Medical Imaging Informatics

Medical Imaging Informatics

Alex A.T. Bui, Ricky K. Taira (eds.)

Springer

Editors
Alex A.T. Bui
Medical Imaging Informatics Group
Department of Radiological Sciences
David Geffen School of Medicine
University of California, Los Angeles
924 Westwood Blvd.
Los Angeles, CA 90024
Suite 420
USA
buia@mii.ucla.edu

Ricky K. Taira
Medical Imaging Informatics Group
Department of Radiological Sciences
David Geffen School of Medicine
University of California, Los Angeles
924 Westwood Blvd.
Los Angeles, CA 90024
Suite 420
USA
rtaira@mii.ucla.edu

ISBN 978-1-4419-0384-6 e-ISBN 978-1-4419-0385-3
DOI 10.1007/978-1-4419-0385-3
Springer New York Dordrecht Heidelberg London

Library of Congress Control Number: 2009939431

Printed on acid-free paper

Springer is part of Springer Science+Business Media (www.springer.com)

For our mentor and friend, Hoosh, who has the wisdom and leadership to realize a vision; and to our students past, present, and future, for helping to pave a path forward.

Foreword

Imaging is considered as one of the most effective – if not the most effective – *in vivo* sampling techniques applicable to chronic serious illnesses like cancer. This simple yet comprehensive textbook in medical imaging informatics (MII) promotes and facilitates two different areas of innovation: the innovations in technology that improve the field of biomedical informatics itself; and the application of these novel technologies to medicine, thus, improving health. Aside from students in imaging disciplines such as radiological sciences (vs. radiology as a service), this book is also very pertinent to other disciplines such as cardiology and surgery. Faculty and students familiar with this book will come to have their own ideas how to innovate, whether it be in core technologies or in applications to biomedicine.

Organizationally, the book follows a very sensible structure related to the process of care, which can in principle be summarized in three questions: *what is wrong*; *how serious is it*; and *what to do?* The first question (what is wrong) focuses mostly on diagnosis (*i.e.,* what studies should be obtained). In this way, issues such as individually-tailored image protocol selection are addressed so that the most appropriate and correct study is obtained – as opposed to the traditional sequential studies. For example, a patient with knee pain and difficulty going up stairs or with minor trauma to the knee and evidence of effusion is directly sent for an MRI (magnetic resonance imaging) study rather than first going to x-ray; or in a child suspected of having abnormal (or even normal) brain development, MRI studies are recommended rather than traditional insurance-required computed tomography (CT). The role of imaging, not only in improving diagnosis but reducing health costs is highlighted. The second question (how serious is it) relates to how we can standardize and document image findings, on the way to providing truly objective, quantitative assessment from an imaging study as opposed to today's norm of largely qualitative descriptors. Finally, the third question is in regard to how we can act upon the information we obtain clinically, from imaging and other sources: how can decisions be made rationally and how can we assess the impact of either research or an intervention?

The textbook has been edited by two scientists, an Associate Professor and a Professor in MII who are both founders of this discipline at our institution. Contributions come from various specialists in medical imaging, informatics, computer science, and biostatistics. The book is not focused on image acquisition techniques or image processing, which are both well-known and described elsewhere in other texts; rather, it focuses on how to extract knowledge and information from imaging studies and related data. The material in this textbook has been simplified eloquently, one of the most difficult tasks by any teacher to simplify difficult material so that it is understandable at all levels.

In short, this textbook is highly recommended for students in any discipline dealing with imaging as well as faculty interested in disciplines of medical imaging and informatics.

Hooshang Kangarloo, MD

Professor Emeritus of Radiological Sciences, Pediatrics, and Bioengineering
University of California at Los Angeles

With the advancement of picture archiving and communications systems (PACS) into "mainstream" use in healthcare facilities, there is a natural transition from the disciplines of engineering research and technology assessment to clinical operations. While much research in PACS-related areas continues, commercial systems are widely available. The burgeoning use of PACS in a range of healthcare facility sizes has created entirely new employment opportunities for "PACS managers," "modality managers," "interface analysts," and others who are needed to get these systems implemented, keep them operating, and expand them as necessary. The field of medical imaging informatics is often described as the discipline encompassing the subject areas that these new specialists need to understand. As the Society of Imaging Informatics in Medicine (SIIM) defines it:

> Imaging informatics is a relatively new multidisciplinary field that intersects with the biological sciences, health services, information sciences and computing, medical physics, and engineering. Imaging informatics touches every aspect of the imaging chain and forms a bridge with imaging and other medical disciplines.[1]

Because the technology of PACS continues to evolve, imaging informatics is also important for the researcher. Each of the areas comprising the field of imaging informatics has aspects that make for challenging research topics. Absent the research these challenges foster and PACS would stagnate.

For the student of medical imaging informatics, there is a wealth of literature available for study. However, much of this is written for trainees in a particular discipline. Anatomy, for example, is typically aimed at medical, dental, veterinary, and physical therapy students, not at engineers. Texts on networks or storage systems are not designed for physicians. Even primers on such topics tend not to provide a cross-disciplinary perspective of the subject.

The authors of *Medical Imaging Informatics* have accepted the challenge of creating a textbook that provides the student of medical imaging informatics with the broad range of topical areas necessary for the field and doing so without being superficial. Unusual for a text on informatics, the book contains a chapter, *A Primer on Imaging Anatomy and Physiology*, subject material this writer knows is important, but is often lacking in the knowledge-base of the information technology (IT) people he works with. Similarly, many informatics-oriented physicians this writer knows do not have the in-depth understanding of information systems and components that IT experts have. Such is the subject matter of the "middle" chapters of the book – *Chapter 3: Information Systems & Architectures*, *Chapter 4: Medical Data Visualization: Toward Integrated Clinical Workstations*, and *Chapter 5: Characterizing Imaging Data*. The succeeding chapters are directed towards integrating IT theory and infrastructure with medical practice topics – *Chapter 6: Natural Language Processing of Medical Reports, Chapter 7: Organizing Observations: Data Models*, *Chapter 8: Disease Models, Part I: Graphical Models*, and *Chapter 9: Disease Models, Part II: Querying & Applications*. Finally, because a practitioner of medical imaging informatics is expected to keep up with the current literature and to know the bases of decision making, the authors have included a chapter on *Evaluation*. With the statistical methods and technology assessment areas covered, the reader will gain the understanding needed to be a critical

---

[1] Society of Imaging Informatics in Medicine website: *http://www.siimweb.org*.

reader of scientific publications and to understand how systems are evaluated during development and after deployment.

Structured in this way, this book forms a unique and valuable resource both for the trainee who intends to become an expert in medical imaging informatics and a reference for the established practitioner.

Steven C. Horii, MD, FACR, FSIIM

Professor of Radiology,
Clinical Director, Medical Informatics Group, and
Modality Chief for Ultrasound
Department of Radiology
University of Pennsylvania Medical Center

Preface

This book roughly follows the process of care, illustrating the techniques involved in medical imaging informatics. Our intention in this text is to provide a roadmap for the different topics that are involved in this field: in many cases, the topics covered in the ensuing chapters are themselves worthy of lengthy descriptions, if not an entire book. As a result, when possible the authors have attempted to provide both seminal and current references for the reader to pursue additional details.

For the imaging novice and less experienced informaticians, in **Part I** of this book, ***Performing the Imaging Exam***, we cover the current state of medical imaging and set the foundation for understanding the role of imaging and informatics in routine clinical practice:

- **Chapter 1 (*Introduction*)** provides an introduction to the field of medical imaging informatics and its role in transforming healthcare research and delivery. The interwoven nature of imaging with preventative, diagnostic, and therapeutic elements of patient care are touched upon relative to the process of care. A brief historic perspective is provided to illustrate both past and current challenges of the discipline.
- **Chapter 2 (*An Introduction to Imaging Anatomy & Physiology*)** starts with a review of clinical imaging modalities (*i.e.*, projectional x-ray, computed tomography (CT), magnetic resonance (MR), ultrasound) and a primer on imaging anatomy and physiology. The modality review encompasses core physics principles and image formation techniques, along with brief descriptions of present and future directions for each imaging modality. To familiarize non-radiologists with medical imaging and the human body, the second part of this chapter presents an overview of anatomy and physiology from the perspective of projectional and cross-sectional imaging. A few systems (neurological, respiratory, breast) are covered in detail, with additional examples from other major systems (gastrointestinal, urinary, cardiac, musculoskeletal).

More experienced readers will likely benefit from starting with **Part II** of this book, ***Integrating Imaging into the Patient Record***, which examines topics related to communicating and presenting imaging data alongside the growing wealth of clinical information:

- Once imaging and other clinical data are acquired, **Chapter 3 (*Information Systems & Architectures*)** tackles the question of how we store and access imaging and other patient information as part of an increasingly distributed and heterogeneous EMR. A description of major information systems (*e.g.*, PACS; hospital information systems, HIS; etc.) as well as the different data standards employed today to represent and communicate data (*e.g.*, HL7, DICOM) are provided. A discussion of newer distributed architectures as they apply to clinical databases (peer-to-peer, grid computing) and information processing is given, examining issues of scalability and searching. Different informatics-driven applications are used to highlight ongoing efforts with respect to the development of information architectures, including telemedicine, IHE, and collaborative clinical research involving imaging.
- After the data is accessed, the challenge is to integrate and to present patient information in such a way to support the physician's cognitive tasks. The longitudinal EMR, in conjunction with the new types of information available to clinicians, has created an almost overwhelming flow of data that must be fully understood to

properly inform decision making. **Chapter 4 (*Medical Data Visualization: Toward Integrated Clinical Workstations*)** presents works related to the visualization of medical data. A survey of graphical metaphors (lists and tables; plots and charts; graphs and trees; and pictograms) is given, relating their use to convey clinical concepts. A discussion of portraying temporal, spatial, multidimensional, and causal relationships is provided, using the navigation of images as an example application. Methods to combine these visual components are illustrated, based on a definition of (task) context and user modeling, resulting in a means of creating an adaptive graphical user interface to accommodate the range of different user goals involving patient data.

Part III, ***Documenting Imaging Findings***, discusses techniques for automatically extracting content from images and related data in order to objectify findings:

- In **Chapter 5 (*Characterizing Imaging Data*)**, an introduction to medical image understanding is presented. Unlike standard image processing, techniques within medical imaging informatics focus on how imaging studies, alongside other clinical data, can be standardized and their content (automatically) extracted to guide medical decision making processes. Notably, unless medical images are standardized, quantitative comparisons across studies is subject to various sources of bias/artifacts that negatively influence assessment. From the perspective of creating scientific-quality imaging databases, this chapter starts with the groundwork for understanding what exactly an image captures, and commences to outline the different aspects encompassing the standardization process: intensity normalization; denoising; and both linear and nonlinear image registration methods are covered. Subsequently, a discussion of commonly extracted imaging features is given, divided amongst appearance- and shape-based descriptors. With the wide array of image features that can be computed, an overview of image feature selection and dimensionality reduction methods is provided. Lastly, this chapter concludes with a description of increasingly popular imaging-based anatomical atlases, detailing their construction and usage as a means for understanding population-based norms and differences arising due to a disease process.
- Absent rigorous methods to automatically analyze and quantify image findings, radiology reports are the sole source of expert image interpretation. In point of fact, a large amount of information about a patient remains locked within clinical documents; and as with images, the concepts therein are not readily computer understandable. **Chapter 6 (*Natural Language Processing of Medical Reports*)** deals with the structuring and standardization of free-text medical reports via natural language processing (NLP). Issues related to medical NLP representation, computation, and evaluation are presented. An overview of the NLP task is first described to frame the problem, providing an analysis of past efforts and applications of NLP. A sequence of subtasks is then related: structural analysis (*e.g.,* section and sentence boundary detection), lexical analysis (*e.g.,* logical word sequences, disambiguation, concept coding), phrasal chunking, and parsing are covered. For each subtask, a description of the challenges and the range of approaches are given to familiarize the reader with the field.
- Core to informatics endeavors is a systematic method to organize both data and knowledge, representing original (clinical) observations, derived data, and conclusions in a logical manner. **Chapter 7 (*Organizing Observations: Data Models*)** describes the different types of relationships between healthcare entities, particularly focusing on those relations commonly encountered in medical imaging. Often in

clinical practice, a disease is studied from a specific perspective (*e.g.*, genetic, pathologic, radiologic, clinical). But disease is a phenomenon of nature, and is thus typically multifaceted in its presentation. The goal is to aggregate the observations for a single patient to characterize the state and behavior of the patient's disease, both in terms of its natural course and as the result of (therapeutic) interventions. The chapter divides the organization of such information along spatial (*e.g.*, physical and anatomical relations, such as between objects in space), temporal (*e.g.*, sequences of clinical events, episodes of care), and clinically-oriented models (*i.e.*, those models specific to representing a healthcare abstraction). A discussion of the motivation behind what drives the design of a medical data model is given, leading to the description of a phenomenon-centric data model to support healthcare research.

Finally, in **Part IV**, ***Toward Medical Decision Making***, we reflect on issues pertaining to reasoning with clinical observations derived from imaging and other data sources in order to reach a conclusion about patient care and the value of our decision:

- A variety of formalisms are used to represent disease models; of these, probabilistic graphical models have become increasingly popular given their ability to reason in light of missing data, and their relatively intuitive representation. **Chapter 8 (*Disease Models, Part I: Graphical Models*)** commences with a review of key concepts in probability theory as the basis for understanding these graphical models and their different formulations. In particular, the first half of the chapter handles Bayesian belief networks (BBNs), appraising past and current efforts to apply these models to the medical environment. The latter half of this chapter addresses the burgeoning exploration of causal models, and the implications for analysis and positing questions to such networks. Throughout, a discussion of the practical considerations in the building of these models and the assumptions that must be made, are given.
- Following the discussion of the creation of the models, in **Chapter 9 (*Disease Models, Part II: Querying & Applications*)**, we address the algorithms and tools that enable us to query BBNs. Two broad classes of queries are considered: belief updating, and abductive reasoning. The former entails the re-computation of posterior probabilities in a network given some specific evidence; the latter involves calculating the optimal configuration of the BBN in order to maximize some specified criteria. Brief descriptions of exact and approximate inference methods are provided. Special types of belief networks (naïve Bayes classifiers, influence diagrams, probabilistic relational models) are covered, illustrating their potential usage in medicine. Importantly, issues related to the evaluation of belief networks are discussed in this chapter, looking to standard technical accuracy metrics, but also ideas in parametric sensitivity analysis. Lastly, the chapter concludes with some example applications of BBNs in medicine, including to support case-based retrieval and image processing tasks.
- **Chapter 10 (*Evaluation*)** concludes by considering how to assess informatics endeavors. A primer on biostatistics and study design starts this chapter, including a review of basic concepts (*e.g.*, confidence intervals, significance and hypothesis testing) and the statistical tests that are used to evaluate hypotheses under different circumstances and assumptions. A discussion of error and performance assessment is then introduced, including sensitivity/specificity and receiver operative characteristic analysis. Study design encompasses a description of the different types of experiments that can be formed to test a hypothesis, and goes over the

process of variable selection and sample size/power calculations. Sources of study bias/error are briefly described, as are statistical tools for decision making. The second part of this chapter uses the foundation set out by the primer to focus specifically on informatics-related evaluations. Two areas serve as focal points: evaluating information retrieval (IR) systems, including content-based image retrieval; and assessing (system) usability.

Contributors

Pablo Abbona, MD
Department of Radiological Sciences
UCLA David Geffen School of Medicine

Denise Aberle, MD
Medical Imaging Informatics &
Department of Radiological Sciences
UCLA David Geffen School of Medicine

Corey Arnold, PhD
Medical Imaging Informatics &
Department of Information Studies
University of California, Los Angeles

Lawrence Bassett, MD
Department of Radiological Sciences
UCLA David Geffen School of Medicine

Kathleen Brown, MD
Department of Radiological Sciences
UCLA David Geffen School of Medicine

Matthew Brown, PhD
Thoracic Imaging Laboratory &
Department of Radiological Sciences
UCLA David Geffen School of Medicine

Suzie El-Saden, MD
Department of Radiology
Veteran's Administration Wadsworth
Los Angeles, California

Ana Gomez, MD
Department of Radiological Sciences
UCLA David Geffen School of Medicine

William Hsu, PhD
Medical Imaging Informatics
UCLA David Geffen School of Medicine

Juan Eugenio Iglesias, MSc
Medical Imaging Informatics
UCLA Biomedical Engineering IDP

Neda Jahanshad, BS
Medical Imaging Informatics
UCLA Biomedical Engineering IDP

Hooshang Kangarloo, MD
Medical Imaging Informatics
UCLA David Geffen School of Medicine

Kambiz Motamedi, MD
Department of Radiological Sciences
UCLA David Geffen School of Medicine

Craig Morioka, PhD
Department of Radiology
Veteran's Administration Wadsworth
Los Angeles, California

Nagesh Ragavendra, MD
Department of Radiological Sciences
UCLA David Geffen School of Medicine

James Sayre, PhD
Departments of Biostatistics &
Radiological Sciences
UCLA David Geffen School of Medicine

Leanne Seeger, MD
Department of Radiological Sciences
UCLA David Geffen School of Medicine

Ilya Shpitser, PhD
School of Public Health
Harvard University

Emily Watt, MLIS
Medical Imaging Informatics
UCLA Biomedical Engineering IDP

Table of Contents

PART I
Performing the Imaging Exam

Wherein an introduction to medical imaging informatics (MII) is provided; as is a review of the current state of clinical medical imaging and its use in understanding the human condition and disease. For new students and the informatician with a minimal background in medical imaging and clinical applications, these chapters help provide a basis for understanding the role of MII, the present needs of physicians and researchers dealing with images, and the future directions of this discipline.

- **Chapter 1** – Introduction
- **Chapter 2** – A Primer on Imaging Anatomy and Physiology

Chapter 1
Introduction

ALEX A.T. BUI, RICKY K. TAIRA, AND HOOSHANG KANGARLOO

Medical imaging informatics is the rapidly evolving field that combines biomedical informatics and imaging, developing and adapting core methods in informatics to improve the usage and application of imaging in healthcare; and to derive new knowledge from imaging studies. This chapter introduces the ideas and motivation behind medical imaging informatics. Starting with an illustration of the importance of imaging in today's patient care, we demonstrate imaging informatics' potential in enhancing clinical care and biomedical research. From this perspective, we provide an example of how different aspects of medical imaging informatics can impact the process of selecting an imaging protocol. To help readers appreciate this growing discipline, a brief history is given of different efforts that have contributed to its development over several decades, leading to its current challenges.

What is Medical Imaging Informatics?

Two revolutions have changed the nature of medicine and research: medical imaging and biomedical informatics. First, medical imaging has become an invaluable tool in modern healthcare, often providing the only *in vivo* means of studying disease and the human condition. Through the advances made across different imaging modalities, majors insights into a range of medical conditions have come about, elucidating matters of structure and function. Second, the study of biomedical informatics concerns itself with the development and adaptation of techniques from engineering, computer science, and other fields to the creation and management of medical data and knowledge. Biomedical informatics is transforming the manner by which we deal and think with (large amounts of) electronic clinical data. *Medical imaging informatics* is the discipline that stands at the intersection of biomedical informatics and imaging, bridging the two areas to further our comprehension of disease processes through the unique lens of imaging; and from this understanding, improve clinical care.

Beyond the obvious differences between images and other forms of medical data, the very nature of medical imaging set profound challenges in automated understanding and management. While humans can learn to perceive patterns in an image – much as a radiologist is trained – the nuances of deriving knowledge from an image still defy the best algorithms, even with the significant strides made in image processing and computer vision. Imaging informatics research concerns itself with the full spectrum of low-level concepts (*e.g.*, image standardization; signal and image processing) to higher-level abstractions (*e.g.*, associating semantic meaning to a region in an image; visualization and fusion of images) and ultimately, applications and the derivation of new knowledge from imaging. Notably, medical imaging informatics addresses not only the images themselves, but encompasses the associated data to understand the context of the imaging study; to document observations; and to correlate and reach new conclusions about a disease and the course of a medical problem.

A.A.T. Bui and R.K. Taira (eds.), *Medical Imaging Informatics*,
DOI 10.1007/978-1-4419-0385-3_1, © Springer Science + Business Media, LLC 2010

The Process of Care and the Role of Imaging

From a high-level perspective, the healthcare process can be seen in terms of three clinical questions (Fig. 1.1), each related to aspects of the scientific method. For a given patient, a physician has to: 1) ascertain what is wrong with the patient (identify the problem, develop a hypothesis); 2) determine the seriousness of a patient's condition by performing diagnostic procedures (experiment); and 3) after obtaining all needed information, interpret the results from tests to reach a final diagnosis and initiate therapy (analyze and conclude). At each point, medical imaging takes on a critical role:

1. What is wrong? Patient presentation, for the most part, is relatively subjective. For example, the significance of a headache is usually not clear from a patient's description (*e.g., my head throbs*). Imaging plays a major role in objectifying clinical presentations (*e.g.,* is the headache secondary to a brain tumor, intracranial aneurysm, or sinusitis?) and is an optimal diagnostic test in many cases to relate symptoms to etiology. In addition, when appropriately recorded, imaging serves as the basis for shared communication between healthcare providers, detailing evidence of current and past medical findings.
2. How serious is it? For many conditions, the physical extent of disease is visually apparent through imaging, allowing us to determine how far spread a problem has become (*e.g.,* is it confined to a local environment or is it systemic?). Moreover, imaging is progressively moving from qualitative to quantitative assessment. Already, we use imaging to document physical state and the severity of disease: tumor size in cancer patients; dual energy x-ray absorptiometry (DXA) scores in osteoporosis; cardiothoracic ratios; arterial blood flow assessment based on Doppler ultrasound; and coronary artery calcification scoring are all rudimentary metrics that quantify disease burden. On the horizon are more sophisticated quantitative imaging techniques that further characterize biophysical phenomena.
3. What to do? Treatment is contingent on an individual's response: if a given drug or intervention fails to have the desired effect, a new approach must be taken to resolve the problem. For many diseases, response assessment is done through imaging: baseline, past, and present studies are compared to deduce overall behavior. By way of illustration, many of today's surgical procedures are assessed on a follow-up imaging study; and the effects of chemotherapy are tracked over time (*e.g.,* is the tumor getting smaller?). Additionally, contemporary image-guided interventional techniques are opening new avenues of treatment.

As the ubiquity and sophistication of imaging grows, methods are needed to fully realize its potential in daily practice and in the full milieu of patient care and medical research. The study of medical imaging informatics serves this function.

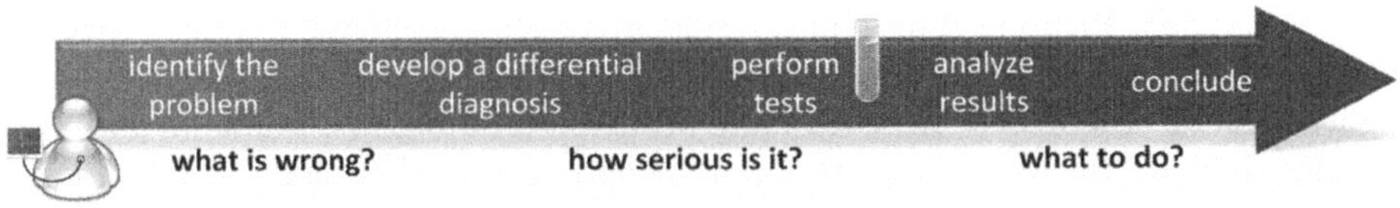

Figure 1.1: The process of care can be roughly summarized in three stages: 1) *what is wrong*, which entails identifying the problem and establishing a differential diagnosis; 2) *how serious is it*, which involves testing the differential diagnosis and determining the extent of the problem; and 3) *what to do*, which based on analysis of test results, concludes with a treatment decision.

Medical Imaging Informatics: From Theory to Application

There are two arms to medical imaging informatics: the development of core informatics theories and techniques that advance the field of informatics itself; and the translation of these techniques into an application that improves health. To demonstrate, we first consider the reasons for the improper use of imaging today, and then how imaging informatics can impact these issues.

Improving the Use of Imaging

The process of providing an accurate, expedient medical diagnosis via imaging can fail for several reasons (Fig. 1.2):

- Sub-optimal study selection. The first potential point of failure arises when an imaging study is requested. Given the fairly rapid changes across all elements of imaging technology, it is unrealistic to believe that a physician can always make up-to-date if not optimal decisions about an imaging exam [9]. Thus, the wrong study may be requested for a given patient. To reduce this problem, practice guidelines have been introduced, but are often generic and do not take into account the specific condition of the patient.
- Poor acquisition. The next potential point of failure occurs during study acquisition. Problems arise due to poor instrumentation (*e.g.,* sensitivity), equipment calibration, poor data acquisition methods, or poor technique. For example, due to the very technical nature of imaging procedures, the average clinician is unable to determine the most specific diagnostic protocol; this process is often left to a technologist or radiologist, who without fully knowing the context of the patient, may not use ideal acquisition parameters.
- Poor interpretation. Study interpretation presents an additional point for potential failure. Poor study interpretation can be due to inadequate historical medical information, poor information filtering/presentation, or poor/mismatched skills by the study reader. Studies have shown that historical clinical information can improve the perception of certain radiographic findings [3]. Poor information presentation often leads to important data being buried within the medical record. Finally, study reading itself can be improved by providing users with the facility to retrieve relevant data from online medical literature, or by choosing the best-matched readers (*i.e.,* generalist vs. specialist) for a particular exam. However, currently available search techniques do not support specific and directed retrievals and no electronic framework exists for efficiently matching a given exam with the most appropriate reader for that exam.

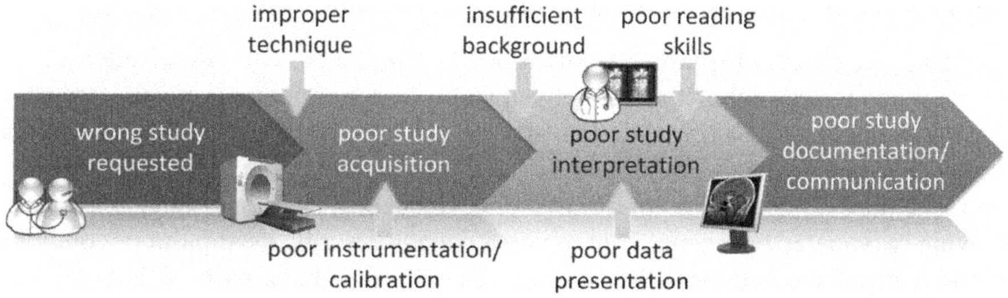

Figure 1.2: Identification of potential problems in the diagnostic process. In emergency cases, the process may also fail due to excessively long times to completion.

- Poor reporting. The last potential point of failure concerns reporting of study results, which is a key concern in the coordination of care as related to the diagnosis and intervention for a given case. This lack of coordination is due to: 1) poor documentation of study results; and 2) difficulties communicating the results of tests to referring healthcare providers. These inefficiencies can lead to problems such as initiating treatment before a definitive diagnosis is established, and duplicating diagnostic studies.

From this perspective, medical imaging informatics aims to improve the use of imaging throughout the process of care. For example, what is the best imaging method to assess an individual's given condition? Are there image processing methods that can be employed to improve images post-acquisition (*e.g.*, histogram correction, denoising, etc.)? These and other questions motivate medical imaging informatics research. Indeed, imaging plays a significant role in the evaluation of patients with complex diseases. As these patients also account for the majority of expenses related to healthcare, by improving the utility of imaging, cost savings can potentially be realized.

Choosing a Protocol: The Role of Medical Imaging Informatics

To further highlight the role of medical imaging informatics, we consider the task of choosing an imaging protocol when a patient first presents in a doctor's office, addressing issues related to sub-optimal study design. When a primary care physician (PCP) decides to obtain an imaging study to diagnosis or otherwise assess a problem, the question arises as to which imaging modality and type of study should be ordered. Furthermore, the ability to make the best decisions regarding a patient is variable across individual physicians and over time. Individual physician biases often creep into decision making tasks and can impact the quality and consistency of healthcare provided [1, 6].

To ground this discussion, we use an example of a 51 year-old female patient who visits her PCP complaining of knee pain. The selection of an appropriate imaging protocol to diagnosis the underlying problem can be thought of in three steps: 1) standardizing the patient's chief complaint, providing a structured and codified format to understand the individual's symptoms; 2) integrating the patient's symptoms with past evidence (*e.g.*, past imaging, medical history, etc.) to assess and to formulate a differential diagnosis; and 3) selecting and tailoring the imaging study to confirm (or deny) the differential diagnosis, taking into account local capabilities to perform and evaluate an imaging study (there is no point in ordering a given exam if the scanner is unavailable or unable to perform certain sequences). We elaborate on each of the steps below, illustrating current informatics research and its application.

Capturing the chief complaint. As mentioned earlier, a patient's description of his or her symptoms is very subjective; for physicians – and computers more so – translating their complaints into a "normalized" response (such as from a controlled vocabulary) is tricky. For instance, with our example patient, when asked her reason for seeing her doctor, she may respond, "My knee hurts a lot, frequently in the morning." Consider the following two related problems: 1) mapping a patient-described symptom or condition to specific medical terminology/disease (*e.g.*, knee hurts = knee pain → ICD-9 719.46, *Pain in joint involving lower leg*); and 2) standardizing descriptive terms (adjectives, adverbs) to the some scale (*e.g.*, Does "a lot" mean a mild discomfort or a crippling pain? Does "frequently" mean every day or just a once a week?).

Several informatics endeavors related to the automated structuring of data are pertinent here. Electronic collections of validated questionnaires are being created, formally defining pertinent positive/negative questions and responses (*e.g.*, see the National Institutes of Health (NIH) PROMIS project [7] and related efforts by the National Cancer Institute, NCI). Such databases provide a foundation from which chief complaints and symptoms can be objectified and quantified with specificity: duration, severity, timing, and activities that either trigger or relieve the symptom can be asked. Likewise, existing diagnostic guidelines intended for non-physicians, such as the American Medical Association Family Medical Guide [5], can be turned into online, interactive modules with decision trees to guide a patient through the response process. Markedly, an inherent issue with such questionnaires is determining how best to elicit responses from patients; aspects of visualization and human-computer interaction (HCI) thus also come into play (see Chapter 4). Apart from structured formats, more complicated methods such as medical natural language processing (NLP) can be applied to structure the statement by the patient, identifying and codifying the chief complaint automatically. Chapter 6 provides an overview of NLP research and applications.

Assessing the patient. The chief complaint provides a basis for beginning to understand the problem, but a clinician will still require additional background to establish potential reasons for the knee pain. For example, does the patient have a history of a previous condition that may explain the current problem? Has this specific problem occurred before (*i.e.*, is it chronic) or did any specific past event cause this issue (*e.g.*, trauma to the knee)? The answers to these questions are all gleaned from questioning the patient further and an exploration of the medical record.

An array of medical and imaging informatics research is ongoing to enrich the electronic medical record's (EMR) functionality and to bring new capabilities to the point of care. A longstanding pursuit of the EMR is to provide an automated set of relevant information and a readily searchable index to patient data: rather than manually inspect past reports and results, the system should locate germane documents, if not permit the physician to pose a simple query to find key points. Informatics work in distributed information systems concentrates on the problems of data representation and connectivity in an increasingly geographically dispersed, multidisciplinary healthcare environment. Patients are commonly seen by several physicians, who are often at different physical locations and institutions. As such, a patient's medical history may be segmented across several disparate databases: a core challenge of informatics is to find effective ways to integrate such information in a secure and timely fashion (see Chapter 3). For imaging, past exams should be made available; but instead of the whole study, only (annotated) sentinel image slices that detail a problem could be recalled. Although manual image capture and markup is presently used, automated techniques are being investigated to identify anatomical regions and uncover potential abnormalities on an image (*e.g.*, CAD); and to segment and quantify disease based on domain knowledge (see Chapter 5). For textual data, such as generated from notes and consults (*e.g.*, a radiology report), NLP techniques are being developed to facilitate content indexing (see Chapter 6). To aggregate the information into a useful tool, a data model that matches the expectations of the clinician must be used to organize the extracted patient data (see Chapter 7), and it must then be presented in a way conducive to thinking about the problem (see Chapter 4).

Specifying the study. Based on the patient's responses and review of her record, the PCP wishes to differentiate between degenerative joint disease and a meniscal tear. If a patient complains of knee pain, then traditionally as a first step an x-ray is obtained. But if the patient's symptoms are suggestive of pain when going up stairs, then a knee magnetic resonance (MR) imaging study is warranted over an x-ray (this symptom being suggestive of a meniscal tear). When asked whether going up stairs aggravates the knee pain, the patient indicated that she was unsure. Thus, her PCP must now make a decision as to what imaging test should be ordered. Furthermore, the selection of the imaging exam must be tempered by the availability of the imaging equipment, the needed expertise to interpret the imaging study, and other potential constraints (*e.g.*, cost, speed of interpretation, etc.).

First, supporting the practice of evidence-based medicine (EBM) is a guiding principle of biomedical informatics, and hence medical imaging informatics. The development and deployment of practice guidelines in diagnosis and treatment has been an enduring effort of the discipline, suggesting and reminding physicians on courses of action to improve care. For instance, if the patient's clinician was unaware of the sign of a meniscal tear, the system should automatically inform him that an MR may be indicated if she has knee pain when climbing stairs; and supporting literature can be automatically suggested for review. Second, formal methods for medical decision-making are central to informatics, as are the representation of medical knowledge needed to inform the algorithms [10]. Techniques from computer science, ranging from rudimentary rule-bases to statistical methods (*e.g.*, decision trees); through to more complex probabilistic hidden Markov models (HMMs) and Bayesian belief networks (BBNs) are finding applications in medicine (see Chapter 8). For example, the evidence of the patient's medical history, her response to the physician's inquiries, the availability of imaging, and the relative urgency of the request can be used in an influence diagram to choose between the x-ray and MR (see Chapter 9). Such formalizations are providing new tools to model disease and to reason with partial evidence. Essential to the construction of many of these models is the compilation of large amounts of (observational) data from which data mining and other computational methods are applied to generate new knowledge. In this example, these disease models can be used: to identify further questions that can be asked to further elucidate the patient's condition (improving the likelihood of choosing an optimal imaging exam); and to select the type of imaging study, and even its acquisition parameters, to best rule in/out elements of the differential diagnosis.

Ultimately, an electronic imaging infrastructure that expedites accurate diagnosis can improve the quality of healthcare; and even within this simple example of choosing an imaging protocol, the role of informatics is apparent in enhancing the process of care. When used appropriately, medical imaging is effective at objectifying the initial diagnostic hypothesis (differential diagnosis) and guiding the subsequent work-up. Given a chief complaint and initial assessment data, one can envision that specialists or software algorithms would select an imaging protocol for an appropriate medical condition even before a visit to the PCP. The PCP can then access both objective imaging and clinical data prior to the patient's visit. Medical imaging informatics research looks to improve the fundamental technical methods, with ensuing translation to clinical applications.

Cost Considerations

Some have targeted the cost of imaging as a major problem in healthcare within the United States: one 2005 estimate by the American College of Radiology (ACR) was that $100 billion is spent annually on diagnostic imaging, including computed tomography (CT), MR, and positron emission tomography (PET) scans [2]. While acknowledging that many factors are contributing to these high costs it is, however, important to separate out two issues: the healthcare cost savings generated as a result of imaging, in light of earlier diagnoses and quality of life; and the *true* cost of performing an imaging study (*i.e.,* versus what is charged).

An "appropriate" process of care that disregards issues related to utilization review and approvals required for imaging studies can be very effective for care of the patient as well as cost-effective. In one study performed by us for a self-insured employer group, we removed all of the requirements for (pre-)approval of imaging studies and allowed primary care physicians to order imaging based on their diagnostic hypothesis and the need of the patient. The imaging costs were instead capitated for the employer group. The number of cross-sectional images, particularly CT and MR, more than doubled and the number of projectional images decreased. However, the net effect was not only significant cost savings to the employer group but also much higher quality and satisfaction by patients [12]. A follow-up study further showed improved health (lowered incidence of chronic disease, decreased number of hospitalizations and emergency room visits, etc.), continued high levels of patient satisfaction, and lowered expenditures within the cost-capitated imaging environment relative to a control group [4]. All of this is to suggest that it is not necessarily the overuse of imaging that is inherently costly, and that there are in fact cost-savings introduced through the unrestricted use of imaging. Of course, a capitated cost agreement with unfettered usage of imaging is not the norm. Unfortunately, the cost of imaging studies is rarely the true cost of performing the study. As an example, presently charges for a brain MR imaging study with and without contrast are in excess of $7,000 at some institutions – largely because of professional fees and attempts to recoup costs (*e.g.,* from non-paying and uninsured individuals). Yet in one internal study we conducted in the 1990s to understand the real cost of CTs and MRs, it was concluded that the price of an MR study is no more than $200 and the price of a CT less than $120. These costs included technologists time, materials used (*e.g.,* contrast) and the depreciation of the scanning machines over five years. Even adjusting for inflation and a moderate professional fee, one can argue that the charges seen today for imaging largely outpace the true cost of the exam. Hence, a current practical challenge for medical imaging informatics is to develop new paradigms of delivery that will encourage the use of imaging throughout the healthcare environment while still being cost-effective.

A Historic Perspective and Moving Forward

Medical imaging informatics is not new: aspects of this discipline have origins spanning back over two or more decades [14]. As such, it is useful to consider this field's interdisciplinary evolution to understand its current challenges and future. Below, we consider four different eras of technical research and development.

PACS: Capturing Images Electronically

Concurrent to the progress being made with respect to CT and MR imaging, initial efforts to create an electronic repository for (digital) imaging in the 1980s led to the creation of picture archive and communication systems (PACS). [8, 11] provide some perspective on the early development of PACS, which focused on linking acquisition devices (*i.e.,* scanners), storage, intra-site dissemination of studies, and display technologies (soft and hard copy). With the introduction of PACS, some of the physical limitations of film were overcome: images were now available anywhere within an institution via a display workstation, and multiple individuals could simultaneously view the same study. Preliminary work also highlighted the need to integrate PACS with other aspects of the healthcare environment and for common data standards to be adopted. Development of the latter was spearheaded by a joint commission of the ACR in conjunction with the National Electrical Manufacturer's Association (NEMA), later leading to establishment of the now well-known DICOM (Digital Imaging and Communication in Medicine) standard. While some academic research in PACS is still being performed today, arguably much of this work has transitioned to industry and information technology (IT) support.

Teleradiology: Standardizing Data and Communications

In 1994, DICOM version 3.0 was released, setting the stage for digital imaging and PACS to be embraced across a broader section of the healthcare arena. At the same time, MR and CT scanners were becoming widespread tools for clinical diagnosis. Recognizing early on the potential for data networks to transmit imaging studies between sites, and partly in response to a shortage of (subspecialist) radiologists to provide interpretation, the next major step came with teleradiology applications. [18] describes the genesis of teleradiology and its later growth in the mid-1990s. Key technical developments during this era include the exploration of distributed healthcare information systems through standardized data formats and communication protocols, methods to efficiently compress/transmit imaging data, and analysis of the ensuing workflow (*e.g.,* within a hospital and between local/remote sites). Legal policies and regulations were also enacted to support teleradiology. From a clinical viewpoint, the power of teleradiology brought about consolidation of expertise irrespective of (physical) geographic constraints. These forays provided proof positive for the feasibility of telemedicine, and helped create the backbone infrastructure for today's imaging-based multi-site clinical trials. Although DICOM provided the beginnings of standardization, there was a continued need to extend and enhance the standard given the rapid changes in medical imaging. Moreover, researchers began to appreciate the need to normalize the meaning and content of data fields as information was being transmitted between sites [15]. Newer endeavors in this area continue to emerge given changes in underlying networking technology and ideas in distributed architectures. For instance, more recent work has applied grid computing concepts to image processing and repositories.

Integrating Patient Data

Alongside teleradiology, medical informatics efforts started to gain further prominence, launching a (renewed) push towards EMRs. It became quickly evident that while many facets of the patient record could be combined into a single application, incorporating imaging remained a difficultly because of its specialized viewing requirements (both because of the skill needed to interpret the image, and because of its multimedia format). Conversely, PACS vendors encountered similar problems: radiologists using imaging workstations needed better access to the EMR in order to provide proper assessment. Hence in this next major phase of development, processes that were originally conceived of as radiology-centric were opened up to the breadth of healthcare activities, sparking a cross-over with informatics. For example, the Integrating the Healthcare Enterprise (IHE) initiative was spawned in 1998 through HIMSS and RSNA (Healthcare Information and Management Systems Society, Radiological Society of North America), looking to demonstrate data flow between HL7 and DICOM systems. Additionally, drawing from informatics, researchers began to tackle the problems of integration with respect to content standardization: the onset of structured reporting; the creation and use of controlled vocabularies/ontologies to describe image findings; and the development of medical natural language processing were all pursued within radiology as aids towards being able to search and index textual reports (and hence the related imaging). Though great strides have been made in these areas, research efforts are still very active: within routine clinical care, the process of documenting observations largely remains *ad hoc* and rarely meets the standards associated with a scientific investigation, let alone making such data "computer understandable."

Understanding Images: Today's Challenge

The modern use of the adage, "*A picture is worth ten thousand words*," is attributed to a piece by Fred Barnard in 1921; and its meaning is a keystone of medical imaging informatics. The current era of medical imaging informatics has turned to the question of how to manage the content within images. Presently, research is driven by three basic questions: 1) what is in an image; 2) what can the image tell us from a quantitative view; and 3) what can an image, now correlated with other clinical data, tell us about a specific individual's disease and response to treatment? Analyses are looking to the underlying physics of the image and biological phenomena to derive new knowledge; and combined with work in other areas (genomics/proteomics, clinical informatics), are leading to novel diagnostic and prognostic biomarkers. While efforts in medical image processing and content-based image retrieval were made in the 1990s (*e.g.*, image segmentation; computer-aided detection/diagnosis, CAD), it has only been more recently that applications have reached clinical standards of acceptability. Several forces are driving this shift towards computer understanding of images: the increasing amount and diversity of imaging, with petabytes of additional image data accrued yearly; the formulation of new mathematical and statistical techniques in image processing and machine learning, made amenable to the medical domain; and the prevalence of computing power. As a result, new imaging-based models of normal anatomy and disease processes are now being formed.

Knowledge creation. Clinical imaging evidence, which is one of the most important means of *in vivo* monitoring for many patient conditions, has been used in only a limited fashion (*e.g.*, gross tumor measurements) and the clinical translation of derived quantitative imaging features remains a difficulty. And, in some cases, imaging remains the only mechanism for routine measurement of treatment response. For example, a recent study suggests that while common genetic pathways may be uncovered for high-grade primary brain tumors (glioblastoma multiforme, GBM), the highly heterogeneous nature of these cancers may not fully lend themselves to be sufficiently prognostic [17]; rather, other biomarkers, including imaging, may provide better guidance. In particular, as the regional heterogeneity and the rate of mutation of GBMs is high [13], imaging correlation could be important, providing a continuous proxy to assess gene expression, with subsequent treatment modification as needed. In the short-term, the utilization of imaging data can be improved: by standardizing image data, pre- and post-acquisition (*e.g.*, noise reduction, intensity signal normalization/calibration, consistent registration of serial studies to ensure that all observed changes arise from physiological differences rather than acquisition); by (automatically) identifying and segmenting pathology and anatomy of interest; by computing quantitative imaging features characterizing these regions; and by integrating these imaging-derived features into a comprehensive disease model.

One can assume that every picture – including medical images – contain a huge amount of information and knowledge that must be extracted and organized. Knowledge can be conveniently categorized twofold [16]: *implicit*, which represents a given individual's acumen and experience; and *explicit*, which characterizes generally accepted facts. Clearly, implicit knowledge is advanced through current informatics endeavors, as employed by the individual scientist and clinician. But informatics can further serve to create explicit knowledge by combining together the implicit knowledge from across a large number of sources. In the context of healthcare, individual physician practices and the decisions made in routine patient care can be brought together to generate new scientific insights. That is to say that medical imaging informatics can provide the transformative process through which medical practice involving imaging can lead to new explicit knowledge. Informatics research can lead to means to standardize image content, enabling comparisons across populations and facilitate new ways of thinking.

References

1. Aberegg SK, Terry PB (2004) Medical decision-making and healthcare disparities: The physician's role. J Lab Clin Med, 144(1):11-17.
2. American College of Radiology (ACR) (2005) ACR chair tells House Committee unnecessary and inferior medical imaging lowers quality of care, costs taxpayers. American College of Radiology. http://www.acr.org. Accessed April 23, 2009.
3. Berbaum KS, Franken EA, Jr., Dorfman DD, Lueben KR (1994) Influence of clinical history on perception of abnormalities in pediatric radiographs. Acad Radiol, 1(3):217-223.
4. Bui AA, Taira RK, Goldman D, Dionisio JD, Aberle DR, El-Saden S, Sayre J, Rice T, Kangarloo H (2004) Effect of an imaging-based streamlined electronic healthcare process on quality and costs. Acad Radiol, 11(1):13-20.
5. Clayman CB, Curry RH (1992) The American Medical Association Guide to Your Family's Symptoms. 1st updated pbk. edition. Random House, New York.
6. Croskerry P (2002) Achieving quality in clinical decision making: Cognitive strategies and detection of bias. Acad Emerg Med, 9(11):1184-1204.
7. DeWalt DA, Rothrock N, Yount S, Stone AA (2007) Evaluation of item candidates: The PROMIS qualitative item review. Med Care, 45(5 Suppl 1):S12-21.
8. Dwyer III SJ (2000) A personalized view of the history of PACS in the USA. Medical Imaging 2000: PACS Design and Evaluation: Engineering and Clinical Issues, vol 3980. SPIE, San Diego, CA, USA, pp 2-9.
9. Edep ME, Shah NB, Tateo IM, Massie BM (1997) Differences between primary care physicians and cardiologists in management of congestive heart failure: Relation to practice guidelines. J Am Coll Cardiol, 30(2):518-526.
10. Greenes RA (2007) A brief history of clinical decision support: Technical, social, cultural, economic, and governmental perspectives. In: Greenes RA (ed) Clinical Decision Support: The Road Ahead. Elsevier Academic Press, Boston, MA.
11. Huang HK (2004) PACS and Imaging Informatics: Basic Principles and Applications. 2nd edition. Wiley-Liss, Hoboken, NJ.
12. Kangarloo H, Valdez JA, Yao L, Chen S, Curran J, Goldman D, Sinha U, Dionisio JD, Taira R, Sayre J, Seeger L, Johnson R, Barbaric Z, Steckel R (2000) Improving the quality of care through routine teleradiology consultation. Acad Radiol, 7(3):149-155.
13. Kansal AR, Torquato S, Harsh GI, Chiocca EA, Deisboeck TS (2000) Simulated brain tumor growth dynamics using a three-dimensional cellular automaton. J Theor Biol, 203(4):367-382.

14. Kulikowski C, Ammenwerth E, Bohne A, Ganser K, Haux R, Knaup P, Maier C, Michel A, Singer R, Wolff AC (2002) Medical imaging informatics and medical informatics: Opportunities and constraints. Findings from the IMIA Yearbook of Medical Informatics 2002. Methods Inf Med, 41(2):183-189.
15. Kulikowski CA (1997) Medical imaging informatics: Challenges of definition and integration. J Am Med Inform Assoc, 4(3):252-253.
16. Pantazi SV, Arocha JF, Moehr JR (2004) Case-based medical informatics. BMC Med Inform Decis Mak, 4:19.
17. The Cancer Genome Atlas Research Network (2008) Comprehensive genomic characterization defines human glioblastoma genes and core pathways. Nature, 455(7216):1061-1068.
18. Thrall JH (2007) Teleradiology: Part I. History and clinical applications. Radiology, 243(3):613-617.

Chapter 2
A Primer on Imaging Anatomy and Physiology

DENISE ABERLE, SUZIE EL-SADEN, PABLO ABBONA, ANA GOMEZ, KAMBIZ MOTAMEDI, NAGESH RAGAVENDRA, LAWRENCE BASSETT, LEANNE SEEGER, MATTHEW BROWN, KATHLEEN BROWN, ALEX A.T. BUI, AND HOOSHANG KANGARLOO

An understanding of medical imaging informatics begins with knowledge of medical imaging and its application toward diagnostic and therapeutic clinical assessment. This chapter is divided into two sections: a review of current imaging modalities; and a primer on imaging anatomy and physiology. In the first half, we introduce the major imaging modalities that are in use today: projectional imaging, computed tomography, magnetic resonance, and ultrasound. The core physics concepts behind each modality; the parameters and algorithms driving image formation; and variants and newer advances in each of these areas are briefly covered to familiarize the reader with the capabilities of each technique. From this foundation, in the second half of the chapter we describe several anatomical and physiologic systems from the perspective of imaging. Three areas are covered in detail: 1) the respiratory system; 2) the brain; and 3) breast imaging. Additional coverage of musculoskeletal, cardiac, urinary, and upper gastrointestinal systems is included. Each anatomical section begins with a general description of the anatomy and physiology, discusses the use of different imaging modalities, and concludes with a description of common medical problems/conditions and their appearance on imaging. From this chapter, the utility of imaging and its complexities becomes apparent and will serve to ground discussion in future chapters.

A Review of Basic Imaging Modalities

The crucial role of imaging in illuminating both the human condition and disease is largely self-evident, with medical imaging being a routine tool in the diagnosis and the treatment of most medical problems. Imaging provides an objective record for documenting and communicating *in vivo* findings at increasingly finer levels of detail. This section focuses on a review of the current major imaging modalities present in the clinical environment. As it is beyond the ability of a single chapter to comprehensively cover all aspects of medical imaging, we aim only to cover key points: references to seminal works are provided for the reader. Also, given the scope of this field, we omit a discussion of nuclear medicine, and newer methods such as molecular and optical imaging that are still largely seen in research environments.

Projectional Imaging

The genesis of medical imaging and radiography started in 1895 with the discovery of x-rays by Roentgen. Today, the use of x-ray projectional imaging comes only second to the use of laboratory tests as a clinical diagnostic tool.

Core Physical Concepts

A thorough handling of x-ray physics can be found in [15, 19]. X-rays are a form of electromagnetic (EM) radiation, with a wavelength ranging from 0.1-10 nm, which

A.A.T. Bui and R.K. Taira (eds.), *Medical Imaging Informatics*,
DOI 10.1007/978-1-4419-0385-3_2, © Springer Science + Business Media, LLC 2010

translates to photons with an energy level of 0.12-125 keV. Above a certain energy level (~12 keV), x-rays are able to penetrate different materials to a varying degree: it is this phenomenon that is taken advantage of in projectional x-ray imaging. Recall from basic physics that when a photon hits an atom, there is a chance of interaction between the photon and any electrons. There are essentially three different ways that an x-ray can interact with matter within the diagnostic energy range:

1. Photoelectric effect. The well-known *photoelectric effect* involves the interaction of a photon with a low-energy electron. If the photon has sufficient energy, then the electron is separated from the atom, with any excess energy from the photon being transformed into the electron's kinetic energy (Fig. 2.1a). The emitted electron is referred to as a *photoelectron*. Given the absence of an electron in the lower energy levels, an electron from a higher energy level moves down to take its place; but in order to do so, it must release its extra energy, which is seen in the form of a photon (characteristic radiation). Thus, the photoelectric effect generates three products: a photoelectron; a photon (characteristic radiation); and an ion (the positively charged atom, hence the phrase *ionizing radiation*). This type of interaction typically occurs with the absorption of low-energy x-rays.
2. Compton effect. Rather than being absorbed, when a high-energy photon collides with an electron, both particles may instead be deflected. A portion of the photon's energy is transferred to the electron in this process, and the photon emerges with a longer wavelength; this effect is known as *Compton scattering* (Fig. 2.1b). This phenomenon is thus seen largely with higher-energy x-rays. Compton scattering is the major source of background noise in x-ray images. Furthermore, Compton scattering is a cause of tissue damage.
3. Coherent scattering. Lastly, an x-ray can undergo a change in direction but no change in wavelength (energy) (Fig. 2.1c). Thompson and Rayleigh scatter are examples of this occurrence. Usually < 5% of the radiation undergoes this effect.

A fourth type of interaction is possible, known as *pair production*. Pair production involves high energy x-rays and elements of high atomic weight. When a high-energy

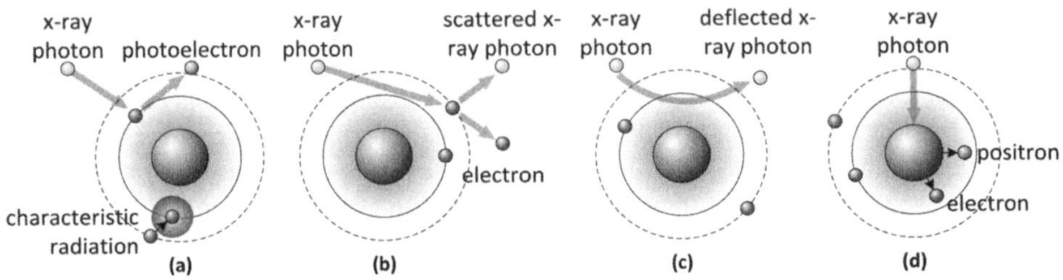

Figure 2.1: Interaction of x-rays with matter, envisioning an atom and its electrons in terms of a nucleus and orbitals. **(a)** The photoelectric effect results in the complete transfer of the energy from an x-ray photon to an electron, which leaves the atom as a photoelectron. Another electron then moves from a higher to lower orbit and in the process emits characteristic radiation. **(b)** The Compton effect results in scattering of the x-ray photon with a portion of the photon's momentum transferred as kinetic energy to the electron. **(c)** Coherent scattering involves the deflection of the x-ray photon in a new direction. **(d)** Pair production occurs when the x-ray photon interacts with the nucleus, its energy being transformed into two new particles, an electron and position.

photon comes close to a nucleus, its energy may be transformed into two new particles: an electron and a positron (excess energy from the photon is transferred as kinetic energy to these two particles) (Fig. 2.1d). For the most part, pair production is rare in medical x-rays given the high level of energy needed.

The degree to which a given substance allows an x-ray to pass through (versus absorbing or scattering the x-ray) is referred to as *attenuation*. Denser materials, particularly comprised of larger atoms, such as the calcium in bone, will absorb more x-rays than soft tissue or fluids. Indeed, photoelectric effects are proportional to the cube of the atomic number of the material. A projectional image is thus formed by capturing those x-ray photons that are successfully transmitted from a source through an object to a detector that is designed to capture the photons.

Dosage. We briefly touch upon the issue of ionizing radiation and patient exposure. Typically, we speak of radiation dosage to describe the amount of radiation absorbed by tissue. The amount of radiation absorbed by tissue is measured in terms of energy absorbed per unit mass; this unit is called a *gray* (Gy), and is defined as: 1 Gy = 1 J/kg. A *dose equivalent* is a weighted measure that accounts for the fact that some types of radiation are more detrimental to tissue than others; the unit for this measure is called a *sievert* (Sv). A sievert is defined as: 1 Sv = 1 J/Kg x *radiation weight factor*, where the radiation weight factor (RWF) depends on the type of radiation. For example, the RWF for x-rays is 1; for neutron radiation, the RWF is 10; and for α-particles, the RWF is 20. The average dose of radiation that a person receives annually from natural sources is ~360 μSv. Regulations state that the maximal allowable maximal amount for most individuals is 1 mSv/year; and for those individuals working closely with radiation, 50 mSv/year. As a point of comparison, a single chest x-ray provides ~500 μSv. Ultimately, a key drive of imaging technology is to minimize the total amount of ionizing radiation exposure to the patient while balancing the ability of the modality to provide diagnostic images.

Imaging

Fig. 2.2 outlines the rudimentary idea behind using x-rays as a means to create medical images. A controlled and focused source of x-rays is allowed to pass through the anatomy of interest; a *detector* is then responsible for quantifying the amount and pattern of x-ray photons, converting the information into a visual image. Detailed discussions of projectional image formation can be found in [36, 39].

X-ray generation. X-rays are generated when electrons of sufficient energy hit certain materials. Generally speaking, a source of electrons is generated by heating a metal cathode (filament) made of tungsten coil; an electrical current is used to induce *thermionic emission*. These released free photoelectrons are then accelerated toward a rotating target anode, usually made of tungsten, copper, or molybdenum. On hitting this surface, the photoelectrons decelerate, leading to the emission of x-ray radiation and thermal energy. In particular, the x-rays are created when the accelerated photoelectrons release some of their energy in interacting with an atom. Two processes generate these x-rays: 1) *bremsstrahlung* (German for "breaking radiation"), where the electron collides with a nucleus and its kinetic energy is completely converted into x-ray photons; and 2) *K-shell emission*, in which the accelerated electron hits another lower-energy bound electron resulting in the same outcome as the photoelectric effect (a photoelectron and characteristic radiation are generated). X-rays produced by the

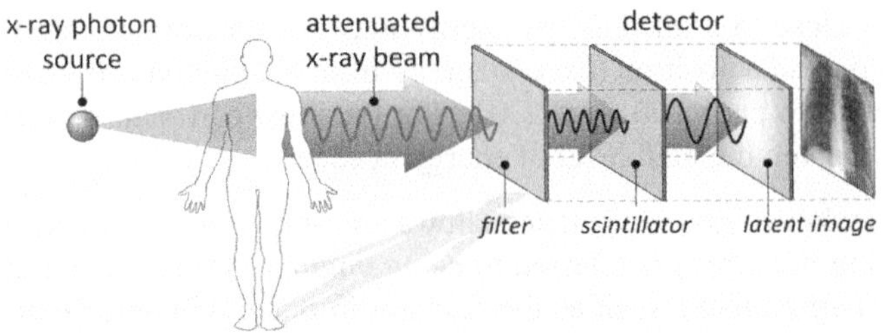

Figure 2.2: An x-ray source is focused into a beam that penetrates the patient, resulting in attenuated x-rays. A filter then removes scatter generated from photon-electron interaction, and the x-rays are detected by a scintillating material that transforms the signal (*e.g.,* into light or an electrical current). The result is a detectable latent image.

former phenomenon are the most useful, and are sometimes referred to as *white radiation.* Fig. 2.3a shows the structure and components of an x-ray tube. A voltage is applied to produce a current across the cathode/anode; and as the voltage increases, the current also increases until a maximal point is reached, the *saturation current*, in which current is limited by the cathode temperature. An x-ray beam's "intensity" is thus measured in terms of milliamperes (mA). Note that the number of x-ray photons generated by the tube is dependent on the number of electrons hitting the anode; this quantity is in turn ultimately controlled by the cathode material's saturation current. Changing the cathode material will therefore result in a different beam intensity. Additionally, the x-rays are of varying energy levels (*i.e.,* polychromatic); for medical imaging, we typically want to use only a portion of this spectrum. For example, there is no reason to expose a patient to non-penetrating x-rays (< 20 keV). The glass encasing the vacuum in which the cathode/anode apparatus exists within an x-ray tube helps to remove some low-energy x-rays. Further filters constructed of thin aluminum can also be placed in the path of the x-ray photons: for instance, a 3 mm layer of aluminum will attenuate more than 90% of low-energy x-rays. This filtering process to remove the lower-energy x-rays is called *beam hardening*. Similarly, copper layers are also sometimes used as filters in order to block high-energy x-rays. The choice of material and the thickness of the filter will determine preferential removal of high- and low-energy x-rays. The x-ray photons generated from this process emanate in all directions; therefore, the x-ray tube is encased in (lead) shielding, with a small aperture to permit some of the x-rays to escape. A *collimator* is used to further refine the beam, limiting its size and controlling the amount permitted to pass through to the patient.

Grids. As the x-rays pass through an object, photons generated as a result of scattering effects occur (*e.g.,* Compton effect), thus resulting in signal noise that degrades end image quality (the consequence is sometimes called *radiographic fog*). To minimize this effect, a (anti-scatter) grid made of high attenuation material is typically placed in front of the detector to block scatter: regularly spaced gaps (or x-ray transmitting material) allow select rays through based on directionality (Fig. 2.3b). By way of illustration, the grid may consist of alternating strips of aluminum and lead, the former material transmitting and the latter absorbing the x-rays. The geometry of the grid ultimately affects the degree of scatter that impacts image formation.

Image contrast. In x-ray images, contrast refers to the difference in visible grayscales seen as a result of differences in attenuation. Given the process of generating a projectional image, there are in general four variables that control the contrast seen in a latent image: 1) *thickness*, in which two objects of the same composition, but one

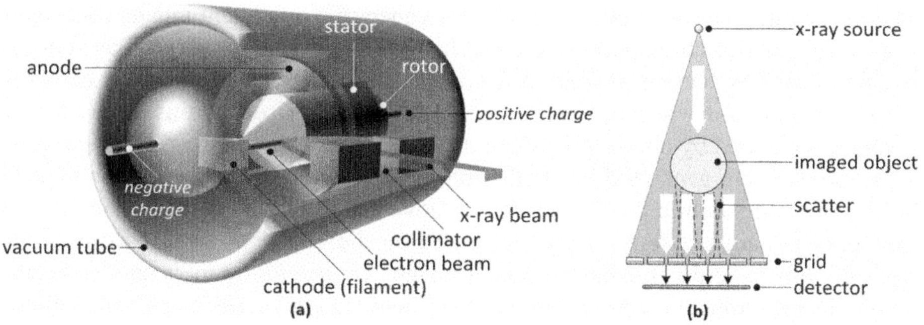

Figure 2.3: **(a)** Cutaway illustration of a x-ray vacuum tube and its components. A potential difference is created between a cathode/anode, resulting in electrons hitting a metal surface. The result is x-ray photons, which are emitted through a collimator. The entire assembly is encased in a vacuum tube and typically shielded. **(b)** A grid is used to remove smatter arising from the Compton effect.

thicker than another, when imaged together the thinner object will produce more contrast; 2) *density*, where more dense materials (*e.g.*, a solid vs. a liquid) will produce higher x-ray attenuation; 3) *material*, where the effective atomic number and attenuation curve dictate interaction with x-ray photons; and 4) *x-ray tube voltage*, which controls the energy of the photons and hence the degree of penetration (higher voltage increases contrast). The first three of these variables can be explained by examining an x-ray's intensity as it passes through a material. X-ray intensity, I, through a material is given by the following equation: $I = I_0 e^{-\mu t}$ where I_0 is the incident x-ray intensity, μ is the *linear attenuation coefficient*, and t is the thickness of the material. μ reflects the removal of x-ray photons from a beam through the interaction of electrons in the material: the higher the electron density, the more likely an interaction between an electron and x-ray photon.

Conventional image formation. Photographic films coated with materials sensitive to x-rays are still perhaps the most commonly used means of forming images. The procedure of exposing film to photons generates a *latent image* that can then be processed to create a visible image. The film itself is usually a transparent plastic sheet that is covered with a radiation-sensitive emulsion; silver halide (*i.e.*, a compound formed by silver and a halogen, such as silver bromide) crystals in gelatin is often used for this purpose. In brief, when a silver halide crystal absorbs x-ray photons, imperfections in the crystal (so-called sensitivity specks) will turn into regions of metallic silver. If a sufficient number of silver atoms are present in an area, the crystal is rendered developable so that the use of a developing solution will change the entire crystal into silver. Hence, those areas that are exposed to more photons will be developed more. On film, developed regions are shown as black. Because of the relatively low effective atomic number of the film, only 3-5% of the x-rays will actually react with the emulsion (the rest pass directly through). Lower-energy light photons are actually easier for film to capture. Based on this fact, an *intensifying screen* is used to enhance the interaction between the film and the x-ray photons. One intensifying technique is to use a fluorescent screen made up of a layer of phosphor that absorbs the x-rays and re-emits visible radiation that is picked up by the silver halide crystals. Current screens can achieve intensification of up to 250x. Thus, combined screen-film systems can reduce the exposure time – but at the cost of some loss of detail due to diffusion effects from fluorescence.

Other techniques have also been explored for generation of a latent image. *Ionography* is one means of detection predicated on a chamber filled with a gas (such as xenon) at high pressure (~5-10 atmospheres). A high potential difference is generated across the chamber, resulting in a strong electric field. The chamber also contains electrodes, one of which is covered by a thin foil. When the x-ray photons interact with the gas molecules, ion pairs are generated inside the ionization chamber. The ions are attracted to the chamber sides, while the free electrons move toward the electrodes. The electrons thus form a charge pattern on the foil based on the concentration of x-ray photon exposure; this pattern is the desired latent image. *Xeroradiography* is another method of x-ray image formation: a plate formed of layers of aluminum, selenium, and aluminum oxide is charged and subsequently exposed to the x-rays. When the x-ray photons impinge on the selenium, a positive charge discharges in proportion to the amount of x-ray exposure; this technique exploits the principle of photoconduction. The selenium surface thus forms a latent image. For the most part, ionography and xeroradiography are less common today given the advent of digital detectors (see below).

Computed radiography. Unlike film, which must act both as an image receptor and as the image display medium, computed radiography (CR) systems separate the task of photon detection and image display. *Photostimulable luminescent phosphor plates* (PSL phosphor plates) are used as the primary image receptor material in CR. These "imaging plates" are similar in concept to conventional radiographic intensifying screens. The major difference between CR and conventional intensifying screens is in the luminescence process. Conventional screens are designed so that the x-ray photon energy absorbed within the phosphor results in prompt fluorescent emission. PSL imaging plates on the other hand are designed so that a large portion of the absorbed x-ray energy is stored within the PSL phosphor material as trapped excited electrons. This stored energy gives rise to a sort of latent image in the PSL plate itself. As such, computed radiography systems are often referred to as *storage phosphor systems*.

Once the PSL plate is exposed, the next stage is CR image formation. At a high level, this process is based on the use of a laser to stimulate the trapped electrons to emit visible light, and a photomultiplier tube (PMT) that captures the light signal and transforms it into a detectable current that quantifies the degree of x-ray photon exposure. The image formation process can be broken up into three major steps:

1. Image pre-read. Before the imaging plate is actually scanned, a *pre-read* of the plate is performed. A survey of points is made using a low power laser to determine the minimum, maximum, and mean exposure values on the plate. These values are used to optimize the high voltage and amplifier gain settings on the photomultiplier and signal conditioning circuits. The minimum, maximum, and mean exposure values are also used to determine the appropriate digital transformation tables (see below) for optimal image display. The pre-read stimulates only a small percentage of the total trapped electrons on the plate so that the latent image is relatively unaltered. A pre-read is analogous to exposure readings performed on autofocus/auto-exposure cameras that automatically set shutter speed and aperture size based on survey information about the light intensity in various image zones.
2. Image main read. Given the information from the pre-read, the main read samples the imaging plate at several points (~4 million over an 8 x 10" area). Each sampled point details the number of trapped electrons in a particular area of the imaging plate. When a point on an exposed plate is stimulated by the laser beam (spot size between 80-200 μm), the trapped electrons in the spot are released and return to a

ground state; in doing so, luminescence radiation with intensity proportional to the absorbed x-ray energy is liberated. This visible light is then amplified by a PMT, which detects the photons and generates a current signal proportional to the number of x-ray photons. The laser scans the PSL plate in a raster format, and detection of the PMT current signal is accordingly synchronized. The analog PMT signal is then transformed into a digital value, thereby allowing a digital matrix representation of the visible image to be computed.

3. Image processing. The digital image is then optimized, usually via tonal conversion and edge enhancement. The goal of the *tonal conversion* is to optimize the image contrast based on: 1) display transfer characteristics; 2) anatomical region of interest; and 3) review medium. In the first stage, a transformation curve is employed based on the capture of the x-rays to the PSL plate is employed. Photostimulable phosphors have a much wider dynamic range than x-ray film: the former's exposure range is about 10,000:1, meaning that the linear response for PSL intensity can go from 5 microroentgens to 50 milliroentgens. In the second stage, the PMT signal digitization to a value (usually a range of 2^{10} values is selected) is transformed dependent upon algorithms that attempt to optimize image contrast based on the anatomical region under examination. Preferred lookup tables and the minimum, maximum, and average PSL values obtained during the pre-read operation are used to compute a transform curve. In the third stage, the transformation goes from a digital pixel value to an analog value (*e.g.*, optical density or luminance), depending upon the characteristic curve of the film or monitor on which the final image is viewed. *Edge enhancement* is often performed on CR images to present an image to the radiologist that is "sharper." This operation is often useful for bone imaging to accentuate fine lines and boundaries between structures. The algorithm used in most CR systems to create an edge enhanced image is called *unsharp masking*.

Digital radiography. Not to be confused with CR, digital radiography (DR; also referred to as *direct radiography*[1]) forgoes the use of the cartridge containing the PSL and separate reader to process the latent image. Instead, digital x-ray sensors are used and the latent image data are transferred directly from a detector to a review monitor. [17, 88] provide earlier descriptions of DR systems; and a more recent general discussion is found in [50]. The different technologies that have been developed as digital detectors can be grouped twofold:

1. Indirect conversion. In this first category, a scintillator is used as an intermediate between the detection of x-ray photons and the generation of visible light, much like an intensifying screen. A *scintillator* is a material that exhibits the property of luminescence when excited by ionizing radiation. The visible light is then detected in a number of ways. Charge-coupled device (CCD) cameras are one method. A CCD camera is a relatively small image sensing device (~3-5 cm^2) that contains a photoactive region. One or more of CCDs are combined in a digital detector, with the image from the scintillator downscaled using an optical lens or fiber optics to project onto the smaller area of the CCD's light-sensitive area. From the CCD, the image can then be read. Alternatively, an amorphous silicon

[1] There remains some ambiguity in terminology, as digital radiography is sometimes used to refer to an umbrella term for both computed radiography and direct radiography. Here, we refer to digital radiography as a separate entity from computed radiography.

photodiode circuitry layer integrated with a thin-film transistor (TFT) array can be used. Using this method, a large flat-panel sensor is constructed by the deposition of silicon with an array of photodiodes, subsequently covered with a scintillator. When this scintillator emits light, the photodiodes are activated and generate an electric charge; the TFT array then records this information to generate an image, with this array correlating directly to the image's pixels.

2. Direct conversion. A range of solid state materials can be used to detect x-rays. For instance, lithium-doped germanium or silicon can detect x-rays. Photons hitting these materials cause the formation of an electron hole pair that can be detected. Current direct conversion methods use a photoconductor, such as amorphous selenium, to convert x-ray photons directly into electrical charges. As in xeroradiography, the charge pattern on the selenium is proportional to the incident x-ray exposure; however, a TFT array is instead used to record the electric charge and to create a digital representation.

Flat-panel detectors (both direct and indirect) have resolution only dependent on the recording device (*i.e.,* the TFT array). Present pixel sizes are on the order of 200 μm for use in thoracic imaging, and 50-100 μm for mammography. A key advantage of direct radiography over CR is the almost immediate generation of the visual image while still preserving the ability to perform digital image processing to improve image quality and reduce noise/artifacts.

Fluoroscopy. The fundamental idea behind fluoroscopy is to use x-rays to provide real-time display of anatomy/physiology, allowing visualization of movement. The original design entailed the use of a fluorescent screen as the detector, allowing an observer to view images directly. Unfortunately, such images are often dim: significant improvement was brought about when x-ray *image intensifier tubes* were introduced, providing much brighter images. The simplest image intensifier tubes are composed of an input phosphor layer coupled with a photocathode, a vacuum tube, and a smaller output phosphor layer integrated with an output window for display. X-rays photons hit the input phosphor (usually cesium iodide, doped with sodium onto an aluminum substrate), which in turn scintillates and emits light photons picked up by the photocathode. The photocathode, often made of an antimony-cesium alloy, produces electrons that are accelerated through a vacuum and focused onto the output phosphor layer. This final phosphor layer luminesces brightly in response to the concentrated electrons, showing the latent image. The image intensifier thus results in a brighter image through two effects: 1) *flux gain*, where the electrons accelerated through the vacuum produce more light as they strike the output phosphor; and 2) *minification*, as the number of light photons is concentrated in a smaller display. The input image's brightness is enhanced by a factor of approximately 10^5 times. Frequently used with image intensifiers are video cameras to record the images.

More recently, digital fluoroscopy has come about with the introduction of flat-panel detectors; additionally, the video camera has been replaced with a CCD-based camera. This change introduces new capabilities, such as *road mapping*, which allows the viewer to temporarily store and display the most recent fluoroscopic image on a screen (*e.g.,* for interventional procedures, such as the placement of guide wires). *Digital subtraction angiography* is another ability, wherein pre- and post-contrast images are combined together to examine the distribution of contrast. And like computed radiography, an advantage of digital fluoroscopy is the ability to perform further image processing to optimize image visualization. For example, frame averaging and edge enhancement can be used to improve image presentation.

Projectional image artifacts. In medical imaging, one often talks of *spatial resolution* – that is, the ability to visually resolve fine details. Higher spatial resolution implies better discrimination of smaller objects. In projectional imaging, there are four sources of *unsharpness* that decrease spatial resolution:

1. Motion blur. Although radiologic image exposure times are relatively short, they are not instantaneous. During this time, a patient may move and/or physiologic processes occur (*e.g.*, normal cardiac motion), therefore causing a blurring artifact as the boundaries of an object are spread over the detector's field.
2. Geometric blur. In reality, an x-ray source is not an idealized point source of photons. Thus, geometric unsharpness occurs due to the physical geometry of image acquisition and image formation, and is influenced by factors such as the size of the x-ray source, the distance between the source and the patient, and the distance from the patient to the detector. Regions at the edge of an object will be formed such that x-ray intensity will gradually increase/decrease, causing unsharpness. These regions are called *penumbra*.
3. Absorption blur. X-rays are not uniformly absorbed by an object; rather, there is a gradated change in x-ray absorption across its boundary. Consider, for instance, the difference between an object who's edges are parallel to the cone of an x-ray beam, versus a perfect sphere: the former will have sharply defined edges as absorption will be uniform, whereas the different points of the sphere will encounter varying amounts of x-ray photons (the center will see maximal amounts, the outer regions the minimum).
4. Detector blur. Lastly, the detector itself can introduce certain phenomena that will create image blur. For instance, the use of an intensifying screen will result in a finite amount of diffusion given the distance between the screen and the film.

[19, 36] provides further details on the geometry of the radiographic image and the reasons for unsharpness.

Computed Tomography

Key work during the mid-20th century in x-ray reconstruction and the theory behind axial tomography led to the development of the first commercial computed tomography (CT) scanner in 1971 for clinical purposes [37]. Relative to conventional projectional x-ray imaging where subtle differences in attenuation (less than 5%) are often lost, CT provides much improved subject contrast with discrimination less than 1% and the current generation of multi-slice CT scanners provide sub-millimeter resolution. We note here that the core physical concept behind CT, x-ray attenuation, is described prior; below, we focus on the principles that enable image formation. The reader is referred to [15] for a complete handling of CT imaging.

Imaging

The projection of an x-ray through an object can be defined through a set of line integrals, representing the total attenuation of the beam as it travels through the different materials composing the object. Recall from the discussion of projectional imaging that x-ray attenuation through one material is described by the equation, $I = I_0 e^{-\mu t}$. The attenuation effect is cumulative so that the transmission of an x-ray through multiple substances is given by the formula:

$$I = I_0 e^{-\int \mu(x,y)ds}$$

where $\mu(x,y)$ is the attenuation coefficient at point (x, y) along the beam's path. Given an *xy*-plane through the object, let r represent the path of an x-ray beam in this plane. Then the above equation can be rewritten in terms of the total attenuation, p:

$$p(r,\theta) = \ln\left(\frac{I}{I_0}\right) = -\int \mu(x,y)ds = \iint \mu(x,y)\delta(x\cos\theta + y\sin\theta - r)dxdy$$

where θ is the angle formed between r and the x-axis (*i.e.*, $r = x\cos\theta + y\sin\theta$). The function, $p(r, \theta)$, is referred to as the *Radon transform*. By recovering $\mu(x, y)$ via an inverse Radon transform, a cross-sectional image of the object in the *xy*-plane is possible: this process is the premise behind *tomographic reconstruction*. In theory, given an infinite number of measurements, one can reconstruct $\mu(x, y)$ perfectly; CT thus uses multiple narrow x-ray beams through the same point in order to collect enough data to sufficiently approximate $\mu(x, y)$ and reconstruct an image. A *sinogram* is the raw data obtained from a tomographic reconstruction, a visual representation of the Radon transform. Each row of the sinogram represents a different projection through the object (Fig. 2.4b).

Reconstruction algorithms. Central to CT imaging is a means to efficiently perform the inverse Radon transform. Several algorithms exist for this purpose, and can be categorized threefold [36]:

1. Simple back-projection. The most straightforward method to reconstruct a 2D image starts by assuming an empty, equally-spaced image matrix. As each x-ray beam contributes to the estimation of μ, the algorithm aims to sum the attenuation from each beam for point (x,y) in the image matrix. The relative contribution of each x-ray path (ray) through the object can be determined knowing the angle at which the ray is transmitted. This procedure is known as *simple back-projection.* The back-projection is created by "smearing" a ray back through the image in the direction it was originally acquired. Conceptually, one can think of this algorithm as adding the value of μ to each pixel based on the rays going through the pixel. While simple to implement, simple back-projection tends to blur image features (Fig. 2.4c) as the point spread function of a back-projection is circularly symmetric, decreasing as an inverse function of the radius ($1/r$).
2. Filtered back-projection. To overcome the blurring in simple back-projection, each ray can be filtered or convolved with a *kernel* prior to the back-projection. In *filtered back-projection,* the filter has the effect of weighting the center of a ray while underweighting the periphery, thus counteracting the blur. Mathematically, the convolution operation in the spatial domain is represented by $p'(x) = p(x) \otimes k(x)$ where $p(x)$ is the original projection data, $k(x)$ is the kernel, $p'(x)$ is the resultant filtered data, and $\otimes$ represents the integral convolution operation. Alternatively, this same operation can be considered in the frequency domain using a Fourier transform (FT), $p'(x) = \mathrm{FT}^{-1}(\mathrm{FT}(p(x)) \times K(f))$, where $K(f) = \mathrm{FT}(p(x))$. This transformation is often called the *Fourier slice theorem.* The advantage of considering this process in the frequency domain is that the convolution operation is transformed into a multiplication. Various kernels exist dependent on the imaging application

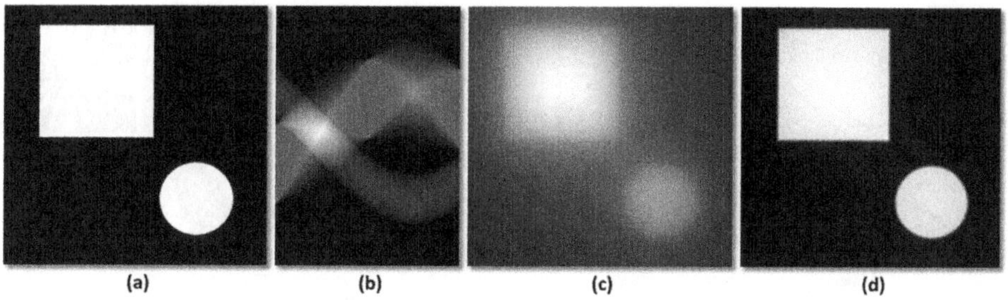

Figure 2.4: Demonstration of CT image reconstruction process. **(a)** Example black and white image source image with two shapes is shown. **(b)** A sinogram is a visual representation of the Radon transform, where each row/column of the sinogram represents projection information. A sinogram with 180 samples for the image in (a) is shown. **(c)** Simple back-projection results in a blurring of the image. **(d)** Different kernels can be applied in filtered back-projection to handle the blurring caused by a point-spread function. Here, a Hamming filter is used to improve the reconstruction; although improved, there are still subtle imaging artifacts relative to the original image.

(*e.g.,* soft tissue or bone visualization), and will affect the in-plane resolution and noise seen in the final image. For instance, in the frequency domain, one can apply the Ram-Lak filter to undo the $1/r$ blurring phenomena; however, this method is highly sensitive to noise, especially at higher frequencies. More robust techniques include the use of a Shepp-Logan, cosine, or Hamming filters that compensate with high-frequency roll-off (Fig 2.4d; Fig. 2.5). For the most part, filtered back-projection is the method used today in clinical CT scanners.

3. Series expansion. Both simple and filtered back-projection algorithms can be run while raw image data is acquired, allowing for more immediate reconstruction. In comparison, *series expansion techniques* (also known as *iterative techniques* and *algebraic reconstruction*) require that all x-ray attenuation data be available before reconstruction commences. Series expansion techniques involve solving large systems of linear equations based on the observed attenuations from each ray; the linear system represents the target image for reconstruction. Examples of these methods include the algebraic reconstruction technique (ART); iterative least-squares technique (ILST); and simultaneous iterative reconstruction technique (SIRT). [43] provides a thorough discussion of these methods.

We note here that the image grid used in reconstruction can be seen as a discretization superimposed on a physical region; as such, the mapping between a given pixel (voxel) to a single point in space is imperfect. For instance, what if the pixel boundary encompasses two different types of materials (*e.g.,* bone and soft tissue; air-surface boundaries)? The resultant attenuation for that pixel is an intermediate value of the two substances. This artifact is referred to as *partial voluming*[2]. Under CT, partial volume averaging often happens when structure boundaries are almost parallel to the CT slice.

Hounsfield units. The results of a reconstruction algorithm are a measure of the attenuation for a given pixel location (x,y). These values are normalized to *Hounsfield units* (HU) prior to generation as final image. Specifically:

---

[2] In point of fact, the partial voluming effect is not unique to computed tomography, but also occurs with other imaging modalities, such as magnetic resonance.

Figure 2.5: (a) Different filters. A ramp function can be used to undo blurring, but is sensitive to noise. Other filters attempt to compensate for higher frequency values. **(b)** Example of different reconstruction filters. From left to right: original image, non-filtered image from simple back-projection, ramp filter, Shepp-Logan filter, cosine filter. The bottom row shows a magnified region in the lower-left the image, where subtle reconstruction artifacts are removed.

$$CT(x,y) = 1000 \times \frac{\mu(x,y) - \mu_{water}}{\mu_{water}}$$

where $\mu(x,y)$ is the computed attenuation value and μ_{water} is the attenuation coefficient of water at the given x-ray beam energy level. For most CT scans, this transformation results in a scale of -1,000 to 3,000 Hounsfield units. These values are principally a representation of a given substance's interaction with x-ray beams due to its density and Compton scattering. Air, therefore, has an HU value of -1,000; and water is 0 HU. Although ranges vary, bone and contrast agent are typically more than 1,000 HU; and soft tissues (including fat) range from -300 to 100 HU. [15] notes that the presence of hydrogen has some influence on the CT value of a substance, allowing for visualization of hydrogenous/fatty tissues on computed tomography; but the predominant factor in determining image contrast between substances remains density.

Display: Windowing and leveling. Based on Hounsfield units, anatomical CT images have a typical grayscale value range of 2^{12} bits (4,096 values). Early image displays were only capable of displaying a grayscale range of 2^8 (256 values); and psychometric studies of the human eye have shown that we are only capable of discerning 40-100 shades of gray simultaneously (dependent on the viewing environment). Hence, CT values are often "downscaled" to a range of 8 bits using a mapping function. This process is referred to as *window and leveling*, and establishes a (linear) transformation between Hounsfield units and visual grayscale: by altering the mapping, alternate tissues and phenomena become more prominent through improved visual contrast (Fig. 2.6a). The *window* refers to the size of the subset of the HU scale that is mapped to grayscale: the values below the window are mapped to 0 (black) and values above the window are mapped to 255 (white). This parameter therefore governs the degree of contrast seen in the image: narrower (smaller) windows are more contrasted. The *level* refers to the location of the midpoint of the window (*i.e.*, where on the HU scale is the window located). Default window-level ranges are often associated with anatomical regions (*e.g.*, window-level settings exist for lung, liver, bone, etc.). Fig. 2.6b-c illustrate the effect of different window-level settings on the same CT data.

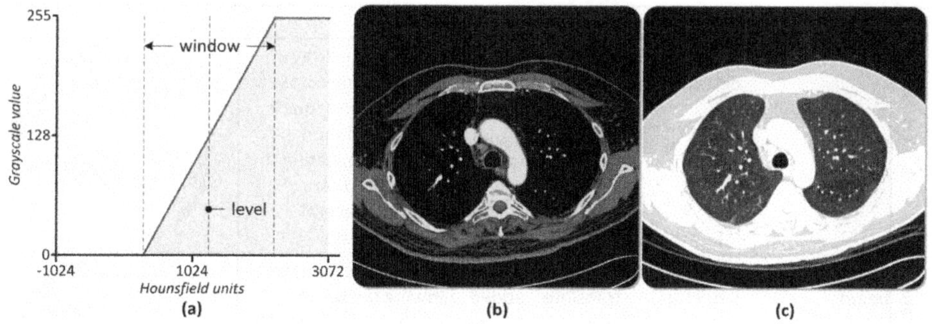

Figure 2.6: (a) Graphical depiction of a linear transform for window and leveling. The Hounsfield units are mapped to grayscale values dependent on the given ramp function. The window defines the degree of contrast, while the level specifies what portion of the spectrum is seen. **(b)** A thoracic CT image shown at a bone window-level setting (window = 2000, level = 500). **(c)** The same image, seen using a lung window-level setting (window = 1500, level = -650). As shown, different anatomical features become apparent using different window-levels.

CT scanner design. The design of CT scanners has gone through several generations, changing the shape of the emitted x-ray beam; the relative motion of the x-ray source, detectors, and patient; and the number of detectors. For instance, the first generation of CT scanners employed a narrow beam x-ray source, a single detector, and a translate-rotate acquisition paradigm wherein the emitter and detector would move parallel to each other for a short distance (*i.e.*, translate), and then would rotate around the gantry and repeat the process. The second generation increased the number of detectors and introduced a fan beam x-ray geometry, thus minimizing translation; and ensuing generations would completely eliminate translational motion. The resultant progression of CT architectures has brought about faster image acquisition: whereas the first generation would require minutes to acquire a single slice, by the third generation a single slice could be obtained in 0.5-1.0 sec. using hundreds of detectors. Table 2.1 summarizes the first four generations of scanners.

A significant improvement in CT scanners was the incorporation of *slip rings* in 3rd and 4th generation scanners. Prior to the use of slip rings, power to the x-ray tubes generating the beams was supplied through wired cabling; and data acquired by the detectors were transmitted back similarly. This design practically limited the amount of gantry rotation to 400-600 degrees before the cables would have to be reversed, thus completely stopping the patient movement through the scanner. With slip rings, power could be provided and data received without cables, allowing continuous rotation of the x-ray source. The consequence was the development of *helical (spiral) CT scanning*, in which the patient could be moved continuously through the scanner with the x-ray beams tracing out a spiral-like path through the body (Fig. 2.7a). Helical CT scanning requires reprocessing of the data into series of planar (axial) image datasets. The benefits of helical CT include: 1) faster scanning of a larger anatomical volume, and hence decreased amounts of contrast media needed to visualize regions; 2) decreased motion artifacts from patient and/or physiological movement (*e.g.*, faster scan times can permit single breath-hold inspiration); and 3) continuous slice acquisition and reconstruction, allowing image slice interpolation (*i.e.*, there are no gaps between image slices). In particular, the feasibility of volumetric acquisition paved the way for

	Geometry	Detectors	Motion	Comment	
1st generation	Pencil-beam	Single detector	Rotate-translate around patient	The detector and the x-ray source moved linearly across a field of view (FOV) to acquire projection data, and then rotated 1 degree, repeating the process across a 180° arc. A single slice took several minutes to obtain data.	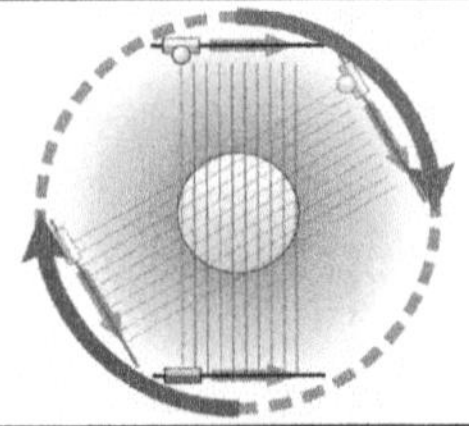
2nd generation	Narrow fan beam	Array of detectors	Rotate-translate around patient	Additional detectors (~30) were added, allowing fan beam geometry to be introduced. This provided some speed-up, but at the cost of additional scatters radiation detection.	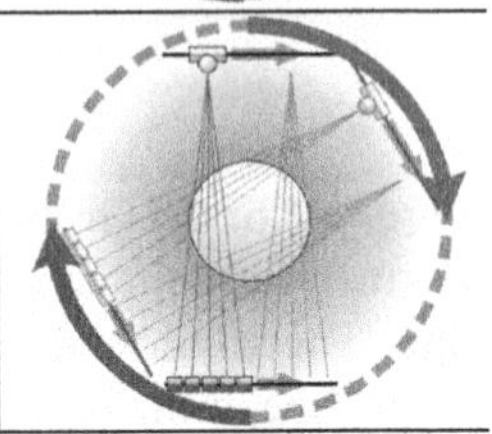
3rd generation	Wide fan beam, with the apex of the fan being the x-ray tube	Array of detectors	Rotate around patient	The array was increased to include hundreds of detectors, removing translational movement; and the angle of the x-ray beam was increased. Unfortunately, this architecture is subject to *ring artifacts* if the detectors are not properly calibrated.	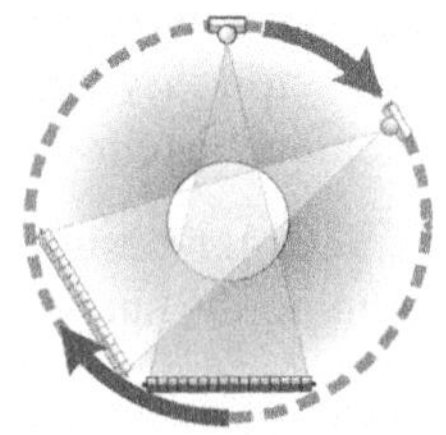
4th generation	Wide fan beam, normalizing the apex of the fan to be the detector	Ring of detectors	Rotate x-ray source only	Thousands of stationary detectors (4,000) are placed in a 360° ring around the patient, removing the need for detector movement and canceling ring artifacts.	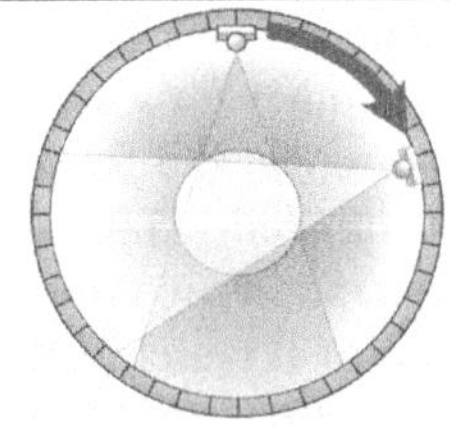

Table 2.1: Summary of the first four generations of CT scanner designs, illustrating the evolution of change in x-ray beam geometry, and the number and placement of detectors. The relative motion and position of the x-ray source and detectors are illustrated in the rightmost column, where the object to be scanned lies in the middle.

CT angiography (see below) and 3D image processing and visualization techniques, including multi-planar reformations (*i.e.*, interpolating viewing in other planes, such as coronal or sagittal) and maximum intensity projections (MIPs).

Varied methods have been explored to quicken CT image acquisition speeds with respect to scanner design:

- Electron-beam (ultrafast) CT scanning. One method for speeding CT image acquisition was seen with the development of *electron-beam computed tomography* (EBCT; also sometimes referred to as ultrafast CT and cine CT). In EBCT, the gantry itself is a large x-ray tube; and instead of moving the x-ray source as in conventional CT, the electron beam focal point (and hence, the x-ray focal point) is swept across the tube's anode and around the patient using magnetic fields. As in 4th generation scanners, a stationary bank of x-ray detectors is used to capture the resultant attenuation data. As EBCT removes all mechanical motion, a scan can be completed extremely quickly, with ~20 images per second (presently, some EBCT commercial systems claim a single image sweep in 0.025 seconds).

As such, EBCT is useful in cardiac imaging. Despite its speed advantage, the cost of EBCT and other technical issues have hampered widespread adoption and newer spiral CT scanner designs are now providing similar capabilities.

- Multi-detector CT scanners. Recognizing that a time-limiting step in earlier generations of CT scanners was the physical speed with which an x-ray beam could be produced, engineers looked for ways to better utilize those x-rays that are already produced. Recall from projectional imaging that a collimator is used to control the size of an x-ray beam; by opening the collimator in a CT scanner, the beam broadens, allowing more x-rays through but at the expense of increasing slice thickness (and therefore, decreasing spatial resolution). To surmount this problem, multiple detector arrays (or rings) are instead used, changing the image formation process so that slice thickness is dependent on the physical size of the detectors rather than the x-ray beam. A single rotation of the x-ray source can therefore generate multiple image slices. Using multiple rows, the coverage in the *z*-axis is anywhere from 24-40 mm. This technique is known as multi-detector computed tomography, or more commonly, *multislice CT* (MSCT). In a *fixed array detector*, the rows of detectors in an MSCT system are of equal size in the *z*-axis; in contrast, in an *adaptive array detector*, the rows are of assorted dimensions, often with the smallest in the center and growing in size toward the outer rows (Fig. 2.7b). Multislice helical scanners allow slice thickness to be specified as part of the (spiral) reconstruction process, combining the information from different rows. As a result, both narrow slices for high-resolution detail and for 3D image post-processing can be derived from the same acquisition dataset. [83] provides a recent discussion of the underlying technology and current trends. As of the end of 2008, commercial MSCT systems had upwards of 320 detector rows and can provide sub-millimeter resolution in the *z*-axis, with nearly isotropic voxel sizes. New techniques have been developed for multislice CT sampling. For example, *double z-sampling* uses periodic motion of the x-ray beam's focal point in the longitudinal direction to enhance data sampling along the *z*-axis. Given the rapid acquisition of imaging data across multiple detectors, a challenge lies in moving the data from the gantry to the computer: for instance, a 64-slice CT system can produce up to 180 to 200 MB/s.

CT x-ray detectors. The first few generations of CT scanners employed xenon-based detectors, using the ionization of the gas by an x-ray photon to induce a current and measure attenuation (much like the use of ionography in projectional imaging). But the limitations of this approach – particularly given its low x-ray detection efficiency and the rapid millisecond response rate needed for modern CT systems – has led to the use of solid-state detectors. Materials for solid-state detectors include light-scintillating compounds like cadmium tungstate ($CdWO_4$), cesium iodide (CsI), and ceramic materials doped with rare-earth oxides; principally, the choice of materials is driven by high detection efficiency and low fluorescence decay time (allowing for fast gantry rotations and the production of artifact-free images). The emitted light is then detected by a silicon photodiode, amplified, and converted into a digital signal (see also the prior discussion with respect to digital radiography).

CT acquisition parameters. Apart from the reconstruction algorithm and kernel, several variables influence how a CT study is obtained and the level of detail and appearance of the images; major parameters are below:

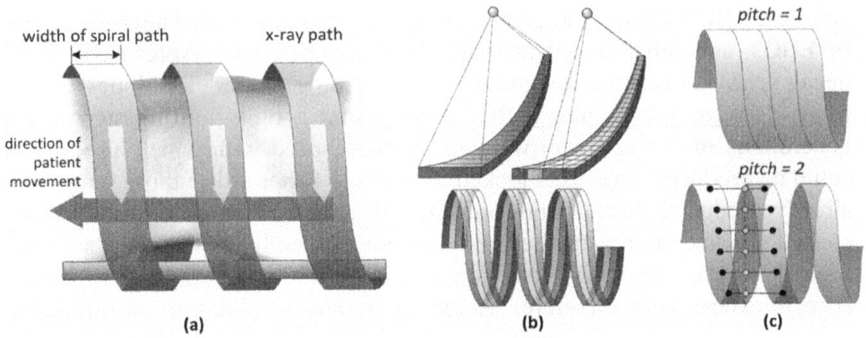

Figure 2.7: (a) Example of helical (spiral) CT scanning process. As the body moves through the scanner, the x-ray tube rotates continuously around the gantry, tracing out a helical path. The width of the spiral path is controlled by the x-ray tube collimation. **(b)** The top shows the difference between a single detector, where collimation drives slice thickness, and the use of multiple detectors. An example of the pathway when multiple detectors are used in a helical CT scanner is shown below. **(c)** Pitch describes the rate of movement of the table relative to the gantry rotation. A pitch value of 1.0 is equal to contiguous axial scanning. Higher pitches decrease radiation dosage, but can introduce require interpolation across a larger spatial volume. For instance, the arbitrary image plane shown in the bottom figure uses projection data from farther away than a pitch of 1.0.

- Slice thickness and collimation. The size of the x-ray beam, and thus the amount seen across detectors is controlled by the collimator opening. In single detector scanners, collimation therefore controls the slice thickness (*i.e.*, *z*-axis dimension). In MSCT, slice thickness is specified during reconstruction by combining information from several detectors. For conventional CT, with all other parameters being equal (*e.g.*, equal kV and mA settings), increasing the slice thickness linearly increases the number of detected x-ray photons and increases the signal-to-noise (SNR) ratio. Increasing slice thickness in these systems will result in improved contrast resolution, but decreased spatial resolution and increased partial voluming effects. An oft discussed concept is a *slice sensitivity profile*, which is the line spread function in the z-axis. The slice sensitivity profile is driven by the finite width of the x-ray focal spot, the penumbra of the collimator, and the finite number of attenuation projections taken from around the patient.
- Table feed/tube rotation. This parameter specifies the rate of table movement (*e.g.*, 5 mm/sec) through the scanner for a given rotation of the x-ray tube. For conventional CT, a volume is scanned section by section. Table movement between rotations can therefore be used to generate overlapping acquisition data (with an increase in radiation dosage to the patient); or conversely, discontinuous information (*e.g.*, for detecting gross findings, such as trauma). For spiral and multislice CT, the table feed rate is used to control pitch[3].
- Pitch. The concept of *pitch* comes into play when considering spiral CT. For single detector spiral CT, *collimator pitch* is defined as the ratio of table feed per rotation to the collimation (*i.e.*, slice thickness), whereas for multi-detector systems, *detector pitch* is given as the ratio of table feed per rotation to detector width:

---

[3] Although for some commercial multi-detector scanners, the slice thickness is independent of table speed because of the helical re-interpolation algorithm.

$$collimator\ \ pitch = \frac{\text{table feed per rotation}}{\text{collimator width}} \qquad detector\ \ pitch = \frac{\text{table feed per rotation}}{\text{detector width}}$$

In 3rd generation CT scanners, detector pitch determines ray spacing; in 4th generation scanners, detector pitch affects view sampling. The relation between detector and collimator pitch is given by dividing the former by the number of detector arrays. In general, a (collimator) pitch < 1.0 produces overlap between sections (and more radiation dosage); pitch = 1.0 indicates no gaps, and for single detector systems, is equivalent to a contiguous axial scan from a conventional CT scanner; and if pitch > 2.0, there is an introduction of gaps in the acquisition and possible artifacts due to an under-sampled volume. Fig. 2.7c visualizes the differences between pitches and the generation of image slices from the acquisition data: as the pitch increases, the sampling points to reconstruct the image become farther apart.

- Tube potential (kVp) and current (mAs). As in projectional imaging, the x-ray tube's voltage and current impacts the production of x-ray photons. The higher the voltage, the more higher-energy x-ray photons will be produced, boosting image contrast. The tube current affects the total number of emitted x-ray photons, and thus affects exposure time and the overall darkness of the image; increasing mAs will decrease noise and increase contrast. Both parameters must be balanced accordingly to minimize the radiation dose to the patient.

Image quality and artifacts. The challenge of a CT protocol is to minimize radiation dosage while maximizing spatial and contrast resolution to provide clinically viable images. CT image noise and artifacts occurs for any number of reasons, including the above acquisition parameters and the choice of reconstruction method [7, 38]. For instance, partial voluming was described earlier, and can result in a loss of detail. And as with other modalities, patient movement, either voluntarily or due to normal physiologic processes (*e.g.*, cardiac movement, breathing) is another source of imaging artifacts. More specific to CT is *beam hardening,* a phenomenon in which lower-energy x-rays are attenuated more than higher-energy x-rays when passing through tissue, thus skewing the average energy distribution toward the latter: the beam is said to "harden." Denser materials, like bone, cause a higher degree of beam hardening. The extent of this shift differs dependent on the projection angle of the beam through the material, which ultimately affects the reconstruction algorithm. Visually, beam hardening can result in *cupping artifacts*, where for a given uniform object, x-rays are attenuated more in the center than in the edges of the object; and *streaking* in the region between two dense objects. Most scanners typically have algorithms that will correct for beam hardening artifacts given the relative attenuation of each x-ray. There are two other causes for streaking. First, *photon starvation* can cause streaking. Attenuation is at its greatest when an x-ray beam travels horizontally through a material; and in the case of a high attenuation substance (*e.g.*, bone), an insufficient number of photons may reach the detector, causing noisy projection data. When reconstructed, this phenomena tends to be magnified and results in horizontal banding across the image. Again, most modern scanners incorporate methods to correct for photon starvation by modulating the x-ray tube current dynamically during acquisition, or through image post-processing. Second, the presence of metal objects (*e.g.*, surgical clips, dental fillings, prosthetic devices) will also cause streaking: the high density and attenuation of such materials is beyond the normal range. Algorithms to minimize the streaking caused by metal are an active area of research (*e.g.*, by re-interpolating the attenuation profile).

Helical CT scanning is prone to additional imaging artifacts due to the method of interpolation and reconstruction. For example, for multislice CT, helical interpolation can produce a *windmill artifact*, which arises as rows of detectors intersect the reconstruction plane [75]. To minimize this occurrence, z-filter helical interpolators can be used. And a *cone beam effect* occurs with wider collimation: the x-ray beam begins to take on a cone-shape, as opposed to the assumed flat plane of the detectors. The outcome is akin to partial voluming, and tends to be more pronounced for data on the outer rows of the detector array. As the number of detector rows and longitudinal space covered by detectors increases, cone beam effects worsen overall. As a result, newer multislice CT systems have altered the reconstruction algorithms to accommodate cone-beam geometries.

Radiation dosage. The radiation dosage of a CT scan is controlled by several factors, including the beam energy, collimation, and pitch. [15] provides the following equation with regard to the relationship between SNR, pixel dimensions (Δ), slice thickness (T), and radiation dosage (D):

$$D \propto \frac{SNR^2}{\Delta^3 T}$$

In general, the radiation dosage from a CT scan is somewhat higher than that of an equivalent radiographic study. Table 2.2 provides approximated amounts of radiation for different anatomical regions. Unlike projectional imaging, the radiation exposure from CT differs in several ways: 1) CT generates a continuous field around the patient, resulting in radially symmetric exposure, whereas radiography exhibits a higher radiation field at the point of entry than at exit; 2) CT x-ray beams are highly collimated to a much smaller target volume as compared to a radiograph; and 3) CT x-rays typically entail higher kVp and mAs to increase contrast and improve SNR, therefore increasing overall radiation. As such, specific parameters of radiation dosage due to CT have been created. For instance, the *multiple scan average dose* (MSAD) takes into consideration the fact that although x-ray beams are highly focused per slice, there is an accumulative effect to neighboring slices as the radiation profile is not perfect: instead, the periphery receives additional exposure. Thus, for instance, when a slice at position x is acquired, the volume at position $(x - 1)$ and position $(x + 1)$ will also receive some radiation (*e.g.*, due to scattering). MSAD thus computes the mean dose to tissue given a large series of scans:

$$\text{MSAD} = \frac{1}{I} \int D_{series}(z)dz$$

Parameter	Head	Thorax	Abdomen	Pelvis
Scan range (cm)	15	31	24	15
Slice thickness (mm)	5	5	5	3
Scan time (sec)	32	32	40	40
Tube current (mA)	200	150	250	250
Organ of interest	Eye lens	Lung	Liver	Bladder
Organ dose (mSv)	22.2	22.1	21.7	19.1
Effective dose (mSv)	0.9	6.3	6.8	3.9

Table 2.2: Effective dose estimates for different anatomical regions for conventional axial CT and spiral CT with pitch value 1.0.

where I represents the interval of the scan length, $D_{series}(z)$ is the dose at position z resulting from the series of CT scans; and the integral is evaluated from $^{-1}/_{2}$ to $^{1}/_{2}$. An approximation to MSAD is given by the *CT dose index* (CTDI). [59] provides a formal review of these radiation dose metrics for CT.

Additional CT Applications

Computed tomography technology is continually evolving, both with respect to the underlying scanner hardware and the software techniques used to reconstruct the images. Furthermore, CT is being seen in an increasing number of cancer screening applications (*e.g.*, lung, virtual colonoscopy). Several adaptations and applications of CT imaging are described below.

Positron emission tomography (PET)/CT. A well-known nuclear medicine imaging technique is PET, which looks to detect gamma rays emitted by an injected short-life radioactive compound. This compound, called a *radiopharmaceutical*, *radiotracer,* or *radionuclide*, is given intravenously and dependent on what type of isotope is injected, emissions occur either as a single photon, which forms the basis for single photon emission tomography (SPECT); or with a higher energy level, positrons, which form the basis for positron emission tomography. Typically, the tracer is incorporated into a biologically active molecule that will be absorbed by a tissue of interest; fluorodeoxyglucose (FDG) is commonly used for this purpose. The concentration of tracer in a given tissue is then measured by scintillating-based detectors. SPECT/PET therefore gives some sense of physiological and metabolic processes, and is often used to ascertain the degree of activity in a given region (*e.g.*, a tumor, brain function, etc.). Unfortunately, the resolution of PET is much lower than other modalities; hence it is often used in conjunction with CT. PET/CT scanners are increasingly common, allowing for simultaneous acquisition of both channels of image data; and ensuing registration and image fusion of metabolic and spatial information (Fig. 2.8a).

CT perfusion. Broadly, perfusion is defined as the steady-state delivery of blood to tissue, assessing the volume per unit time. The underlying theory behind CT perfusion is that given intravenous contrast bolus, the attenuation of tissue will change based on its delivery through the bloodstream and subsequent uptake. Images taken pre- and post-contrast can provide quantitative maps of the change in attenuation. Notably, regions of tissue not receiving sufficient blood (and therefore, oxygen) are made

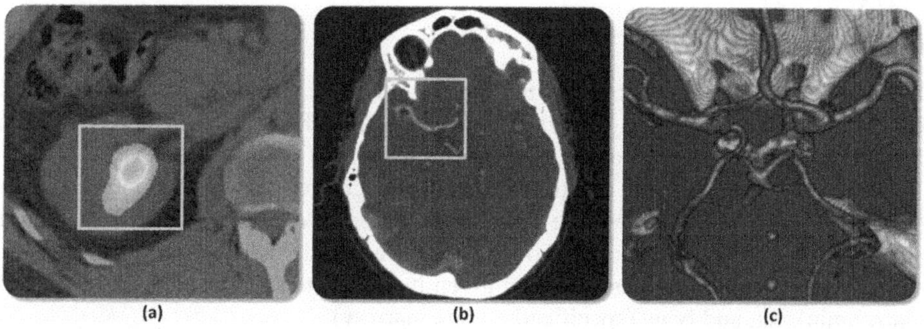

Figure 2.8: **(a)** Example of fused PET/CT data. Here, tracer uptake is seen in the calyx of the kidney as it circulates through the blood; color coding is used to indicate concentration. **(b)** Demonstration of CT angiography. The highlighted region shows a cerebral artery demarcated by contrast. **(c)** A 3D reconstruction of the CTA region.

detectable. For this reason, CT perfusion has been studied with respect to cerebrovascular flow; for instance, the method has been used in the assessment of stroke and other neuro-pathology [23, 48, 57].

CT angiography. With the advent of helical CT, computed tomography angiography (CTA) was introduced and is now commonly used to visualize arterial and venous blood vessels throughout the body, including in the brain, heart, lungs, and abdomen. CTA can be used to detect aneurysms, arteriovenous malformations, vessel narrowing (*e.g.,* atherosclerotic disease, stenosis), and thrombosis (blood clots in veins). A bolus of contrast agent is injected into a periphery vein, and CT images are subsequently taken of the region. Relative to traditional catheter angiography, CTA offers a less invasive, faster means of imaging blood vessels. Multi-detector CT has further improved CT angiography (MD-CTA), providing faster scanning (and less patient motion artifacts), improved contrast, and reduced need for contrast material (for arterial phase studies) [46, 68]. For example, MD-CTA provides high spatial resolution contrast casts of intracranial aneurysms, with the advantages of characterizing the intramural and peri-aneurismal environment, and the accurate depiction of the location and extent of intramural calcification, intraluminal thrombus, and of impinging surrounding bony structures (Fig. 2.8b).

Dual source and dual energy CT. Rather than use a single x-ray tube, a dual source CT (DSCT) scanner incorporates a second x-ray tube and set of detectors, offset at 90° from the first tube/detector arrangement. One tube/detector pair covers the entire scan field, whereas the second provides a more limited scope, central to the field of view. DSCT provides better temporal resolution while also reducing the radiation dosage. Introduced in 2005, initial demonstrations of this technology have been for coronary CTA [25] and other cardiac scans. Running the two x-ray tubes at different energy levels (*e.g.,* one at 80 kVp and another at 140 kVp) presents *dual energy* CT. A limitation of conventional CT is that tissues of different chemical compositions may have similar x-ray attenuation and thus appear similarly on images. This problem can potentially be solved by examining the behavior of the material at two different energies, providing spectral information within a single scan [42]. Information from the detector-tube pairs are reconstructed separately, and then post-processing is applied to compute dual-energy information. Potential applications for dual energy CT include the generation of iodine contrast maps (including its separation from high-attenuation tissue, such as bone) and lung perfusion.

Magnetic Resonance

Developed for clinical purposes in the 1970s, magnetic resonance (MR) imaging is based on the use of a strong magnetic field to align the nuclear magnetization of hydrogen atoms in water molecules. The application of a radiofrequency (RF) field alters the atom's alignment, creating a rotating magnetic field that is measurable by sensors; subsequent changes in the RF field (and hence, detected magnetic field) permit an image to be constructed. Today, MR is a standard cross-sectional imaging modality that is useful for visualization given its ability to image soft tissue (muscles, joints, brain), fat, and bone (specifically, bone marrow).

Core Physical Concepts

[55] provides a more detailed handling of basic MR physics, which we review here. All nuclei possess *intrinsic angular momentum*, also known as *nuclear spin*, which

affects all dynamical nuclear properties (*e.g.*, the interaction of nuclei with magnetic fields, etc.). Additionally, spin is a fundamental property of all elementary particles: electrons, neutrons, and protons all possess intrinsic spin, as do particles with a zero rest-mass (*e.g.*, photons) (Fig. 2.9a). Nuclear spin has two components: 1) the contribution of spin by individual nucleons (*i.e.*, neutrons and protons); and 2) the orbital angular momentum of the nucleons. This orbital motion is actually caused by the spinning of the collective nucleons forming a nucleus, rather than by the independent orbital motion of the individual nucleons. Nuclear spin comes in multiples of ½ and is either positive or negative. A nucleus' net nuclear spin is dependent on the number of protons and neutrons present: particles with opposite spins will cancel out each other's contributions so that the net spin is dependent on the number of unpaired protons and neutron (each adding ½ spin to the net spin). Thus, hydrogen nuclei (^{1}H, 1 proton, no neutrons) exhibit a non-zero spin and as a result there is a *magnetic dipole moment* – similar to how in classical electrodynamics, a rotating electrically charged body generates a magnetic moment. Recall from basic physics that a magnetic moment is a vector quantity (*i.e.*, it has both a quantitative component and a direction).

Spins and external magnetic fields. When spins are placed in a strong external magnetic field (such as emitted by an MR scanner), the nucleons precess around an axis along the direction of the field[4] (Fig. 2.9b, Fig. 2.9c). The frequency of precession is governed by the *Larmor equation*, $\omega_0 = \gamma B_0$, where ω_0 is the angular frequency; γ is the gyromagnetic ratio, a constant specific to the nuclei; and B_0 is the strength of the applied magnetic field. Specifically, protons will align themselves in either a low- or high-energy state, corresponding to the relative parallel or opposing alignment of the spin to the external magnetic field (Zeeman effect). The number of nuclei that move into the lower energy state is only slightly greater that those that move into the higher energy state – the difference in numbers is less than 0.001% but is enough to be exploited in MR imaging, with the exact proportion determined by the strength of the externally applied magnetic field and temperature. The sum of the magnetization vectors from all of the spins is the termed the *net magnetization*. The net magnetization vector, M, consists of two parts: 1) the *longitudinal magnetization*, M_z, which arises from the excess of protons in the lower energy state and causes net polarization parallel to the external magnetic field (*i.e.*, in the z-axis); and 2) the *transverse magnetization*, M_{xy}, which is due to coherence forming between two protons (*i.e.*, becoming in phase), causing net polarization perpendicular to the external magnetic field (*i.e.*, in the xy-plane).

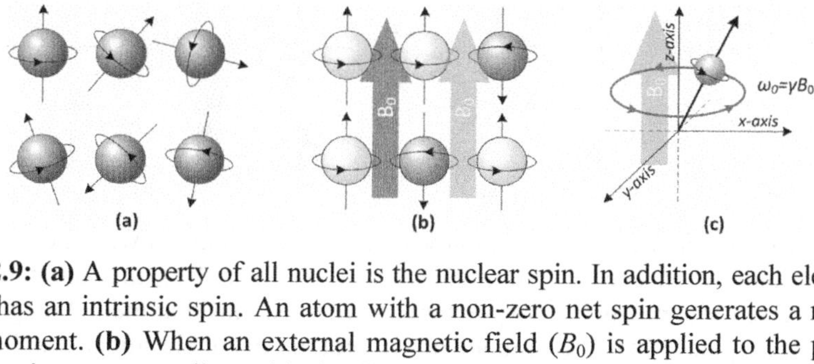

Figure 2.9: **(a)** A property of all nuclei is the nuclear spin. In addition, each elementary particle has an intrinsic spin. An atom with a non-zero net spin generates a magnetic dipole moment. **(b)** When an external magnetic field (B_0) is applied to the particles, the magnetic moments align with the field, being either parallel or opposite in direction. **(c)** The alignment to B_0 is not exact; instead, the nucleons precess around the axis with an angular frequency, ω_0, given by the Larmor equation.

[4] By convention, the direction of the field is taken to be the z-axis.

Applying a radiofrequency pulse. When the system is stable, the transverse magnetization component is zero; but when energy is introduced into this system, both the longitudinal and transverse components change. If an electromagnetic wave, such as an RF pulse, is introduced to a stable spin system at the same angular frequency as ω_0, then the energy is absorbed via *resonance* and the net magnetization moves away from the direction of the main magnetic field: the spins are excited such that the longitudinal magnetization tips away from the *z*-axis, and the protons become in phase (Fig. 2.10a). Note that the angle at which the RF pulse is applied relative to B_0 determines the extent to which net magnetization vector is affected. This angle is commonly referred to as the *flip angle*, *α*.

Once the RF pulse is terminated, the protons begin to re-radiate the absorbed energy at the same resonance frequency; this signal is detected via the scanner's RF coil (by induction of a current). The initial amplitude of this signal is given by the amount of "tipping" from the *z*-axis toward the *xy*-plane. A flip angle of 90° produces maximal tipping into the *xy*-plane. As the absorbed energy dissipates, both M_z and M_{xy} return to an equilibrium state: recovery of excited spins to the longitudinal magnetization to the *z*-axis is called *T1 relaxation* or *recovery* (also called spin-lattice relaxation); and the restoration of the transverse magnetization to zero is called *T2 relaxation* (also called spin-spin relaxation). T1 is a specific measure of how long it takes for M_z to return to 63% of its original value and is associated with the enthalpy of the spin system (the amount of spins in parallel/anti-parallel state). T2 is quantified by the amount of time it takes M_{xy} to return to 37% of its original value and is associated with a system's entropy (the amount of spins in phase). In particular, the T2 phenomenon results from *spin-spin interactions* where two nearby protons can cause each other to "flip" so that one changes from anti-parallel to parallel alignment, while the other changes from parallel to anti-parallel (*i.e.,* one gains the excitation energy from the other). Phase coherence with other excited protons is lost during this exchange and the end result is a relaxation of the transverse magnetization. T1 and T2 will vary dependent on the material (Fig. 2.10b).

The T2 signal generated from spin-spin relaxation decays quickly due to the loss of coherence. This signal is referred to as *free induction decay* (FID). T2 assumes a perfect

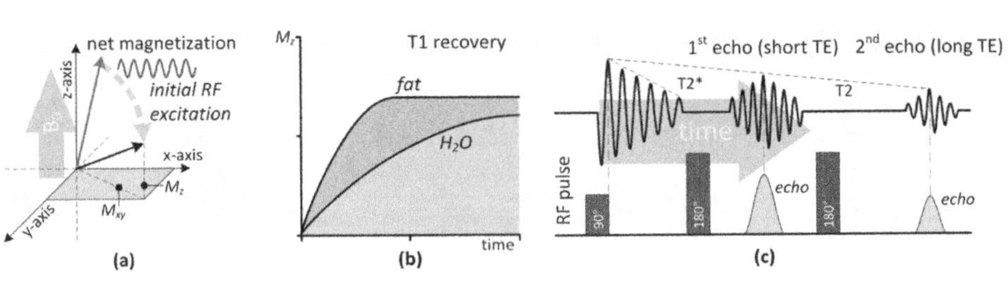

Figure 2.10: (a) The net magnetization vector is closely aligned with the *z*-axis when in equilibrium. When energy is introduced (*e.g.,* an RF pulse), the longitudinal (M_z) and transverse (M_{xy}) magnetization vectors are affected, pushing the net magnetization toward the *xy*-plane. **(b)** The amount of ^{1}H protons available alters the T1 and T2 properties of a substance. For example, the T1 relaxation for fat is shorter than that of water. **(c)** Example of a spin echo and the difference between T2 and T2*. A 90° RF pulse initiates the sequence; subsequently, a 180° RF pulse is used to flip spins, creating another signal peak.

external magnetic field – but in reality, fields often have inhomogeneities and external sources of interference exist, causing a further loss of phase coherence. *T2** therefore reflects the additional compounding effect of imperfections in the external magnetic field. To recover some of the some signal and information about the environment, a technique called *spin echo* is sometimes used: a refocusing RF pulse is used so that the spins are flipped 180°, inverting the phase-position of each spin. In flipping the spins, those protons precessing faster are effectively made "behind" spins that were processing at a slower rate. After some time, the spins will have caught up with each other and a "spin echo" is created: a signal peak is seen when this occurs (Fig. 2.10c). The time taken for this signal peak to be seen is referred to as the *echo time* (TE).

Imaging

So, what do T1 and T2/T2* mean in terms of imaging human tissue? The protons that generate MR signals are primarily those in cellular fluids and lipids (*i.e.,* the nuclei of hydrogen atoms that are relatively free to move within their environment). The hydrogen protons in tightly-bound environments such as within proteins or DNA usually do not contribute to MR signals, and the same situation exists for those in solid structures such as bone. We can first of all consider a water molecule moving through its environment within tissue as a result of local chemical and physical forces. The magnetic properties of its two hydrogen protons generate a small magnetic field of ~1 mT and the molecule's motion is therefore also influenced by the magnetic properties of the other water molecules in its vicinity (and reciprocally influences their motion). When excited protons are involved following RF excitation, it is the interactions with their local environment that cause them to lose their excess energy and return to the lower energy state with the emission of RF radiation (*i.e.,* this can be seen as the basis for re-establishing longitudinal magnetization during relaxation). The rate at which molecules can move within their environment is related to their size and thus small molecules have a low probability for interaction. Hence liquids such as cerebral spinal fluid (CSF) have long T1 values, for instance (Table 2.3). Medium-sized molecules (*e.g.,* lipids), in contrast, move more slowly, have a greater probability for interaction as a result, and exhibit relatively short T1 values. In contrast, T2 relaxation reflects spin-spin interactions, which tend to happen much more rapidly than T1 relaxation; T2 values are therefore generally less than T1 values (Table 2.3).

As T2 arises mainly from neighboring protons, a higher interaction probability exists with larger molecules over smaller molecules. Macromolecular environments will therefore display shorter T2 values than water-based fluids (*e.g.,* CSF). A final point to note is that both T1 and T2 measurements in a small volume of tissue result from the integrated motional effects of all compounds that contain hydrogen protons in that volume, be they those of small molecules, lipids or macromolecules.

Tissue	T1 (msec)	T2 (msec)	Tissue	T1 (msec)	T2 (msec)
Muscle	870	47	Gray matter (brain)	920	100
Liver	490	43	White matter (brain)	790	92
Kidney	650	58	CSF	2400	160
Lung	830	80			

Table 2.3: Examples of average T1 and T2 times for different types of tissues. T1 relaxation times are much higher than T2 relaxation times.

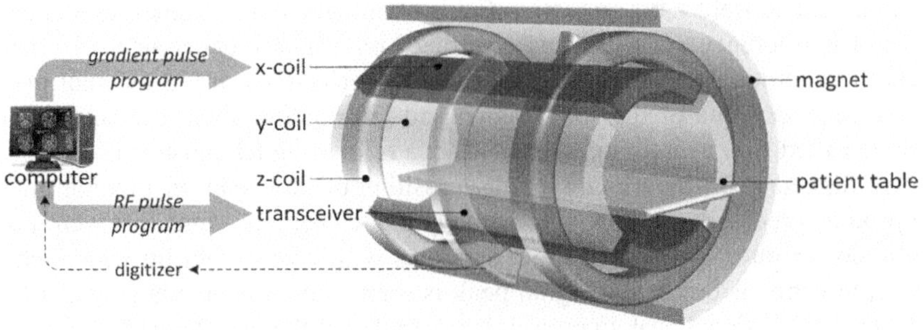

Figure 2.11: Cross-section of an MR scanner, showing the relationship of the gradient coils (x-, y-, and z-coil), the RF transceiver (*i.e.,* RF pulse generator and receiver), the external magnet, and the patient table. The z-coil surround the entire table and permit slice selection. The y-coil is to the left/right of the table and is responsible for phase encoding. The x-coil is atop and below the patient, and is used for frequency encoding. A computer is responsible for coordinating the execution of the pulse sequence across the magnets/coils and taking information from the digitizer to create an image.

Gradients and k-space. Thus far, discussion has only centered about the idea that we can detect changes in nuclear spins – but this information must be spatially localized in order to construct an image. To this end, MR scanners use *gradients*. Gradients are linear variations of the magnetic field strength in a selected region. Typical gradient systems are capable of producing gradients from 20-100 mT/m (for a 2.5T MR scanner). Three types of gradients are applied, according to the axis of imaging (x-, y-, or z-axis; Fig. 2.11). We first consider the z-axis: for instance, the magnetic field strength may be weakest at the top of the patient's head, and increase in strength to the strongest strength at the patient's feet. The consequence is that the Larmor frequency changes gradually along the z-axis, and a mapping is created between field strength, position, and angular frequency: by emitting the corresponding ω, only the matched portion along the z-axis will respond. The selection of the region in the z-axis is often referred to as *slice selection.* A shallow gradient will produce a thicker slice, while a steeper gradient produces a thinner slice. Next, we examine the problem of determining the (x,y) position of a signal within the slice, which is referred to as *spatial encoding*. Spatial encoding consists of two steps: *phase encoding* and *frequency encoding*. For phase encoding, a magnetic gradient is applied in the y-axis after the original RF pulse. Spins higher in this linear gradient are affected more and become more in phase relative to spins lower in this new y-axis gradient. This change in phase therefore induces a unique phase to be associated with each point in the y-axis; from this phase, the y-position can be computed. For spatial encoding, a third linear gradient is applied, this time in the x-axis, with the effect of shifting the Larmor frequencies of the spins such that along the direction of the x-axis gradient, the spins become slower. This alteration in the frequency permits the x-position to also be identified by examining the frequency of a given point. The full mathematics of this process is beyond the scope of this chapter; the reader is instead referred to [15] for the particulars.

The (x,y) information obtained from phase and spatial encoding are stored in *k-space*, a 2D or 3D matrix (dependent on whether a single image slice or image volume is being considered). The y-axis of k-space represents phase information, while the x-axis captures angular frequency information. Note that k-space coordinates therefore have no correspondence to an image's coordinates. From this k-space representation, a 2D/3D (discrete) Fourier transform is applied to transform the frequency data into a spatial domain. For 2D slices, the current clinical MR scanner resolution is about 1 mm^3, while research models can exceed 1 μm^3.

Pulse sequences and image contrast. The contrast in MR images are dominated by three inherent tissue characteristics: proton density, ρ; T1; and T2. By varying the nature and timing of the RF pulses, different tissues and pathologies can be emphasized to take advantage of differences in these characteristics. An MR *pulse sequence* describes a timed set of RF and gradient pulses, wherein the interval between the RF pulses and the amplitude and shape of the gradients is given. In general, an MR pulse sequence specifies (in order): 1) the slice selection gradient; 2) the excitation RF pulse; 3) the phase encoding gradient; 4) the generation of an echo (*e.g.*, secondary RF pulse); and 5) the frequency encoding gradient. There are three categories of common MR pulse sequences: spin echo (SE); inverse recovery; and gradient echo. MR pulse sequence design is an ongoing research area; we briefly cover these sequences below and refer to [11, 85] for additional information.

As mentioned earlier, *spin echo imaging* is characterized by the use of a slice-selective 90° RF pulse, after which transverse magnetization decays with T2*; followed by a 180° RF pulse to refocus the spins. Two parameters help to characterize the RF component of these pulse sequences: the *time of repetition* (TR) indicates the amount of time that elapses between 90° RF pulses, which in turn determines how much of M_z is recovered by a tissue; and TE impacts how much measurable signal in M_{xy} is lost due to dephasing. With variations on TR and TE, three MR contrast images are defined:

1. Proton (spin) density images. The first type of image uses a long TR and short TE, which controls for both T1 and T2 relaxation. As a result, the contrast in proton density images is due primarily to variations in the concentrations of mobile ^1H.
2. T1-weighted images. T1-weighted images use a short TR and short TE. The short TR only allows tissues with a short T1 relaxation to fully recover (*e.g.*, fat) – substances with longer T1 properties will only partially recover. For example, at short TRs the difference in relaxation time between fat and water can be detected. The use of a short TE minimizes the loss of the transverse signal due to T2 relaxation.
3. T2-weighted images.T2-weighted images employ a long TR and a long TE. The long TR value allows all tissues to reach full longitudinal magnetization, while the long TE highlights any tissue with a long traverse magnetization (*e.g.*, fluid).

In general, spin echo images proffer excellent image quality due to recovery from the echo signal; however, this improvement in image quality comes at the cost of a longer scan time (and ensuing motion artifacts). Also, T1-weighted images are better at visualizing anatomy, while T2-weighted images are more sensitive to pathology.

An *inverse recovery* (IR) pulse sequence is a variation of spin echo that uses an additional 180° RF pulse before the first 90° RF pulse. The effect is that longitudinal magnetization is completely flipped (*i.e.*, inverted), but leaving M_{xy} unaffected (*i.e.*, it remains zero). Thus, only T1 relaxation occurs, at which point the 90° RF pulse is applied. The interval between the first 180° and 90° RF pulses is called the *inversion time* (TI). By changing TI, image contrast is affected by varying the degree of longitudinal magnetization. Two common IR pulse sequences are short TI inversion recovery (STIR) and fluid-attenuated inverse recovery (FLAIR). STIR is often employed to suppress the signal arising from fat. In comparison, FLAIR uses a long TI that results in almost complete suppression of high T1 materials (*e.g.*, CSF) while providing visualization of fat, edema, and tumors (particularly in neurological studies, such as with brain tissue).

Lastly, *gradient echo* (GRE) pulse sequences use a scanner's gradient coils instead of additional RF pulses to generate an echo signal. The frequency encoding gradient (see above) is changed so that it is first applied with a negative polarity, which removes phase coherence between spins; and then the gradient is reversed, thereby causing the spins to rephase, which in turn forms a gradient echo. The advantage of GRE sequences is that as there is no need for a 180° RF pulse, a very short TR can be achieved, which in turn results in faster imaging that is more resistant to motion artifact. The tradeoff, however, is that GRE sequences do not permit a long T1 recovery; and furthermore, there is no recovery from T2*. Given the relatively fast nature of these pulse sequences, there is a potential for T1 signals to remain between cycles (*i.e.,* the system does not return to equilibrium within the time allotted). Thus, some GRE pulse sequences add a signal to purposely cause dephasing before the next RF pulse; this process is called *spoiling* and sequences using this technique are referred to as *spoiled GRE.* Two clinically popular spoiled GRE sequences are spoiled gradient echo (SPGR) and FLASH (fast low shot angle). Different image contrasts in spoiled GRE sequences are similar to SE, but introduce the use of the flip angle: proton-density images are obtained with a long TR, low flip angle, and short TE; T1-weighted images are given by decreasing TR and increasing the flip angle; and T2*-weighted imaging occurs with long TR and long TE.

Signal-to-noise ratio. As might be imagined, the quality of an MR image is beholden to a number of practicalities, ranging from hardware concerns such as the strength and homogeneity of the external magnetic field, and limitations of the RF coils (*e.g.,* thermal noise); through to motion artifacts (patient movement in the scanner); and ultimately the image reconstruction parameters. At each stage, noise is introduced into the process, resulting in image degradation. In MR, the ratio of this noise to the MR signal is referred to as the *signal-to-noise ratio* (SNR), where the desire is to have a high SNR. For example, SNR increases with field strength, permitting higher resolution and faster scanning. However, higher field strengths have increased safety concerns and are more costly. Likewise, increasing the number of scan repetitions or increasing pixel resolution (*i.e.,* larger pixels) will improve SNR, but at the expense of time and the ability to see finer details within the image. [85] provides a summary of the parameters that affect SNR. Alternatively, it is possible to employ denoising algorithms as part of the post-processing of an MR image to improve quality.

Additional MR Imaging Sequences

The versatility of MR is demonstrated by the numerous adaptations of the core technology and the continued research in the field. A few of the more prominent uses and applications of MR that are now clinically seen are described below.

Diffusion MRI. Diffusion techniques are based on the measurement of Brownian motion in water; by understanding how water moves, one can infer information about the local (anatomical) environment [31]. Instead of a homogeneous magnetic field, a linear gradient is imposed, resulting in varying rates of precession. A *pulsed gradient spin echo* sequence extends a standard SE with two additional diffusion gradient pulses to create a diffusion-encoding gradient. The diffusion-encoding gradient causes phase shift to vary with position: all spins that remain in the same location will return to their initial state; however, any spins that have moved (*i.e.,* due to diffusion) will experience a different total phase shift. The result is that the diffusion of water molecules along a field gradient reduces the MR signal: the higher the degree of diffusion, the greater the loss of signal. Images will therefore show low signal intensity where diffusion along the applied diffusion gradient is high. The amount of diffusion-weighting in an image is given by the amount of expected signal loss, calculated as:

$$\frac{S}{S_0} = e^{-\gamma^2 G^2 \delta^2 \left(\Delta - \frac{\delta}{3}\right) D} = e^{-bD}$$

where S is the resultant signal intensity, S_0 represents the original signal intensity, γ is the gyromagnetic ratio, G is the strength of the gradient, δ is the duration of the pulse, Δ is the time between the two diffusion-gradient pulses, and D is the diffusion constant. Typically, this equation and degree of diffusion-weighting is summarized in terms of the *b-value*.

The displacement of water molecules follows one of two basic patterns: *isotropic diffusion*, where the net random displacement in any direction is equal (*e.g.,* free diffusion, like in a fluid such as CSF with no constraints on spatial direction); and *anisotropic diffusion*, where the movement of water is confined to certain directions due to high-density regions (*e.g.,* such as along a fiber). *Diffusion-weighted* MRI (DWI) visualizes the variation in water molecule mobility, irrespective of these displacement patterns. Based on the above equation, if the same volume is repeatedly scanned but with different diffusion gradients, one can quantitatively assess the diffusion constants of the local environment. This calculation of the diffusion constants is known as the *apparent diffusion coefficient* (ADC). ADC values can then be mapped spatially to create an image with diffusion serving as the basis for visual contrast. Such ADC-based images are referred to as *ADC maps*. Dissimilar to DWI, *diffusion tensor* MRI (DTI) tries to quantify anisotropy by ascertaining the direction of water molecule movement. By varying the direction of the diffusion-gradient field, different values of D are obtained for a given point in space, providing information on the local structures that restrict water movement. The directionality of water movement at a point can then be represented in 3D by a tensor – in this case, a 3 x 3 symmetric positive matrix, with the overall direction corresponding to the matrices' main eigenvector. Because of the difficulty in visualizing tensors, another method of quantifying the degree of anisotropy is *fractional anisotropy* (FA), which transforms the tensor to a scalar representation that is more readily presented. Presently, DTI is used for fiber tracking (*e.g.,* such as in the brain). Diffusion-weighted images are sensitive to motion artifact. DTI, in particular, because of the number of gradient changes needed, suffers from long scan times and/or noise. However, both DWI and DTI are active areas of research and development, given their unique ability to illuminate anatomical structures.

MR angiography (MRA). MR angiography is primarily used for visualization of vasculature, including aneurysms and cardiovascular function. Perhaps the most common MRA technique today, *contrast-enhanced magnetic resonance angiography* (CE-MRA) is based on T1 values for blood and the use of an MR contrast agent to affect this value (*i.e.,* a relaxation-enhancing medium). By reducing blood's T1 value, image formation is no longer dependent on the flow of blood, but rather, the detection of the contrast. CE-MRA uses a spoiled GRE sequence with a short TR to have low signal (due to the longer T1) from the stationary tissue and short TE to minimize T2* effects. CE-MRA hence proffers fast acquisition and high-quality images over large regions of vasculature. Other modes of MRA take advantage of the flow of blood and other fluids to induce a range of imaging artifacts that can be used to visualize surrounding tissue:

1. Time of flight angiography (TOF). Spins flowing into a slice selection region are initially unaffected by the magnetic field; on entering the area, they emit a higher MR signal. 2D TOF uses this fact (Fig. 2.12a) with a flow-compensated gradient echo sequence to reconstruct (*e.g.*, via maximum intensity projection) from multiple images a 3D image of the vessels akin to that seen with to conventional angiography.
2. Phase contrast angiography (PCA). Spins moving parallel to a magnetic field develop a phase shift proportional to the velocity of the spin. Knowing this, PCA uses a bipolar gradient that dephases (and hence encodes) the spins in proportion to their velocity. This information can then be used to calculate the relative magnitude of flow and direction, generating flow velocity maps.

More recently, tagged MRI velocimetry using proton spins that have been magnetically tagged by various methods have shown potential as a noninvasive technique for measuring flow volume and velocity (see below).

Perfusion MR. Unlike MRA, perfusion MR imaging is designed to quantify microcirculatory tissue perfusion rather than the gross flow from larger vascular axes. Though the chief use of perfusion MR has been in cerebral studies (*e.g.*, stroke), more recently work has been done to apply this modality to assess myocardial perfusion [41] and cancer. MR perfusion studies utilize some type of intravascular tracer, which can be categorized twofold: 1) exogenous materials, such as in injected contrast; and 2) endogenous materials, such as tagged ^{1}H nuclei. *Dynamic susceptibility contrast* (DSC) MR is an example of the former, measuring the decrease in T2/T2* during the first pass of the tracer through the capillary bed. An example of the latter is *arterial spin labeling* (ASL; also referred to as arterial spin tagging), which marks inflowing water proton spins in arterial blood to visualize flow, and has been used for assessment of tumor response (*e.g.*, for anti-angiogenesis agents) and cerebral blood flow [20, 86].

Magnetic resonance spectroscopy (MRS). Although MR focuses on the use of hydrogen atoms in water, other biological compounds contain ^{1}H and other nuclei also naturally respond to the presence of an external magnetic field (*e.g.*, phosphorous,

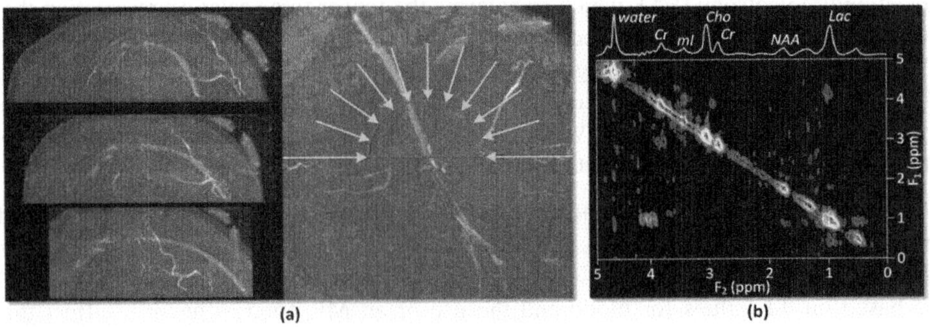

Figure 2.12: (a) Example of TOF MR angiography for treatment of an intracranial aneurysm. **(b)** Example of 2D magnetic resonance spectroscopy for primary brain tumors. Detected metabolites peaks are labeled in the spectrum at the top.

fluorine)[5]. MRS uses these insights to provide biochemical information on a region and is beginning to see clinical usage. The underlying principle of MRS is *chemical shift*. The presence of electrons around a nucleus dampens the effect of a magnetic field. Based on the configuration of a molecule, therefore, the magnetic field will be experienced to lesser or greater extent: ^{1}H signals from compounds will exhibit slightly different frequencies dependent on their local chemical environment. This last point implies that the Lamor frequencies of different compounds will therefore vary. To take advantage of this fact, MRS uses a water-suppressing pulse sequence to remove the predominant signal from H_2O; a chemical shift selective (CHESS) technique or an inversion recovery sequence akin to that used for fat suppression is used. Detected MR signals therefore arise from metabolites in the body. As in standard chemical spectroscopy, a spectrum is plotted. Each metabolite has a known frequency, which can be used to identify its peaks in the spectrum and its relative concentration based on the area under the peak. Early work on 1D ^{1}H MR spectroscopy demonstrated the ability to detect several metabolic intermediates (NAA, lactate, choline, creatine, myo-inositol, glutamine, and glutamate) [45, 65, 80, 81]. However, under conventional 1D ^{1}H MRS, the spectral peaks of several key compounds overlap (*e.g.*, methyl, GABA). Alternatively, 2D MRS (Fig. 2.12b) enables the unambiguous resolution of overlapping peaks of 1D MRS, allowing a more detailed map of the chemical environment of protons, thus complementing the metabolic information obtained from other modalities such as PET (positron emission tomography). Applications of MRS include assessment of different cancers, including brain and prostate [74, 79], and neurological disorders.

Functional MRI (fMRI). The details of fMRI are largely outside of the scope of this text, and we only briefly describe it here; the reader is referred to [16] for a comprehensive discussion. The association between hemodynamics and neural activity has been long known: active cells consume oxygen, so portions of the brain that exhibit higher levels of oxygen are likely activated. This principle serves as the foundation for fMRI, a now standard tool for neurological mapping. fMRI uses the *blood oxygenation level dependent* (BOLD) effect as the basis for spatially highlighting brain activity: the difference in magnetic susceptibility between oxygenated and deoxygenated blood is used to establish MR T2* signal variation. As such, a T2*-weighted GRE pulse sequence is often used for fMRI studies. In theory, repeated measurements provide a statistical basis for mapping those regions that are associated with thought processes and/or a given activity. Unlike other methods for establishing brain activity, fMRI does not require injections of radioactive isotopes, scan times can be short, and resolution is relatively good (currently on average 2.5 x 2.5 mm). Although there is some recent debate in regard to the interpretation of the BOLD effect and fMRI [54], its utility as a tool has spawned new insights into neurological function [84].

Ultrasound Imaging

Descriptions of medical applications based on the principles of sonar (sound navigation and ranging) began to appear shortly after the end of World War II. Donald *et al.* developed the first prototype ultrasound scanner and reported its diagnostic utility in identifying normal abdominal organs and several common pathological processes [22].

---

[5] However, arguably only ^{1}H and ^{31}P exist at sufficient levels of concentration *in vivo* to be detectable presently.

Since the original description of its diagnostic utility, ultrasonography has emerged as the most frequently utilized imaging technique in the world.

High frequency sound waves, ranging from 1-20 MHz, are employed to generate cross-sectional images of the human body. The *ultrasound transducer*, which contains one or more piezoelectric crystals, is used both as a sender and a receiver of sound waves. An electrical current passing through the piezoelectric crystal causes it to vibrate and generate sound waves that propagate through the body tissues (Fig. 2.13a). The reflected echoes returning to the crystal are then converted back into electrical pulses. These signals are processed by the *scan converter* into shades of gray and displayed on a monitor as a cross-sectional image [69, 70]. The strength of the returning echoes is dependent on the physical characteristics of the particular biological tissue. The ability of a material to return an echo is referred to as *echogenicity*. As not all biological tissues behave similarly, organ delineation and disease detection are achievable via ultrasound imaging. Today's ultrasound image displays provide several capabilities to help assess structure and fluid flow:

- A (amplitude) mode. The characteristics of a single ultrasound beam are analyzed as the sound wave travels through the body and returns to the source transducer. The display of A-mode information can be likened to that of an oscilloscope.
- B (brightness) mode. Also known as *gray scale ultrasound* imaging, B-mode is currently the standard method to display diagnostic images. Tissue differences are recorded in shades of gray depending on their strength and distance from the source transducer. A strong reflector of sound waves is displayed as a bright dot.
- M (motion) mode. M-mode is a type of B-mode imaging whereby a series of B-mode dots are displayed on a moving time scale. This technique is presently used to record fetal heart rate and perform other cardiac imaging procedures.
- Real-time B scan. This modality provides a cinematic view of the area being evaluated, displaying sequential images in real-time.
- Doppler sonography. The direction of blood flow and velocity within blood vessels are assessed in real-time using Doppler principles. Clinical applications include the evaluation of arterial and venous flow in normal and diseased organs, as well as blood vessels. There are two types of Doppler ultrasound commonly used in clinical practice: *Color Doppler* and *power Doppler*. Color Doppler is used to visualize the velocity and direction of blood flow within a vessel (Fig. 2.13c) [33]. Power Doppler is a newer technique that overcomes some of the limitations of the conventional color Doppler (but does not assess the direction of blood flow).

In addition, 3D/4D ultrasound systems are now becoming viable, allowing for 2D ultrasound images to be rendered in 3D and motion displayed in real-time. In view of its low cost to acquire and maintain, portability, pain free application, and lack of ionizing radiation, ultrasonography is now a widely used technique for evaluation of diseases in both pediatric and adult populations, including its use in pregnant women to assess the fetal well-being. Real-time imaging capabilities provide instant and continuous visual feedback of organ structures and blood flow patterns. This characteristic is particularly important when performing ultrasound-guided interventional procedures. In addition, there are endoscopic, intra-cavitary, intra-operative, and intravascular ultrasound systems available for use in specific clinical conditions. Acquiring good quality diagnostic images, however, is dependent upon the skill of the operator. Poorly trained personnel tend to create poor quality images by inferior scanning techniques and improper selection of transducers.

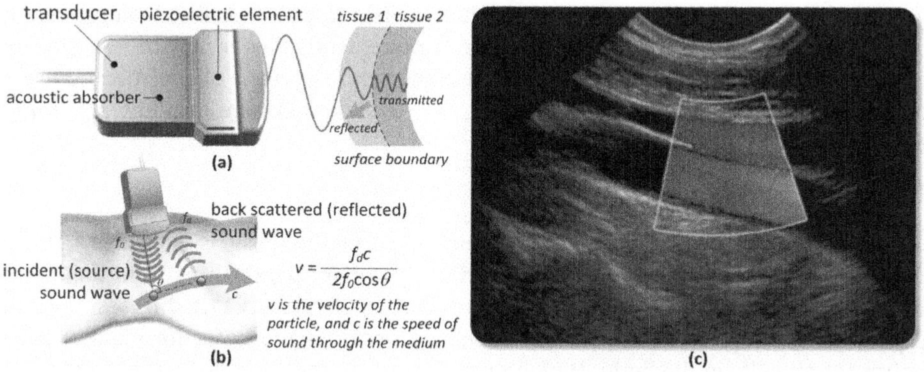

Figure 2.13: (a) Basic principles of ultrasound. A transducer produces sounds waves via a piezoelectric element stimulated by electrodes. As the sound waves pass through the surface boundary of different materials, a portion of the energy is absorbed, and the rest reflected. **(b)** Based on the change in a reflected sound wave from a moving object, Doppler ultrasound can provide flow velocity data. **(c)** Color Doppler longitudinal ultrasound image of the right upper extremity demonstrates normal brachial artery and vein. Flow away from the transducer is displayed in red (right brachial artery), while flow towards the transducer is displayed in blue (right brachial vein).

More recent technical innovations are contributing significantly to the advancement of the field of diagnostic ultrasound imaging:

- Ultrasound microbubble contrast agents. These bubbles provide augmentation of the ultrasound signal, enhancing Doppler and gray scale images. The microbubble particles are less than 10 μm in diameter and therefore cross capillary beds easily. Microcirculation and vessel flow can be demonstrated by using microbubble contrast agents using harmonics (see below). Drug delivery to a specific target in the body, such as a tumor, has been achieved by incorporating ligands on the surface of the microbubble to recognize receptors on the cell membrane. Other therapeutic options include *sonoporation*, or the transient increase in cell membrane permeability during intravascular ultrasound.
- Harmonic imaging. Images are constructed from returning echoes having 2-3x the frequency of the original transmitted frequency (*i.e.,* harmonic waves), resulting in images with a higher spatial resolution, reduced SNR, and improved image contrast [34, 53]. Harmonic imaging has been used to improve the resolution of ultrasound images in conjunction with microbubble ultrasound contrast agents.
- High intensity focused ultrasound (HIFU). This technique uses high intensity focused ultrasound waves to heat and destroy tissue at previously determined specific depths. This technology is currently used for the ablation of uterine fibroids.
- Elastography. This technique is used to measure the mechanical stiffness of organs and has been shown to be helpful in identifying tumors of the prostate, thyroid gland, breast and liver [34, 66]. For example, diseased tissues such as found in tumors and inflammatory lesions, tend to be harder than surrounding normal tissue; based on this observation, elastrography can help distinguish malignancies.

An Introduction to Imaging-based Anatomy & Physiology

Given this basic background on core imaging modalities, we now turn to the use of imaging to understand human anatomy and physiology. This primer will serve to ground our examples and discussions in the chapters to come. Rather than provide an extensive review of each anatomical region, a more detailed overview of two systems, neurological and respiratory, are provided along with coverage of the specific use of imaging for mammography; subsequently, we review other core anatomy/physiology and the role of imaging.

Respiratory System

The respiratory system has the critical function of facilitating gas exchange between the air and circulating blood. At this interface, oxygen (O_2) diffuses from the alveoli into the respiratory alveolar capillaries and carbon dioxide (CO_2) diffuses from the capillaries into the alveoli. This gas exchange provides the basis for cellular metabolism of all cells in the body. Beyond providing a surface area for gas exchange, the respiratory system must also: move air to and from this exchange surface; protect the respiratory surfaces from exposure to the environment, inhaled substances, and potential pathogens; and produce sounds. The respiratory system is divided into the *upper respiratory system* extending from the nose to the pharynx and the *lower respiratory system*, which extends from the larynx (voice box) to the pulmonary alveoli. The upper respiratory system conducts air to the ultimate units of gas exchange; during which it filters, warms and humidifies incoming air to protect the lower, more delicate gas exchange surfaces. The respiratory tract has both conducting and respiratory portions: the conducting portions transport air to the deeper parts of the lungs; the respiratory portions participate in gas exchange. The upper respiratory system is purely conductive; whereas the lower respiratory system has both conducting and respiratory components. The lower respiratory system is the focus of this section.

The Larynx and Trachea

Air from the upper respiratory system enters the *larynx* through a small opening called the *glottis*. The passage of air through the glottis vibrates the *vocal cords* (the walls of the glottis) and produces sound waves. The larynx is essentially a cylinder with incomplete cartilaginous walls supported by ligaments and muscles (Fig. 2.14a). Three large cartilages form the larynx:

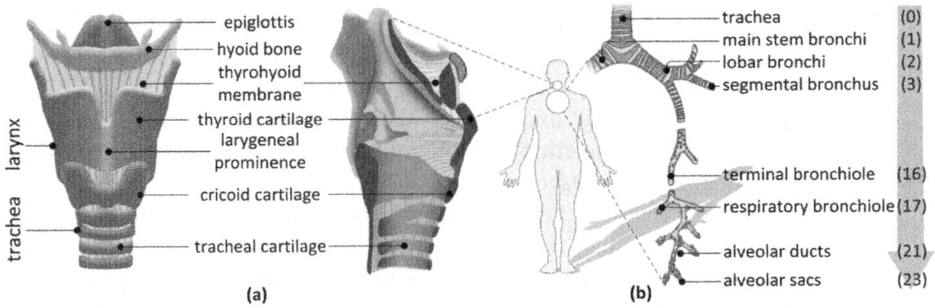

Figure 2.14: (a) Anterior view of the larynx (left) and lateral view (right) with anterior surface facing right. **(b)** The trachea spans roughly 11 cm from the cricoid cartilage to the carinal bifurcation, then divides to form the right and left main stem bronchi. The trachea and subsequent bronchi to the level of the terminal bronchioles contain cartilage within their walls and form conducting airways. Beginning with the respiratory bronchioles, which contain no cartilage, both gas transport and exchange can occur. The airway tree has 23 generations that divide into successive branches that are narrower, shorter, and more numerous; the levels of different areas are shown to the right.

1. The *thyroid cartilage*, a large U-shaped cartilage that forms the anterior and lateral walls of the larynx. The prominent anterior portion of the thyroid cartilage is the Adam's apple, or thyroid prominence.
2. The *crycoid cartilage*, located below the thyroid cartilage, is larger posteriorly and provides support in addition to the thyroid cartilage.
3. The *epiglottis* is a shoe-horned shaped elastic cartilage with ligamentous attachments to the thyroid cartilage below and the hyoid bone above. The epiglottis projects superior to the glottis and form a lid over it during swallowing to prevent the movement of food or liquid into the lower respiratory tract.

The *trachea* is a tube extending from the cricoid cartilage to the origins of the main stem bronchi. It is roughly 2.5 cm in diameter and extends ~11 cm in length. The trachea is lined by a pseudostratified ciliated columnar epithelium, as with the nasal cavity. A band of connective tissue, the submucosa, resides under the mucosa and contains mucous glands that communicate with the tracheal epithelium via secretory ducts. The trachea is comprised of 15-20 C-shaped tracheal cartilages separated by annular ligaments (Fig. 2.14b). The tracheal cartilages provide support for the trachea to prevent collapse or overexpansion during changes in respiratory pressures. The C-shaped tracheal cartilages form the anterior and lateral walls of the trachea. The posterior tracheal wall is intimate to the esophagus, contains no cartilage, and is formed by an elastic membrane and the *trachealis muscle*, a smooth muscle that contracts in response to sympathetic stimulation. Absent a rigid cartilage, the posterior tracheal wall is deformable and can distort during swallowing to allow a food bolus to pass through the esophagus. Contraction of the trachealis muscle alters the diameter of the trachea; with sympathetic stimulation the tracheal diameter enlarges to accommodate greater airflow. The bifurcation of the trachea in the right and left main stem bronchi is called the *carina*.

The Lungs and Airways

The *right* and *left lungs* occupy the *right* and *left pleural cavities*, respectively, and are separated by the *mediastinum* within which lie the heart; aorta and great vessels; the esophagus; lymph nodes, the thoracic duct, and lymphatic vessels; and various nerves, including the vagus, phrenic, and recurrent laryngeal nerves. The right and left lungs subtend the right and left main stem bronchi, then further subdivide into *lobes*. The lobes of the lungs are anatomic units completely encased by *visceral pleura*. The right lung has three lobes: right upper, right middle, and right lower; the left lung has two lobes: left upper and left lower (Fig. 2.15a). The *lingula* of the left upper lobe occupies roughly the same anatomic region in the left chest as does the right middle lobe in the right chest; the lingula is anatomically a part of the left upper lobe because it lacks its own visceral pleural envelope.

Within the lung parenchyma, the airways and pulmonary arteries run together in an interstitial envelope and generally have the same cross-sectional diameter, while the pulmonary veins run independently. The right and left main stem bronchi (the primary bronchi) and their corresponding main pulmonary arteries exit the mediastinum at the *right* and *left hila*, respectively. The primary bronchi further subdivide into lobar (secondary) bronchi along with their corresponding pulmonary arteries. The pulmonary veins, formed by the convergence of venules and veins of the lungs, converge at the left atrium.

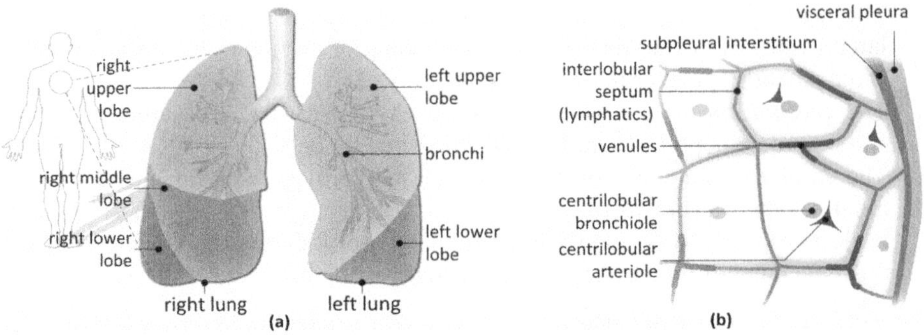

Figure 2.15: (a) The lobes of the lung are discrete anatomic units surrounded by their own visceral pleura. The right lung has three lobes whereas the left lung has two lobes. **(b)** The secondary lobule is the smallest unit of lung surrounded by a fibrous partition. A terminal bronchiole and arteriole arise from the center of the lobule, surrounded by the bronchovascular interstitium that contains pulmonary lymphatics. The interlobular septa form the lobular margins and contain pulmonary venules and lymphatics.

Segments of the lobes. Each lobe further subdivides into *bronchopulmonary segments*. The lung segments lack discrete connective tissue boundaries, but are defined by their cartilaginous segmental airway (tertiary bronchus) and artery, called *bronchovascular bundles*. Although there are frequent anatomic variations, the adult right lung typically has 10 segments and the left lung has 8 segments. The bronchovascular bundles repeatedly divide into smaller and smaller units: the pulmonary arteries divide into arterioles; the bronchi into bronchioles. In total, there are 23 generations of airways (Fig. 2.14b). The conducting zone consists of the trachea and the first 16 generations of airways, ending in the terminal bronchioles. As the bronchi subdivide, their cartilaginous walls become progressively thinner, being replaced by smooth muscle. The last 7 airway generations comprise the respiratory zone beginning with respiratory bronchioles, which measure 0.3-0.5 mm in cross-section (Fig. 2.16a). These are the first anatomic units in which gas exchange occurs. The walls of the respiratory bronchioles have smooth muscle but lack cartilage. Autonomic nervous stimulation of their smooth muscle walls regulates bronchiolar diameter and airway resistance: sympathetic stimulation produces bronchodilatation and parasympathetic stimulation produces bronchoconstriction. The final airway tree generation is made up of the *alveolar sacs*.

Secondary pulmonary lobule. The smallest unit of lung structure surrounded by its own interstitial envelope (fibrous partition) is the *secondary pulmonary lobule* (Fig. 2.15b). The secondary pulmonary lobule is an irregularly polyhedral structure measuring 1-2.5 cm in diameter. The secondary pulmonary lobule is subtended by a central terminal bronchiole and arteriole surrounded by an interstitial envelop that also contains pulmonary lymphatics. Within the lobule, the centrilobular bronchiole further branches into 2-3 terminal bronchioles, each subdividing into 2-3 respiratory bronchioles; the arteriole repeatedly subdivides to the level of the pulmonary capillaries. The walls of the pulmonary lobule, called *interlobular septa*, contain the pulmonary venules and lymphatics and are continuous with the visceral pleura. Although rarely visible under normal conditions, the secondary pulmonary lobule becomes visible on chest radiographs and on computed tomography in disease states in which the pulmonary interstitium

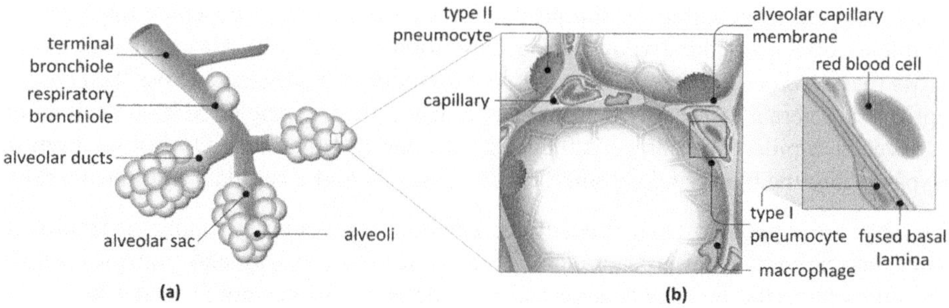

Figure 2.16: **(a)** Lower airway anatomy, including the alveoli. **(b)** The alveolus is composed of Type I pneumocytes (alveolar epithelial lining cells). Small pores connect adjacent alveoli. Type II pneumocytes are responsible for the production of surfactant. Alveolar macrophages move freely within the alveoli and phagocytize debris. The alveolar capillary membrane represents the fusion of the basal lamina of the alveolar epithelium and capillary endothelium and contain gas on one side and blood on the other.

becomes thickened by fluid, fibrosis, or cells (Fig. 2.17). Radiographic signs of interstitial disease include the presence of prominent septal lines that demarcate the polyhedral secondary lobules, prominence of the visceral pleura and fissures separating the lungs, and thickening of the airways, called *cuffing*, due to thickening of the bronchovascular interstitium.

Alveolar ducts and alveoli. *Alveolar ducts* and *alveoli* branch directly off of the respiratory bronchioles. Alveolar ducts are gas exchange units from which alveolar sacs and alveoli branch. The *alveolar sacs* are common chambers from which individual *alveoli* extend (Fig. 2.16a). There are ~150 million alveoli in each adult lung; the total surface area available for gas exchange is 75 m^2, about the size of a tennis court. The alveolus is suspended within an extensive network of elastic tissue responsible for the elastic properties of the lung and the capacity of the alveoli to expand and contract during ventilation (Fig. 2.16b). The lining cells of the alveolar epithelium are called *pneumocytes*. *Type I pneumocytes* are the primary epithelial cells of the alveoli; *Type II pneumocytes* are present in equal numbers but occur at the corners between adjacent

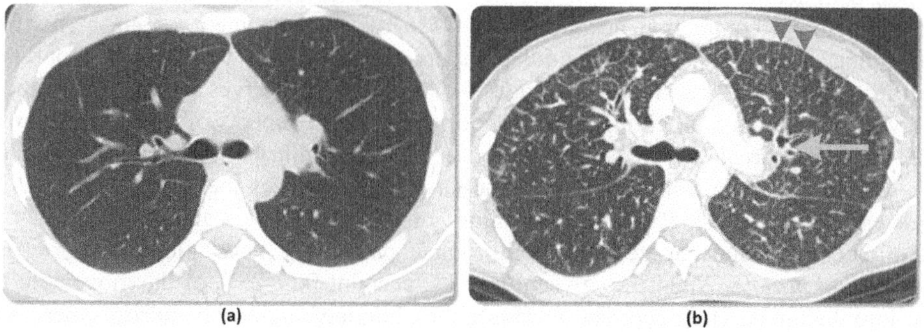

Figure 2.17: **(a)** Axial CT images of normal lung and **(b)** abnormal lung with peribronchial cuffing (arrow) and marked prominence of interlobular septa (arrowheads) in a patient with tumor infiltration of the interstitium (lymphangitic carcinomatosis). Normally, secondary pulmonary lobules are rarely visible on CT, but become apparent when edema, tumor cells, or fibrosis thicken the lung interstitium.

alveoli and are responsible for the production of *surfactant*, a substance that prevents complete collapse of the alveoli during expiration (Fig. 2.16b). Gas exchange occurs across the *alveolar-capillary membrane*. The total diffusion distance across the alveolar-capillary membrane is roughly 0.5 μm. The transit time of a single red blood cell across the alveolar-capillary membrane is 0.3-0.75 sec., during which time O_2 diffuses from the airspaces into the blood and carbon dioxide passes from the capillary into the alveolus.

The lungs are supplied by two circulations. The pulmonary arteries supply the respiratory zone. One pulmonary arterial branch accompanies each airway and branches with it. The lungs have the most extensive capillary network of any organ in the body. The pulmonary capillaries occupy 70% to 80% of the alveolar surface area. This capillary network receives the entire cardiac output and is the basis for gas exchange. The conductive zone is supplied separately by the bronchial circulation. These vessels originate from the descending aorta and provide nutrients to the conducting airways themselves. Both the systemic bronchial arterial and pulmonary arterial circulations ultimately drain into the pulmonary veins and left atrium.

The Pleura, Chest Wall, and Respiratory Muscles

The chest cavities within which lie the right and left lungs are the *pleural cavities*, so named for the serous pleural membrane that lines the cavities, the *parietal pleura*, and the membrane that lines the lung surfaces, the *visceral pleura*. The parietal pleura is adherent to the ribs and *intercostal muscles* of the inner chest wall as well as the *diaphragm*, the dome-shaped skeletal muscle separating the thoracic and abdominal cavities. The visceral pleura is adherent to the lung surface, continuous with the interlobular septa, and forms the fissures that separate the lobes of the lung. The parietal and visceral pleura are normally closely apposed, separated by a thin layer of pleural fluid produced by both pleural surfaces. The fluid provides a smooth coating and allows movement of the two pleural surfaces during breathing. The intrapleural fluid also creates surface tension between the two pleural surfaces such that they remain adherent. This fluid bond is responsible for a negative intrapleural pressure that, as will be discussed below, is important for pulmonary ventilation.

A number of muscles contribute to quiet and forced breathing. The primary muscle of inspiration is the dome-shaped *diaphragm*, divided into right and left *hemidiaphragms*. Upon contraction, the hemidiaphragms descend to increase the volume of the chest. Diaphragmatic contraction accounts for ~75% of the movement of air in normal breathing. The right and left hemidiaphragms are innervated by the paired phrenic nerves, supplied by cervical nerve roots 3-5. The phrenic nerves descend from their cervical origins to the hemidiaphragms along the lateral aspects of the mediastinum. Injury to the phrenic nerve can result from spinal cord injury at or above the level of the cervical roots or when trauma or neoplasm compromises the phrenic nerve as it descends in the chest. Phrenic nerve injury results in paralysis and elevation of the hemidiaphragm and can severely compromise pulmonary ventilation. The *external intercostal muscles*, situated between the ribs, account for ~25% of the movement of air into the lungs. Their contraction causes the obliquely angled ribs to rise superiorly, expanding the width and depth of the chest cavity. Under normal conditions, inspiration (inhalation) is an involuntary, but active process that requires contraction of the diaphragm and external intercostal muscles. A number of accessory muscles of inspiration become active when the depth and frequency of ventilation must increase, as during exercise or in individuals with various forms of lung disease. Expiration (exhalation) is normally involuntary and passive, results from relaxation of the muscles of inspiration, and depends on the natural elasticity of the lungs rather than muscular contraction. Active expiration involves the use of several accessory muscles that contract to depress the ribs or force the diaphragm up, as during exercise or with singing, when precise regulation of flow across the vocal cords is required.

Pulmonary Ventilation: Inspiration and Expiration

Pulmonary ventilation refers to the movement of air into and out of the respiratory tract. Pulmonary ventilation is necessary in order to maintain alveolar ventilation, the movement of air into and out of the alveoli. Two physical principles are the basis for pulmonary ventilation: 1) the pressure of a gas varies inversely with its volume in a closed environment (Boyle's Law); and 2) if gas containers of different pressures are connected, gas will flow from the area of higher to lower pressure until the systems have equal pressure. During inspiration, the diaphragm and the external intercostal muscles contract, increasing the size of the thoracic cavity (Fig. 2.18a). When the glottis is open, continuity is established between the atmosphere and chest cavity. As the chest cavity enlarges, intra-alveolar pressure drops and air flows from the atmosphere to the alveoli. When the lungs can no longer expand, inspiration stops and atmospheric and intra-alveolar pressures equalize. During expiration, passive relaxation of the diaphragm and external intercostal muscles causes the thoracic cavity to decrease in size, raising intra-alveolar pressure and forcing air out of the lungs into the atmosphere. Expiration ceases when atmospheric and intra-alveolar pressures equalize.

Pressure Relationships during Inspiration and Expiration

The direction of airflow during breathing is determined by the relationships between atmospheric and respiratory pressures. Atmospheric pressure is normally 760 mm Hg. Respiratory pressures are so small they are normally measured in cm H_2O. Respiratory pressures are typically described relative to atmosphere (*e.g.*, a respiratory pressure of -4 cm H_2O is 4 cm H_2O below atmospheric). *Intrapulmonary* (*intra-alveolar*) *pressure* (P_A) is the pressure in the alveoli. P_A rises and falls with breathing, but always eventually equalizes with atmospheric pressure. *Intrapleural pressure* (P_{ip}) is the pressure within the pleural space. Intrapleural pressure also fluctuates with breathing, but is normally an average of -4 cm H_2O lower than atmospheric pressure and can reach -18 cm H_2O during a powerful expiration. A negative intrapleural pressure is normally

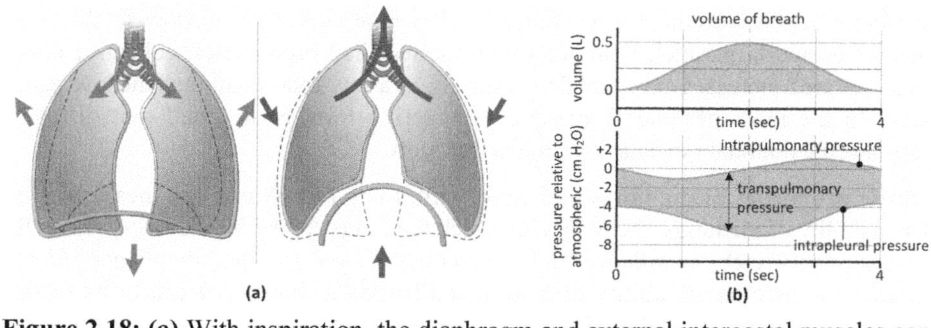

Figure 2.18: (a) With inspiration, the diaphragm and external intercostal muscles contract, causing the lungs to increase in both diameter and length. Intrapulmonary pressure falls and air moves in. During expiration, the muscles relax and the chest volume decreases, raising intrapulmonary pressure and causing air to move out. **(b)** Changes in pulmonary pressures with inspiration and expiration.

maintained for several reasons: 1) the surface tension from the thin film of pleural fluid secures the pleural surfaces together; 2) lung elastic recoil tends to decrease the volume of the lungs; and 3) the elasticity of the chest wall tends to pull the thorax outward to increase the volume of the chest wall. The *transpulmonary pressure* (P_L), also called distending pressure, is the pressure that maintains inflation of the lungs and prevents collapse. Transpulmonary pressure is $P_L = P_A - P_{ip}$. The more positive the transpulmonary pressure, the greater the distension (inflation) of the lungs. For example, at end inspiration, P_{ip} = -7 cm H_2O and P_A = 0 cm H_2O, therefore transpulmonary pressure is 7 cm H_2O. Respiratory pressure changes during ventilation are illustrated in Fig. 2.18b. With inspiration, the chest wall enlarges and intrapleural pressure decreases from -4 to -7 cm H_2O. Transpulmonary pressure increases, the lungs increase in volume, and air enters the lung until intrapulmonary and atmospheric pressures become equal. With expiration, the chest wall decreases in size; intrapleural pressure rises, transpulmonary pressure decreases, and the lungs deflate. Lung deflation reduces alveolar volume, raising intrapulmonary pressure. Air flows from the airspaces to the atmosphere until intrapulmonary and atmospheric pressures reach equilibrium.

Factors Influencing Airflow

The major determinants of airflow are airway resistance and the elastic properties of the lungs and chest wall.

Airway resistance. For airflow to occur, a pressure difference must exist between the mouth and the alveoli. The pressure difference is determined by both the rate and pattern of flow. Airflow resistance is directly proportional to airway length and inversely proportional to the 4th root of airway radius:

$$R = \frac{8Ln}{\pi r^4}$$

where R is the resistance, L is the length of the tube, n is the gas viscosity, and r is the radius of the tube. As such, doubling airway length will double resistance, while decreasing airway radius by half will increase resistance by 16-fold. The major site of airflow resistance is in the large and medium sized bronchi up to the 7th generation. One might expect the greatest resistance to be at the level of the bronchioles because of their small radius; however, only 10-20% of total airway resistance normally occurs at the bronchiolar level. This seeming paradox is due to the fact that although airflow resistance at the individual bronchiolar level is relatively high, airflow in the small airways occurs in parallel. Beginning with the terminal bronchioles, collective cross-sectional area increases substantially, resulting in a low total combined airflow resistance. In the most peripheral airways, convective flow is so small that diffusion becomes the dominant mechanism of gas transport.

Elastic properties of the lungs and chest wall. Lung *compliance* is a measure of the distensibility of the lungs, the ease with which the lungs can be inflated. *Stiffness* is the resistance to stretch, or inflation and is inversely related to lung compliance. *Elastic recoil* is defined as the ability of a stretched (inflated) lung to return to its resting volume. Elastic recoil is directly related to lung stiffness and is inversely related to compliance: the greater the stiffness, the greater the elastic recoil and the lower the compliance. Lung compliance, C_L, is represented by the ratio:

$$C_L = \frac{\Delta volume}{\Delta pressure}$$

where the change in lung volume is compared to the change in P_{ip}. However, compliance is not the same across all lung volumes: it is high at low lung volumes and low at high lung volumes. The major determinates of lung compliance are the following:

- Lung structure. Lung compliance results from the elastin and collagen fibers that enmesh the alveolar walls, airways, and pulmonary vasculature. Elastin fibers are highly distensible and can be stretched to nearly twice their resting length, while collagen fibers resist stretch and limit lung distensibility at high lung volumes.
- Surfactant effects. Surface tension at the alveolar air-liquid interface also significantly affects lung compliance. The surface forces in the alveolus tend to minimize surface area, promoting alveolar collapse, which creates positive pressure within the alveolus. The pressure (P) that develops within alveoli correlates directly with surface tension (T) and inversely with radius (r), as reflected by the Laplace equation, $P = 2T/r$. Alveoli are interconnected but vary in size. If surface tension were uniform throughout the lung, smaller alveoli (having a smaller radius) would tend to collapse and deflate into larger alveoli, causing *atelectasis* in the smaller airspaces and over-distension of larger alveoli. Fortunately, surface tension is not uniform between alveoli because of surfactant. Surfactant works by reducing surface tension at the gas-liquid interface. Moreover, surfactant lowers surface tension more at smaller surface volumes, which promotes alveolar stability by preventing the deflation of smaller alveoli into larger alveoli.
- Chest wall compliance. The chest wall also has elastic properties. The elastic recoil of the chest wall is such that if the chest were unopposed by lung elastic recoil, the chest would enlarge to ~70% of total lung capacity. If the chest wall is expanded beyond 70%, it recoils inward. At volumes <70% of total lung capacity, chest wall recoil is directed outward. The outward elastic recoil of the chest wall is greatest at residual volume; the inward elastic recoil of the chest wall is greatest at total lung capacity. The volume at which the elastic recoil of lung and chest wall are in equilibrium, in opposing directions, is functional residual capacity.

Measures of Lung Function

Measures of lung volumes and expiratory flow rates provide diagnostic information about pulmonary function. *Lung volumes* are defined volumes of air inspired or expired during the respiratory cycle. *Lung capacities* are specific combinations of lung volumes (Fig. 2.19). Most volumes can be measured with *spirometry*, in which a subject breathes into a closed system. However, because the lungs do not empty completely following a forced expiration, both residual volume (RV) and *functional residual capacity* (FRC) are measured using other methods.

A sudden or forceful blow to the chest can render an individual extremely breathless and with the sensation that he cannot inspire air for several seconds. What causes this effect? The condition results from the loss of RV. The airspaces collapse beyond what is physiologic and the alveoli must overcome high surface tension forces to re-expand. An analogy would be the force required to begin to inflate a balloon.

Minute and alveolar ventilation. Ventilation can be considered as the total volume of air brought into or out of the lungs. *Minute ventilation* is the product of tidal volume (V_T) and breaths/minute (500 ml x 12 breaths/minute = 6 L/min). Not all inspired air

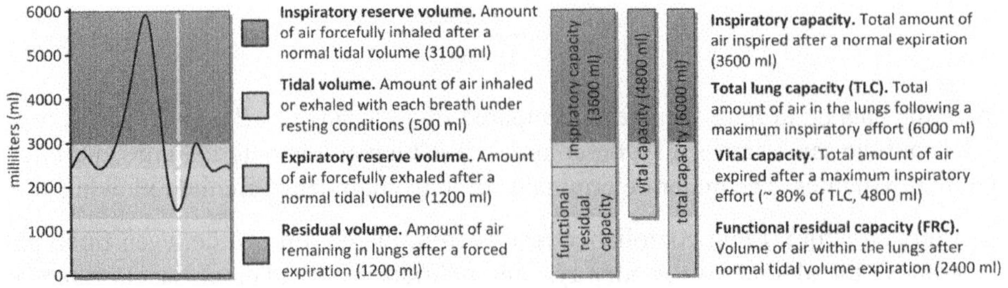

Figure 2.19: The relationships between lung volumes and capacities. Definitions of the lung volumes and capacities are provided. Amounts given assume an average-sized young, healthy male.

contributes to gas exchange because it is distributed between gas exchange units and conducting airways. That portion of V_T in the conducting airways that does not contribute to gas exchange is called *anatomic dead space*, and is normally about 150 ml or 30% of V_T. *Alveolar ventilation* refers the total air available for gas exchange, and is the product of (V_T – dead space) and breaths/minute ((500 - 150) x 12 = 4200 ml/minute). Dead space volume increases when alveoli are not perfused; in that setting, the airspaces not contributing to gas exchange represent alveolar dead space. The sum of alveolar and anatomic dead space is call *physiologic dead space*. The relationships between dead space volume, alveolar ventilation, and minute ventilation are very important. For the same minute ventilation, different breathing patterns provide different volumes of alveolar ventilation. Alveolar ventilation is greater when increasing the depth of breathing rather than the rate of breathing.

Expiratory flow rates. Spirometry is also used to determine rates of airflow and flow volume relationships. During spirometry, the subject exhales maximally and as forcefully as possible into the spirometer, while the volumes of air exhaled are measured over time. The most commonly measured expiratory flow rates are: 1) FVC, the forced vital capacity; 2) FEV_1, the forced expiratory volume in one second; 3) the ratio of FEV_1/FVC (normally greater than 0.70 to 1.00); and 4) the FEF_{25-75}, the forced expiratory flow between 25% and 75% of the FVC, which provides a measure of airflow in the small airways. In the normal setting, at least 70% or more of the FVC is expired in the first second of a forced expiratory maneuver. Diseases of airflow obstruction represent conditions of inadequate expiration. In *obstructive lung disease*, expiratory flow rates are decreased, resulting in a decrease in the ratio of FEV_1/FVC. Because of expiratory airflow obstruction, air is trapped in the lung at the end of a forced expiration, resulting in increased residual volumes. In contrast, *restrictive ventilatory defects* result in limitations of inspiratory airflow and are characterized by stiff lungs with decreases in all lung volumes and capacities. Although FVC is reduced, expiratory flow rates and the FEV_1/FVC ratio are preserved or may increase. These pathophysiologic changes in pulmonary function are summarized in Table 2.4.

Basic Respiratory Imaging

Several modalities are used to image the lung. The most commonly used medical imaging examination is projectional chest radiography (CXR), in which frontal (or frontal and lateral) projections are obtained with the patient at suspended maximal inspiration. A considerable amount of information can be gleaned from the CXR: the lung volumes and texture of the lung parenchyma; the presence of focal nodules, masses,

Measure	Restrictive disease	Mild obstruction	Severe obstruction
FVC	↓↓	Normal	↓
FEV_1	↓↓	↓	↓↓
FEV_1/FVC	Normal	↓	↓↓
FEF_{25-75}	↓↓	↓	↓↓
TLC	↓	Normal	↑
RV	↓	Normal	↑
FRC	↓	Normal	↑
FEV_1 post-bronchodilator	No change	> 15%*	> 15%*

Table 2.4: Alterations in pulmonary function in restrictive and obstructive lung diseases. (*) FEV_1 improves more than 15% in asthma.

or lung consolidations; the status of the pulmonary circulation; abnormalities of the pleura; cardiac size and shape; alterations of the bony chest wall; and potential masses or other abnormalities in the mediastinum. The chest radiograph is often the first examination to suggest pulmonary pathology. More advanced imaging techniques such as CT, MR, and ultrasound are typically used to better characterize respiratory pathology because they provide a cross-sectional perspective and eliminate the superimposition of structures that is characteristic of projectional imaging. CT is the most commonly used advanced imaging technique to further characterize pulmonary parenchymal, pulmonary vascular, and pleural pathology, owing to its high spatial resolution and the high native contrast of aerated lung relative to soft tissue. For the assessment of the lung parenchyma, high spatial resolution is desirable, and helical CT sequences are used that allow for reconstructions of contiguous or overlapping sub-millimeter axial image series. High-resolution imaging is routinely performed to assess for focal and infiltrative lung diseases, emphysema, and abnormalities of the airways. Intravenous contrast is used during chest CT in order to optimally distinguish normal vascular structures from non-vascular soft tissues as well as to assess for abnormalities of the pulmonary arteries themselves, such as with pulmonary embolism, when blood clots migrate from systemic veins (typically the lower extremity) to occlude portions of the pulmonary circulation. Intravenous contrast can also help to characterize airless lung, such as atelectasis (collapse), pneumonic consolidation, or other processes in which the lung has become diseased and airless; different enhancement patterns of consolidated lung may suggest specific conditions and better delimit lung parenchyma when there is concomitant pleural disease. There is increasing interest in CT techniques that use low radiation exposure (low dose) in order to minimize radiation during this commonly acquired procedure. Although CT is typically the modality of choice for assessing the pleura, ultrasound is non-ionizing and is a common alternative used to detect and localize abnormal collections of fluid in the pleural spaces (effusions). Abnormalities of diaphragm motion are also commonly examined using ultrasound because this technology enables continuous imaging during respiration. The failure of the diaphragm to move properly on ultrasound during certain respiratory maneuvers can establish the existence of diaphragm paralysis. MR imaging is quite commonly used to characterize the pulmonary circulation and abnormalities related to the heart. Most normal lungs show little or no MR signal above background; however, conditions resulting in consolidation or infiltration of the lung are observable, albeit without the level of spatial quality as is attainable with CT. In some patients, MR may afford optimal characterization of soft tissue abnormalities, such as in patients with cancers involving the chest or mediastinum. Finally, there are several experimental applications of MR using oxygen as a paramagnetic agent or hyperpolarized ^{3}He ventilation, which provides a source of nuclear magnetic resonance (NMR) signals. Among potential clinical applications are the opportunity to image the air spaces of human lungs, to non-invasively investigate human lung ventilation, and to study the dynamics of inspiration/expiration and functional imaging.

Imaging Analysis of Pulmonary Pathophysiology

Expiratory airflow obstruction can result from either parenchymal lung disease or from intrinsic airway disease. The major parenchymal lung disease is emphysema. The primary airway causes of fixed airflow obstruction are chronic bronchitis and constrictive bronchiolitis. Asthma is also a disease of airflow obstruction, but differs from chronic bronchitis in that the airflow obstruction is not fixed and is reversible with bronchodilators. The mechanisms of airflow obstruction can be distinguished with CT, but result in similar obstructive profiles on spirometry.

Asthma. *Asthma* is characterized by chronic airway inflammation, bronchial wall thickening, airway remodeling, and reversible airflow obstruction due, in part, to airway smooth muscle contraction [21]. During an asthma attack, an inciting event causes the smooth muscles of the small airways to contract and eventually hypertrophy, causing bronchoconstriction and distal air-trapping. Airway inflammation is a primary feature of asthma, and can cause airway edema, hypersecretion of thick mucous that plugs the airways, inflammatory cell infiltration of airways, narrowing of the airway lumen, and expiratory air-trapping. Drugs that minimize bronchoconstriction (bronchodilators) and that inhibit the immune response (systemic or inhaled corticosteroids) are mainstays in the treatment of asthma [24]. Pulmonary function tests (PFTs) demonstrate decreases in all measures of expiratory airflow, which normalize with administration of a bronchodilator.

CT obtained during suspended maximal inspiration and suspended expiration can detect subtle changes in expiratory airflow obstruction that may not be detectable by global PFTs. On expiratory CT in an individual with normal lung function, the lung

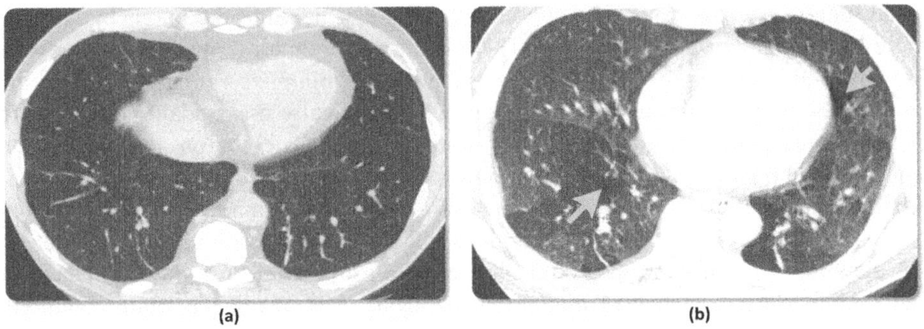

Figure 2.20: Axial CT images in a patient with mild asthma show **(a)** normal lung attenuation (density) at suspended inspiration, while **(b)** in suspended expiration, the lungs have characteristic features of airtrapping at the level of the bronchioles (small airways), visible as lobular and multi-lobular areas of low lung attenuation (arrows). Normally, expiration would appear as an increase in lung attenuation in a uniform gradient that is greatest in the dependent lungs.

volumes decrease and show a smooth gradient of increased attenuation (density) that is greatest in the gravity-dependent bases. In the asthmatic patient, inspiratory images are typically normal and the lung parenchyma is uniform in attenuation (Fig. 2.20), although bronchial wall thickening, luminal narrowing, or mucous may be present with more advanced disease. However, expiratory imaging shows lobular and multilobular regions of low lung attenuation. These regions correspond to airtrapping at the bronchiolar level, resulting from smooth muscle hyperreactivity, luminal narrowing, and premature airway closure. These imaging features may provide the first evidence of small airways disease.

Chronic bronchitis. *Chronic bronchitis* is also characterized by chronic inflammation in the bronchial walls, hypertrophy of the mucous glands and mucous hypersecretion, typically from smoking-related injury. In contrast to asthma, airway smooth muscle hyperreactivity is not a major contributing factor and the airway changes may be irreversible. The result is increased airflow resistance and airflow obstruction from luminal narrowing and mucous hypersecretion [10, 35].

Emphysema. *Emphysema* is characterized by enlargement of airspaces distal to the terminal bronchioles due to destruction of alveolar walls (Fig. 2.21a), typically from smoking-related inflammatory mediators and enzymatic degradation of lung elastin and collagen [73, 78, 87]. The lungs lose elasticity, become highly distensible (having increased compliance) and are easy to inflate. Importantly, expiratory narrowing of the airways results from a loss of the normal radial traction on the airways by elastic lung tissue, causing premature airway collapse and air trapping in large emphysematous holes (Fig. 2.21b). The increased alveolar dead space results in an increased RV and compromises inspiratory reserve volumes. Both chronic bronchitis and emphysema are visible on chest CT; in the case of emphysema, the extent of lung destruction can be quantified *in vivo* using advanced image analysis techniques [6, 27].

Idiopathic interstitial pneumonias. Idiopathic interstitial pneumonias are a heterogeneous group of diffuse, parenchymal lung disorders resulting from injury to the lung parenchyma and are associated with varying degrees of inflammation and fibrosis. One of these entities, *idiopathic pulmonary fibrosis* (IPF), has distinct clinical, radiographic,

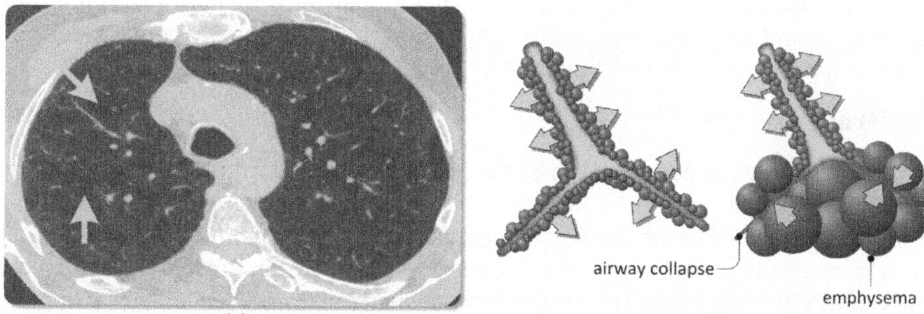

Figure 2.21: **(a)** Axial CT through the upper lobes of a smoker shows emphysema resulting from destruction of alveolar walls and the development of large cyst-like holes (arrows). **(b)** The normal elasticity of lung created by its fibrous and collagen structure provide radial traction on small airways during expiration, maintaining the patency of the small airways (left). With loss of the normal lung architecture by emphysema, this radial traction is lost, resulting in premature airway closure on expiration (right).

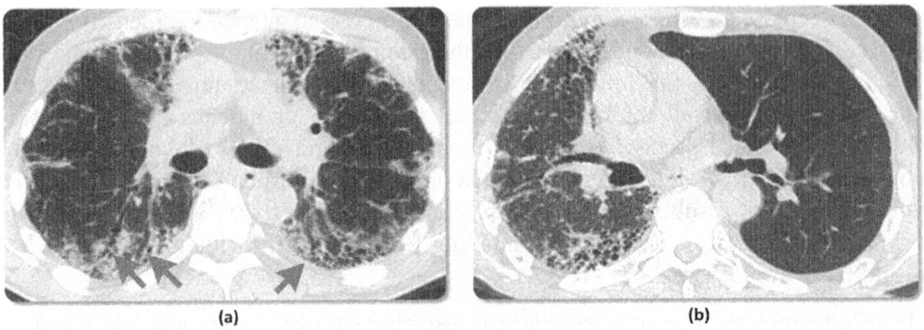

Figure 2.22: **(a)** Typical appearance of end-stage pulmonary fibrosis in which the normal lung architecture is replaced by fibrous honeycomb cysts (right arrow). There is traction bronchiectasis on the subtending airways due to increased radial traction by the stiff lungs (left arrows). **(b)** The same patient following successful left single lung transplantation shows the dramatic difference in the volumes of the fibrotic (right) and normal transplant (left) lungs. The fibrotic lung is stiff, noncompliant and has increased elastic recoil, causing it to decrease in volume, whereas the normal transplant lung is compliant and distensible. The differences in compliance between the two lungs causes the mediastinum to shift rightward.

and pathological features and is characterized by heterogeneous areas of normal lung, active fibrosis, and endstage honeycomb fibrosis [1]. The lungs become stiff, noncompliant, and decrease in volume (Fig. 2.22). On spirometry, maximal expiratory airflow may be higher than normal at a given volume due to increased elastic recoil, as well as from the increased radial traction on the airways, increased airway diameters, and decreased airway resistance. On CT, *interstitial fibrosis* appears as replacement of the normal lung architecture by irregular fibrous bands, small fibrous cysts (called *honeycombing* because of their resemblance to a beehive), and enlarged airways (traction bronchiectasis) resulting from increased radial traction on the airway walls due to the stiff lungs (Fig. 2.22) [56, 62]. The process begins in the subpleural lung, extending more centrally along the bronchovascular bundles, and is most severe in the lower lobes. Image processing techniques have been developed to quantify the degree of fibrosis as well as to estimate the amount of residual inflammation (alveolitis) that may be amenable to immunosuppressive therapy [89]. CT has become a defining feature in the characterization of interstitial fibrosis.

The Brain

The central nervous system consists of the brain and spinal cord. The peripheral nervous system is composed of the nerves that enter and exit these structures. We shall confine this discussion to the brain. The brain contains *gray matter* and *white matter*. The gray matter is made up of the cell bodies, or *neurons*, that reside in the cerebral cortex and within the deep nuclei. The white matter is made up of the axons that extend from these cell bodies to various target tissues; the axons are *myelinated*, meaning they are sheathed in a fat containing substance called myelin that speeds up the transmission of electrical impulses traveling along the axon. On gross pathology, gray matter and white matter appear gray and white, respectively. The CT and MR appearance of these tissues is also distinct on imaging. The density of the highly compact gray matter full of cell bodies attenuates the CT x-ray beam, making the tissue appear whiter. Alternatively, the white matter, consisting of myelinated axons, contains fat thereby making its appearance on CT darker. In summary, on CT the gray matter appears whiter and the white matter appears blacker (Fig. 2.23a). Recall that MR depends on the differing relaxation times of water within various tissues. With white matter behaving more like fatty tissue, it is brighter on T1-weighted images and darker on T2-weighted images; and the gray matter, being densely cellular with relatively little water content, is conversely darker on T1-weighted images and brighter on T2-weighted images (Fig. 2.23b-c).

Cerebral Hemispheres

On a gross anatomic scale the brain is divided into two hemispheres, the right and the left. Each hemisphere is then divided into four lobes: the *frontal*, *parietal*, *temporal*, and *occipital* lobes (Fig. 2.24a). Each lobe is comprised of cerebral cortex and the connected axons that project to form the *white matter tracts*. There is right/left governance, meaning that the right side of the brain is responsible for the left side of the body, and therefore the descending white matter tracts typically cross somewhere below the cerebral cortex to reach their contralateral body part.

Along the surface of the brain are multiple convolutions that are known as the *gyri*, and multiple crevices in between, called *sulci*. Larger sulci are called *fissures* and several of these fissures divide the hemisphere into the different lobes. The lateral horizontally oriented fissure is also known as the *Sylvian fissure* and it separates the frontal and parietal lobes from the temporal lobe. The *central sulcus* or *Rolandic fissure* is located superiorly and separates the frontal and parietal lobes. Posteriorly, the *parieto-occipital fissure* separates the occipital from the parietal lobe (Fig. 2.24a). The brain can also be broken down into the functional areas, known as *Brodmann areas* (Fig. 2.24b), which are responsible for discrete activities such as voluntary motor function and sensory function, along with their adjacent association areas that are responsible for integrating the information into useful and purposeful action, further discussed below. Table 2.5 summarizes the different lobes/regions and their primary functions.

Cerebral cortex. The *cerebral cortex* contains the neuron cell bodies and is responsible for consciousness. The gray matter cortex is only 2-4 millimeters thick, but it constitutes 40% of the brain mass. This portion of the cortex is highly convoluted, a configuration that triples the surface area of the brain without increasing the volume. These convolutions are easily appreciated on gross examination of the brain.

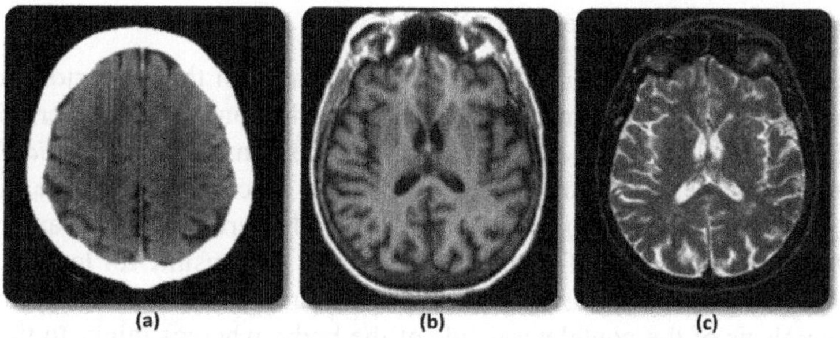

Figure 2.23: (a) An example of a brain CT, where gray matter appears brighter than white matter due to different in the x-ray attenuation of fat in the latter. The skull, because of the bone, appears as a bright ring around the brain. **(b)** A typical T1-weighted axial normal brain (cerebrospinal fluid (CSF) in the ventricles appears dark). **(c)** A normal T2-weighted axial brain image (CSF appears bright).

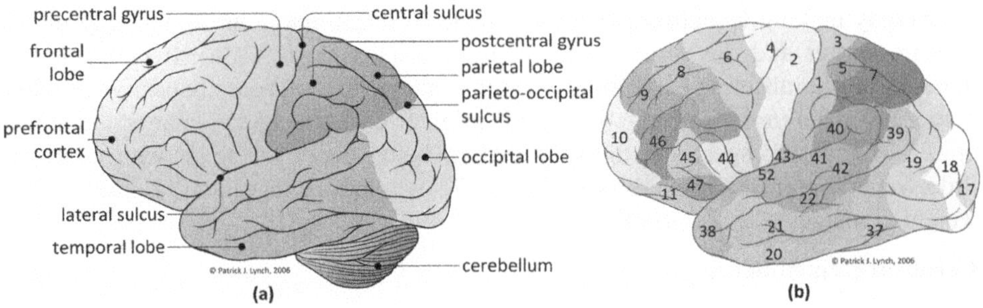

Figure 2.24: **(a)** Lateral schematic of the brain depicting the lobes in different colors. Anterior is on the left of the image, posterior is on the right. **(b)** Approximate Brodmann's areas are shown. For instance, Area 44 is responsible for motor speech function on the left side, whereas Area 2 controls contralateral body motor function. Areas 39 and 40 are involved in speech comprehension. Area 1 is the primary sensor strip. *Brain templates adapted from Patrick J. Lynch.*

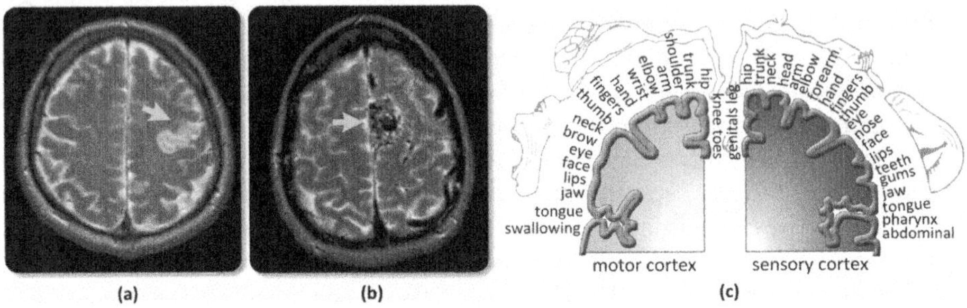

Figure 2.25: **(a)** Axial T2-weighted image demonstrating a stroke in the patient's left frontal lobe cortex. The lesion is located in the left primary motor strip (arrow) in a patient who presented to the emergency room with acute onset of right arm weakness. Note that the right side of the image represents the left side of the patient's brain. **(b)** An MRI in a patient with right leg motor seizures. This image demonstrates an arteriovenous malformation (AVM, a congenital tangle of blood vessels) near the vertex along the portion of the homunculus expected to govern leg motor function. **(c)** Coronal schematic of the homunculus with motor function depicted on the right and sensor function on the left. Note that the amount of cortex devoted to the functions may not always be the same between the left/right sides.

Motor function. The *primary motor cortex* is located in the posterior frontal lobe immediately anterior to the central sulcus (Fig. 2.24a), and is also referred to as the precentral gyrus. This cortex contains specialized neurons called *pyramidal cells* that generate the impulses for voluntary movement that descend through the main motor tract (the corticospinal tract) to the contralateral body part. The gyrus anterior to this cortex is the *premotor cortex*. The premotor cortex is responsible for learned repetition such as typing or playing an instrument. As such, damage to the primary motor cortex results in paralysis of the contalateral side of the body, whereas injury to the premotor cortex will often result in loss of the learned skill while strength in the affected limb is preserved (Fig. 2.25a-b). The area referred to as *Broca's area* (Brodmann's area 44) is responsible for motor speech (Fig. 2.24b) and is located in the left hemisphere near the Sylvian fissure in most right-handed people and also in most left-handed people as

Lobe/Region	Function
Front lobe (primary motor cortex)	Voluntary control of skeletal muscle
Parietal lobe (primary sensory cortex)	Responsible for the perception of touch, pressure, vibration, temperature, and taste
Occipital lobe (visual cortex)	Handles perception of visual stimuli
Temporal lobe (auditory, olfactory cortex)	Handles perception of sounds and smells
All lobes (association areas)	Integration and processing of sensory data; processing and initiation of motor activities

Table 2.5: Summary of brain lobe/regions and primary functions.

well. Occasionally a left-handed person may have primary language function in the right hemisphere, referred to as right hemisphere dominance.

The large portion of the frontal lobes known as the *prefrontal cortex* (Fig. 2.24a) is dedicated in humans to cognition, complex learning, mood and executive functions such as reasoning and judgment. This area develops slowly (it does not fully mature until the late teens/early 20s) and depends on positive and negative feedback.

The senses. The primary somatosensory cortex is located in the anterior parietal lobe just posterior to the central sulcus and is known as the post central gyrus or *primary sensory strip*. This region is responsible for conscious perception of sensation. The cell bodies in the gray matter of the sensory strip are specialized receptor neurons that receive input from skin and skeletal muscle via the white matter tract known as the *spinothalamic tract*. Damage to this portion of the cortex will result in a loss of sensation in the affected limb. The gyrus posterior to the post central gyrus is the *somatosensory association cortex* and is responsible for integrating and analyzing sensory input such as temperature, pressure, and size. For example, with this additional input, one not only feels the presence of loose change in the pocket but one can recognize a quarter or dime or nickel by feel alone [30]. Impairment of the association cortex would therefore affect one's ability to recognize objects based on feel – but without affecting the ability to feel the object.

The distribution of motor and sensory function along the cerebral cortex is quite orderly and regular between individuals. This distribution is referred to as the *homunculus*. As seen in Fig. 2.25c, motor/sensory function for the leg occurs near the vertex or top of the brain, and as one travels inferiorly functional zones for the arm and the face are encountered, with motor and sensory function for the face residing just superior to the Sylvian fissure.

Within the occipital lobes are the *primary visual cortex* and the *visual association cortex*. The primary visual cortex is located at the occipital pole and represents the largest of the sensory areas, receiving information from the retina through the optic tracts and like the previously mentioned areas, there is left-right governance of the visual fields (*i.e.,* the right occipital cortex is responsible for the left visual field of each eye and the left occipital cortex is responsible for the right visual field of each eye). Problems with one visual cortex results in blindness of the contralateral visual field; damage to both occipital cortices results in total cortical blindness. The visual association cortex is responsible for comprehension of vision. Damage to the association cortex results in retained sight, but lacking the capacity to recognize objects (Fig. 2.26a). Along the lateral temporal cortex are the primary and association auditory areas. The

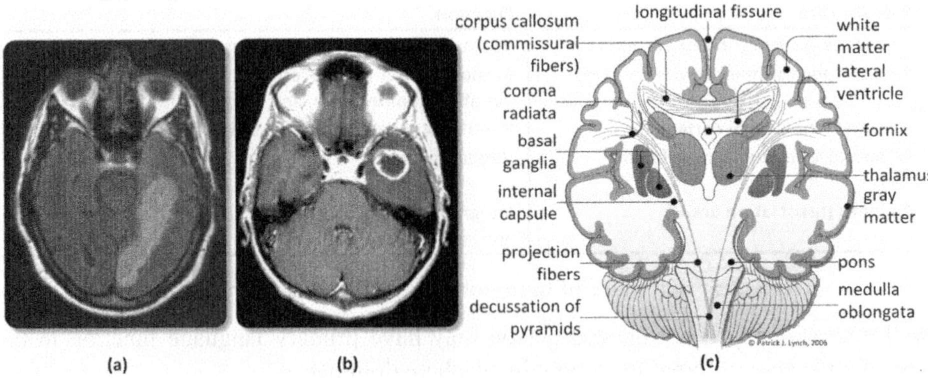

Figure 2.26: (a) MR image showing damage to the association cortex. **(b)** Contrast-enhanced MR of a temporal lobe tumor in a patient presenting with seizures preceded by the sense of smell of burning rubber. **(c)** Coronal schematic of the brain, showing some of the fiber tracks, including the corpus callosum and projection fibers that cross over in the decussation of pyramids. *Brain template adapted from Patrick J. Lynch.*

primary auditory cortex receives information from the cochlea creating the sensation of sound while the *association auditory cortex* is responsible for the recognition of the sound as music, thunder, clapping, etc. [30]. Memories of sound are also stored here.

Deep along the medial margin of the temporal lobe is the *olfactory cortex* that provides for conscious awareness of odors. This region has further evolved in vertebrates into storage for memories and emotions. Thus, for instance, seizures that originate in the medial temporal lobe are often preceded by an olfactory aura (Fig. 2.26b).

Cerebral White Matter

The white matter is well organized and divided into tracts. These tracts may run vertically to and from the cerebral cortex to the spinal cord (*projectional fibers*), from one side of the cerebral cortex to the other side (*commissural fibers*), or they may project from one gyrus to the next (*association fibers*) (Fig. 2.26c). The largest intracranial white matter tract, the *corpus callosum*, is a commissure that connects the right and left hemispheres. As it travels from side to side, it is best seen in cross-section on sagittal and coronal images of the brain. The other tracts cannot be resolved individually on routine clinical imaging; however, *tractography* is now possible using diffusion tensor imaging (see earlier) (Fig. 2.27a-b). The axons in white matter deliver electrical impulses that may release chemical neurotransmitters at the terminus of the axon that, in turn, act on target tissue such as muscle, gut, or glandular tissue. These neurotransmitters cannot be directly imaged by conventional MR or CT imaging.

Basal Nuclei

The basal nuclei, also called the *basal ganglia* or *corpus striatum*, consist of the caudate nucleus, the putamen, and the globus pallidus. These largely gray matter structures receive input from the entire cortex, and are important in starting and stopping motor function, monitoring the rhythm of movement (like the swinging of arms while walking). Insult to these structures causes tremors, slowness, and Parkinsonism. The *thalamus* is another of the deep gray matter nuclei. It is comprised of two parts, on

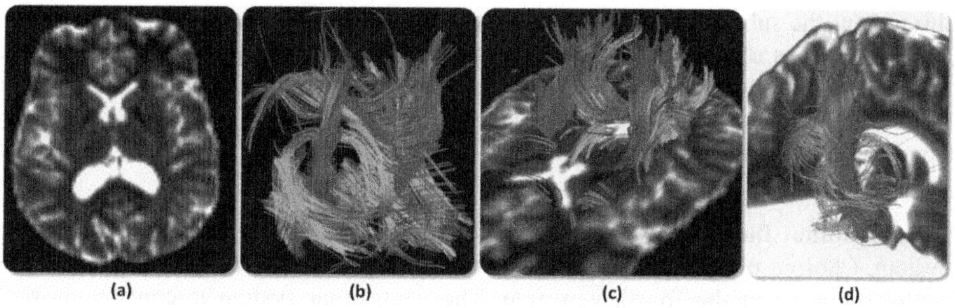

Figure 2.27: Examples of diffusion tensor imaging and tractography. **(a)** An axial image created under a DTI sequence. **(b)** Visualization of the fiber tracks in the volume, under a 3D perspective; note the fibers crossing between hemispheres (commissural fibers). **(c)** The same fiber tracks, shown in conjunction with an axial slice. **(d)** a 3D rendering of the fibers relative to the corpus callosum. *Rendered using MedINRIA.*

either side of midline, with a central connection across the third ventricle known as the *massa intermedia*. The thalamus is responsible for mediating sensation, arousal, learning, and memory. Damage to the thalamus can cause contralateral sensory loss; and if both hemi thalami are damaged, coma may occur.

Brainstem

The brainstem serves as a conduit for the white matter tracts that pass from the cerebral cortex to the body and extremities; and also houses the cranial nerves and their nuclei. The brainstem is divided into the *midbrain*, the *pons*, and the *medulla*. The midbrain is the most cranial of the brainstem levels, with the nuclei for cranial nerves III and IV exiting the midbrain to reach the orbits where they innervate the extraocular muscles. The dorsal portion of the midbrain contains the superior and inferior colliculi. The *superior colliculi* function as a visual reflex center for head and eye movement and the *inferior colliculi* serve as the auditory relay for the startle reflex that results in head turning. The pons contains the nuclei for cranial nerves V, VI and VII. Cranial nerve (CN) V is largely responsible for facial sensation and motor innervation to the muscles of mastication. Cranial nerve VI innervates a solitary extraocular muscle, the lateral rectus muscle. Cranial nerve VII is responsible for motor function of the muscles of facial expression. Finally, the medulla is a tightly packed region and the smallest in cross-sectional diameter. The nuclei for CN VIII, IX, X, XI and XII are found in the medulla. CN VIII consists of the superior and inferior vestibular nerves (balance) and the cochlear nerve (hearing).Cranial nerve IX is the glossophayngeal nerve, partially responsible for the muscles that coordinate swallowing. CN X, the vagus nerve, innervates the vocal cords. CN XI innervates the trapezius muscles allowing shrugging of the shoulders, and CN XII innervates the tongue.

The medulla is also largely responsible for homeostasis: the center for cardiovascular force and rate of ventricular contraction is located here. The medulla is also home to the respiratory center, regulating the rate and depth of breathing. The centers for vomiting, hiccupping, coughing, sneezing, and swallowing are also located here.

Meninges

The brain is covered by three protective layers known as the *meninges*. The toughest outermost layer is located along the inside of the skull and is called the *dura mater*. The second layer is called the arachnoid mater. The space deep to the arachnoid membrane is the *subarachnoid space*. The cerebrospinal fluid (CSF; see below) travels throughout the subarachnoid space down around the spinal cord and up over the surface convexities of the brain until it is reabsorbed by the arachnoid granulations and returned to the venous system (superior sagittal sinus). This fluid surrounds the brain and serves as a shock absorber. The deepest layer of the meninges that lies directly upon the brain surface is called the *pia mater*. Together, the meninges serve as a barrier to infection and trauma.

Cerebrospinal fluid. The CSF is produced in the choroid plexus of the ventricular system. Choroid plexus exists in the lateral, third and fourth ventricles and within the outlet foramina of the fourth ventricle. The ventricular system is continuous with the central canal of the spinal cord. Fluid produced by the choroid plexus travels from the lateral to the third and then the fourth ventricles. The fluid then passes to the subarachnoid space. If there is a blockage along the pathway, then the affected ventricles will dilate and the patient will develop hydrocephalus.

Cerebral Vascular Anatomy

Blood-brain barrier. Unique to the brain is the *blood-brain barrier* (BBB). This is a filter formed by the specialized endothelial cells that line the walls of the capillaries of the brain. These capillary walls contain tight junctions that allow some substances to pass easily such as glucose, amino acids, and some electrolytes. Other substances that are easily passed include alcohol, nicotine, some anesthetics as well as gases such as oxygen and carbon monoxide.

Cerebral arteries. The four main arteries that supply the brain are the two *internal carotid arteries* (ICAs) that ascend anteriorly in the neck and the two vertebral arteries that ascend posteriorly in the neck. The internal carotid arteries begin their course in the neck at approximately the level of the C3-4 disc where the *common carotid artery* (CCA) bifurcates into the ICA and the *external carotid artery* (ECA). The ICA enters the skull through its own carotid canal and then bifurcates into the anterior (ACA) and middle (MCA) cerebral arteries, supplying the majority of blood flow to one hemisphere. The vertebral arteries enter the skull through foramen magnum and join to form the basilar artery. The basilar artery then travels up the ventral surface of the brainstem supplying blood flow to the brainstem and the cerebellum and finally to the occipital lobes at the posterior aspect of the cerebrum. There are two small posterior communicating arteries that connect each posterior cerebral artery to its ipsilateral ICA, and there is a small anterior communicating artery that connects the two ACAs together. This arterial network is known as the *Circle of Willis* (Fig. 2.28a). Because of the "circular" configuration of this blood supply, there is collateral flow available in the event of occlusion of one of the major arteries. For instance, if the right ICA were to shut down, the blood could flow from the left-sided carotid system through the anterior communicating artery to the right MCA. Similarly, the basilar artery can assist in the delivery of blood flow through the right posterior communicating artery to the right distal ICA (which is usually preserved in the setting of ICA occlusion in the neck) and then to the right MCA.

Pathophysiology of a stroke. A cerebral infarction (*i.e., stroke*) occurs when there is a lack of blood flow to a portion of the brain and the brain tissue dies. This event is often secondary to a blood clot or piece of atheromatous plaque that breaks loose and travels with the blood flow into a major cerebral artery, preventing blood and oxygen from reaching the brain tissue. When the neurons are deprived of oxygen, the cell wall

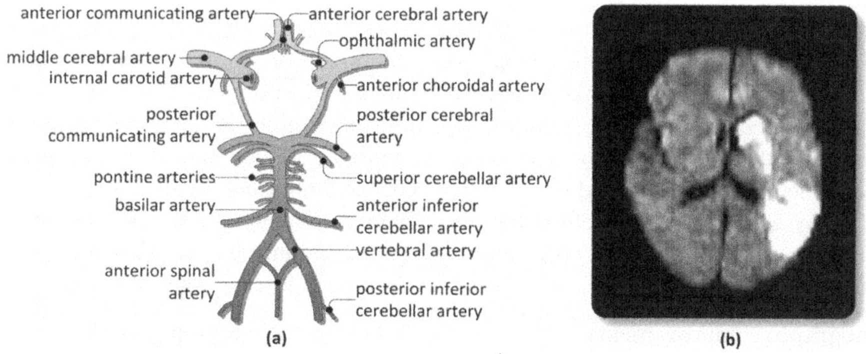

Figure 2.28: **(a)** Illustration of the Circle of Willis, showing the different arteries. **(b)** Axial diffusion weighted image (DWI) demonstrating restricted diffusion in the left basal ganglia and left parietal cortex within the distribution of the left MCA (middle cerebral artery) that was occluded by embolus.

can no longer maintain the necessary ionic gradients across the membrane and sodium and water enter into the cells from the surrounding interstitium, causing cellular swelling or cytotoxic edema. Diffusion-weighted MR can detect this shift of water and is an excellent tool in the diagnosis of early cerebral infarction (Fig. 2.28b).

Breast Anatomy and Imaging

In the adult woman, the breasts are milk-producing glands located over the pectoral muscles. Ligaments on both sides of the breast bone and sternum support and attach the breasts to the chest wall (Fig. 2.29a). Breasts contain no muscle tissue; a layer of fatty tissue (subcutaneous adipose tissue) surrounds the mammary glands and extends throughout the breast. 15-20 *lobes*, arranged in a circular fashion, comprise each gland; and each lobe in turn is made up of small bulb-like *lobules* that produce milk in response to hormones (*i.e.,* estrogen, progesterone, prolactin). *Ducts* transport the milk from lobules to the nipple. The blood supply to the breast region is drawn from the *axillary artery* and the *internal mammary artery*. Also, *lymphatic vessels* run throughout breast tissue; these vessels connect with a network of *lymph nodes* located around the breasts' edges and in surrounding tissues in the armpit and collarbone.

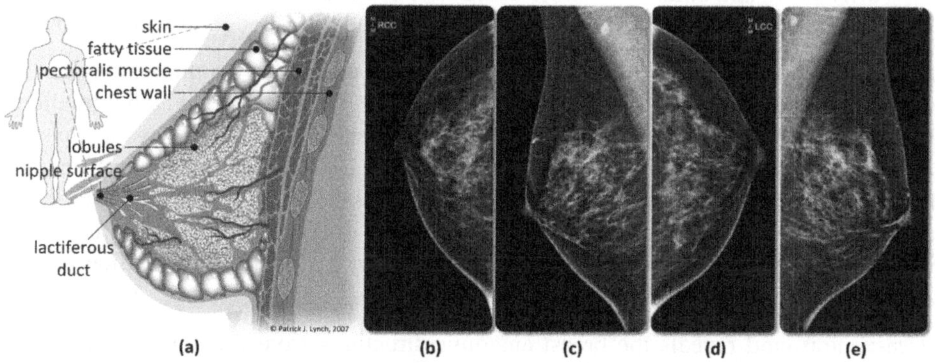

Figure 2.29: **(a)** Anatomy of the adult female breast, sagittal view. *Drawing adapted from Patrick J. Lynch.* Right breast mammograms, craniocaudal (CC) view **(b)** and mediallateral oblique (MLO) view **(c)**. Left breast mammograms, CC view **(d)** and MLO **(e)**.

Breast Imaging

Mammography continues to be the primary imaging modality for breast cancer screening and diagnosis, with digital mammography being the most important recent technologic improvement in this area. Additionally, ultrasonography is commonly used in conjunction with mammography. Advances in imaging-guided breast biopsy techniques have led to the widespread use of stereotactic- and ultrasound-guided breast core needle biopsy as the primary method for breast biopsy of abnormal imaging findings. Beyond these core modalities, other more advanced modalities are used including magnetic resonance imaging (breast MRI) and radiolonuclide imaging of the breast.

Mammography. Improvements in the overall quality of mammography performance are related to the efforts of programs established both by professional societies and government agencies. The introduction of the American College of Radiology (ACR) Mammography Accreditation Program in 1987 and the Mammography Quality Standards Act in 1994 are among the most significant of these efforts [58]. Standardized documentation of mammographic findings via the ACR Breast Imaging Reporting and Data System (BI-RADS) [5] has also played a major role in the evolution and adoption of mammography.

Today, mammography exams can be divided twofold:

1. Screening. *Screening mammography* is an examination of an asymptomatic woman to detect clinically occult breast cancer [8]. The standard screening examination includes two views of the breast: a mediolateral oblique (MLO) and a craniocaudal (CC) (Fig. 2.29b-e) [4]. The effectiveness of screening mammography for mortality reduction from breast cancer is related to earlier cancer detection (Stage 1) and has been confirmed by evaluations of randomized clinical trials [77].
2. Diagnostic. *Diagnostic mammography* is indicated when there are clinical findings such as a palpable lump, localized pain, nipple discharge, or an abnormal screening mammogram that requires additional work up [3, 28, 32, 61]. To correlate the clinical and imaging findings, a marker (*e.g.,* radiopaque BB or other) is often placed over the skin in the area of clinical concern prior to performing the mammograms (Fig. 2.30a). The diagnostic workup may include MLO, CC, and mediolateral (ML) views and additional views using spot compression and magnification techniques, correlative clinical breast examination, and ultrasonography (see below).

Breast ultrasound. Breast ultrasound is an essential adjunct to mammography for the workup and diagnosis of palpable and mammographically-detected abnormalities. Historically breast ultrasound was used to differentiate solid and cystic masses. Advances in ultrasound technology have led to high-resolution ultrasound imaging helping differentiate benign and malignant solid masses [26, 76]. In addition to lesion characterization, breast ultrasound is used to guide interventional breast procedures, including cyst aspiration, core needle biopsy (see below), fine needle aspiration (FNA), and ultrasound-guided preoperative needle localization.

Breast ultrasound reveals the breast anatomic structures from the skin surface to the chest wall (Fig. 2.32). Normal skin measures < 3 mm and is composed of two parallel echogenic (white) lines separated by a thin, hypoechoic (dark) band. Just under the skin lies the subcutaneous fat followed by the interwoven bands of fibroglandular tissue and breast fat. Both subcutaneous and breast fat are mildly hypoechoic (gray), whereas the

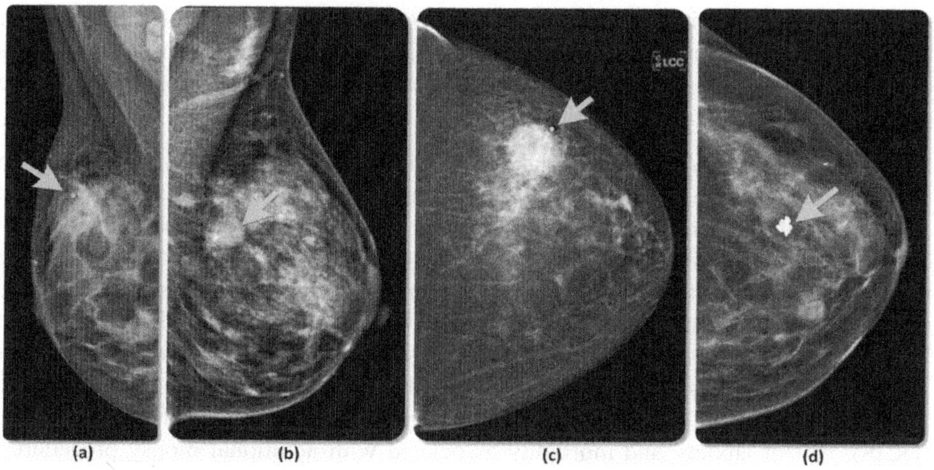

Figure 2.30: **(a)** Right breast MLO view shows a BB marker overlying a breast mass at the site of clinically palpable lump (arrow). **(b)** Left breast MLO shows an oval circumscribed mass (arrow). Ultrasound demonstrated a simple cyst. **(c)** Left breast CC view with a round mass with indistinct and microlobulated borders (arrow). Pathology results of the excised mass showed invasive ductal carcinoma. **(d)** Example of a benign, popcorn-like calcification typical of a calcified fibroadenoma.

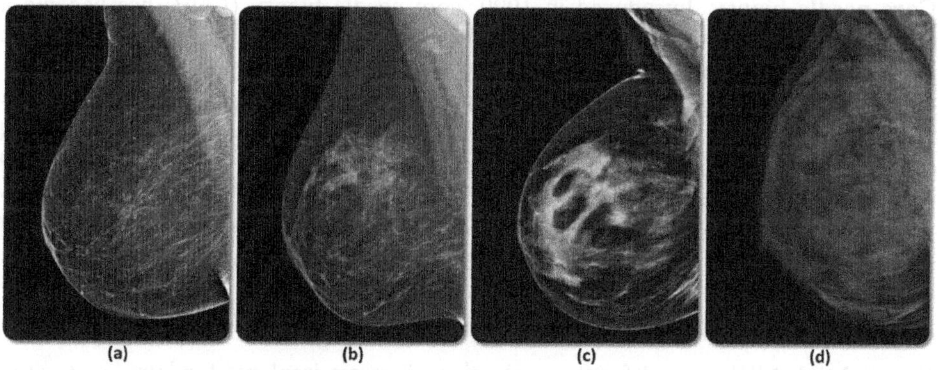

Figure 2.31: BI-RADS breast tissue composition categories. **(a)** Category 1, almost entirely fatty. **(b)** Category 2, scattered islands of fibroglandular tissue. **(c)** Category 3, heterogeneously dense. **(d)** Category 4, extremely dense.

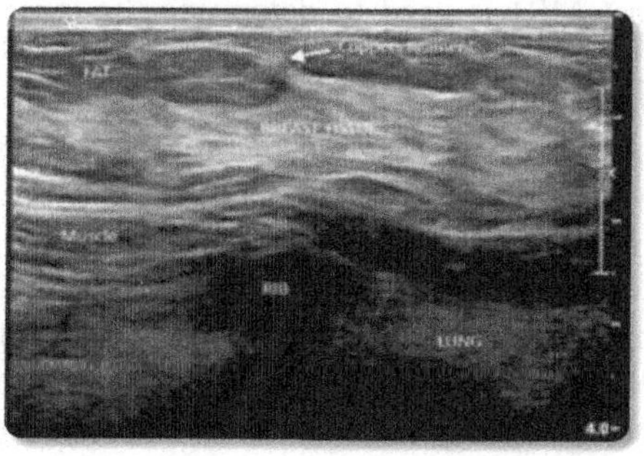

Figure 2.32: Normal ultrasound anatomy of the breast.

fibroglandular tissue is hyperechoic (light gray to white). Deep to the fibroglandular tissue is the retroglandular fat, which lies against the chest wall. The chest wall is composed of the more superficial band of the pectoralis muscle, the ribs laying deep to the pectoralis muscle, and the parietal pleura.

A *breast screening ultrasound* is defined as bilateral whole breast ultrasound of an asymptomatic woman with normal mammograms. Several studies have shown that small, clinically and mammographically occult breast cancers may be detected with screening ultrasound in women with dense breast tissue [14, 18, 29, 44, 49]. Despite the encouraging results from these studies, many potential drawbacks are associated with ultrasound screening of the breast. Of particular concern is the high number of incidental benign masses encountered during screening ultrasound for which either biopsy, aspiration or short interval follow-up ultrasound is recommended. Additional problems include an extremely limited ability to detect ductal carcinoma *in situ* (DCIS), patient anxiety and morbidity associated with additional biopsy procedures, added cost, lengthy exam times, and highly variable inter-operator performance with ultrasound. A large study of screening breast ultrasound in high risk women with dense breast breasts conducted by the American College of Radiology Imaging Network (ACRIN) and the Avon Foundation independently evaluated screening ultrasound compared to screening mammography. The study concluded that a single screening ultrasound will yield an additional 2.1-7.2 cancers per 1,000 high-risk women, but will also substantially increase the number of false positives [9].

Core needle biopsy. Introduced in 1990, core needle biopsy (CNB) has become a desirable alternative to excisional surgical biopsy as it is less costly, results in less morbidity, and minimizes scarring. Notably, CNB of the breast overcomes the limitations of FNA cytology because insufficient samples are less frequent, the interpretation can be performed by a pathologist without special training in cytopathology, and CNB can differentiate invasive from *in situ* breast cancer [40, 64]. CNB are performed with imaging guidance to sample a clinical or imaging identified abnormality. Imaging guidance can be provided by ultrasound or mammography (stereotactic). The choice of ultrasound vs. stereotactically-guided CNB is based on which modality best demonstrates the abnormality and the location of the abnormality in the breast. However, ultrasound is usually preferred as it is faster and more comfortable for the patient.

Breast MRI. MRI is used for the evaluation of breast implants for intra- and extra-capsular rupture. Breast MRI using IV contrast agents show a high sensitivity for the detection of breast cancer as cancers show rapid contrast enhancement. However, specificity varies as numerous benign entities can also show rapid contrast enhancement. Contrast-enhanced breast MRI is used to determine the size and extent of invasive cancers; identifying multifocal and multi-centric lesions; evaluating the ipsilateral breast of a woman with unilateral axillary metastases; and identifying recurrent carcinoma in the conservatively treated breast. A multi-institutional study [52] concluded that women at high-risk for breast cancer would benefit from screening MRI. In that study, high-risk included women 25 years of age or older who were genetically at high risk (BRCA1/2 carriers or with at least a 20% probability of carrying such a mutation). The study found that screening MR imaging led to biopsies with higher positive predictive value and helped detect more cancers than either mammography or ultrasound. As such, the American Cancer Society recently recommended breast MRI screening for women at high risk for breast cancer [71].

Radionuclide imaging. Another area of active investigation involves scanning of the breast after the injection of the radionuclide-labeled substances that concentrate in breast tumors. Technetium-99m (Tc-99m) methoxyisobutyl isonitrile (MIBI) breast scintigraphy (*scintimammography*) has been under investigation for several years. Initial reports indicated high sensitivity (> 90%) and specificity (slightly < 90%) [47]. Later reports, however, indicated a relatively low sensitivity for small cancers, those found only by mammography (56%), and those 1cm or larger (39%) [82]. However, a new technology using this radionuclide agent image specially designed for the breast, *breast-specific gamma imaging*, is undergoing clinical trials [12, 13] with early results showing utility in avoiding biopsies of palpable breast masses with indeterminate mammographic and ultrasonographic features. In addition, Tc-99m sulfur colloid has been proven useful and is now widely used for the identification of sentinel nodes [72]. Prior to surgery, the isotope is injected into the breast in the vicinity of a biopsy proven breast cancer. The injected isotope in theory drains through the same lymphatic chain as the tumor. At surgery, the sentinel nodes draining the site of the cancer are identified using a radioisotope probe. The sentinel nodes are removed and evaluated histologically. If the sentinel nodes are negative for tumor, axillary node dissection can be avoided.

Tumor uptake also has been identified on PET after the injection of ^{18}F 2-deoxy-2-fluoro-D-glucose [2]. This agent also accumulates in axillary nodes, providing information about nodal status. These methods will require additional studies to determine sensitivity, specificity, and cost-effectiveness.

Breast Cancer and other Findings

Masses. A mass is any space-occupying lesion that is seen on at least two mammographic projections. If a density is seen on only one view, it is described as an *asymmetry*. Masses are described by their shape and margins (Fig. 2.30b-c). The shape can be round, oval, lobular, or irregular. Oval and round masses are usually benign, whereas an irregular shape suggests a greater likelihood of malignancy. The margins of masses are the most reliable indicator of the likelihood of malignancy. The margins can be described as circumscribed, microlobulated, obscured (partially hidden by adjacent tissue), indistinct (ill-defined), or spiculated. *Circumscribed margins* favor a benign etiology with a likelihood of malignancy probably < 2% (10-12). Ultrasound is often necessary to determine whether a round or oval circumscribed mass is cystic or solid. *Microlobulated margins* increase the likelihood of malignancy. If the mass is directly adjacent to fibroglandular tissue of similar density, the margin may be obscured, and additional imaging should be done in an attempt to show the margins as completely as possible. A finding of *indistinct margins* is suspicious for malignancy. A mass with *spiculated margins* has lines radiating from its border, and this finding is highly suggestive of malignancy. An area of spiculation without any associated mass is called an *architectural distortion*. The density of a mass compared with normal fibroglandular tissue provides another clue as to its etiology. In general, benign masses tend to be lower in density than carcinomas; however, the density of a mass is not always a reliable sign as to whether it is benign or malignant.

Cystic masses. Breast ultrasound can reliably identify cystic masses. BI-RADS describes three types of cystic masses:

1. Simple cysts. A *simple cyst* is round or oval shaped, anechoic (black with no internal echoes) mass with smooth margins, an imperceptible wall, and increased posterior acoustic echoes (Fig. 2.33a). This last feature means it appears as though a flashlight is shining through the back of the cyst. Because cysts develop within the terminal duct lobular unit of the breast, it is not uncommon to see clusters of cysts or coalescing cysts.
2. Complicated cysts. Sometimes cysts with echogenic interiors are seen, such as a debris-filled cyst. These cystic masses are called *complicated cysts* and further evaluation may be needed. Ultrasound-guided aspiration can be performed to verify its cystic nature, to exclude a solid mass, and to confirm complete resolution of the mass after aspiration.
3. Complex mass. A *complex mass* is defined as a mass with both cystic and solid components. The solid component is usually seen as a mural nodule or an intra-cystic mass. A complex mass can also be composed of thick walls and anechoic center. A cyst with a solid component is suspicious for a malignancy, such as a papillary carcinoma or a necrotic infiltrating carcinoma. Benign papillomas can also present as a complex mass. The diagnostic evaluation of a complex mass is ultrasound-guided CNB of the solid component or surgical excision.

Solid masses. Several studies have defined criteria to aid in the distinction of benign and malignant solid breast masses [26, 76]. Although no single or combination of sonographic features is 100% diagnostic for a benign mass, careful use of established criteria can help differentiate benign and malignant solid masses and avoid biopsy of certain solid masses.

Mass shape, margins, orientation relative to the skin surface, echogenicity, and posterior echoes are the minimum preliminary characteristics that should be assessed in solid masses. Typically benign sonographic features of solid masses include an ellipsoid or oval shape, width greater than anteroposterior diameter (orientation parallel to the skin surface), three or fewer gentle lobulations, circumscribed margins, a pseudocapsule, echogenicity hyperechoic to fat (*i.e.,* whiter than fat), and absence of any malignant features (Fig. 2.33b). In comparison, malignant sonographic features of solid masses include an irregular or angular shape; more than three lobulations; ill-defined, spiculated or microlobulated margins; width smaller than anteroposterior diameter; markedly hypoechoic (dark) echogenicity; a surrounding thick, echogenic (white) halo; posterior shadowing (black shadows posterior to the mass), duct extension; and associated calcifications (Fig. 2.33c) [67]. There appears to be overlap in these features, and some malignant masses may have features suggesting they are benign, which could lead to false-negative interpretations of malignant solid masses.

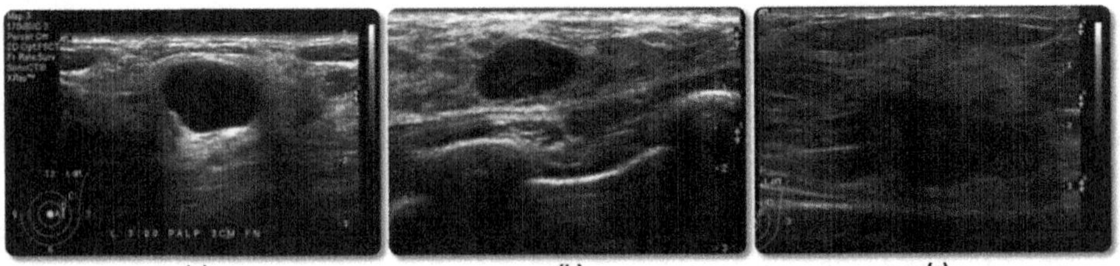

(a) (b) (c)

Figure 2.33: **(a)** Breast ultrasound, simple cyst. **(b)** Breast ultrasound, benign mass (fibroadenoma). **(c)** Breast ultrasound, malignant mass (invasive carcinoma).

Calcifications. Calcifications are described on mammograms by their morphology and distribution. The calcifications can be placed into three general categories: 1) typically *benign calcifications* (Fig. 2.30d) can be usually identified by their mammographic features and include skin, vascular, coarse, large rod-like, round, egg-shell, and milk-of-calcium types; 2) *intermediate concern calcifications* are described as amorphous or indistinct (these are tiny or flake-shaped calcifications that are small or hazy in appearance so that a more specific morphologic classification cannot be made; and 3) *higher probability of malignancy calcifications* (Fig. 2.34) can be described as plemorphic, heterogeneous, or fine, linear, and branching. Calcifications are also characterized in mammography reports by their distribution. *Grouped* or *clustered calcifications* include more than five in a small area (< 2cc) and can be benign or malignant. *Linear calcifications* are in a line and may have small branch points. When linear calcifications are in a line and branching their distribution is duct-like and suspicious for malignancy. *Segmental calcifications* are distributed in a duct and its branches with the possibility of multifocal carcinoma in a lobe (or segment) of the breast. A segmental distribution tends to be "triangular" with the apex towards the nipple. *Regional calcifications* are in a larger volume of breast tissue, and usually do not indicate suspicious calcifications. Finally, *diffuse* or *scattered calcifications* are distributed randomly through both breasts and are almost always benign.

Musculoskeletal System

The musculoskeletal system develops from mesenchyme (embryonic connective tissue) and includes bones and cartilage; muscles, ligaments, tendons; and other tissue types, collectively called "soft tissues." The musculoskeletal system is primarily responsible for locomotion and upright gait. Musculoskeletal imaging deals with a wide variety of pathology affecting numerous individual parts of the system.

The base musculoskeletal unit is the bone-muscle-joint complex with its stabilizing ligaments, tendons and joint capsule (Fig. 2.35). The bones provide the solid central portion of an extremity. The bone is surrounded by complex compartmentalized groups of muscles, each with a specific task. Muscle fibers blend into tendon fibers at either end of each muscle, and the tendon ultimately attaches to the bone, usually spanning over one and occasionally two joints. The musculotendinous junction is the weakest link of this base unit. As such, the majority of tears and strains occur at this junction. *Ligaments* are composed of dense connective tissue that spans between bones, providing substantial

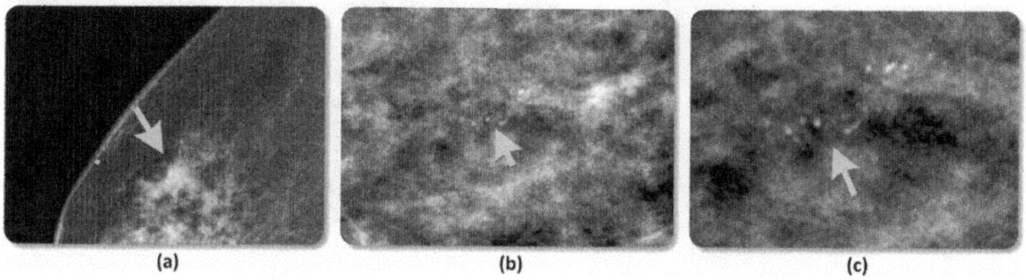

Figure 2.34: (a) MLO view of the right breast shows an area of arquitectural distortion of the breast tissue. Pathology demonstrated an invasive ductal carcinoma. **(b-c)** Magnification views demonstrating plemorphic microcalcifications (arrows). Pathology reveled microcalcifications in association with ductal carcinoma *in situ* (DCIS).

stabilization to the joint. Even though ligaments often blend in with the capsule of the adjacent joint, they can be anatomically (and often radiologically) distinguished from the capsule.

Imaging of the Musculoskeletal System

Each imaging modality, as described in the beginning of this chapter, provides specific information for certain tissue types of the musculoskeletal system. The following is a review of respective imaging information provided by each modality.

Plain radiography. Plain radiography (Fig. 2.35-2.37) should usually be the initial modality for assessing musculoskeletal pathology. It is an inexpensive yet essential tool for diagnosis of fractures, arthritis, bone tumors, infections, and many other entities. In many situations, plain radiography provides sufficient information to determine the appropriateness of conservative vs. surgical management. Radiographs are usually obtained in two perpendicular plains for long bones, and commonly with a third oblique view for evaluating joints. Pathology seen on one view is typically confirmed on a second view. The second view provides further information about localization and 3D orientation of pathology, rendering more detailed characterization (Fig. 2.38a-b; fracture displaced on one view).

Plain radiography is routinely used to characterize *fractures* (Fig. 2.38c-d: femur fracture). The diagnostic information needed by the treating physician includes the location of the fracture along the bone, orientation of the fracture (transverse, oblique, spiral, complex, etc.), the direction of the distal fragment displacement (anterior, posterior, etc.), the degree and orientation of angulation between the fragments, and the amount of diastasis (*i.e.,* separation), impaction, or overriding of the fragments. Impaction or overriding may require traction for treatment. For bone tumors, radiographs provide information on the location of the tumor along the bone. The appearance of the tumor margin reflects the aggressiveness of the lesion. Radiographs also display the mineralized matrix of a bone tumor (osteoid or cartilage) and the amount and nature of periosteal new bone formation, if present. The sum of this information often aids the radiologist in developing a differential diagnosis, and establishes if a lesion needs biopsy and/or treatment. In the setting of *arthritis*, plain radiography demonstrates

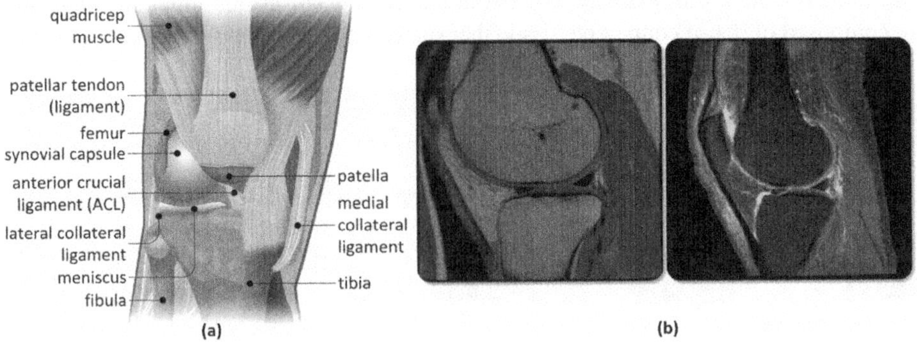

Figure 2.35: (a) Coronal depiction of the typical knee. Example of the musculoskeletal system in the knee, showing the muscles, ligaments, tendons, and bones. **(b)** Examples of the knee on a sagittal view using MR.

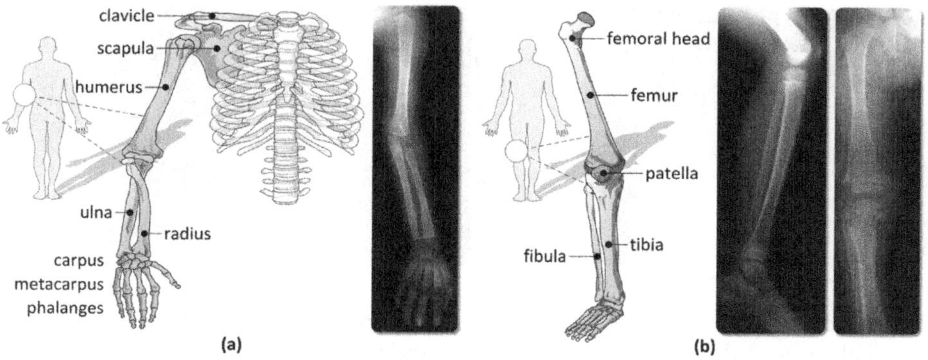

Figure 2.36: Projectional imaging of the extremities, shown alongside schematic diagrams labeling the key bones. **(a)** The shoulder, arm, and major groups of bones in the hands (carpus, metacarpus, phalanges) are shown relative to the ribcage and portion of the spine. **(b)** The upper and lower leg are shown, with the rightmost x-ray showing the knee, femur, and femoral head.

changes in the joint space, bone remodeling including formation or erosion, and often, joint effusions. By evaluating the character of bone and soft tissue changes and the joints involved, a specific diagnosis of the type of arthritis can often be made.

Fluoroscopy. Fluoroscopy provides imaging in real time, and allows visualization of movement; as such, it is widely used in musculoskeletal imaging for joint injections. Joint injections may be performed to instill contrast material for conventional x-ray *arthrography* such as in wrist arthrography, which is used to evaluate the integrity of ligaments and the triangular fibrocartilage, major stabilizers of the wrist (Fig. 2.38e; wrist arthrogram with leak). Fluorscopy-guided injections may be used in conjunction with advanced imaging such as MR imaging (*e.g.*, MR arthrography). A joint injection may also be performed to administer medication (corticosteroids or anesthetics) for therapeutic pain control or to diagnose if the joint is the source of a patient's pain. Fluoroscopy can also be used to guide a joint aspiration to obtain a fluid sample for laboratory analysis in cases of suspected joint infection (*e.g.*, septic arthritis).

Computed tomography. CT is superior to other modalities for imaging bone detail, and is routinely used to evaluate the extent of fractures in complex bony structures such as the bony pelvis, cervical spine, ankle, and wrist. Information provided by CT over plain radiography in a patient with a complex fracture includes not only a far more detailed display of fracture anatomy, but also entrapment of bony fragments in

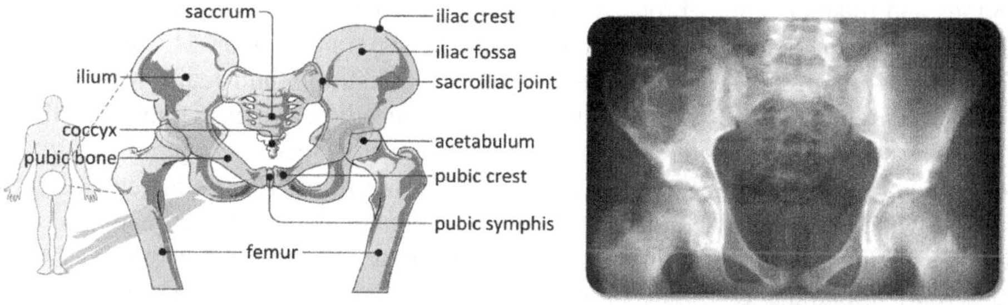

Figure 2.37: Plain film radiograph of the pelvic bone. The left diagram labels the major bones/regions of the region.

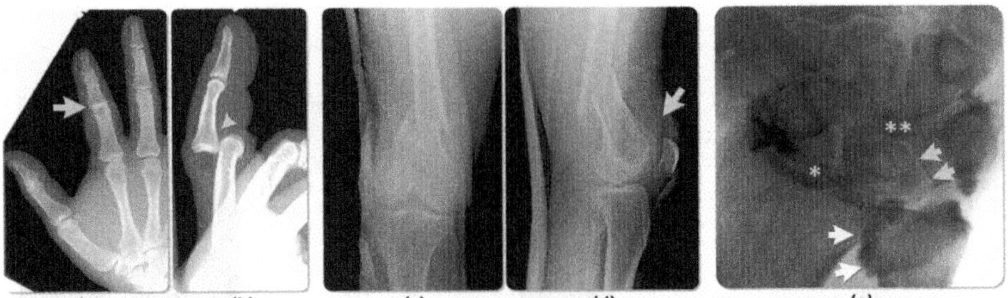

Figure 2.38: **(a)** Frontal radiograph of the digit suggests only narrowing of the proximal interphalangeal (PIP) joint (arrow). **(b)** The lateral view reveals a posterior dislocation of the PIP joint (arrowhead). **(c)** Frontal and **(d)** lateral views of the knee demonstrate a predominantly transverse fracture of the distal femur with half-shaft lateral displacement, apex posterior angulation and mild impaction. There is a large hemarthrosis (bloody joint effusion, arrow). **(e)** Frontal view of the wrist following injection of contrast (black material) into the wrist joint (*) demonstrates leak of the contrast through the lunotriquetral interval (rightmost arrows) into the midcarpal compartment (**) and through the central disk of the triangular cartilage complex (arrowhead) into the distal radioulnar joint (white arrows).

joint spaces or intraarticular extension of a fracture. This knowledge aids the trauma surgeon in properly planning an appropriate action. For bone tumors, CT provides a detailed look at the tumor margins, matrix, and extent superior to plain radiography. The wide availability of multi-detector CT provides a seamless reformation from axial imaging into coronal and sagittal planes, which has facilitated CT's surrogate role when an MRI is contraindicated in the patient. Occasionally CT may shed additional light on the nature of a bone or soft tissue lesion by demonstrating mineralization in or about soft tissues that may have not been evident on plain radiography or MR. This ability is helpful in cases of certain sarcomas (*e.g.*, malignant mesenchymal neoplasms), tendon tug lesions (trauma), and calcific tendinosis (*i.e.*, overuse). CT is also widely used for preoperative planning in joint and limb replacements, providing a suitable tool for accurate measurements and choosing the correct size for a replacement. Though lower in sensitivity than MR, CT arthrography can be also used to diagnose internal derangements of joints when MRI would be associated with unacceptable artifact (Fig. 2.39a, CT arthrography knee with screws).

Magnetic resonance imaging. Of all the modalities used for musculoskeletal imaging, MRI provides unparalleled soft tissue contrast. Its sensitivity and specificity for common intraarticular pathology has been repeatedly proven in several investigations. Tendons and ligaments are fibrous structures that are dark on all MR pulse sequences. A disruption of the tendon or degeneration (tendinosis) translates on MR to characteristic architecture and/or signal changes. With muscle strains and tears, MRI readily demonstrates edema. With a complete tear of a musculotendineous junction, fibers will be diastatic. MRI provides a road map for the orthopedic surgeon should the patient require arthroscopy. Knee MRI aids in diagnosis of internal derangements such as tears of the mensiscus; cruciate and collateral ligaments; joint cartilage; and surrounding tendons. Shoulder MRI is a valuable tool for diagnosis of pathology of the rotator cuff, including complete or partial tears, impingement, muscle atrophy, and joint degeneration. The addition of intraarticular contrast (via fluoroscopic guidance) aids in diagnosis of labral tears. Similar in composition to the meniscus in the knee,

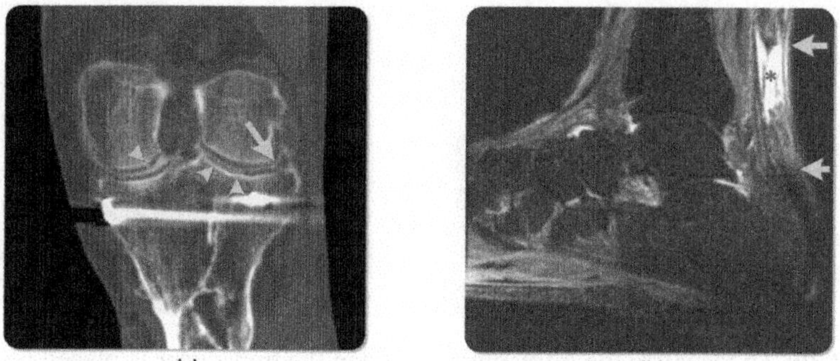

(a) (b)

Figure 2.39: **(a)** Coronal CT arthrogram of the knee demonstrates several cartilage defects (arrowheads, white contrast material filling the gaps in dark gray cartilage layer) and blunted lateral meniscus (arrow) consistent with a free-edge tear. **(b)** Sagittal inversion recovery (fluid-sensitive) sequence of the ankle shows a large fluid filled gap (*) in the Achilles tendon with retracted tendon ends (arrows).

the *glenoid labrum* is a fibrocartilageneous rim structure that enlarges the articulating bony surfaces of the glenohumeral joint. Labral tears are commonly seen in a younger patient engaged in throwing sports such as tennis, baseball or basketball. Hip MRI is commonly used to evaluate for bone marrow edema and fractures (both from overuse and direct trauma) that are suspected clinically but not seen on radiographs. In case of *avascular necrosis* of the femoral head (an ischemic process resulting in dead bone) MR demonstrates edema with adjacent reparative granulation tissue, providing a very specific picture to hone down the diagnosis. Like the shoulder, the hip is a ball-and-socket joint. The socket in the shoulder is the glenoid portion of the scapula whereas in the hip it is the acetabular portion of the pelvic bone. As in the shoulder, there is a fibrocartilageneous rim around the rim of the acetabulum, also called the *labrum*. Acetabular labral tears are also best evaluated with fluoroscopic guided addition of intraarticular contrast to the hip joint. Ankle MRI is another commonly utilized diagnostic that aids in the evaluation of ligament and tendon injuries about the ankle joint, such as the Achilles tendon (Fig. 2.39b), or in the evaluation of fractures and cartilage defects of the bony structures about the ankle joint.

Ultrasound. Musculoskeletal ultrasound is an important adjunct in diagnosis of various pathologies of the musculoskeletal system. Ultrasound is inexpensive and readily available. Musculoskeletal ultrasound is, however, operator dependent and has a rather flat learning curve. This modality is most commonly used for evaluation of periarticular superficial structures. Ultrasound requires a high frequency transducer to obtain adequate resolution of the fine architecture of the tendons and ligaments. This modality is valuable for evaluation of the rotator cuff tendons of the shoulder, which is conducted by a focused ultrasound examination of each tendon while the patient is instructed to perform various maneuvers to bring in the respective tendon into view (Fig. 2.40; with both patient and ultrasound picture). Chronic shoulder tendon abnormalities, acute tears, and periarticular fluid collections are readily visualized on ultrasound. The examination has to be focused as this modality often does not provide an overview of the entire joint. The possibility of dynamic imaging is a very special property of ultrasound. Documentation of tendon snapping (which occasionally occurs around the hip), or tendon impingement (which is common at the shoulder) is easily accomplished.

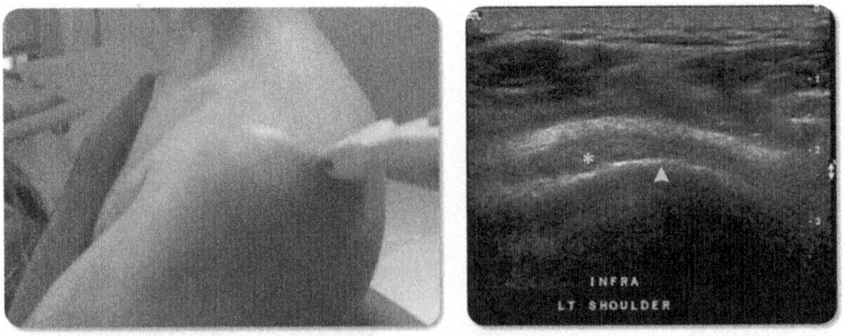

Figure 2.40: To visualize the infraspinatus tendon the transducer is placed posterior to the shoulder and the patient is instructed to place the ipsilateral hand flat over the opposite chest wall (internal rotation and mild adduction of the shoulder). The figure on the right shows the intact infraspinatus tendon as an isoechoic (gray) stripe of tissue (*). The humeral head cortex (arrowhead) prevents any further penetration of the ultrasound beams.

A cine ultrasound sequence is obtained while the tendon of interest is imaged with the patient performing the maneuver that would cause the snapping sensation or pain. Ultrasound can also be used to guide many musculoskeletal interventions, such as biopsies in the hand and feet, and targeted aspiration and/or injections of periarticular fluid collections. Ultrasound has gained substantial popularity for diagnosis of rheumatologic diseases, as it provides an inexpensive tool to document *synovitis* (inflammation of the joint lining), joint effusion, and bony erosions. However, deep intraarticular structures are not amenable to ultrasound examination. Deeper penetrating transducers with lower frequencies provide images of lesser resolution, and ultrasound cannot penetrate bony cortex. At times when MRI is contraindicated or ferromagnetic material would obscure the area of interest due to susceptibility artifact, ultrasound can often provide a diagnostic alternative.

Cardiac System

The heart has four chambers: the *left* and *right ventricles*, and the *left* and *right atria* (Fig. 2.41). Formed of cardiac muscle (myocardium), the heart is responsible for pumping blood throughout the body: the right side of the heart collects de-oxygenated blood and moves it to the lungs' alveoli, where O_2/CO_2 gas exchange occurs; the newly oxygenated blood then flows back to the left side of the heart via the pulmonary veins, from where it then circulates to the remainder of the body.

Embryologically, the heart develops from a single solid tube, the *primitive cardiac tube*. This tube grows at a faster longitudinal rate than the rest of the embryo, causing it to fold/bend to the right. The cardiac tube is canalized, creating the heart's chambers. The left ventricle forms before its right counterpart, explaining the asymmetric size of these chambers. Generally, the right ventricle is located more anterior. The atria are subsequently formed from a confluence of veins: a union of pulmonary veins forms the left atrium; and the superior and inferior vena cava veins join to create the right atrium. The middle of the heart, that is the lowest portion of intra-atrial septum (*i.e.*, the partition between the atriums) and the highest portion of the intra-ventricular septum, are formed by the *endocardial cushion*. The endocardial cushion also helps create the *tricuspid valve*, which separates the right atrium from the right ventricle; and, the *mitral valve,* which separates the left atrium from the left ventricle.

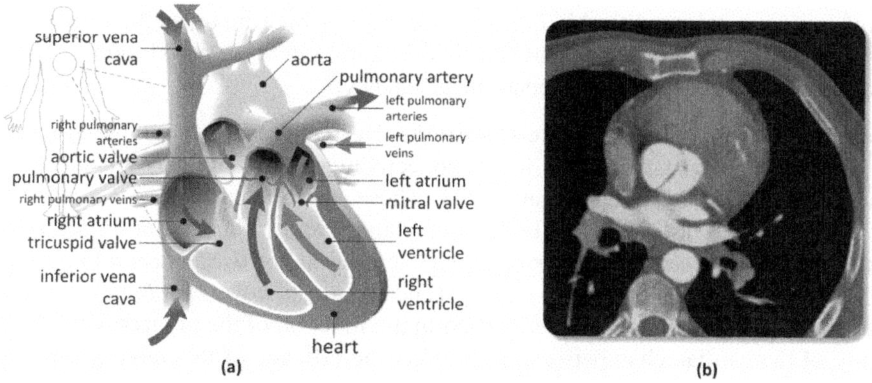

Figure 2.41: (a) Schematic diagram of the heart, illustrating the flow of blood through the chambers, arteries, and veins. **(b)** Example of axial cardiac CT image slice.

Cardiac Medical Problems

Congenital heart disease. Congenital heart defects are the most common birth defect, affecting about 1 in 125 babies in the United States [63]. Types of defects include:

1. Septal defects. Problems with the development of the endocardial cushion can give rise to *atrial septal defects* (ASD) and/or *ventricular septal defects* (VSD). The most common form of congenital heart disease is VSD[6], accounting for about ⅓ of all defects. Left to right shunts (*i.e.,* holes through which fluid can move), including VSD and ASD, cause an increase in pulmonary flow and *cardiomegaly* (an enlarged heart).
2. Obstructions. Blockages in blood flow due to improper growth of the valves, arteries, or veins (including narrowing) can occur. For instance, *aortic valve stenosis* refers to the incomplete closure of the aortic valve, resulting in left ventricle hypertrophy. Similarly, *pulmonary valve stenosis* affects the pulmonary valve, resulting in right ventricle hypertrophy. *Pulmonary stenosis* is the narrowing of the pulmonary artery, reducing the flow of blood from the right ventricle to the lungs.
3. Cynanotic defects. Abnormalities on the right side of the heart including *pulmonary atresia* (the malformation of the pulmonary valve between the right ventricle and pulmonary artery), *tricuspid atresia* (an absence of the tricuspid valve), and tetralogy of Fallot cause hypertrophy of the right ventricle – but no radiographic cardiomegaly. *Tetralogy of Fallot*, in particular, is the most common form of cyanotic congenital heart disease, and is defined by pulmonary stenosis, an *overriding aorta* (an aortic valve malpositioned such that the aorta is connected to both the right and left ventricles), and VSD that together cause the right ventricular hypertrophy. *Truncus arteriosus*, a rare condition, occurs when a key structure fails to properly divide into the pulmonary artery and aorta, thereby forming only one large vessel leading out of the heart, rather than two separate vessels per ventricle. *Transposition* of the great arteries describes the reversed positioning of the aorta and the pulmonary artery.

---

6 In premature infants, the most likely problem is instead *patent (ductus) ateriosus*, the failure of the ductus arteriosus to close. The ductus arteriosus is a shunt that connects the pulmonary artery to the aortic arch in the fetus, allowing blood from the right ventricle to bypass the fetal lungs. Normally, this shunt is closed at birth.

Another congenital condition is *Ebstein's anomaly*, which involves the displacement of the tricuspid valve into the right ventricle, resulting in an enlarged right atrium and smaller right ventricle with pulmonary insufficiency.

Heart disease and other cardiac conditions. Heart disease is an umbrella term for medical conditions affecting the heart; collectively, these problems are the leading cause of death in the United States [51]. These interrelated conditions encompass coronary heart disease, with the buildup of plaques and inflammation in the arteries that supply the myocardium (*atherosclerosis*), leading to *cardiac ischemia* (a restriction in blood supply resulting in tissue damage) and ultimately *myocardial infarction* (heart attack); and *cardiomyopathy* (deterioration in the function of the myocardium). Hypertrophy of the cardiac chambers can also affect the efficacy of the valves, resulting in leaking: for example, dilation of the right ventricle often leads to *tricuspid regurgitation*, a disorder where the tricuspid fails to close properly, allowing blood to flow backwards from the right ventricle to the right atrium; similarly, *mitral regurgitation* affects the left side of the heart. *Cor pulmonale* is a change in function of the right ventricle (usually from hypertrophy) due to a (chronic) respiratory disorder causing long-term high blood pressure in the pulmonary artery.

Basic Cardiac and Vascular Imaging

Several modalities are used to evaluate the heart and surrounding vessels. Projectional imaging includes the use of x-ray. Projectional x-ray imaging can be used to see the heart, great vessels, as well as vasculature in the lungs. As a heuristic, the normal *cardiothoracic ratio* (*i.e.*, the width of the heart to the chest cavity) is about 50%. Examination of the vessels beneath the diaphragm on x-ray can help to determine if there is increased or decreased flow: for instance, large vascular channels below the diaphragm suggest that pulmonary vasculature is increased. Cross-sectional imaging of the heart includes the use of CT, MR, and ultrasound. Cardiac CT is useful in assessing the myocardium, pulmonary veins, aorta, and coronary arteries. Cardiac disease can be evaluated by CT: for instance, *coronary calcium screening* attempts to quantify the extent of calcification in the coronary arteries as a corollary to coronary artery disease. Magnetic resonance is also increasingly used in cardiac imaging. Specifically, MRI can help locate myocardial areas receiving insufficient blood from the coronary arteries. For example, coronary MR studies (with contrast) can identify damaged muscle due to an infarct. MR is also used to diagnosis pulmonary artery disease, ventricular problems, ischemic heart disease, and cardiac tumors (Fig. 2.42a-b). Perhaps the most common cardiac imaging mode, ultrasound-based *echocardiography* is used to evaluate the heart and the origin of the great vessels: cardiac motion and blood flow can be evaluated via Doppler with ischemic areas (being less mobile) readily identified. Echocardiograms also permit evaluation of the cardiac walls for prolapse and other insufficiencies. Finally, *coronary angiography* is a noninvasive imaging technique that results in high resolution 3D visualization of the moving heart and great vessels. It can be performed via CT (*i.e.*, computed tomography angiography, CTA) or MR imaging (*i.e.*, MR angiography, MRA), although the former is presently the clinical standard. The test presently takes about ten minutes, with newer generation scanners (multi-detector or multi-slice scanning) provide faster and finer resolution angiographic imaging. More recently, dual energy source CT has been developed for angiography, using two sources and two detectors simultaneously; this modality provides full cardiac detail with significantly lower radiation exposure (~½ the dosage).

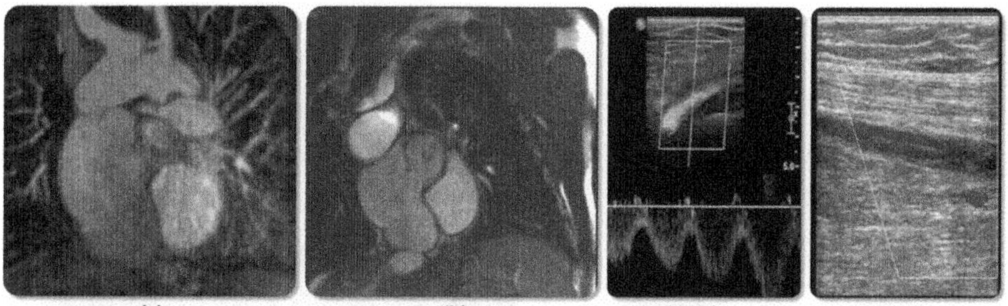

(a) (b) (c) (d)

Figure 2.42: (a) Example coronal slice from cardiac MR, allowing visualization of the different chambers. **(b)** A sagittal view of the same patient's heart on MR. **(c)** Color duplex ultrasound image through the right femoral vein, shows a normal venous flow with respiratory variance (see waveform in lower half). **(d)** Color Doppler ultrasound images through the right femoral vein demonstrate a mixed echogenicity filling defect within the vessel lumen compatible with acute deep vein thrombosis. Some flow is demonstrated in the vessel periphery (blue signal).

Vascular imaging. The carotid arteries in the neck are assessed using a combination of gray scale ultrasound, Doppler, and color Doppler techniques to detect atherosclerotic disease and stenosis (*i.e.,* vessel narrowing) [60]. The internal carotid arteries are a low-resistance system and therefore are characterized by a broad systolic peak and gradual down slope into a diastolic portion of the cardiac cycle (Fig. 2.42c). The external carotid arteries have a higher resistance as they supply the muscles of the scalp and the face. The waveform is characterized by a narrow systolic peak and absent or decreased diastolic flow. Clinically, hemodynamically significant stenosis is present when there is vessel narrowing of 50% or more of the vessel lumen. To assess the degree of stenosis several criteria are available, the most common being an abnormal peak systolic velocity (> 125 cm/sec), end diastolic velocity (> 40 cm/sec), and the ratio of peak systolic velocity in the internal carotid artery (ICA) to the peak systolic velocity in the ipsilateral common carotid artery (> 2.5) [60].

Gray scale and Doppler ultrasound techniques are also used for imaging the femoral and popliteal veins in patients with suspected deep vein thrombosis (DVT). Under normal circumstances these veins are easily compressible and color Doppler images show blood flow filling the entire lumen; waveform shape and amplitude vary with respiration. In DVT, the vessel lumen cannot be completely obliterated or compressed and contains a hypodense soft tissue mass (requisites) (Fig. 2.42d).

Urinary System

The kidneys are retroperitoneal solid organs of the abdomen responsible for maintaining fluid and electrolyte homeostasis; creating urine; and excreting waste products. Located posteriorly in the abdominal cavity, the kidneys develop as a paired structure, one on the right and one on the left, as a result of union between the *uretic bud* (the origin of the urinary collecting system, including calyces, renal pelvis, and ureter) and the *primitive kidney* (metanephrogenic blastema). If the uretic bud and primitive kidney are not united in the usual way, then one kidney will develop, with ensuing *unilateral renal agenesis*. In adults, healthy kidneys vary in size with age and sex, averaging 9-13cm in length. Blood enters the kidneys through the *renal artery* (Fig. 2.43a); filtered blood leaves through the *renal vein*. The *renal cortex* covers the innermost

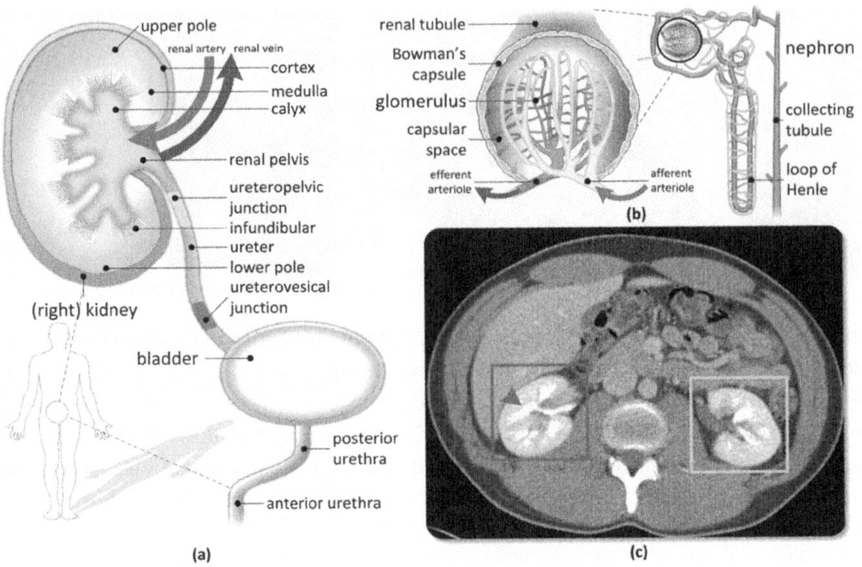

Figure 2.43: **(a)** Schematic diagram of a portion of the genitourinary system. The right kidney, its relation to the bladder via the ureter, and the urethra are shown. The kidney's blood supply enters from the renal artery and exits through the renal vein. **(b)** In the medulla, the glomerulus is a rounded mass of capillaries where blood filtering occurs; a glomerulus is part of the nephron structure.

portion of the kidney, termed the *medulla*. The medulla contains a matrix of blood vessels, *glomeruli*, and urine tubules. As blood flows through the kidney, it progresses through capillaries bundled into ball-shaped structures known as glomeruli, where filtering primarily occurs (Fig. 2.43b). A glomerulus is considered a part of a *nephron*, the functional unit of the kidney, and is surrounded by a *Bowman's capsule*. The urine tubules gather first into a *minor calyx*, and then into *major calyces* that subdivide the kidney. The calyces then empty into the *renal pelvis*, the kidney's main conduit for collecting urine, directing it to the *ureter*. The ureter is a narrow tube (~4 mm) that transports urine from the kidney to the bladder via peristalsis. The ureters drain into the bladder, which in turn is emptied during urination via the urethra. Unlike the kidneys, the bladder has a different origin, the *cloaca*, which divides into the anterior of the bladder and the posterior of the rectum.

Basic Imaging of the Urinary System

An *intravenous pyelogram* (IVP) is an x-ray examination of the kidneys, ureters, and urinary bladder that uses contrast injected into the veins.

Renal function and suspected obstructions to the collection system can be evaluated by injection of iodinated contrast material. An *intravenous pyelogram* (IVP) is an x-ray examination of the kidneys, ureters, and urinary bladder that uses contrast injected into the veins. Radiographic images taken soon after contrast uptake into the kidneys is called a *nephrogram*; and when obtained after some delay as the contrast is filtered into the bladder, a *urogram*. Specifically, these types of evaluation can occur using CT (Fig. 2.44a): contrast is injected and an immediate CT scan of the renal area is obtained, showing the nephrogram phase and function of the kidneys; a delayed CT cut of the abdomen through to the bladder is then acquired, showing the calyces, pelvis,

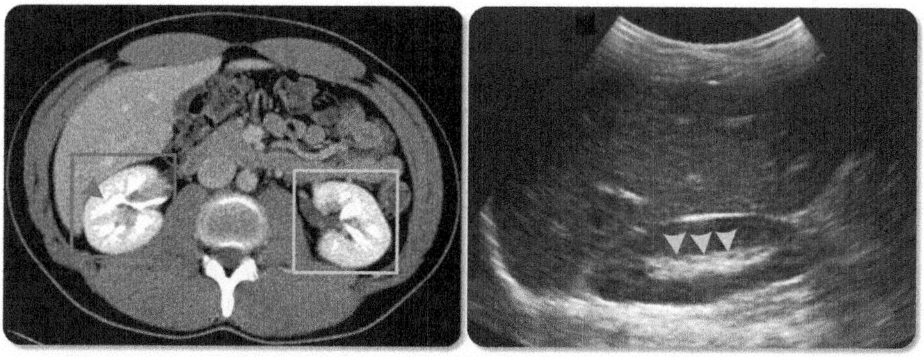

Figure 2.44: (a) A contrast-enhanced CT study of a normal adult kidney. The right and left kidneys are identified (boxes). High attenuation areas (arrowhead) inside the kidneys are the renal pelvis and ureter. **(b)** Longitudinal grayscale ultrasound image through the right upper quadrant show a normal kidney with dark renal cortex and an echogenic central sinus composed of fibrofatty tissue (arrowheads).

and ureters. Rather than intravenous administration, contrast can also be directly injected into the kidney and collecting system in an antegrade fashion, puncturing the kidney to reach a dilated region (*e.g.*, hydronephrosis): this procedures generates an *antegrade urogram*. Conversely, injection in a reverse manner from the urethra towards the bladder is termed a *retrograde urethrogram*. For suspected urethral obstructions, the CT urogram can be augmented with a delayed traditional projectional image to visualize the bladder. For direct injections into the bladder, the imaging study is called a *cystogram*; if the patients voids after the contrast study, the exam is known as a *voiding cystourerthrogram*. Lastly, the kidney can also be imaged via ultrasound, as showin in Fig. 2.44b.

Urinary Medical Problems

As the kidneys are responsible for maintaining proper levels of several substances within the bloodstream, including glucose and water, damage to the kidney or obstructions to the collection process can lead to a range of medical consequences ranging from the subtle (*e.g.*, hypo- or hypertension) to the serious (*e.g.*, renal failure). On imaging, four broad categories of issues exist with the GU system: congenital problems resulting in location or fusion abnormalities; cysts; obstructions; and cancer.

Location and fusion abnormalities. In rare circumstances, during its developmental phase a kidney may ascend past the usual L2-L3 lumbar vertebrae levels and progress into the chest, creating a *thoracic kidney*. This condition is generally asymptomatic and can be easily diagnosed by ultrasound or a CT urogram. If the lower poles of the kidneys (right or left) are fused together then the appearance is similar to a U-shape and is called a *horseshoe kidney*. Ectopia, the displacement or malposition of an organ, can also occur with the kidneys. *Abdominal crossed fused ectopia* occurs when the upper pole of an ectopic kidney and the lower pole of its opposite normal kidney are joined: a kidney thus appears absent on one side of the body. In *pelvic ectopia*, the lower levels of the two kidneys are fused together in the pelvis. This phenomenon may be mistaken for an abdominal mass, though again CT urography or ultrasound can differentiate it from other masses readily.

Cysts. *Simple (renal) cysts* are probably the most common form of a kidney cyst and can become quite large, leading to a distortion of the collecting system. Such cysts are diagnosed by a combination of CT urography and ultrasonography. *Polycystic disease* is a childhood malady (it can also occur in adults, but with much less frequency), and is a two-organ disease involving the liver and the kidneys in a reverse ratio of severity – that is to say, that if the liver is heavily affected, then the kidneys are less severely impacted, and vice versa. A rule of thumb for differentiating polycystic disease from multiple simple cysts is that in the former, the cysts are too numerous to count. On ultrasound, cysts present as a basic fluid collection with sound enhancement passing through the object (*i.e.,* the sound reflection becomes stronger, traveling much faster than through the surrounding medium): the region is empty, being absent of any echogenic material. The presence of echogenic material within the area is instead suggestive of a neoplasm.

Obstructions. Obstruction of the renal collecting system by stones or masses can produce dilation, also known as *hydronephrosis* (Fig. 2.45). Prolonged dilation can progressively result in atrophy of the kidney. Renal stones (nephrolithiasis) are easily detected both under ultrasound and CT. An acute urethral obstruction, such as by a stone, generally does not cause significant proximal dilatation[7]. Patients present with renal colic pain and the diagnosis of urethral stone can be made with an abdominal non-contrast CT. Urethropelvic junction (UPJ) obstruction, a common entity presenting as a neonatal mass, is the result of lack of peristalsis of the proximal ureter and the dilatation of the renal pelvis and calyces. Urethrovesical junction (UVJ) obstruction is similar: the adynamic segment (*i.e.,* the non-peristalatic section) occurs in the distal ureter and is referred to as a *primary megaureter*. Diagnosis of primary megaureter can be made by fluoroscopy, observing the aperistaltic segment of the ureter. Megaureter can also occur as part of prune belly syndrome, wherein the anterior abdominal wall musculature is absent or hypoplastic. Severe obstruction at the level of the pelvis and infundibulum, *pelvoinfundibular atresia* results in multi-cystic disease (*i.e.,* there are multiple cysts not communicating with each other) and no kidney function (Fig. 2.46). A milder form of obstruction with some kidney function and distortion of normal architecture of the collecting system, *pelvoinfundibular stenosis* is more difficult to diagnosis.

Obstructions can further occur due to blockage in the posterior or urethral valves (the latter also known as *urethral diverticulum*). Strictures of the urethra as a result of secondary infections, like gonorrhea, can result in obstruction of the urethra, abnormal drainage of the bladder, and ultimately proximal dilatation.

Cancers. In children, cancers of the kidney are often *Wilms' tumors* (nephroblastomas). Although rare, Wilms' tumors are the fifth most common pediatric cancer and are believed to arise from embryological cells that were destined to form the kidneys, but that instead fail to develop normally and instead multiply in their primitive state, becoming a tumor. On ultrasound, Wilms' tumors typically appear as inhomogeneous mass within the kidney, with multiple areas of decreased echogenicity compatible with necrosis on larger lesions. CT of Wilms' tumors present with mixed, low attenuation areas with bands of enhancing tissue surrounding cystic and necrotic regions. Bladder cancer in children is also relatively uncommon, and usually impact connective tissues

---

[7] In contrast, congenital obstructions cause a significantly larger amount of dilation in the area of the urinary tract proximal to the obstruction site.

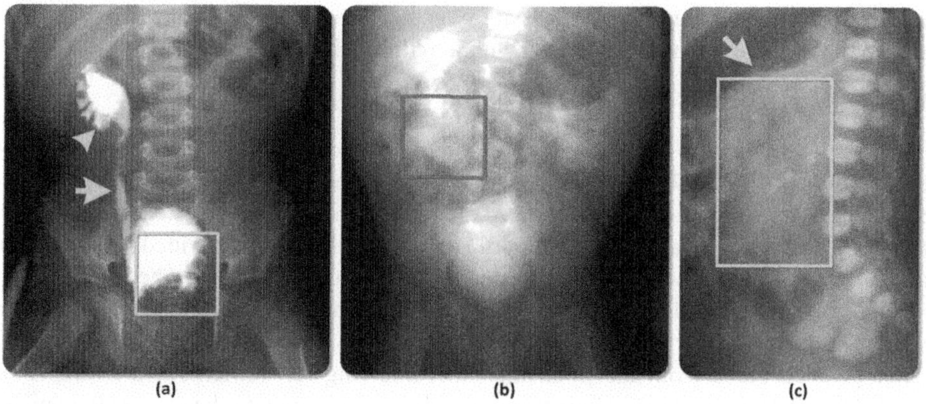

Figure 2.45: **(a)** Hydronephrosis can occur when there is a bladder outlet obstruction, including tumors as in this case with the grape-like appearance (box). This image shows a childhood rhabdomyosarcoma of the bladder, which in turn causes obstruction and hence dilatation of the right ureter (arrow), renal pelvis, and calyses (arrowhead). **(b)** Enlarged hydronephrosis of the right kidney, one of the most common reasons in neonates being a mass. Here, the mass is seen displacing the stomach anteriorly. The mass itself is seen with not much contrast because of dilution of contrast material in the hydronephrotic kidney. The image also shows dilation of the renal pelvis and ureter on the right side (box) and a normal functioning left kidney. **(c)** Saggital view of previous image, with the mass shown inside the box The arrow indicates the stomach.

(*i.e.*, sarcomas); *rhabdomyosarcoma* is the most common. Like Wilms' tumors, both ultrasound and CT can detect these growths, which appear as heterogeneous masses.

In adults, cancers involving the kidney parenchyma are *hypernephromas*. These lesions present as solid masses that disturb the renal contour; the margin being lobulated or irregular in appearance. On CT, these masses will appear as (large) heterogeneously attenuating entities. Dependent on the extent of the tumor, urograms may show involvement and/or destruction of the infundibulum and calyces. Approximately 85% of solid renal lesions are *renal cell carcinomas* (RCC), which can be further distinguished on

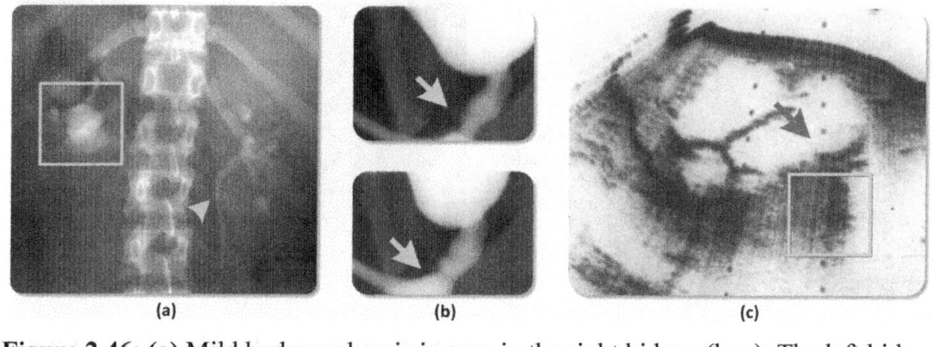

Figure 2.46: **(a)** Mild hydronephrosis is seen in the right kidney (box). The left kidney and ureter show signs of pelvoinfundibular stenosis (arrowhead), and a mild beginning to multicystic disease. **(b)** A contrast study showing megaureter, with dilation of the ureter in the posterior urethral vessel. **(c)** Ultrasound of multicystic disease (arrow). The boxed region shows sound enhancement as a result of the fluid in the cysts.

CT from normal parenchyma post contrast administration. Unlike pediatric cases, tumors of the ureter, bladder, and renal pelvis are typically *transitional cell carcinomas* (TCC) and frequently present as obstructions with a hydronephrotic kidney.

Upper Gastrointestinal (GI) System

The liver is the largest abdominal organ and is located in the right upper quadrant of the abdomen (Fig. 2.47a). It is responsible for a number of major physiologic functions, including the production of bile to aid in digestion, glycogen storage, lipid metabolism, and detoxification of the blood. Physically, the liver is divided into the *right* and *left hepatic lobes*, which are separated by the *interlobar fissure*. The liver is characterized by its dual blood supply, which comes from the *hepatic artery* and *portal vein*. The hepatic artery provides oxygenated blood, whereas the portal vein draws deoxygenated blood from the small intestine, allowing for nutrients and toxins to be extracted. On ultrasound examination, the liver parenchyma has an echogenicity greater or equal to the kidney, and lower than the pancreas and spleen (Fig. 2.47b). Pathological processes such as fatty infiltration generally cause a uniform or focal increase in the echogenicity of the liver [60, 69] (Fig. 2.48a). Other focal hepatic lesions such as simple cysts; and benign and malignant solid tumors are easily detected. Doppler is of great value for the evaluation of waveform analyses of the portal vein, hepatic vein, and hepatic artery in patients with liver cirrhosis or in assessing liver transplants.

The gallbladder is a small pear-shaped structure located beneath the liver that stores and concentrates bile from the liver. Ultrasound plays an important role in imaging of the gallbladder in pathological processes such as cholelithiasis (gallstones), acute cholecystitis or gallbladder carcinoma [60] (Fig. 2.48b).

The pancreas (Fig. 2.49) is a retroperitoneal organ that generates and stores a variety of compounds: its endocrine functions include the production and release of several key hormones directly into the bloodstream including insulin and glucagon (made by β- and α-islet cells); and its exocrine functions encompass the release of a variety of digestive enzymes into the small intestine. Divided into the *uncinate process*, *head*,

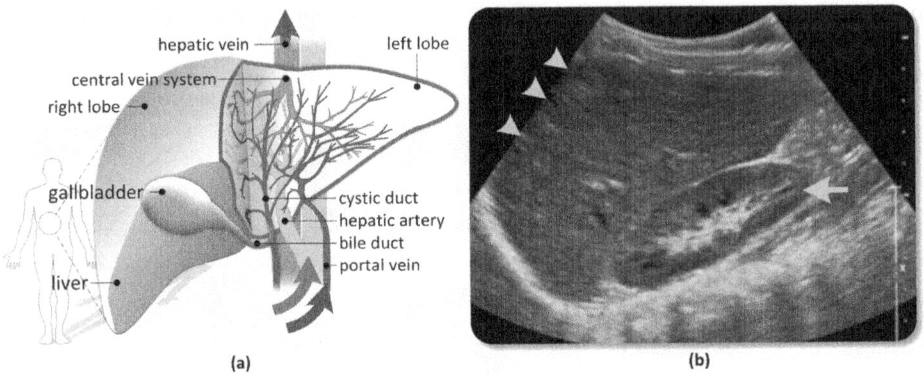

Figure 2.47: (a) Anatomical schematic of the liver and gallbladder, showing a cutaway view of the former. **(b)** Longitudinal grayscale ultrasound image through the right upper quadrant demonstrates a normal liver parenchyma (arrowheads) with an echogenicity equal to the kidney (arrow).

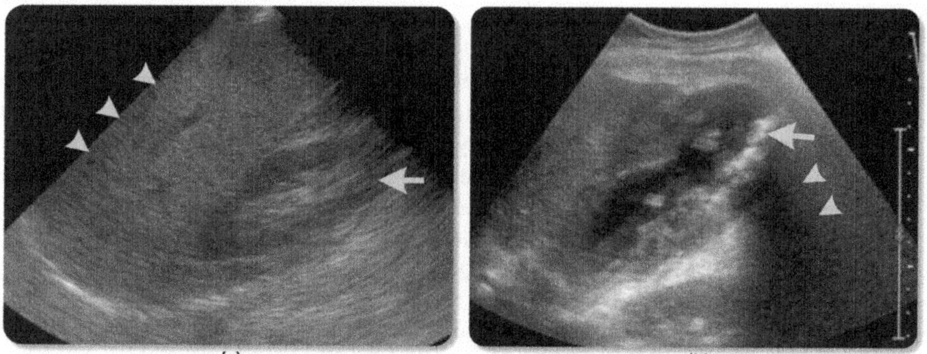

(a) (b)

Figure 2.48: **(a)** Longitudinal grayscale ultrasound images through the liver shows diffuse increased liver echogenicity (arrow heads) consistent with fatty infiltration. Notice the marked difference with the hypoechoic kidney (arrow). **(b)** Longitudinal grayscale ultrasound image through the right upper quadrant demonstrates a distended gallbladder with diffusely thickened wall. Multiple echogenic gallstones (arrow) are seen within the gallbladder. Due to their calcium content the gallstones generate acoustic shadowing (arrowheads).

neck, *body*, and *tail*, the pancreas' normal thickness for the head, body and tail is around 3.0, 2.5 and 2.0 cm respectively. The head of the pancreas is located to the right of the mesenteric vessels, and the neck and body are located anterior to these vessels. Due to its deep location the pancreas is often difficult to image by ultrasound techniques. Its echogenicity is usually variable and depends on the amount of fatty replacement within the pancreas. Normally the pancreas appears equal to or more echogenic than the liver. Ultrasound imaging is useful in assessing complications of pancreatitis, a diffuse inflammatory process of the pancreas. Solid tumors of the pancreas, such as pancreatic adenocarcinoma and islet cell tumors, are seen as focal pancreatic masses [60].

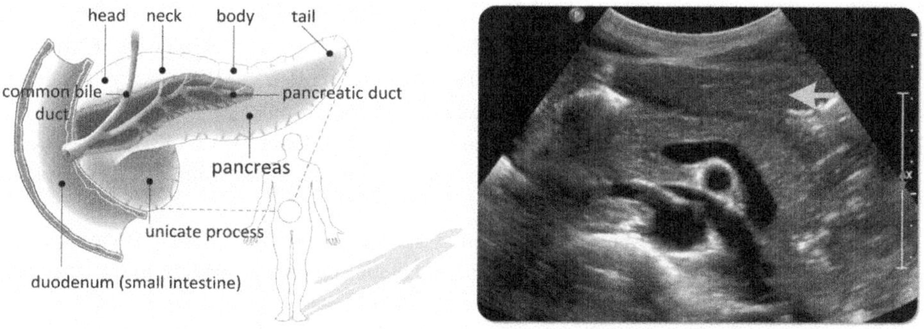

(a) (b)

Figure 2.49: **(a)** Schematic of pancreas. The pancreatic duct joins with the common bile duct to inject digestive enzymes into the small intestine. **(b)** Ultrasound of normal pancreas.

References

1. (2002) American Thoracic Society/European Respiratory Society International Multidisciplinary Consensus Classification of the Idiopathic Interstitial Pneumonias. This joint statement of the American Thoracic Society (ATS), and the European Respiratory Society (ERS) was adopted by the ATS board of directors, June 2001 and by the ERS Executive Committee, June 2001. Am J Respir Crit Care Med, 165(2):277-304.
2. Adler LP, Crowe JP, al-Kaisi NK, Sunshine JL (1993) Evaluation of breast masses and axillary lymph nodes with [F-18] 2-deoxy-2-fluoro-D-glucose PET. Radiology, 187(3):743-750.
3. American College of Radiology (ACR) (1994) Clinical Practice Guideline for the Performance of Diagnostic Mammography and Problem-solving Breast Evaluation.
4. American College of Radiology (ACR) (1994) Standards for the Performance of Screening Mammography.
5. American College of Radiology (ACR) (2003) Breast Imaging Reporting and Data Systems (BI-RADS), 3rd ed.
6. Aziz ZA, Wells AU, Desai SR, Ellis SM, Walker AE, MacDonald S, Hansell DM (2005) Functional impairment in emphysema: Contribution of airway abnormalities and distribution of parenchymal disease. AJR Am J Roentgenol, 185(6):1509-1515.
7. Barrett JF, Keat N (2004) Artifacts in CT: Recognition and avoidance. RadioGraphics, 24(6):1679-1691.
8. Bassett LW, Hendrick RE, Bassford TL (1994) Quality Determinants of Mammography: Clinical Practice Guideline No. 13 (AHCPR 95-0632). Agency for Health Care Policy and Research, US Dept Health and Human Services.
9. Berg WA, Blume JD, Cormack JB, Mendelson EB, Lehrer D, Bohm-Velez M, Pisano ED, Jong RA, Evans WP, Morton MJ, Mahoney MC, Larsen LH, Barr RG, Farria DM, Marques HS, Boparai K (2008) Combined screening with ultrasound and mammography vs. mammography alone in women at elevated risk of breast cancer. JAMA, 299(18):2151-2163.
10. Berger P, Perot V, Desbarats P, Tunon-de-Lara JM, Marthan R, Laurent F (2005) Airway wall thickness in cigarette smokers: Quantitative thin-section CT assessment. Radiology, 235(3):1055-1064.
11. Bitar R, Leung G, Perng R, Tadros S, Moody AR, Sarrazin J, McGregor C, Christakis M, Symons S, Nelson A, Roberts TP (2006) MR pulse sequences: What every radiologist wants to know but is afraid to ask. Radiographics, 26(2):513-537.
12. Brem RF, Floerke AC, Rapelyea JA, Teal C, Kelly T, Mathur V (2008) Breast-specific gamma imaging as an adjunct imaging modality for the diagnosis of breast cancer. Radiology, 247(3):651-657.
13. Brem RF, Petrovitch I, Rapelyea JA, Young H, Teal C, Kelly T (2007) Breast-specific gamma imaging with 99mTc-Sestamibi and magnetic resonance imaging in the diagnosis of breast cancer–a comparative study. Breast J, 13(5):465-469.
14. Buchberer W, Niehoff A, Obrist P, et al. (2002) Clinically and mammographically occult breast lesions: Detection and classification with high resolution sonography. Semin Ultrasound CT MR, 21:325-336.
15. Bushberg JT, Seibert JA, Leidholdt Jr EM, Boone JM (2002) The Essential Physics of Medical Imaging. 2nd edition. Lippincott Williams & Wilkins, Philadelphia, PA.
16. Buxton RB (2002) Introduction to Functional Magnetic Resonance Imaging: Principles and Techniques. Cambridge University Press, New York, NY.
17. Chotas HG, Dobbins JT, III, Ravin CE (1999) Principles of digital radiography with large-area, electronically readable detectors: A review of the basics. Radiology, 210(3):595-599.

18. Crystal P, Strano SD, Shcharynski S, Koretz MJ (2003) Using sonography to screen women with mammographically dense breasts. AJR Am J Roentgenol, 181(1):177-182.
19. Curry TS, Dowdey JE, Murry RC, Christensen EE (1990) Christensen's physics of diagnostic radiology. 4th edition. Lea & Febiger, Philadelphia, PA.
20. De Bazelaire C, Rofsky NM, Duhamel G, Michaelson MD, George D, Alsop DC (2005) Arterial spin labeling blood flow magnetic resonance imaging for the characterization of metastatic renal cell carcinoma. Academic Radiology, 12(3):347-357.
21. de Jong PA, Muller NL, Pare PD, Coxson HO (2005) Computed tomographic imaging of the airways: Relationship to structure and function. Eur Respir J, 26(1):140-152.
22. Donald I, Macvicar J, Brown TG (1958) Investigation of abdominal masses by pulsed ultrasound. Lancet, 1(7032):1188-1195.
23. Eastwood JD, Lev MH, Provenzale JM (2003) Perfusion CT with iodinated contrast material. Am J Roentgenol, 180(1):3-12.
24. Fanta CH (2009) Asthma. N Engl J Med, 360(10):1002-1014.
25. Flohr T, McCollough C, Bruder H, Petersilka M, Gruber K, Süß C, Grasruck M, Stierstorfer K, Krauss B, Raupach R, Primak A, Küttner A, Achenbach S, Becker C, Kopp A, Ohnesorge B (2006) First performance evaluation of a dual-source CT (DSCT) system. European Radiology, 16(2):256-268.
26. Fornage BD, Lorigan JG, Andry E (1989) Fibroadenoma of the breast: Sonographic appearance. Radiology, 172(3):671-675.
27. Gietema HA, Schilham AM, van Ginneken B, van Klaveren RJ, Lammers JW, Prokop M (2007) Monitoring of smoking-induced emphysema with CT in a lung cancer screening setting: Detection of real increase in extent of emphysema. Radiology, 244(3):890-897.
28. Gold RH, Montgomery CK, Rambo ON (1973) Significance of margination of benign and malignant infiltrative mammary lesions: Roentgenologic-pathologic correlation. AJR Am J Roentgenol, 118:881-894.
29. Gordon PB, Goldenberg SL (1995) Malignant breast masses detected only by ultrasound. A retrospective review. Cancer, 76(4):626-630.
30. Guyton AC (1987) Basic Neuroscience: Anatomy and Physiology. Saunders, Philadelphia, PA.
31. Hagmann P, Jonasson L, Maeder P, Thiran J-P, Wedeen VJ, Meuli R (2006) Understanding diffusion MR imaging techniques: From scalar diffusion-weighted imaging to diffusion tensor imaging and beyond. RadioGraphics, 26(S1):S205-223.
32. Hall FM, Storella JM, Silverstone DZ, Wyshak G (1988) Nonpalpable breast lesions: Recommendations for biopsy based on suspicion of carcinoma at mammography. Radiology, 167(2):353-358.
33. Hamper UM, DeJong MR, Caskey CI, Sheth S (1997) Power Doppler imaging: Clinical experience and correlation with color Doppler US and other imaging modalities. RadioGraphics, 17(2):499-513.
34. Harvey CJ, Pilcher JM, Eckersley RJ, Blomley MJ, Cosgrove DO (2002) Advances in ultrasound. Clin Radiol, 57(3):157-177.
35. Hasegawa M, Nasuhara Y, Onodera Y, Makita H, Nagai K, Fuke S, Ito Y, Betsuyaku T, Nishimura M (2006) Airflow limitation and airway dimensions in chronic obstructive pulmonary disease. Am J Respir Crit Care Med, 173(12):1309-1315.
36. Hendee WR, Ritenour ER (2002) Medical Imaging Physics. 4th edition. Wiley-Liss, New York, NY.
37. Hounsfield GN (1973) Computerized transverse axial scanning (tomography): Part 1. Description of system. Br J Radiol, 46(552):1016-1022.
38. Hsieh J (2003) Image artifacts: Appearances, causes, and corrections. Computed Tomography: Principles, Design, Artifacts and Recent Advances. SPIE Press.

39. Huang HK (1987) Elements of Digital Radiography: A Professional Handbook and Guide. Prentice-Hall, Inc.
40. Jackson VP, Bassett LW (1990) Stereotactic fine-needle aspiration biopsy for nonpalpable breast lesions. AJR Am J Roentgenol, 154(6):1196-1197.
41. Jerosch-Herold M, Muehling O, Wilke N (2006) MRI of myocardial perfusion. Seminars in Ultrasound, CT, and MRI, 27(1):2-10.
42. Johnson TRC (2009) Dual-energy CT – Technical background. Multislice CT, pp 65-73.
43. Kak AC, Slaney M (2001) Algebraic reconstruction algorithms. Principles of Computerized Tomographic Imaging. Society for Industrial and Applied Mathematics (SIAM), Philadelphia, PA.
44. Kaplan SS (2001) Clinical utility of bilateral whole-breast US in the evaluation of women with dense breast tissue. Radiology, 221(3):641-649.
45. Kato T, Inubushi T, Kato N (1998) Magnetic resonance spectroscopy in affective disorders. J Neuropsychiatry Clin Neurosci, 10(2):133-147.
46. Katz DS, Hon M (2005) Multidetector-row CT angiography: Evolution, current usage, clinical perspectives and comparison with other imaging modalities. In: Catalano C, Passariello R (eds) Multidetector-row CT Angiography. Springer.
47. Khalkhali I, Mena I, Jouanne E, Diggles L, Venegas R, Block J, Alle K, Klein S (1994) Prone scintimammography in patients with suspicion of carcinoma of the breast. J Am Coll Surg, 178(5):491-497.
48. Koenig M, Klotz E, Luka B, Venderink DJ, Spittler JF, Heuser L (1998) Perfusion CT of the brain: Diagnostic approach for early detection of ischemic stroke. Radiology, 209(1):85-93.
49. Kolb TM, Lichy J, Newhouse JH (1998) Occult cancer in women with dense breasts: Detection with screening US - Diagnostic yield and tumor characteristics. Radiology, 207(1):191-199.
50. Korner M, Weber CH, Wirth S, Pfeifer K-J, Reiser MF, Treitl M (2007) Advances in digital radiography: Physical principles and system overview. RadioGraphics, 27(3):675-686.
51. Kung HC, Hoyert DL, Xu J, Murphy S (2008) Deaths: Final data for 2005. National Vital Statistics Reports, 56(10):1-121.
52. Lehman CD, Isaacs C, Schnall MD, Pisano ED, Ascher SM, Weatherall PT, Bluemke DA, Bowen DJ, Marcom PK, Armstrong DK, Domchek SM, Tomlinson G, Skates SJ, Gatsonis C (2007) Cancer yield of mammography, MR, and US in high-risk women: Prospective multi-institution breast cancer screening study. Radiology, 244(2):381-388.
53. Lencioni R, Cioni D, Bartolozzi C (2002) Tissue harmonic and contrast-specific imaging: Back to gray scale in ultrasound. Eur Radiol, 12(1):151-165.
54. Logothetis NK (2008) What we can do and what we cannot do with fMRI. Nature, 453(7197):869-878.
55. Lufkin RF (1998) The MRI Manual. Mosby, St. Louis, MO.
56. Lynch DA, Travis WD, Muller NL, Galvin JR, Hansell DM, Grenier PA, King TE, Jr. (2005) Idiopathic interstitial pneumonias: CT features. Radiology, 236(1):10-21.
57. Mayer TE, Hamann GF, Baranczyk J, Rosengarten B, Klotz E, Wiesmann M, Missler U, Schulte-Altedorneburg G, Brueckmann HJ (2000) Dynamic CT perfusion imaging of acute stroke. AJNR Am J Neuroradiol, 21(8):1441-1449.
58. McLelland R, Hendrick RE, Zinninger MD, Wilcox PA (1991) The American College of Radiology Mammography Accreditation Program. AJR Am J Roentgenol, 157(3):473-479.
59. McNitt-Gray MF (2002) AAPM/RSNA Physics tutorial for residents: Topics in CT: Radiation dose in CT. RadioGraphics, 22(6):1541-1553.
60. Middleton WD, Kurtz AB, Hertzberg BS (eds) (2004) Ultrasound, The Requisites. Mosby St. Louis, MO.

61. Moskowitz M (1983) The predictive value of certain mammographic signs in screening for breast cancer. Cancer, 51(6):1007-1011.
62. Mueller-Mang C, Grosse C, Schmid K, Stiebellehner L, Bankier AA (2007) What every radiologist should know about idiopathic interstitial pneumonias. Radiographics, 27(3):595-615.
63. National Heart Lung and Blood Institute (NHLBI) (2007) What are congenital heart defects? National Institutes of Health. http://www.nhlbi.nih.gov/health/dci/Diseases/chd/chd_what.html. Accessed January 26, 2009.
64. Parker SH, Lovin JD, Jobe WE, Luethke JM, Hopper KD, Yakes WF, Burke BJ (1990) Stereotactic breast biopsy with a biopsy gun. Radiology, 176(3):741-747.
65. Passe TJ, Charles HC, Rajagopalan P, Krishnan KR (1995) Nuclear magnetic resonance spectroscopy: A review of neuropsychiatric applications. Progress in Neuro-Psychopharmacology and Biological Psychiatry, 19(4):541-563.
66. Ragavendra N, Ju J, Sayre JW, Hirschowitz S, Chopra I, Yeh MW (2008) In vivo analysis of fracture toughness of thyroid gland tumors. J Biol Eng, 2:12.
67. Rahbar G, Sie AC, Hansen GC, Prince JS, Melany ML, Reynolds HE, Jackson VP, Sayre JW, Bassett LW (1999) Benign versus malignant solid breast masses: US differentiation. Radiology, 213(3):889-894.
68. Rubin GD, Shiau MC, Leung AN, Kee ST, Logan LJ, Sofilos MC (2000) Aorta and iliac arteries: Single versus multiple detector-row helical CT angiography. Radiology, 215(3):670-676.
69. Rumack CM, Wilson SR, Charboneau JW, Johnson JAM (2005) Diagnostic Ultrasound. 3rd edition. Elsevier Mosby, St. Louis, MO.
70. Sanders RC (ed) (1997) Clinical Sonography: A Practical Guide. Lippincott Williams & Wilkins Philadelphia, PA.
71. Saslow D, Boetes C, Burke W, Harms S, Leach MO, Lehman CD, Morris E, Pisano E, Schnall M, Sener S, Smith RA, Warner E, Yaffe M, Andrews KS, Russell CA (2007) American Cancer Society guidelines for breast screening with MRI as an adjunct to mammography. CA Cancer J Clin, 57(2):75-89.
72. Schwartz GF, Guiliano AE, Veronesi U (2002) Proceeding of the consensus conference of the role of sentinel lymph node biopsy in carcinoma or the breast April 19-22, 2001, Philadelphia, PA, USA. Breast J, 8(3):124-138.
73. Sharafkhaneh A, Hanania NA, Kim V (2008) Pathogenesis of emphysema: From the bench to the bedside. Proc Am Thorac Soc, 5(4):475-477.
74. Sibtain NA, Howe FA, Saunders DE (2007) The clinical value of proton magnetic resonance spectroscopy in adult brain tumours. Clinical Radiology, 62(2):109-119.
75. Silver MD, Taguchi K, Hein IA, Chiang B, Kazama M, Mori I (2003) Windmill artifact in multislice helical CT. Medical Imaging 2003: Image Processing, vol 5032. SPIE, San Diego, CA, USA, pp 1918-1927.
76. Stavros AT, Thickman D, Rapp CL, Dennis MA, Parker SH, Sisney GA (1995) Solid breast nodules: Use of sonography to distinguish between benign and malignant lesions. Radiology, 196(1):123-134.
77. Tabar L, Fagerberg CJ, Gad A, Baldetorp L, Holmberg LH, Grontoft O, Ljungquist U, Lundstrom B, Manson JC, Eklund G, et al. (1985) Reduction in mortality from breast cancer after mass screening with mammography: Randomised trial from the Breast Cancer Screening Working Group of the Swedish National Board of Health and Welfare. Lancet, 1(8433):829-832.
78. Taraseviciene-Stewart L, Voelkel NF (2008) Molecular pathogenesis of emphysema. J Clin Invest, 118(2):394-402.

79. Thomas M, Lange T, Velan S, Nagarajan R, Raman S, Gomez A, Margolis D, Swart S, Raylman R, Schulte R, Boesiger P (2008) Two-dimensional MR spectroscopy of healthy and cancerous prostates in vivo. Magnetic Resonance Materials in Physics, Biology and Medicine, 21(6):443-458.
80. Thomas MA, Huda A, Guze B, Curran J, Bugbee M, Fairbanks L, Ke Y, Oshiro T, Martin P, Fawzy F (1998) Cerebral 1H MR spectroscopy and neuropsychologic status of patients with hepatic encephalopathy. AJR Am J Roentgenol, 171(4):1123-1130.
81. Thomas MA, Ke Y, Levitt J, Caplan R, Curran J, Asarnow R, McCracken J (1998) Preliminary study of frontal lobe 1H MR spectroscopy in childhood-onset schizophrenia. J Magn Reson Imaging, 8(4):841-846.
82. Tolmos J, Cutrone JA, Wang B, Vargas HI, Stuntz M, Mishkin FS, Diggles LE, Venegas RJ, Klein SR, Khalkhali I (1998) Scintimammographic analysis of nonpalpable breast lesions previously identified by conventional mammography. J Natl Cancer Inst, 90(11):846-849.
83. Ulzheimer S, Flohr T (2009) Multislice CT: Current technology and future developments. Multislice CT, pp 3-23.
84. Weiller C, May A, Sach M, Buhmann C, Rijntjes M (2006) Role of functional imaging in neurological disorders. J Magnetic Resonance Imaging, 23(6):840-850.
85. Weishaupt D, Köchli VD, Marincek B (2006) How does MRI work? An Introduction to the Physics and Function of Magnetic Resonance Imaging. 2nd edition. Springer, Berlin, Germany.
86. Wong EC, Cronin M, Wu WC, Inglis B, Frank LR, Liu TT (2006) Velocity-selective arterial spin labeling. Magnetic Resonance in Medicine, 55(6):1334-1341.
87. Wouters EF, Groenewegen KH, Dentener MA, Vernooy JH (2007) Systemic inflammation in chronic obstructive pulmonary disease: The role of exacerbations. Proc Am Thorac Soc, 4(8):626-634.
88. Yaffe MJ, Rowlands JA (1997) X-ray detectors for digital radiography. Phys Med Biol, 42(1):1-39.
89. Zavaletta VA, Bartholmai BJ, Robb RA (2007) High resolution multidetector CT-aided tissue analysis and quantification of lung fibrosis. Acad Radiol, 14(7):772-787.

PART II
Integrating Imaging into the Patient Record

Wherein we consider how imaging and the electronic medical record continue to evolve and the methods for presenting this information to users. Issues are examined pertaining to the communication and retrieval of images; and also to the associated clinical information. The growing number of medical databases and the means to access this information are described. Simple access to this wealth of information is insufficient though, as physicians must be able to make sense of the data in order to reach diagnostic and therapeutic conclusions. Appropriate methods to visualize the data must be in place for users to search, analyze, and understand medical datasets.

- **Chapter 3** – Information Systems & Architectures
- **Chapter 4** – Medical Data Visualization: Toward Integrated Clinical Workstations

Chapter 3
Information Systems & Architectures

ALEX A.T. BUI AND CRAIG MORIOKA

Since the advent of computers in medicine, the objective of creating an electronic medical record (EMR) has been to transcend the traditional limitations of paper-based charts through a digital repository capable of quickly organizing patient data, and ultimately, aiding physicians with medical decision-making tasks. This chapter introduces concepts related to the EMR, and covers the development of information systems seen in today's clinical settings. The data and communication standards used by these systems are described (*e.g.*, Digital Imaging and Communications in Medicine, DICOM; Health Level 7, HL7). But as healthcare progressively moves from a centralized practice to a more distributed environment involving multiple sites, providers, and an array of different tasks (both clinical and research), the underlying information architectures must also change. A new generation of informatics challenges has arisen, with different frameworks such as peer-to-peer (P2P) and grid computing being explored to create large scale infrastructures to link operations. We highlight several ongoing projects and solutions in creating medical information architectures, including teleradiology/telemedicine, the integrated healthcare enterprise, and collaborative clinical research involving imaging.

The Electronic Medical Record

The development of the electronic medical record[1] has been a longstanding pursuit of the medical informatics community. At its core, the purpose of an EMR is to provide computerized access to patient information. A spectrum of data elements comprises the EMR, capturing an individual's medical history and current status: demographics, vital signs, lab results, reports, medications, and imaging (*e.g.*, radiology, endoscopy) are principal components of the record. From an operational viewpoint, a 2003 Institute of Medicine (IOM) report details the essential functions of an electronic health record [57], including: patient information and results management; computerized physician order entry (CPOE); decision support; electronic communication; administrative processes (*e.g.*, billing, patient scheduling, utilization reviews); patient education, and public health reporting. This IOM report and others document the potential benefits of the EMR, which can be categorized threefold:

1. Improved quality of healthcare. EMR advocates point to better quality care through enactment of automated reminders, such as with childhood vaccinations and cancer screenings; implementation of evidence-based medicine and enhanced physician performance through decision support tools [32]; computer-aided diagnosis (CAD) to improve disease detection and standardize assessment; and continuity of care. In particular, the latter is important in the context of chronic disease management,

[1] The term, *electronic health record* (EHR), is also commonly used in the literature. While some authors use EHR to refer to a more global construct, we use both terms interchangeably here.

A.A.T. Bui and R.K. Taira (eds.), *Medical Imaging Informatics*,
DOI 10.1007/978-1-4419-0385-3_3, © Springer Science + Business Media, LLC 2010

allowing for closer monitoring of a patient's condition and coordination of multi-disciplinary clinical teams through the sharing of information.

2. Improved efficiencies. In theory, the ready availability of computerized results and the ability to quickly find critical information in the record (relative to paper charts) can generate time savings for physicians [9, 62]. Further efficiencies are brought about through the reduction of lost data and redundant/unnecessary tests (*e.g.*, re-ordering of labs) – effectively decreasing resource utilization [11]. Another area of improvement includes automating simple tasks (*e.g.*, re-ordering medications). The caveat is that the EMR and its interface must be conducive to such workflow: showing too much (*i.e.*, unnecessary) data is counterproductive; and the data organization must match the healthcare provider's expectations. For instance, one meta-analysis of time efficiency studies found that while the use of EMRs may decrease review time, it can increase the time needed by physicians to document cases [91].
3. Reduction in errors. The seminal 1999 IOM report, *To Err is Human*, catalyzed interest in methods to reduce the number of accidental deaths brought about through medical mistakes [56] and to improve overall safety. Evaluations of CPOE and alert systems involving drug prescriptions have shown their utility in this regard, minimizing uncertainties (*e.g.*, unclear handwriting) and detecting adverse events (*e.g.*, drug interactions, allergies) [10].

An important consequence of the EMR is the *longitudinal patient record*, also referred to as the virtual health record [101], wherein the distributed electronic records of an individual's healthcare episodes are effectively linked together, following a person over the course of his/her lifetime. Notably, each of the above areas results in possible cost savings [16, 44, 114].

Despite these advantages and the push towards EMR deployment in the United States (US) [105], studies between 2001-2006 show that EMRs have been adopted by only 20-30% of all medical practices [60]. By some estimates, this statistic is optimistically high as the survey definition of EMRs has been broadly encompassing of an assortment of computer-based solutions not necessarily concerned with clinical data (*e.g.*, scheduling); some newer analyses shows that only about 12% of US physicians routinely use fully functional EMRs [46]. Adoption has been slow for a variety of reasons: monetary costs, privacy issues, lack of means to transfer existing (paper) records, lack of functionality (*e.g.*, poor user interfaces, required workflow changes), and lack of (data) standards are often cited. In point of fact, the potential of the EMR has come under recent scrutiny given its failure to realize the expected benefits [18, 72, 102, 115]. Critics note that the majority of positive evidence for the electronic medical record and its applications comes from a handful of institutions and/or are self-reported by the developer – it is unclear whether such benefits can be translated elsewhere. However, much of the problem may lie in managing immediate expectations and follow-through on implementation plans [23]. Nonetheless, because of the relative nascence of the area and the limited degree of adoption, many agree that long-term studies of the EMR are needed to truly evaluate its impact.

EMR Information Systems

Much of the early development in EMRs started in niche areas, addressing data and workflow needs in specialized fields of medicine; current clinical information systems are a generalization and convergence of these efforts. We present two categories of systems: *hospital information systems* (HIS) and *picture archive and communication systems* (PACS). The former group includes databases containing demographics, clinical notes, labs, pharmacy, and administrative information (*e.g.*, insurance, billing, and financial data). The latter group, PACS, deals with the storage and communication of medical images.

Hospital Information Systems

A discussion of EMRs is closely intertwined with hospital information systems, the actual infrastructure for storing, viewing, and communicating data. Initially used for financial services in the 1960s, HIS capabilities have grown into the EMR core functions laid out in [57] (Fig. 3.1): clinical care (*e.g.*, medical chart review, CPOE, clinical protocol and guideline implementation, alerts and reminders); administrative management (*e.g.*, scheduling, billing, admission/discharge/transfer tracking); and in the case of academic medical centers, teaching and research (*e.g.*, discovery of trends in a population via data mining; creation of teaching files for instruction). [6, 7, 37, 67] provide some historical perspective on HIS and the efforts to define its structure and scope. Early HIS efforts for clinical usage involved the development and use of MUMPS (Massachusetts General Hospital Utility Multi-Programming System), a programming language geared toward (hierarchical) database applications. As HIS evolved, the architecture has changed from largely centralized databases to *n*-tier frameworks within an institution: to handle the number of users and emergent interrelated services; and to provide an abstraction layer to older (centralized) information systems. A simple example of this architectural change is in the deployment of web-based applications with today's HIS serving as the medium for an integrated view of the patient record. But noticeably, many current commercial EMRs and HIS still support derivatives of MUMPS.

Efforts have been made to establish reference models and frameworks for describing HIS. For example, $3LGM^2$ is a meta-model that divides a HIS into three inter-related layers of abstraction [119]: 1) the domain, detailing at a high-level the relationships between users and entities of interest in an enterprise; 2) the logical tool layer, wherein applications in concert with other resources (*e.g.*, databases) act upon the entities defined in the domain; and 3) the physical layer, which describes the hardware (*e.g.*, networks, servers) and constructs used operationally by the applications.

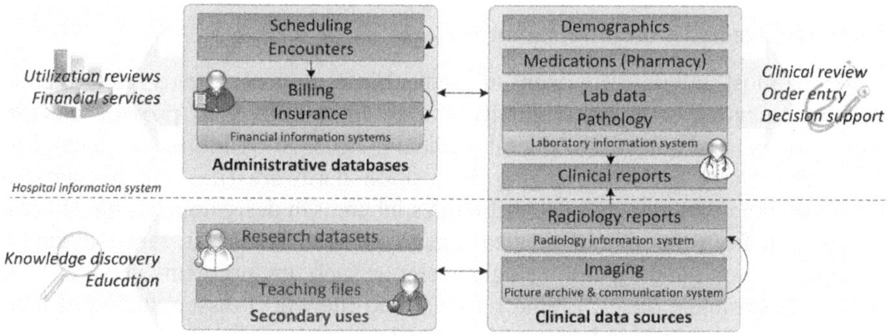

Figure 3.1: A high-level view of the data sources and functions addressed by the electronic medical record and hospital information systems (HIS).

Department-specific information systems. Although the boundary between different medical information systems is becoming blurred, applications have been developed for handling the particularities of a given clinical department. Two prime examples include *radiology information systems* (RIS) and *laboratory information systems* (LIS). Both RIS and LIS can be viewed as subsets of HIS with dedicated functionality:

- Radiology information systems. Arguably, RIS are a microcosm of the HIS with the added requirement of interfacing with an institution's PACS (see below) and the workflow seen in a radiology department. Specific functions attributed to RIS include scheduling (*e.g.*, patient studies/scanner allocation), patient and imaging study management (*e.g.*, study tracking), and reporting (*e.g.*, dictation support, review of previous study results). A distinguishing aspect of RIS *vs.* PACS is that the former is typically constrained to non-imaging data, whereas PACS is primarily concerned with the storage/manipulation of the imaging data. There is a tight coupling of the information in RIS with that of PACS. By way of illustration, RIS patient scheduling information is used to drive imaging study prefetching algorithms (from online archives) in anticipation of needed comparisons between studies for an individual; conversely, once the new study is acquired from the scanner and within a PACS, the radiologist's PACS/RIS work list is updated and an interpretative report must be generated for the corresponding imaging series. As imaging has become an integral part of the healthcare process, integration between HIS and RIS has been a concern: 1) to provide radiologists with a comprehensive context of the patient's history and presentation to reach a proper interpretation; and 2) to ensure that radiology results are quickly disseminated to referring physicians within a medical enterprise.
- Laboratory information systems. Analogous to RIS for radiology, LIS deal with the requirements of laboratory and pathology departments, providing interfaces to the various instruments used to assess chemistry, hematology, immunology, microbiology, genetic, and other histopathologic markers. Fundamental to any LIS is specimen tracking, result validation, and the incorporation of assay results into electronic reports that are communicated to requesting clinicians. Simple automated alert mechanisms are often implemented as part of LIS applications to warn of unexpected or critical test values outside of reference ranges (*e.g.*, a blood creatinine level of 2.0 mg/dl). For histopathology involving expert interpretation, LIS often support structured data entry and/or the dictation of reports, along with the integration of genomic/proteomic information and microscopy images.

Picture Archive and Communication Systems

Perhaps the most unique component of medical information systems, picture archive and communication systems originally started in radiology to provide electronic capture, storage, and retrieval of digital medical images. At a rudimentary level, PACS is the intermediary between imaging acquisition hardware (*i.e.*, the scanners) and applications involving images. The difficulties inherent in designing and implementing a PACS stem from the size of medical image studies, which can range from 2-800 MB, depending on the nature of a single exam; as such, the problems include: 1) long term storage of image data; 2) rapid retrieval and (network) dissemination of images for clinical review; and 3) indexing of image data.

Fig. 3.2 illustrates the different components comprising a PACS and the high-level flow of information; [51] provides a discussion of these constituent elements:

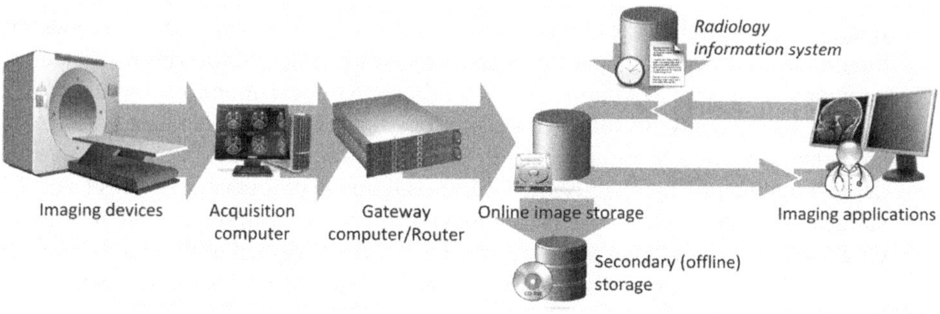

Figure 3.2: Components of a picture archive and communication system (PACS). Arrows between the different parts illustrate data flow.

- Imaging device and acquisition computer. With the introduction of PACS, medical imaging has gone from being an analog, film-based process to an ecosystem of digital acquisition devices. Digital modalities now include computed tomography (CT), magnetic resonance (MR), computed radiography (CR), positron emission tomography (PET), digital x-rays, mammography, and ultrasound. In many cases, and especially for the cross-sectional modalities (*i.e.,* CT, MR), the acquisition devices provide raw signals that must be reconstructed into the image slices/ volumes; such processing is handled by a dedicated acquisition computer that is coupled with the scanner. The acquisition computer maintains a small amount of local storage for recently acquired datasets, but sends the studies to the PACS for long-term storage via the gateway.
- Gateway computer/router. The gateway computer is responsible for two tasks: 1) receiving the images from the acquisition computer, and 2) forwarding the images to an archive and/or secondary device (*e.g.,* a viewing workstation). The logic for routing images in the network to and from the archive and viewing workstations can be embedded within the gateway; aspects of this functionality are often considered a part of the PACS controller. For example, based on study header data, a chest CT would be sent to the workstations in a thoracic reading room (vs. neuroradiology). More sophisticated management (*e.g.,* load balancing) can also be a part of the gateway/router.
- Online image storage (archive) and secondary storage. The archive itself consists of a database that maintains (meta) information about images stored in the PACS, immediate, online storage for recent images, and an "offline" secondary storage for older studies. The database is used to organize and index imaging studies (*e.g.,* by patient ID, date), and is updated as new images are sent from the gateway. Thus, this database is also capable of responding to queries about available studies, and initiating retrieval of studies (*e.g., find and send available MR liver studies for a given patient to a given PACS station*). The online storage is usually a fast network area storage (NAS) or storage area network (SAN) device, with petabytes (PB) of available disk space. Secondary storage employs slower devices, and for permanent storage and archive, may use arrays of optical disks (*e.g.,* CD-RW). Rules indicating when images should be moved from online to offline storage (and its opposite) are a part of the archive. For instance, based on an institution's online capacity and anticipated volume of imaging, only six months of guaranteed storage may be available, in which case studies outside of this period are automatically sent to secondary storage and marked for removal.

- Imaging applications. The final component of PACS is the applications requesting images. In clinical radiology settings, these applications are the viewing workstations, equipped with high resolution, high brightness/contrast monitors for diagnostic review. Increasingly, web-based applications for enterprise-wide access to PACS are being supported: intermediate web servers between the browser and archive are deployed to handle these activities. Additional image processing services (*e.g.*, 3D surface rendering, specialized simulations) also interact with the archive to retrieve data for analysis. Hanging protocols have also been designed to provide layout and presentation for common display patterns, and is discussed further in Chapter 4.

Data Standards for Communication and Representation

Given the nature of medical data and the functions of an EMR, its implementation is complex and multifaceted. The heterogeneity of information poses a problem in creating systems capable of accessing and integrating the diversity of databases maintaining patient information. Significant efforts have been made in the past two decades to provide standards upon which data between different components of the EMR can communicate. These endeavors entail definitions for the network protocols, the data structures (*i.e.*, representation and format), and to some extent a shared set of definitions (*e.g.*, data dictionaries). Three key standards at present are: 1) DICOM (Digital Imaging and Communications in Medicine), the *de facto* standard for representing and sharing medical imaging data, and now implemented as part of industry practice; 2) HL7 (Health Level 7), a collection of standards addressing data exchange within hospital applications, data models, and document structures; and 3) LOINC (Logical Observation Identifiers Names and Codes), a codification for clinical laboratory values and common observations. Momentum to integrate the different EMR data sources, both within a single institution and across sites, continues to refine and address these standards; their current state is described below.

DICOM (Digital Imaging and Communication in Medicine)

The DICOM 3.0 standard, established in 1993, grew out of earlier ACR-NEMA (American College of Radiology; National Electrical Manufacturers Association) efforts to provide data and communication interoperability between the components and vendors making up a PACS (*e.g.*, scanners, computers, storage systems, printers, etc.). Specifically, the standard facilitates interoperability of medical imaging equipment by specifying: 1) a set of protocols for network communication followed by devices conformant to the DICOM standard; 2) a syntax and semantics for commands and associated information that can be exchanged using these protocols; and 3) a set of media storage services to be followed by standard compliant devices, as well as a file format and a directory structure to facilitate access to images, waveform data, and related information. The full DICOM standard is available online [84]: it does not specify any implementation details, but only provides guidance and structure to the communication and storage of digital medical images. DICOM has continually updated and adapted itself to keep pace with the changes in the imaging environment and the needs of users. For example, newer parts of DICOM handle grayscale (display) calibration, compression, security, and web-based services. As of 2008, there were 16 official parts to the DICOM standard, summarized in Table 3.1; and additional supplements extend the standard with additional functionality.

Part	Description	Part	Description
3.1	Introduction and Overview	**3.10**	Media Storage and File Format
3.2	Conformance	**3.11**	Media Storage Application Profiles
3.3	Information Object Definitions	**3.12**	Media Formats and Physical Media
3.4	Service Class Specifications	**3.14**	Grayscale Standard Display
3.5	Data Structures and Encoding	**3.15**	Security and System Management Profiles
3.6	Data Dictionary	**3.16**	Content Mapping Resource
3.7	Message Exchange	**3.17**	Explanatory Information
3.8	Network Communication Support	**3.18**	Web Access to DICOM Persistent Objects

Table 3.1: Current parts of the DICOM standard, covering image representation and transmission. Parts 3.9 and 3.13 are retired.

The DICOM Model

Constructs in the DICOM standard revolve around two ideas:

- Object classes. In DICOM, all data is represented within an information object class. Thus, any entity such as a patient's demographics, image acquisition variables, and the image data itself is specified by an object class. DICOM distinguishes between *normalized* object classes (which basically are atomic entities) versus *composite* objects that are constructed from two or more normalized classes (akin to a *struct* in a programming language like C).
- Service classes. A service class refers to a process upon which data is generated, operated (transformed), or communicated. Examples of services are storage, querying, retrieval, and printing of images. Like its object class counterpart, services classes are divided between normalized and composite services.

Data model. The core DICOM data model uses entity-relationship (ER) and object-oriented concepts as the foundation for the object classes given in Parts 5 (Data Structure and Encoding) and 6 (Data Dictionary) of the standard. Fig. 3.3 illustrates the base hierarchy of patient and imaging data concepts, which is often referred to as DICOM's "model of the real-world." At the top of this hierarchy, the patient object is the main entity around which all other data is organized, with demographics and information on the individual relevant to conducting the imaging procedure (*e.g.*, weight, contrast allergies, notable medical history/contraindications, etc.). A patient object is associated with one or more time-stamped imaging studies, with each study encoding data on its nature (*e.g.*, institution, reason for exam, referring physician, etc.). In turn, each imaging study consists of one or more imaging series that describe the acquisition parameters for a given scan sequence (*e.g.*, modality, scanner, CT- and MR-specific values, contrast agent, orientation, etc.). Each imaging series then consists of a set of individual image slices that make up the sequence, with descriptors for the image resolution, (physical) location, and a 2D array of pixel values constituting the image (or waveform values).

Two more recent constructs seen in the DICOM data model are the structured report and the presentation state, which are aimed at linking ancillary data generated as a result of interpreting an imaging study. Acknowledging the potential of structured and form-based data entry to standardize reporting and improve compliance, such as seen with BI-RADS (ACR's breast imaging-reporting and data system), DICOM structured reporting (DICOM SR, Supplement 23) defines object classes for the storage of both structured information and free-text reported as part of a study (*e.g.*, the RIS report dictated by a radiologist) [54]. SR instances can also contain external references (such as to other images or reports), facilitating the integration of information to understand the context of the imaging study. The SR instances are associated per imaging series.

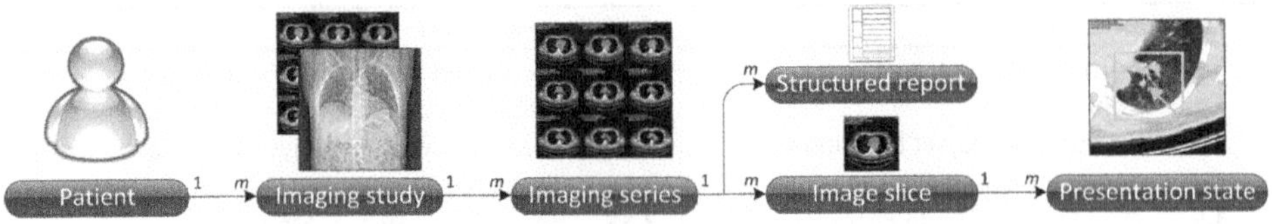

Figure 3.3: Basic DICOM hierarchical data model, which considers the relationship between a patient, an imaging study, and the parts of the study (series, images, reports, presentation states).

The DICOM standard allows different patterns of an SR report, called *SR templates*; different DICOM working groups address different SR specialties to establish these templates. For instance, DICOM Working Group 15 helped create Supplement 50, which provides templates for mammography. Likewise, DICOM breast imaging templates (DICOM Supplement 79) allow capture of diagnostic breast imaging reports and link BI-RAD findings within the impression section of the structured report. Table 3.2 lists additional SR templates. DICOM presentation states (DICOM PS, Supplement 33) further contextualize the image interpretation process by allowing a user to capture a sentinel image along with its visual characteristics (*e.g.*, window/ level, scale) and simple annotations (*e.g.*, lines, boxes); one or more DICOM PS objects can be associated with a given image slice.

From this model, DICOM declares several normalized object classes for patient, study, results, storage resources, and annotations; these base definitions are then glued together in composite object classes to provide specificity. For instance, the CT image class definition aggregates attributes from the patient and study information objects. This precision in defining data objects necessitates the use of element tags to uniquely identify object properties. A tag consists of two parts represented as four-digit hexadecimal codes: a group number and a data element number. For instance, the tag (0008,0020) represents the study date; *0008* is the group number and *0020* is the element number. Those properties sanctioned by DICOM are referred to as *standard data elements*, and have even group numbers; these tags are laid out in the data dictionary. *Information object definitions* (IODs) state those elements that are required as a part of a conformant object definition. For example, a CT image slice has certain agreed upon properties that must be present to be considered DICOM compliant. IODs are detailed in Part 3 of the DICOM standard. To provide for extensions and

DICOM Supplement	Structured Report Template	Working Group (WG)
Supplement 26	Ultrasound obstetrics and gynecology	DICOM WG 12
Supplement 50	Mammography and thoracic CAD	DICOM WG 15
Supplement 66	Catheterization laboratory	DICOM WG 1
Supplement 71	Vascular ultrasound	DICOM WG 12
Supplement 72	Adult echocardiography	DICOM WG 12
Supplement 76	Quantitative ateriography and ventriculography	DICOM WG 1
Supplement 77	Intravascular ultrasound	DICOM WG 1
Supplement 78	Fetal and pediatric echocardiography	DICOM WG 12
Supplement 79	BI-RADS	DICOM WG 8

Table 3.2: Sampling of DICOM structured reporting (SR) templates. A variety of supplements have been published by modality/anatomy working groups.

vendor-specific information, *private data elements* are allowed and are specified with odd group numbers; however, this degree of flexibility is often problematic for developers using DICOM images, as there is no published reference list for these additional fields.

Services. Part 4 of the DICOM standard describes service classes. Conceptually, there are two parts: DICOM message service elements (DIMSEs), which provide lower-level processing; and more complex operations that combine DIMSEs. The DICOM service classes are further divided based on the type of object operated upon: *normalized services* operate on normalized object classes; and *composite services* handle composite object classes (Fig. 3.4). There are six normalized DIMSEs for atomic events and notifications involving normalized information objects, which support management of patient, study, and results objects along with printing functions. Additionally, there are five composite DIMSEs, comprising core operations: C-ECHO (for confirming network connections); C-STORE (transmission of data, *i.e.,* information objects); C-FIND (responding to queries about finding information objects on a storage device); C-GET and C-MOVE (responsible for initiating requests for copying of information objects from one device to another). To illustrate the composition of higher-level services, consider a query/retrieve operation, which involves the use of a C-FIND (to find the imaging study on a given device), a C-MOVE (to move the study from the storage device), and a C-STORE (to save the study to the target device). An implemented subset of a service class' methods is referred to as a DIMSE *service group*.

Service classes consider the role of a device in a requested operation: for example, is the device invoking the service, or is the device performing the operation? The former is referred to as a service class *user* (SCU), the latter a service class *provider* (SCP). For a given service class, a device may be capable of acting as the SCU, the SCP, or both dependent on its function in an overall PACS.

Communication. The interchange of data in DICOM is governed by the concept of service-object pair (SOP) classes, which aggregate an information object definition and service class together – this completes the object-oriented perspective in that data and methods coalesce into one entity. To demonstrate, an MR scanner would implement the SCP for the MR image storage SOP class, meaning that it is capable of sending MR images to a given destination. A device wanting to receive the MR images

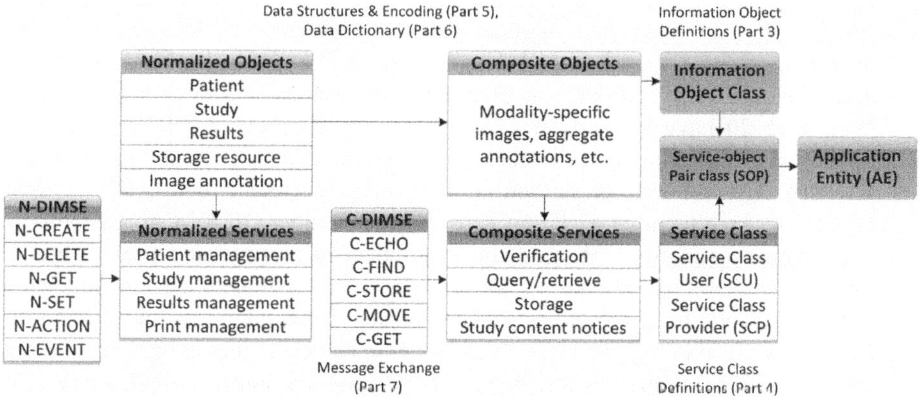

Figure 3.4: Interplay of the DICOM information object classes and service classes. Functionality is divided into normalized and composite operations and data. The relationship between the constructs is shown by connecting arrows.

would need to implement the corresponding SCU for this SOP class. Collectively, the implemented information object definitions and service classes for a device compose a DICOM application entity (AE), which expresses the capabilities of a device within a PACS network. The communication process can be encapsulated in three steps, and [12, 51] provide further discussion:

1. Association. To start the process, an initiating AE first creates an association between itself and a target device (another DICOM AE), sending information on its capabilities (*i.e.,* the SOPs it implements) and the requested service. In return, the target AE also declares its SOP classes. If compatible SCUs/SCPs for the service are found, then both the user and the provider are notified and the service request is accepted; a valid association is formed.
2. Provider processing. The AE implementing the SCP constructs the requested SOP instance. A DICOM message object message (specified in Part 7 of the standard) is constructed with command and data set elements wrapping the SOP instance. The DIMSEs needed to execute the requested service are invoked, and the message object and its contents are thus sent over the network to the associated AE.
3. User processing. The associated AE receives the DICOM message, reconstructing the SOP instance (through a reverse set of DIMSEs), and the SCU completes any additional actions on the data.

DICOM Extensions

DICOM continues to evolve to meet the growing and changing role of imaging in clinical and research environments, as typified by the many technical working groups (WGs) and supplements that have been established over the years: as of late 2008, well over 100 supplements have been specified, many of which have been formally accepted into the standard. Several efforts currently exemplify DICOM's extension:

- Inclusion of new medical disciplines using imaging. DICOM was conceived of to manage clinical radiological imaging. However, the expansion of such imaging to other areas, along with the use of other imaging modalities (*e.g.,* optical, microscopy) has led to the adaptation of DICOM models to other areas, including dentistry (Supplement 92), ophthalmology (Supplements 91, 110), and pathology. For instance, DICOM WG 26 is working on an improved image standard for pathology microscopy images, specimen images, and tissue microarray images.
- Integration with other data standards. The intersection of imaging with other aspects of clinical workflow and data has established synergies between DICOM and other efforts to formalize medical data representation. For example, DICOM WGs 8 & 20 have worked on allowing for the creation of DICOM structured documents that reference or access persistent data objects available outside the DICOM environment. In particular, data objects created under HL7 Clinical Document Architecture (see below) are supported under DICOM supplement 101.
- Extending functionality. Beyond the original need to store and present medical images, DICOM has started to address the need to share data and algorithms associated with imaging. For instance, results from computer aided diagnosis (CAD) are given in Supplements 50 and 65 with respect to mammography and chest imaging. DICOM WG 23 is exploring a "plug-in" architecture and standardized application programmer interface (API) for interactive application services and hosted software; in such a framework, algorithms (*e.g.,* for image processing) can be shared between DICOM systems.

Health Level 7 (HL7)

Started in 1987, Health Level Seven (HL7) is a standard for the exchange of data between healthcare information systems [38]. The standard's somewhat enigmatic name is a reference to the top layer of the Open Systems Interconnection (OSI) Reference Model, a tiered abstract description for communications and computer network protocol design; the seventh level of the OSI Model is the application layer, in which HL7 exists. Like DICOM, the primary purpose of the HL7 standard is to specify the formatting and structure of information exchange: it does not describe the actual technical details of how the information is to be passed from one system to another. HL7 does not try to assume a particular architecture with respect to the placement of data within applications but is designed to support a central patient care system as well as a more distributed environment where data resides in departmental systems. Given its evolution, there are now several components to HL7.

Messaging Protocol

The HL7 messaging protocol, referred to as HL7 v2.x, was first developed to provide a common format to exchange textual healthcare data between information systems. HL7 messages are created in response to some real-world healthcare event that initiates the need for data flow among systems. This real-world event is called a *trigger event*. For example, a trigger event may be for the completion of a patient's hospital admission, or the completion of a lab assay; such occurrences may subsequently need data about that patient to be sent to a number of systems (*e.g.*, updating an occupied bed census; paging a doctor with an irregular lab warning). Trigger events can exist at various levels of granularity and involve one or more entities.

When a message is sent from one system to another, the original message is acknowledged by the receiver (*i.e.*, an *acknowledgment* message is sent), ensuring the receipt and the successful processing of the message. This return message may include data of interest to the originating source (*e.g.*, a response to a lab order may provide a generated lab requisition ID and information about the type of test requested).

HL7 messages may also represent queries from one system to another in order to retrieve needed information. By way of illustration, the arrival of a patient for an imaging exam may trigger an HL7 message from an RIS to the PACS; a PACS prefetching application may require additional information on the type of imaging study in order to retrieve relevant past examinations for comparison. As such, the PACS may issue a request back to RIS for more information. This transaction is a query, in contrast to an unsolicited information update; the response (acknowledgement) contains the results of the query. In this manner, HL7 queries are analogous to a server-client paradigm.

Data encoding. HL7 messages are made up of *segments*, a logical grouping of data fields. Each segment is identified by a unique three character name known as a *segment ID*. A data field is simply a string of characters. Within a segment, the fields comprising the segment are separated by a delimiter (*e.g.*, the pipe character "|"). Individual fields can be further sub-divided into smaller components using another delimiting character (*e.g.*, a carat, "^"). As an example, consider an HL7 admission/ discharge/transfer (ADT) message to transmit portions of a patient's administrative data from one system to another as a continuous character string:

```
MSH|^~\&|HIS|UCLAMedCtr|RIS|UCLAMedCtr|20080518193924||ADT^A01^ADT_A01|0000000101|P|
2.6|EVN|A01|20080518193924|||a021|PID|||123456^^^PH||BRUIN^JO^^^Ms||19880819|F|||100
Westwood Blvd^^VENICE^CA^90066|PV1||I|5W^17^2^^^^^^WardSouth||||TESTDR^DEERE^JOHN^^^
DR|||ORT|||||||||15521|COM|
```

The above ADT message contains the following segments:

- Message header (MSH). The MSH segment provides details about the message itself, such as the sending and receiving systems, timestamps, and message type.
- Event type (EVN). The EVN segment of the message provides information about the triggering event, including the event type, the (real-world) event timestamp, and the originating source of the event (*e.g.,* the patient or some other system).
- Patient ID (PID). The PID segment provides the information specific to the individual patient, with demographics fields.
- Patient visit (PV1). Finally, the PV1 segment specifies fields pertinent to the patient's visit to the hospital, such as the visit number, ward/bed, attending doctor, and financial classification.

As HL7 v2.x is meant to standardize data interchange and not applications, it only requires certain data fields so that different institutions may provide optional information: the standard contains methods to tailor messages for local usage. The only data fields required in a message are those needed to support the logical structure and relationships between messages: many fields are defined but are optional. As EMR system capabilities have matured, the HL7 v2.x protocol has been updated to include new functionality, such as embedding logical rules and statements as part of a message to incorporate clinical guidelines and decision support. Arden syntax, a logical rule grammar for the expression of medical conditions and recommendations, was adopted to encode *medical logical modules* (MLMs) [92]. An MLM represents a standalone knowledgebase that can be used to aid in basic decision processes. An example of the invocation of an MLM through HL7 might be the monitoring of queried labs results for abnormal values, from which guidelines to correct the value may be supplied to a receiving application (and hence passed on to the clinician).

Reference Implementation Model (RIM)

While HL7 v2.x provides the groundwork for communication between systems, it does not actually direct how data should be organized within an information system, or how the messages should be instantiated. As the databases containing patient information grew more complex, models showing how the data should be put together to formulate HL7 messages were needed. In April 1996, HL7 started development of the reference implementation model (RIM), with the aim of creating detailed data types, classes, state diagrams, use case models, and terminology to derive domain-specific information models. In doing so, the intent was to enable the definition of the data needed in different healthcare contexts, the source of the HL7 message fields, and the semantic and lexical relationships between the sources of the field values. Over several subsequent meetings and interim draft releases that progressively harmonized the data model, HL7 released version 1.0 of the RIM in 2001. HL7 RIM is a critical aspect of HL7 v3, being the underlying backbone of all information models and structures that aspires to solve semantic interoperability problems. There are six fundamental classes of the RIM, shown in Table 3.3. Briefly, every healthcare process/event is described by an act (which notably dovetails with the HL7 messaging paradigm of event triggers). Each act may have any number of participants, described

Class name	Definition	Example	Subclass
Entity	Physical thing or organization and grouping of physical things	Person, organization, material, place, medical chart	Yes
Role	The competency of the *Entity* that participates in the *Act*	Doctor, patient, nurse, specimen	Yes
Act	The intentional action(s) that are performed. The *Entity* with a certain *Role* performs an act.	referral, transportation procedure, observation	Yes
Participation	How an entity, in a particular *Role*, functions during the scope of an *Act*. Participation signifies performance	Entity:Doctor-A, Role: Physician, Act: admit a patient, Participation: participation of a physician	No
Act-Relationship	Ability to relate two acts.	The relationship between an order for an event and occurrence of the event	No
Role-Link	The relationships between any two entity roles.	Indicates the physician's relationship with an institution, and the patient's relationship with the institution to express the patient/physician relationship	No

Table 3.3: The six major classes of the HL7 reference implementation model.

by an entity assuming a given role. How such an entity is involved in the act is described through a participation relationship. Acts can be related to other acts, as delineated by an act-relationship. Likewise, roles can be inter-related through role-links. The base data model provides for subclassing of the act, role, and entity objects to specialize definition and properties; this ability creates an inheritance network. For instance, Living Subject is a subclass of entity; and Person is a subclass of Living Subject. Numerous data models based on HL7 RIM have been developed and archived online as part of HL7's efforts. Although criticism of the HL7 RIM has been raised (*e.g.*, [104]), its proponents note that the inherent richness (and complexity) of HL7 RIM has led to differing viewpoints and semantics [100].

Clinical Document Architecture (CDA)

Initiated in 1998, a "next generation" representation for clinical data interchange was being explored by HL7. Initially referred to as the patient reference architecture (PRA), the HL7 clinical document architecture (CDA) is a document markup standard that specifies the structure and semantics of documents [24, 25]. The HL7 CDA is another key component of the HL7 v3 effort, providing a standard for structuring free-text reports generated as part of clinical processes. The need for the CDA can be seen twofold: 1) HL7 messages can only contain a certain type of rudimentary information (*e.g.*, header information, as opposed to deeper semantics, and only text data); and 2) the primary focus up till the late 1990s was on administrative usage of HL7 for process/workflow and simpler content, not clinical information. In 2000, HL7 CDA v1.0 became an approved ANSI (American National Standards Institute) standard.

Key aspects of the CDA standard include: documents are encoded in eXtensible Markup Language (XML); document parts derive their meaning from HL7 RIM and use HL7 v3 data types; and the CDA specification is expressive and flexible. A CDA document can be included in an HL7 message or can exist independently as an XML file. The six characteristics of a CDA document include the following:

1. Persistence. A clinical document continues to exist in an unaltered state, for a time period defined by local and regulatory requirements.
2. Stewardship. A document is maintained by an organization entrusted with its care.
3. Potential for authentication. A clinical document is an assemblage of information that is intended to be legally authenticated.
4. Context. A clinical document establishes the default context for its contents.
5. Wholeness. Authentication of a clinical document applies to the whole and does not apply to portions of the document without the full context of the document.
6. Human readability. A clinical document is human readable.

Fig. 3.5 demonstrates the XML structure for a CDA-based report. A CDA document contains both a *header* and a *body*. The header provides the context for the document's creation, while the body contains the actual content of the document. The header has three main purposes: 1) to facilitate clinical document exchange across and within institutions; 2) to support document management and indexing; and 3) to compile an individual's clinical documents into a longitudinal EMR. The header contains document metadata and encounter data that help provide context to the origin and purpose of the document. The document body is comprised of *sections*, *paragraphs*, *lists*, and *tables*. Each of these structural components can contain captions, text, multimedia components (via external references), and standardized codes. Importantly, each component can be nested (*e.g.*, a sub-section within a section). Additionally, each section can be associated with codifications (*e.g.*, an observation may be linked to a SNOMED (Systematized Nomenclature for Medicine) code) helping enable more useful indexing and searching of medical document corpora.

Three levels of increasing compliance are defined by the CDA standard. Level 1 is the least constrained, only requiring that the CDA document contain a properly structured CDA header. The body section for the CDA document is placed after the section element <structuredBody> as free-text narrative. Level 2 compliance requires section level templates applied to form CDA documents that contain mandatory and optional

```
<clinicalDocument>
  CDA Header
  <structuredBody>
  <section>
    <text>...</text>
    <observation>
      <reference>

  <externalObservation/>
      </reference>
    </observation>
    <observation>
    ...
    </observation>
  </section>
  <section>
    <section>
    ...
    </section>
  </section>
  </structuredBody>
</clinicalDocument>
```

Figure 3.5: Example of the base HL7 clinical document architecture XML structure.

document sections. For instance, an inpatient report may require a historyAndPhysical section as well as a section on vitalSigns, but a cardiovascularExamination block may be optional. Lastly, Level 3 CDA documents contain entry-level templates that allow more complete structuring of the clinical information.

Logical Observation Identifier Names and Codes (LOINC)

In 1994, researchers at the Regenstrief Institute set out to develop a universal, pre-coordinated coding system for laboratory tests. Today, version 2.22 of the Logical Observation Identifier Names and Codes (LOINC) database contains in excess of 50,000 codes for laboratory results, clinical measurements, and findings from other diagnostic tests. LOINC has become the standard coding system for LIS [52, 95]. The primary purpose of LOINC is to codify the range of observations seen in clinical practice. With HL7 the prevailing method for electronic communication amongst health-care institutions, a particular objective of LOINC is to help standardize communication of results and observations within HL7 messages.

Formal name entries in LOINC are derived using six major axes (Table 3.4), with up to four additional minor axes providing further specificity. Table 3.5 depicts examples of LOINC codes and formal LOINC names for laboratory and clinical observations. Notably, LOINC is continually being updated and expanded. Presently, efforts are split along three divisions:

1. Laboratory. The first and most mature aspect of the database covers lab observations. As such, the largest number of entries in LOINC is from this division. More recently, LOINC included supports molecular pathology observations used to identify genetic mutations including substitutions, deletions/insertions, tumor associated genes, and gene deletions.

Axes	Description
1	*Component* (analyte) (*e.g.,* creatinine, glucose, hemoglobin)
2	*Property measured* (*e.g.,* catalytic concentration, entitic length, mass concentration)
3	*Timing* (*e.g.,* an observation may be a moment in time, an average, or over an interval of time)
4	*System* (*e.g.,* type of sample or organ examined, blood arterial, urine, liver)
5	*Scale* (*i.e.,* whether the measurement is quantitative, ordinal, nominal, or narrative)
6	*Method used to produce the observation* (optional)

Table 3.4: The LOINC standard uses six major axes to define name entries.

LOINC Code	LOINC Names
11125-2	Platelets:Morph:Pt:Bld:Nom
11157-5	Leukocytes:Morph:Pt:Bone Mar:Nom
11218-5	Albumin:MCnc:Pt:Urine:Qn
11882-8	Gender:Type:Pt:^Fetus:Nom:US
21612-7	Age:Time:Pt:^Patient:Qn:Reported
29471-0	Blood Flow.systole.max:Vel:Pt:Hepatic Vein:Qn:US.doppler

Table 3.5: Example laboratory and clinical LOINC codes with the formal LOINC names based on five major axes (component:property:timing:system:scale).

2. Clinical observations. The clinical LOINC division is concerned with non-laboratory diagnostic studies, critical care, nursing measures, patient history and physical (H&P), and medical instrument surveys. For instance, included this division are LOINC codes for naming radiology reports; over 2,000 radiology codes exist in LOINC version 2.22. Notably, LOINC codes are in turn used in other standards: LOINC identifiers are being used in DICOM ultrasound reporting. In addition to radiology, LOINC has developed over 400 codes that can be used in tumor registries, capturing information on initial diagnosis, anatomical sites, and tumor features (*e.g.*, size, histopathology).
3. Claims attachments. The third division of LOINC development handles the definition of new LOINC terms and codes to handle claims-related data. In the United States, the HIPAA (Health Insurance Portability Assurance Act) draft rule for claims attachments proposes using HL7 messages with LOINC codes to identify individual observations within attachments. In addition, HIPAA defines six specific types of claims attachments: laboratory results; non-laboratory clinical reports; ambulance transport; emergency room visits; medications; and rehabilitation. This third division of LOINC efforts has managed the creation of the codes needed for these new attachment types.

Distributed Information Systems

Data standards are the vehicle for creating communication interoperability between systems, but do not really specify how an information system and its components should be organized. Client-server and *n*-tier architectures have been used to design many healthcare information systems within a single enterprise; extending beyond this boundary, previous efforts in creating distributed frameworks have often relied on *mediated architectures* to provide data integration across databases. In such frameworks, a software agent (*i.e.*, the mediator) is constructed to provide data mapping/translation and other network-aware services; communication between different systems thus occurs between the agents, enabling a single (disambiguated) syntax for finding and passing information. CORBAMed, the application of the CORBA (Common Object Request Broker Architecture) standard to the healthcare environment, is an example of a mediated framework. Unfortunately, the complexity of such systems impairs the ability of the architecture to support a large number of users and/or sites, often leading to poor performance [40]. As a result, new methods of distributed data access are being examined, taking into consideration scalability, resource utilization (network, hardware), and modularity (*i.e.*, abstraction, extensibility). And like many mission critical operations, clinical information systems must have uninterrupted uptime (*i.e.*, 24 hours/7 days a week) to ensure timely access to patient data. These concerns permeate the design and construction of a newer generation of distributed information architectures in the healthcare environment.

Peer-to-peer Architectures

In the past decade, attention has turned to peer-to-peer (P2P) architectures. Made infamous by MP3 music sharing services, P2P concepts have actually been in existence much longer (a literature search finds that IBM actually used the phrase "peer-to-peer" in a 1984 architecture paper). Although frequently associated with file sharing applications, P2P actually has broader applications and in fact its concepts can be found in a variety of older networking protocols, including the once common Usenet news system and the ubiquitous simple mail transfer protocol (SMTP) used by e-mail servers. Generally speaking, a peer-to-peer topology is any network that does not have fixed clients and servers, but a number of peer nodes that function as both clients and servers to the other nodes on the network. As such, nodes in a P2P network have been termed *servents* as they act (simultaneously) as clients and servers. This network arrangement is in sharp contrast with the client-server model: any node is equally able to initiate or complete any supported transaction. Peer nodes may differ in local configuration, processing speed, network bandwidth, and storage quantity.

P2P networks establish a mesh topology that can grow quite large. Analyses of these networks show a curious result: that the distribution of connections for a given node tends to follow a power law [96], and that most networks tend to naturally develop certain nodes that are "critical" to the overall system (*e.g.,* taking out one of these nodes would break off a large part of the topology). These special nodes are often referred to as super-peers, ultra-peers, or *super-nodes* – such nodes are often characterized by a high degree of connectedness with other nodes. Theoretically, the urban myth of "six degrees of separation" appears to hold true in most P2P networks (and in fact, in communications on the Internet) – that there is a finite number of "hops" that one must make from node-to-node to find targeted data (on average): this phenomenon in P2P is due to the evolution of these highly connected nodes such that most "edge" nodes are only a few hops from these super-nodes.

One way to understand peer-to-peer networks is through their historic development in light of the generations of technology:

- First generation. The first generation of P2P networks involved the use of a centralized resource, such as a database, to maintain information about the content of the network and the specific location of data (*i.e.,* which nodes share which files).
- Second generation. The next generation of peer-to-peer networks were aimed at removing the dependency on a centralized resource, and thus focused on completely new data structures (*e.g.,* distributed hash tables, DHTs) and optimizing decentralized search mechanisms in growing networks.
- Third generation. The third generation of peer-to-peer networks contains improvements upon the first two generations, but tackles questions of quality of service, security, and anonymity in regard to content sharing.

These three levels of P2P are not definitive; many hybrid architectures exist between these layers. Nonetheless, we use this perspective to further illustrate key constructs and issues in regard to P2P systems and their applications to the healthcare environment. To confine the scope of the discussion, we limit our description to the data sharing aspects of peer-to-peer architectures.

First Generation P2P: Centralized Searching

The most straightforward P2P framework to understand and implement, the centralized P2P framework is a hybrid of client-server and basic P2P concepts, where certain services are provided through the client-server mechanism, but the basic data transfer is performed from node to node (*i.e.,* peer-to-peer). In the context of data sharing, this service is a centralized index (such as a database) used to store information on the location of queryable content: a client accesses this server to find the information and then communicates directly with the node containing the data file (Fig. 3.6). When

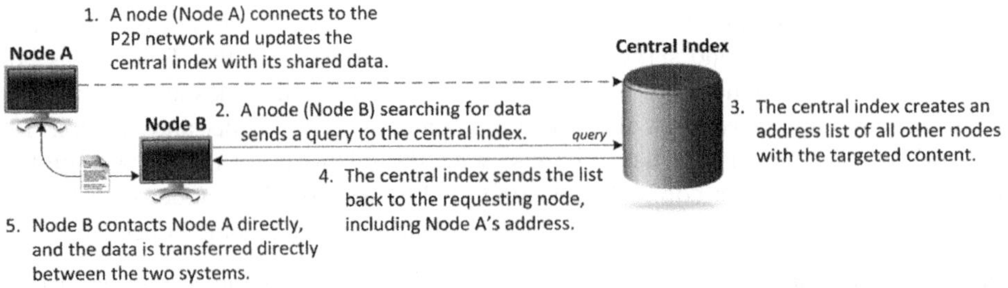

Figure 3.6: Basic data flow for first generation peer-to-peer computing. A centralized index is used to facilitate discovery.

nodes connect to the P2P network, it is responsible for indicating what data it shares so that the index can be updated. The original Napster model is indicative of this architecture. The attractiveness of such systems may be considered fourfold:

1. In theory, the majority of network traffic is spent between each node within the network. Assuming that the amount of data to be transferred is moderately large, the network traffic incurred by a query to the server is minimal in comparison to the amount of data passed in a query.
2. The centralized index allows for access control: users must go through this system in order to find information in the network. If information needs to be removed from the network, its entries in the database can be readily deleted, thus quickly revoking it shared status (*i.e.,* because it is no longer discoverable).
3. As new information is added to the P2P network, all servents will automatically see changes, as the central index is the authority for all information.
4. Querying time is constant and consistent in that a query to the network will always provide the same result from the central index; moreover, the system can assure completeness of results, being aware of all content in the network.

However, the strengths of a centralized approach are also its potential problems:

1. The central server is the bottleneck because all queries must be processed by a single master node. Also, the service's capacity must be scaled up as the number of users increases. A potential solution lies in mirroring the database, but results in propagation update issues (*i.e.,* the replicated indices must be synchronized).
2. The centralized service is also a lynchpin for the entire network: if it fails or goes down, then the entire network will collapse.
3. In order for any information to be seen in the network, nodes sharing data must actively update the central index. Ultimately, the central index is not necessarily aware of whether a given sharing node/client is actually online, or if the shared content is still available. Because of the fluid behavior of P2P networks, a node can enter or leave the network at any time. Thus, if a node goes offline suddenly (purposefully or not) or a node deletes its local copy of a shared data file, querying the index may result in passing references to systems where the data is no longer accessible.

Second Generation P2P: Simple Decentralized Searching (Query Flooding)

The fact that a centralized P2P network can be easily disrupted spawned a new wave of research to overcome the problems of scalability and reliability. Thus, the concept of completely decentralized services epitomizes the second generation of peer-to-peer

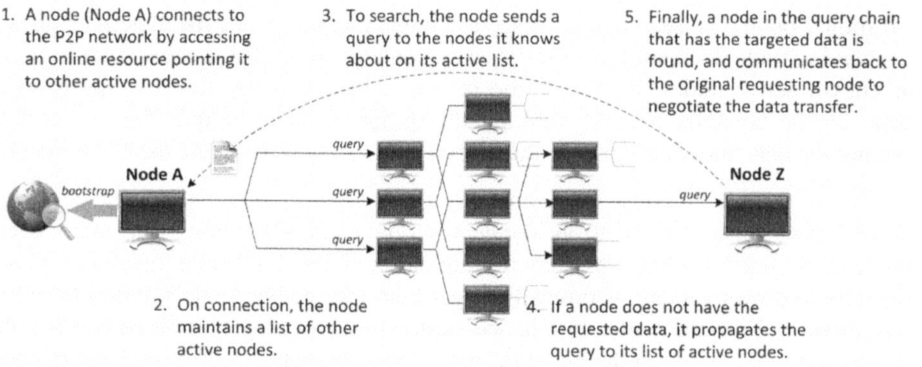

Figure 3.7: Second generation P2P using query flooding. To find a given data file, a node initiates a query sequence that passes a request from node to node. Eventually, the query will be seen by all nodes in the network, discovering the targeted data.

systems. Specifically, these architectures removed the central search index; however, new methods were then needed to enable nodes to find content within the network.

To replace the central index, preliminary efforts used a simple search paradigm: if a given node needed to query the network to find data, it would query a finite number of other nodes to which it is connected; if these neighboring nodes do not have the data, they in turn pass the query on to their neighbors, etc.[2]. Once found, the node containing the desired data communicates back to the original requesting node (Fig. 3.7). While uncomplicated, this approach incurs several technical and practical difficulties: 1) the new nodes joining a P2P network need a mechanism to find other nodes already in the network; 2) the "loopback" query propagation problem (*i.e.*, cycles in a network topology resulting in the original servent initiating a series of queries that comes back to itself or an already queried node in a recursive manner); and 3) the inability to assure timely discovery of targeted content within the network (unlike centralized P2P, where query time is constant, queries can be propagated indefinitely without finding the desired data). Partly being based on the idea of data locality in that the information one needs is typically close by (in terms of the network), naïve implementations of this strategy often fail to scale, as servents can overwhelm a network with queries (*i.e.*, query flooding); in point of fact, some analyses show that the number of queries and the traffic amounts to more than the actual targeted data [97].

Gnutella. Gnutella, the first popular decentralized approach to P2P, exemplifies the simple search method. The original open protocol (version 0.4) defined only five types of messages: ping, to discover hosts on a network; pong, a reply to a ping; query, to search for a file; query hit, a reply to a query; and push, to initiate a download request. To connect to a working node in the P2P network, a range of bootstrapping techniques have been used by Gnutella clients (*e.g.*, permanent IP addresses, web caches of online servents, UDP (user datagram protocol) host caches). On successful connection, the new node updates its working list of online nodes in the P2P network and establishes connections (via pings) to some predetermined number of active nodes (determined by

[2] Some readers may remember a 1970s American television commercial for shampoo that featured the line, "*And she told two friends, and they told two friends, and so on, and so on...*" The simple search mechanism works in a similar fashion to quickly spread the query.

a response pong). To search, a node sends a query message to its list of active nodes; if a receiving node does not have the requested data it forwards the request on to its list of known active nodes. In theory, the request will eventually find its way to every node on the Gnutella network through this propagation technique. On successfully finding the data, the node contacts the original requester to negotiate the file transfer.

To speed download rates, extensions to Gnutella introduced the idea of *swarm downloading* (also referred to as *segmented downloading*). Noting that many users' Internet connections have asymmetric upload/download rates, a receiving node's capacity to download data is typically larger than the sending node's ability to upload information. Swarm downloading hence exploits this fact: when a given file is found on more than one node in the network, the system segments the file so that each node uploads in parallel a different part of the file to the requester, thereby maximizing the requester's download capacity (the data is reconstructed on the requester's node). Other extensions to the base protocol have added a maximal number of node "hops" (*e.g.*, search up to five neighbors away from the requesting node) to the search space and timeouts for query propagation (*e.g.*, if a query is not responded to within 20 minutes, terminate the search) to limit the impact of queries on the network.

Critics of this approach argue that searching on Gnutella networks is often slow and unreliable [97]: nodes are constantly connecting and disconnecting, so the network is never completely stable. Moreover, given the large size of the network (some analyses show the network averages around 100,000 nodes at any given time), search requests take a long time to search and generally do not guarantee discovery. However, given its completely decentralized quality, Gnutella-based networks are highly robust.

Second Generation P2P: Distributed Hash Tables

The problem of finding information in a decentralized P2P topology resulted in two independent research groups coming up simultaneously with variants on the same solution (one at MIT, the other at the University of California at Berkeley), a concept now called a *distributed hash table* (DHT) [93, 106]. Hash tables are a common programming data structure that associate a key with a value, with the primary aim of providing rapid lookup (*i.e.*, given the key, the value can be retrieved in $O(1)$ or constant time). Rather than storing the hash table on a single computer, a DHT spreads the mapping over all the nodes in the network, thereby removing any centralized resource. The DHTs are used to store the locations of shared data files – in essence, a key is generated representing the file, and the value is the file itself; the DHT points to the location of a file in the network.

A DHT starts by choosing a hash function[3] to represent an n-dimensional coordinate space (typically an n-dimensional torus or algebraic ring) that is mapped across the m nodes in a P2P network. From this hash function, the key and value pairs are mapped to this coordinate space. Each node in a network stores a part of the DHT; [93] refers to this part of the table as a *zone*. In addition, each node stores additional information on adjacent zones within the table. Within its mapped space, a node is held responsible for the corresponding set of possible key-value pairs. Requests for an insertion,

[3] For those readers unfamiliar with this term, a *hash function* can be thought of as a (mathematical) operation that for a given object generates a (unique) signature (called a *hash code* or *key*) based on the object's content. Typically the hash function space is much smaller than the object itself, allowing for a compact representation. A basic example of a hash function is the modulo operator.

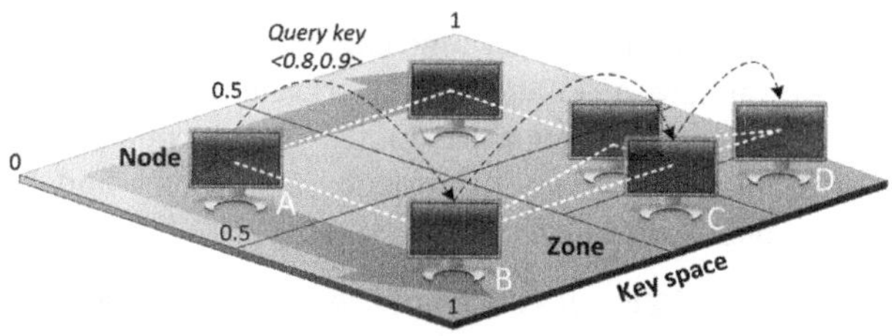

Figure 3.8: Distributed hash tables break up an n-dimensional key space across multiple nodes in a P2P network. Dotted white lines show the mesh topology of the network. In this simplified example, a 2D space is used over six nodes, with values ranging from 0-1. Each node maintains information on adjacent zones, allowing it to propagate a query towards the node responsible for maintaining a requested key. Here, a query from the node in *Zone A* for they key <0.8, 0.9> is progressively forwarded (dashed arrows) to *Zone D*, which is responsible for the range <0.75-1, 0.75-1>.

deletion, or lookup is routed along the different zones (and hence, nodes) until the node responsible for the information is uncovered (Fig. 3.8). For example, to locate a record, a node applies the hash function to get the key. The key is then mapped to the coordinate space to discover which node contains the target entry of interest. In knowing its adjacent neighbors' spaces, a node can forward the request progressively closer to the node responsible for the key. Once the query is received by the responsible node, a lookup occurs to discover if the data is available, and which node holds the actual file. As nodes are not fixed in a P2P network, the DHT technique must handle the continual changes in the network topology to keep the index consistent. As nodes join the network, they split the zone of the node they connect with, becoming a new neighbor. Conversely, as a node leaves the network, its neighbor merges its zone's contents with the node's to take over the mapping space. To provide additional robustness, a DHT-based network may utilize several hash functions (hence overlaying multiple mappings) on the nodes to ensure that if mapping information is lost (*e.g.*, a node does not cleanly leave the network before its zone is consolidated) that another route can be used to find a given piece of data.

DHTs thus provide a decentralized search mechanism, as each node is only locally aware and routes actions based by "pointing" in the right direction; but, unlike simple search, this method guarantees a definitive answer as to whether a given file exists within the network, and within a finite amount of time. The first implementations of DHTs brought about several questions about how to improve P2P search performance [94]. First, because the assignment of adjacent zones has no correspondence to the real-world physical network, a zone's neighbor may actually be located thousands of miles away, resulting in increased network traffic and delays; if geographic proximity could be taken advantage of in the assignment process, faster routing could occur. Second, the framework considers all nodes as equivalent in the DHT; yet it is clear that ultra-peers and nodes with more resources can handle a larger part of the DHT, protocols could distribute the zones in a more optimal manner.

Third Generation P2P

Although the second generation of peer-to-peer overcame centralized search to create robust networks, overall performance and security in these open systems still proved challenging. Efforts in the latest generation improve upon these efforts based on inspection of real-world network behavior, often resulting in hybrid frameworks. For example, research has suggested that superimposing a small random set of connections, akin to the unstructured nature of Gnutella networks, into a DHT map (*i.e.,* having some nodes maintain information on a randomly chosen non-adjacent node), can improve search performance [99]. Two protocols, BitTorrent and Freenet, help to underscore some of the ongoing issues.

BitTorrent. Observation of P2P networks noticed that data that was in widespread demand but available from only a small number of nodes on the network often resulted in substantial slow down of download speeds, as these nodes were forced to split their upload bandwidth across multiple requests. Contributing to this problem was the fact that nodes successfully downloading this data in turn did not share it with other requesters, which would help alleviate the overall demand by spreading the requests over a larger pool of resources. The BitTorrent P2P protocol [20] was developed in response to this issue, leveraging swarm downloading concepts to enable fast dissemination of a given data file to a wide audience of users. The basic idea is that BitTorrent nodes are forced to share the data that is downloaded from the network, thus adding their upload capacity to the distribution effort: each user becomes a member of a swarm, contributing to the spread of a given file's contents.

A file that is to be shared is initially segmented into several blocks, and a hash code is generated for each block using SHA-1 (secure hash algorithm). The hash code is used as a checksum value and to uniquely identify the block. The address of a *tracker*, a server that is responsible for maintaining a running list of nodes that have downloaded a given block, is combined with the hash codes into a single file, called a *torrent*. To download a shared file, a node must obtain the appropriate torrent and contact the designated tracker. The tracker orchestrates the download process, directing requesters to a given node to acquire a block; the tracker thus distributes the requests over the set of nodes within the swarm. Note that a given node, in the process of still downloading the entire data file, may at the same time be uploading a given block to another node, optimizing its participation in the dissemination process. In point of fact, the BitTorrent protocol gives preference to those nodes with higher upload bandwidths by automatically connecting such nodes to more download connections in order to speed sharing. Overall, BitTorrent has the effect of increasing average download speeds when more users connect to download/upload a given file. Markedly, unlike other P2P frameworks, the BitTorrent protocol does not handle searching within the network; a given node must discover the torrent through some other means (*e.g.,* a web search). Because the tracker serves as a centralized resource, successive changes to the original BitTorrent architecture have introduced several trackers and "trackerless" torrents based on distributed hash tables to improve overall reliability.

Freenet. The premise of Freenet is to provide a completely anonymous and secure means of P2P data sharing [19, 109], ensuring the integrity of content and the privacy of its users. The system is designed so that documents stored by Freenet are completely encrypted and replicated across a large number of continuously changing, anonymized computers in the network. Communications between nodes are also encrypted. The discovery of which nodes host a given file is further complicated by the fact that a given file may be broken up into sections spread out over multiple nodes. To support this construct, Freenet participants must allocate a portion of their computer's storage space to hold shared content – however, the user has no control (or direct access) over what data is stored locally, a somewhat controversial feature of the system.

Freenet uses a concept called *key based routing* (KBR), which is similar in nature to distributed hash tables. Each document inserted into a Freenet network has a key that uniquely identifies the object. Two types of keys are supported: *content hash keys* that use SHA-1 as a checksum; or *signed subspace keys* (SSK) based on public key infrastructure to sign a document and validate its contents. Each Freenet node stores in its local cache information not only about its documents and their keys, but also a *routing table* that associates other Freenet nodes with their performance in responding to a given key. These performance statistics are used by a node to decide how best to route requests. Thus, for example, to find a given document, a node first looks in its local cache for the key; if it does not find this key, it uses the routing table to determine the node that will most quickly be able to locate the key and passes along the request. This search process propagates accordingly from node to node. The search algorithm incorporates logic to stop recursive/redundant search paths, allowing a node to retract repeated requests by trying the "next best" entry in its routing table. Searching terminates when either the data is located, or the query passes through a maximal number of nodes; at which point the search path is traversed in reverse, passing back the requested data if found and updating each node's routing table with performance statistics. Data is checked by each node against the key to ensure that the document's content is unchanged. A novel feature of Freenet occurs in the response step, as intermediate nodes (*i.e.,* between the requester and source nodes) may decide to create a local copy of the data based on previous search patterns. This copying step is based on the idea of moving data "closer" to the requester on the assumption that other nodes in the requester's neighborhood may similarly ask for the same data. The results of this intermediate caching are twofold: 1) data replication occurs naturally; and 2) subsequent retrievals of the data may be faster (requiring less node hops to find the data). Freenet document insertion follows the same node traversal pattern described for searching, automatically deciding where to route the document for storage in the network and potentially partitioning the file in the process; in doing so, documents should be found where expected. But as with Gnutella, Freenet's protocol cannot promise that a given document will be found in a timely manner, even if it is stored in the network.

Freenet's design ensures that no tracking information is maintained so that a given node only knows about its neighbors and their requests; the ultimate source of a given document (and conversely, a requester) is hidden. As a node passes a document to its neighbor, it does not know if the neighbor will subsequently forward the document to another node or if the neighbor is the original requester: this policy is intentional, so that anonymity is maintained.

P2P Healthcare Applications

Several systems have adopted P2P concepts in application frameworks for healthcare, focusing on the need to share patient data for clinical and research purposes.

Santa Barbara County Care Data Exchange (SBCCDE). Initiated in 1998, the Santa Barbara County Care Data Exchanged aimed to make a patient's medical record electronically available to any authorized individual, including the patient himself [14]. The intention was to provide access to demographics, imaging, pharmacy, laboratory, insurance (*e.g.,* authorization) and other clinical records through web-based portals. Building a county-wide, public healthcare infrastructure, SBCCDE was also meant to assess if information sharing between healthcare participants was feasible; if such a framework would be sustainable; and whether improved outcomes would ensue (*e.g.*, better patient care, lowered costs, etc.). Acknowledging the fragmented nature of existing technology across the numerous information holders (due to a lack of standards implementation), a "brokered" peer-to-peer system was originally chosen to realize data sharing. P2P agents at each site provided data translation and integration in a mediated manner. Centralized services included a master patient index (MPI)[4], record location, and access controls (*i.e.*, determining authorized system users, and which users would have rights to a given piece of data). A similar mediated P2P architecture is described in [33], which puts forth the idea of a trusted physician (or caregiver) to authorize data access. Unfortunately, although early reports lauded the project and results were encouraging (*e.g.*, a reduction in redundant lab and imaging exams), the SBCCDE experiment was ended in late 2006 for a number of reasons. [78] cites the continual delays in technical deployment and the failure to establish access to key data sources; partly because of these issues, SBCCDE was unable to provide compelling functionality above and beyond existing web portals provided independently by participating major institutions. [48] further argues that in the long run, it was also unclear who should pay to support such an infrastructure.

Shared Pathology Informatics Network (SPIN). In 2000, the National Institutes of Health (NIH) funded research to develop a "virtual database" for locating human tissue specimens across research collections distributed throughout the United States [27, 76]. The idea was to create a single system wherein biospecimens matching certain query criteria could be easily found and then shared. A P2P framework was proposed within the CHIRPS (Consented High-performance Index and Retrieval of Pathology Specimens) project, using the idea of super-nodes to establish a routing meta-layer to pass queries between participating sites. A shared XML syntax for the query and results was developed for this system. Each institution or group of institutions identifies a super-node, responsible for authentication and distributing queries to secondary databases; the super-node also knows how to translate a query to its specific databases and conversely, how to transform the results to the given XML format. A mesh network is created between the super-nodes. Thus, a query from a user of this system would pass from the client to its super-node, who would in turn pass it to other super-nodes, propagating the query to the underlying layer of databases. Results are sent in reverse through the super-nodes, aggregated per site. This approach minimizes the information needed to establish network connections to local systems, instead making authorized super-nodes responsible for connectivity. The strength of this architecture lies in its degree of institutional autonomy to implement specific policies at the super-node level; however, if a super-node fails, this removes all of its underlying databases from a search and can potentially partition the network.

Grid Computing

The prior section viewed P2P as a means to share data; we now turn to the bigger picture of distributed computing and the sharing of resources in general. The phrase, *grid computing*, was termed in the 1990s to describe an environment where a multitude of networked computers are available to handle computationally expensive

---

[4] A *master patient index* is a mechanism to consolidate differing patient identifiers (IDs) between systems. In a simple sense, the MPI is a large table that has columns representing the different information systems, with rows being an individual: a given row entry thus provides a map between systems.

processes[5]. Well-known applications of grid computing encompass a range of areas, from astronomical data analysis (*e.g.*, SETI@home) to bioinformatics (*e.g.*, protein folding and modeling). P2P file sharing has been perceived as a subset of grid computing [30]; however, the former now concentrates on the problems of distributed search and routing, while the latter focuses on application infrastructure. Note that grid computing differs from *parallel computing* – while some concepts overlap, parallel computing typically entails the use of multiple CPUs in a single machine with little dependence on networking for shared memory and resources, whereas grid computing is dependent on network connectivity and further handles issues of multiple users and security.

Grid computing aims to solve large computational problems by establishing clusters of computers connected through an existing network layer (*e.g.*, the Internet). Using as a springboard past work in distributed computing, today's grid computing thus seeks to combine resources into a single "umbrella" and to hide the complexities of the system from the average user. The coming together of different groups and resources into one entity for a mutual, computational purpose is often referred to as a *virtual organization.* The fundamental difficulty lies in how to transparently share a heterogeneous pool of resources (*e.g.*, varying hardware, computing platforms, programming languages, etc.) across different users and environments that may span geographically significant distances. As such, some issues in grid computing include:

- Providing an open and generic application-level framework so that current applications can be readily (if not automatically) transformed to take advantage of a grid. Such a framework would necessitate a system capable of analyzing a program and intelligently splitting data and processes across grid participants.
- A description language for facilitating resource discovery on a given node participating in the grid. With such knowledge, the grid should attempt to mitigate the effects of any slower resources on an overall computation by optimizing execution plans (the so-called "least common denominator" problem, in which the speed of a given computation is limited by the slowest computer).
- Quality of service (QoS) requirements of a given program and/or user for a job submitted to the grid should be guaranteed, along with reliability and consistency.
- The performance threshold for a given application needs to be carefully considered. The use of additional distributed computational power to execute an algorithm does not assure speedup: there is an inherent overhead with the use of a grid (*e.g.*, with management of sub-processes, network traffic, etc.), which can be negligible for "large" problems but may actually penalize simpler programs.
- Given the shared environment of the grid, data and programs should be accessible only to authorized individuals, ensuring security.

A dichotomy has been suggested for describing grids based on their intended functionality: 1) *computational grids*, which tackle the sharing of CPU cycles; and 2) *data grids*, where data is the primary resource distributed amongst nodes and proessed accordingly. However, both grid types share a set of supporting services, including: responding to queries about the status of a grid and its contents; data and session management to distribute files and handle submitted jobs (*e.g.*, scheduling, resource

---

[5] Arguably, the concepts inherent to grid computing have their origins in older areas of computer science, including cooperative computing (1970s) and distributed computing and processing (1980s).

allocation); and load balancing. Several middleware layers have been developed to support grid functionality and the division of labor over a large number of CPUs. For example, BOINC (Berkeley Open Infrastructure for Network Computing) is a generalized client-server architecture for distributing processing tasks in a "volunteer computing" model [4]. We further describe two systems that presently typify general architectures for distributed grid computing: the Globus Toolkit, and Condor.

Globus Toolkit

The *Globus Toolkit* (GT) [29, 110] is an open source implementation of several standards set out by the Open Grid Forum, an international community of grid developers and users. Because of its ongoing standards support and its ability to work with several other high performance computing packages, including MPI (message passing interface, commonly used in parallel computing) and Condor (see below), Globus is frequently used as middleware to allow applications to take advantage of an underlying grid infrastructure. In particular, the Globus Toolkit follows the *Open Grid Services Architecture* (OGSA), a service-oriented model for ensuring interoperability of communication and data between grid participants [87]. First released in 2003, OGSA uses XML and web services as building blocks for grids.

GT has implemented protocols and tools in four areas, which collectively can be used to support a grid and its applications:

1. Resource management. The *grid resource allocation manager* (GRAM) protocol is responsible for remote job execution, allowing communication between job schedulers and other aspects of the grid infrastructure. GRAM itself is not a job scheduler, but rather handles the authorization, submission, monitoring, location, and termination of jobs submitted to grid resources. As of GT v4.0, GRAM supports JSDL (Job Submission Description Language), an XML specification for describing non-interactive tasks for execution on remote systems.
2. Grid information services. The *monitoring and discovery service* (MDS) provides information about resources available through the grid and their respective status.
3. Data management. Several methods have been established by grid developers for quickly moving/replicating data between nodes as part of computational tasks, including GridFTP (a grid-based version of the conventional file transfer protocol), and data replication services (DRS) to facilitate local access.
4. Security. A concern of early developers was the lack of a single method to authenticate grid user across resources: a single sign-on mechanism was needed. The *grid security infrastructure* (GSI), a public key cryptography system for authenticating users and encrypting communications, was conceived partly in response to this issue. GT implements the GSI as part of its security framework.

As alluded to prior in the discussion of data standards, a continual impediment to large scale data analysis is the multitude of data formats that exist, and the subsequent data management that must occur across the collections. To surmount this problem in the grid, OGSA data access and integration (OGSA-DAI) is a web service that facilitates the querying, retrieval, updating, transformation, and delivery of data distributed across the grid (*i.e.,* to grid clients). The objective of OGSA-DAI is hence to enable federation of disparate, heterogeneous data sources into a single logical resource: data in the form of (relational) databases, XML, and other file formats are made accessible on the grid through a uniform set of operations. Access to metadata about grid data sources is supported. In essence, OGSA-DAI attempts to minimize the need for technical knowledge of underlying data sources (*i.e.,* location, formatting) in accessing grid-based data collections, allowing the user to focus on the task of analysis. Closely aligned with OGSA-DAI are web services supporting queries across the grid. OGSA distributed query processing (OGSA-DQP) consists of two parts: 1) the *grid distributed query service* (GDQS) that coordinates and optimizes a query across different data sources on the grid, acting much like a "globalized" database query compiler; and 2) the *query evaluation service* (QES), which executes the determined query plan on a given data source. The implementation of OGSA-DQP intertwines with OGSA-DAI at several points: GDQS uses metadata about a source (*e.g.,* the database schema) to plan the query; and QES employs OGSA-DAI to abstract configuration issues for data access.

Notably, the latest version of the GT supports several programming languages (*e.g.,* Java, C, Python) and is based heavily on the web services resource framework (WSRF). Classic web services are stateless operations: between different sessions, there is no idea of data persistence. WSRF allows web services to maintain a state across different sessions, creating distributed entities that encapsulate both properties and methods. In many ways, this concept can be likened to a CORBA orb.

Condor

Condor is a batch job submission and management system supporting high throughput computing [21, 108]. Although not considered a grid in the same sense as an OGSA-based system, Condor allows for the formation of computer clusters from standard desktop workstations and servers (like a Beowulf cluster) using spare CPU cycles (*i.e.,* when the computer is idle). Indeed, Condor's strength lies in its full framework for job scheduling, queuing, and execution. A high-level description of a Condor network and a job submission is as follows:

1. Each computer participating in the network installs a background daemon process that informs the Condor system of its available resources (*e.g.,* amount of physical memory, disk space allocated for runs, operating system, available program environments, etc.). The system's description is called a machine ClassAd. This cluster of contributing computers is termed a *pool.* The daemon also monitors for idling activity, alerting the pool as to when the computer is available for use.
2. A user submits a job description to the pool. This description, known as a job ClassAd, specifies execution requirements and desired performance (*e.g.,* the location of the program executable and data files, amount of memory needed, etc.).
3. One computer designated in the pool, the *matchmaker*, is responsible for planning and scheduling the job. Planning entails the identification of resources needed to properly execute the job, matching machine and job ClassAds so that adequate resources are available. On finding these resources, scheduling determines which specific computers will be used, and when the procedures will be executed. The job goes into a queue and awaits execution across the different computers; Condor monitors this process, constantly updating its knowledge about the status of each computer in the pool and outstanding jobs.
4. When a resource is ready, the program is run on the machine. Execution of a program via Condor occurs in a *sandbox*, a secured area in which the program runs. This partitioning protects the resource from unintended (or malicious) runtime issues, while supporting a virtual machine and custom environment for

the program to execute. Condor supports a variety of *universes* to facilitate different programming environments within the sandbox; for instance, the Java universe supports class execution in the Java virtual machine.

5. Finally, when all parts of a job are completed, the final run result is returned.

Flexibility exists in supporting data access within a Condor network: a shared file system can be used; Condor can be instructed to copy required resources (program and data) to a sandbox as needed; or Condor can redirect input/output (I/O) requests from the program back to the original (submitting) machine.

Condor is well-suited to computational procedures that can be broken into smaller, independent parallel tasks (*i.e.,* no inter-process communication), running one program concurrently on many machines. However, Condor also enables a richer execution pattern by allowing multiple steps to be composited: a directed acyclic graph (DAG) of serially executed programs can be given to a pool.

Ultimately, Condor and the Globus Toolkit can be seen as complementary technologies. Markedly, the Globus Toolkit does not contain a job scheduler – a function for which Condor is ideal. Conversely, Condor is geared towards clusters of computers within a single institution or enterprise whereas Globus was designed to bridge such boundaries. The synergy between Condor and Globus is demonstrated by *Condor-G*, a Globus-enabled version of Condor that primarily uses an OGSA framework for connecting pools between organizations and handling security and resource management, and the Condor system for job submission and scheduling into Globus.

Grid Computing Healthcare Applications

Grid computing has found its way into the healthcare arena, with numerous projects in the United States and the European Union (EU) using this technology to support biomedical research. Below, several areas of development are highlighted as they pertain to imaging informatics endeavors.

Medical image analysis and dissemination. To deal with both the large amount of data inherent to imaging studies and the complex computations that are employed in analysis, many research groups have exploited grid-based infrastructure to facilitate biomedical image distribution and processing. For example, [113] details methods to bridge secure DICOM protocols to the Globus Toolkit, allowing DICOM compliant devices to become grid resources. The Biomedical Informatics Research Network (BIRN) [64] also employs aspects of the Globus and OGSA frameworks, focusing on shared access to neuroimaging studies and to services for analysis and visualization. Other works have created non-Globus based frameworks. [69] describes the use of a grid to handle images acquired for digital confocal microscopy, wherein a single high resolution image is split into tiles that are distributed across grid nodes for image processing. GridPACS [36, 89] provides a basis for managing online imaging datasets stored on a grid, virtualizing and federating data access across participants to enable rapid data discovery, distribution, and processing. Domain-specific applications using a grid to standardize image assessment (*e.g.,* via CAD) have been prototyped: the eDiaMoND [61] and MammoGrid [3] efforts are demonstrative, for instance, in the area of mammography.

Enabling Grids for E-SciencE (EGEE) II. EGEE II is an EU-funded computational and data grid, which as of early 2008 spanned over 200 sites, boasting 41,000 CPUs and 5 PB of storage for collaborative online processing [1, 31]. A full infrastructure is provided to quickly establish virtual organizations and to take advantage of this collective computing power. Building from past grid efforts, including Globus and Condor, EGEE developers have created *gLite*, a lightweight middleware package designed to further simplify grid application deployment [70]. Used in a broad range of scientific areas, biomedical applications are predominant in EGEE II, and specifically, computationally intensive tools involving medical imaging analyses. Representative programs in this area include: GATE, a Monte Carlo simulator for planning radiotherapy treatments using an individual's imaging studies; SiMRI3D, which simulates 3D MR sequences; gPTM3D, for volumetric 3D reconstructions; and Medical Data Manager (MDM), for secure DICOM management [80]. Components of these imaging-oriented grid applications were explored as part of the overall MEDIGRID project.

Cancer Biomedical Informatics Grid (caBIG). Launched in 2004 by the NIH's National Cancer Institute (NCI), caBIG is an ongoing effort to unify cancer investigations by creating a complete infrastructure connecting teams of researchers, clinicians, and other individuals involved in oncology [82]. The impetus behind caBIG is to allow scientists to answer research questions more rapidly and efficiently, accelerating progress in all aspects of cancer research. A founding principle of caBIG is an open, extensible architecture supporting interoperability between users and systems; for example, an objective early on was to link NCI-designated cancer centers throughout the US. A differentiating point was made between *syntactic* versus *semantic interoperability* [118]: the first reflecting data format, the second data content. caBIG aimed to facilitate both types of interoperability. Thus, standards development for data representation and communication, as represented by caCORE [65] and caGRID [88], have been two important activities of this NIH undertaking. caCORE's design is an *n*-tier architecture, providing a layer for web services for several interacting parts: *enterprise vocabulary services* (EVS), a controlled terminology system; *cancer bioinformatics infrastructure objects* (caBIO), an object-oriented data model defining attributes for common entities; and *cancer data standards repositories* (caDSR), a formal means to relate semantic metadata between a controlled terminology and caBIO classes. caGRID is a Globus-based infrastructure whose functionality is threefold: 1) information and services discovery over heterogeneous resources; 2) data integration and large-scale analysis; and 3) distributed data management, sharing, and coordination (*e.g.*, for multi-site clinical trials). Each resource within caGRID is wrapped as a web service, leveraging WSRF. caGRID provides support for grid workflow management using the web service's business process execution language (BPEL) standard; aggregate/join queries involving different grid data sources; and large dataset retrieval using the concepts of enumeration and database cursors. Inherent to caGRID is an authentication/security framework, and like many computational grids, it has the ability to replicate and distribute services to improve computational throughput. caCORE is the mainstay for a semantic service-oriented architecture promoted as part of caGRID. By way of illustration, semantic resource discovery in caGRID is enabled through a service's metadata description, defining its provider and the expected inputs/outputs in terms of caBIO classes. This metadata is stored in a Globus index service that provides lookup (similar to the yellow and white pages in web service's *universal description, discovery, and integration* (UDDI) framework[6]).

[6] UDDI's terminology of yellow and white pages refers to phone books: yellow pages are used to find a given service based on some higher-level grouping or abstraction.

Like EGEE II, caGRID has developed an intermediary toolkit, *Introduce*, to ease the authoring of grid services and the general use of the system.

Cloud Computing: Beyond the Grid

Grid-based computing establishes a highly structured environment for computing tasks. Developing from this framework, a more general structure, referred to as *cloud computing*, has been popularized. Though consensus on a definition of cloud computing has yet to form, the core idea is that data is permanently stored on servers connected to the Internet (*i.e.*, at data centers) and that a network of software services is available to act on this data; as needed, data is cached temporarily on clients (*e.g.*, desktop computers, handheld devices, etc.), which utilize the online software to facilitate access and processing [43]. Distributed file systems, databases, and job allocation are integral to cloud computing; as such, existing architectures (grids, P2P, mediators) and newer platforms like Apache's Hadoop are absorbed as part of the cloud infrastructure. By taking advantage of the ubiquity of the Web and shared processing power, cloud-based applications aim to enable widespread accessibility (so-called device/location independence, with access to data and applications "anywhere"), to increase system reliability, and to provide scalability in terms of the number of users and the dynamic availability of computational power (*i.e.*, as needed resources). Several frameworks for cloud computing have been released, including Nimbus, a set of tools based on Globus' GT v4.0's grid services to support scientific applications [35]. Thus far, the healthcare environment has yet to exploit cloud computing; however, natural applications include EMR storage/access and more sophisticated processing frameworks for biomedical data and informatics algorithms.

Discussion and Applications

Now more than ever there is an emphasis on sharing clinical and research data. Beyond the obvious reasons of facilitating patient care outside of the conventional single provider/institution model, it is believed that new knowledge about diseases and their treatment will only come about through team science and the pooling of data, computational resources, and domain expertise [123]. Imaging, with its emergent role in all areas of healthcare, has been a particular focus so as to establish normalized datasets for research and validation studies [112]. In response, new information architectures and systems must evolve to address this shift in healthcare delivery and the "collaboratory" paradigm. The feasibility of these approaches must be evaluated, making certain that the infrastructures balance scale with performance, if not being widely deployable to bring together the diverse mix of existing EMR databases, technologies, and standards. And as seen with prior attempts, there must be demonstrated value-added to *all* the individuals vested in the healthcare process: physicians, payors, and patients are all stakeholders that drive acceptance and uptake of changes, if not dictating how such systems will be paid for and maintained in the future. Moreover, with such change come new considerations. Questions of security and ethics are now commingled with our increased ability to access and analyze an ever growing body of patient data [8, 50, 55]. As such, there is a move towards empowering the patient as a

---

(*e.g.*, finding a restaurant); whereas white pages are typically used to find a specific address or phone number, given a piece of information (*e.g.*, the phone number for Vincenti Restaurant in Los Angeles). Likewise, a web service search can be based on its metadata description (yellow pages) or to find its provider (white pages).

gatekeeper to his medical data [75]. For instance, the Personal Internetworked Notary and Guardian (PING) project allows an individual to control access to the contents of his medical record, storing copies of information into a trusted data store [103]. New models of data custody (*i.e.,* who owns the data, who has rights to see the data) should be explored as the usage of clinical data expands. In this light, the issue of personal health and medical records (PHRs, PMRs) versus the EMR must now be considered: as commercial interests have entered the arena (*e.g.,* Microsoft's HealthVault, Google Health Initiative), the boundary between patient, physician, and institutional stewardship and access are progressively blurred. It remains unclear as to whether a PHR separate from the EMR is advantageous: although the end users are different (*i.e.,* patient *vs.* physician), both ultimately draw upon the same data to improve healthcare.

Teleradiology, Telemedicine, and Telehealth

One early area of research and development in distributed EMRs was telemedicine; and in the specific context of imaging, teleradiology. Broadly, the aim of telemedicine is to connect patient data acquisition at a local site (*e.g.,* a rural clinic) with expertise at a remote site (*e.g.,* an academic medical center). By transferring information pertinent to the patient's presentation (*e.g.,* history, signs and symptoms, images, etc.), a medical specialist can render an opinion and answer questions, helping guide a local physician's diagnosis and therapeutic decisions. [86] defines telemedicine as, "*the use of electronic communication and information technologies to provide or support clinical care at a distance*," whereas the newer area of telehealth is wider in scope being, "*the use of electronic information and telecommunication technologies to support long-distance clinical health care, patient and professional health-related education, public health and health administration.*" Both federal and private health insurers have covered some telemedicine services, despite the fact that the costs and benefits of providing many of these services have not been well studied. Indeed, while some debate lingers as to the efficacy of telemedicine (in terms of clinical outcomes and costs) [41], the majority of studies are supportive of both diagnostic accuracy and patient satisfaction [39]. Additionally, the small number of telemedicine studies that do include quality of life (QoL) metrics find a positive impact on patients [59].

Early efforts. Initial works in telemedicine can be categorized threefold, focusing on the use of telecommunications technology for medical diagnosis, monitoring, and therapeutic purposes whenever distance and/or time separates the patient and healthcare provider:

1. In *store-and-forward telemedicine*, clinical data are collected, stored, and then forwarded to be interpreted later. A store-and-forward system eliminates the need for the patient and the clinician to be available at either the same time or place.
2. *Home-based telemedicine* services enable physicians and other healthcare providers to monitor physiologic measurements, test results, images, and sounds, usually collected in a patient's residence or a care facility.
3. *Office/hospital-based telemedicine services* are real-time clinician-patient interactions that substitute for face-to-face encounters between a patient and a physician or other healthcare provider.

Perhaps because of the innate nature of medical imaging in producing a visual representation that requires specialist interpretation (thus making the data amenable to transmission), teleradiology was one of the initial telemedicine efforts. Teleradiology is now the leading application in this area, and was the first to receive full reimbursement under US Medicare provisions. Teleradiology involves the transmission of medical images to a radiologist for final interpretation. The first documented use of teleradiology occurred in 1950 in Philadelphia, PA: [34] developed a system using telephone lines and a facsimile machine to transmit images between two hospitals over 27 miles apart (45 kilometers). In a 2003 survey, of 1,924 professionally active post-training radiologist, results show that 67% of active radiology practices use teleradiology [28]. The use of teleradiology or (distributed) PACS in multi-specialty private practice was significant, with 81% of survey respondents responding positively. For those practices that reported using teleradiology, 82% reported transmitting images to a radiologist's home as the most common destination. Fifteen percent of all practices use teleradiology to send images outside of their own practice facilities.

Current trends. Beyond teleradiology, psychiatry and neurology are two subspecialties that now provide evidence for the effectiveness of telemedicine. Namely, because of the primary dependence on verbal interactions for patient assessment in both of these fields, current technologies are ideal for supporting this mode of communication. Analyses have shown that various psychiatric and neurological diagnostic assessments can be effectively administered through an interactive videoconference system [42]. Likewise, treatments administered in these subspecialties via telemedicine frameworks appear to achieve comparable results with traditional patient-clinician office visits. Several other recent research studies have shown the benefits of home-based telemedicine interventions in chronic diseases (*e.g.*, heart disease, diabetes) and rehabilitation. Two key benefits uncovered in these systems are the enhanced communication between the patient and the healthcare provider; and the closer monitoring of the patient's general health. The one caveat of these studies was the requirement for additional resources and dedicated staff to service these home-based patients: because of the additional "overhead" of interpretation and management of these individuals, it is difficult to assess whether improved outcomes are due to the increased level of care provided by dedicated clinical staff versus the technology intervention itself.

Growing out of the ability to capture and send images over broadband connections, telemedicine applications in dermatology and ophthalmology have grown recently [66], highlighting some issues in telemedicine. For instance, most studies of teledermatology have evaluated store-and-forward techniques; but there continues to be highly variable rates in diagnoses [42]. For tele-ophthalmology, high rates of diagnostic concordance and accuracy are seen, but only for some eye conditions such as diabetic retinopathy: tele-ophthalmology has been less successful at diagnosing cataracts and other lens abnormalities. Other applications have shown viability (*e.g.*, wound care), but limited studies (*e.g.*, small sample sizes, few reviewing clinicians, lack of comparisons to conventional in-person office examinations) have hindered adoption. A burgeoning area of development, several researchers have also proposed the use of telemedicine for oncology – that is, *tele-oncology* [90, 122]. Early works focused primarily on the use of videoconferencing to support multidisciplinary tumor boards [13, 53, 79]. Increasingly, efforts are turning to telemedicine to support oncology consults [116]; symptom and pain management in cancer survivors [26, 98]; and long-term palliative care [22]. In these and other clinical areas, rigorous evaluations are still needed to compare intra- and inter-observer levels of agreement with conventional patient-clinician office visits. Moreover, a careful study of patient outcomes is needed: did the telemedicine encounter provide comparable care for the patient, as compared to a similar live encounter with a clinician? In general, advocacy for an expanded role for telemedicine in specific applications necessitates an examination of the rates of missed diagnoses, incorrect treatments, and under what clinical conditions should a patient schedule an in-person office visit and decline a telemedicine consult.

Toward the future: Wireless health. Traditionally, telemedicine has focused on the transmission of patient data through landlines and fixed networks (*e.g.*, the Internet). But the proliferation of broadband wireless services, along with more powerful and convenient handheld devices, is helping introduce real-time monitoring and guidance for a wide array of patients: low-cost sensors and wireless systems can now create a constantly vigilant and pervasive monitoring capability at home, work, and non-clinical settings, providing continuous data collection via personal/body area networks (PAN/BAN) [73] (Fig. 3.9). Building from earlier works in telemedicine, a nascent research community is connecting medical care with technology developers, wireless and sensing hardware systems, network service providers, and enterprise data management communities. Wireless health prototypes have been demonstrated in a multitude of areas, such as: monitoring of the elderly and the disabled [2, 68, 71, 117]; physiologic monitoring [81, 111]; emergency triage [74]; and rehabilitation [63, 120, 121]. New applications driven by ubiquitous sensing can be imagined. For example, the selection of a medical imaging protocol or other diagnostic test could be driven by an individual's data collection over the past day, optimizing the acquisition; or follow-up therapy could be tailored based on a patient's quality of life and quantified performance in real-world daily activities. Yet while there is potential benefit to patients and healthcare providers, this frontier of technical innovation posits social

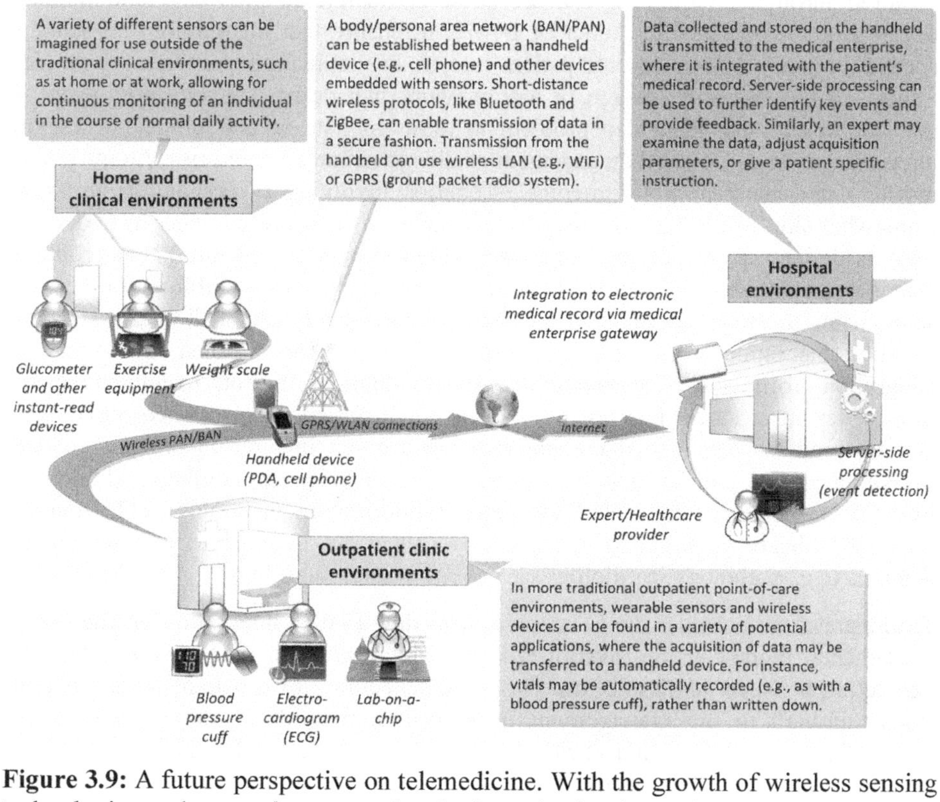

Figure 3.9: A future perspective on telemedicine. With the growth of wireless sensing technologies and networks, a new level of monitoring is made possible in different environments.

implications as to how such data should be used, interpreted, and secured. And as seen with DICOM, success for this new area will depend critically on the development of standards in practice, technology, and information processing that ensures every required degree of interoperability, integrity, and compliance. To this end, several industrial consortia, such as the Continua Health Alliance and the Wireless Health Alliance, have been formed to initiate the standardization process.

Integrating Medical Data Access

Several strategies have been put forth to integrate the diverse set of information and standards seen in healthcare. Comprehensive clinical data models have been created and continually refined to provide a semantic view of data elements and their interaction; HL7 RIM and DICOM are working examples. However, each of these models addresses a finite area within healthcare: the interplay between entities within different domain models is only beginning to be addressed through integrative endeavors. Here, we highlight a few of these efforts; a higher-level perspective, a phenomenon-centric data model, is presented in Chapter 7 as a comprehensive abstraction for coalescing a patient record into a logical observational framework.

Integrating the Healthcare Enterprise (IHE). Started in 1998 through HIMSS and RSNA (Healthcare Information and Management Systems Society, Radiological Society of North America), IHE is an initiative designed to integrate healthcare information systems through the use of recognized standards, such as HL7 and DICOM. Fostered through an ongoing collaboration of healthcare providers and industry developers, the fundamental tenet of IHE is to ensure that all required information needed for medical decisions and patient care are available (and correct) when needed. Central to IHE is in the design of a framework that defines an *integration profile* that addresses some identified problem in terms of data access, clinical workflow, infrastructure, and/or overall management challenge within an enterprise. An integration profile represents real-world medical scenarios involving *actors* and *transactions* (somewhat akin to the idea of use case scenarios in software engineering). Here, an actor is defined as any entity (*e.g.,* information system, application) that produces, manages, or acts on information. Transactions are exchanges of information between actors using messages, defined by the use of a specifically established data standard. Notably, integration profiles can be interdependent. Vendors are then able to take the integration profile and create potential software solutions, leading to implementation and live testing to assess interoperability between different systems; once vetted, commercial products can then document meeting a given integration profile. To date, twelve integration profiles have been generated specifically for radiology; additionally, four profiles have been created for general information technology (IT) domains within healthcare (Table 3.6) and work is concurrent to expand to other clinical domains (*e.g.,* pathology, cardiology).

To illustrate, we consider the Scheduled Workflow (SWF) integration profile [58], a fundamental module in the overall IHE model for radiology. The SWF details the interaction between admission, ordering, and scheduling of a patient for an imaging study; through to the management of worklists, procedural status tracking and notifications; the storage and retrieval of the images; and finally the generation and storage of a report. This profile precisely defines the terminology and transactions necessary to accomplish this operation. Fundamentally, SWF describes communication

	Name	Description
Radiology	*Scheduled workflow*	Complete data flow for an imaging encounter
	Patient information reconciliation	Resolution of (incorrect) patient image data association
	Consistent presentation of images	Ensuring correct image and annotation visualization
	Presentation of grouped procedures	Management of multiple procedures in one study acquisition
	Post-processing workflow	Scheduling of image reconstruction, image processing, CAD
	Reporting workflow	Tracking of reporting events (*e.g.*, interpretation, transcription)
	Evidence documents	Linking non-imaging data (*e.g.*, CAD results) w/reports
	Key image note	Addition of textual notes, pointers to sentinel images
	Simple image and numeric reports	Standard way for generating alphanumeric reports w/images
	Charge posting	Provides procedural information to billing systems
	Basic security	Basic security/HIPAA, consolidation of audit trails
	Access to radiology information	Methods for sharing radiological data across departments
General IT	*Patient-identifier cross referencing*	Master patient index mechanism for a given institution
	Retrieve information for display	Obtaining/display documents and patient-centric information
	Enterprise user authentication	Single-user sign-on for multiple information systems
	Patient synchronized applications	Maintaining patient context across applications (*e.g.*, CCOW)

Table 3.6: Integration profiles supported by IHE.

of information between the HIS, RIS, PACS, and imaging modalities, with eight different actors: ADT patient registration, order placer, department scheduler (order filler), acquisition modality, image manager/archive, performed procedure manager, image display, and evidence creator. The SWF integration profile begins with the registration of a patient by an ADT Patient Registration actor. The patient registration information is passed to the Order Placer and the Order Filler. Additionally, the order for an imaging service is passed from the Order Placer to the Order Filler. The Order Filler assigns an accession number to the image order, which is subsequently used to link associated data together (*e.g.*, connecting the image study to the radiology report). As an image order can result in more than one procedure, the Order Filler is responsible for creating a mapping between one or more scheduled procedure steps. In the IHE model, each scheduled procedure step represents a basic unit of work (*e.g.*, by the technologist or radiologist) and a billable set of acts that can be coded. The scheduled procedure steps are provided by the modalities as part of the DICOM Modality Work List standard. The modalities then send Performed Procedure Step messages back to the Image Manager and the Order Filler so that these actors know the state of the patient during the imaging exam. The Order Filler can query the Image Manager to determine image availability, and the Order Filler can notify the Order Placer when the order has been completed. The modality transfers the patient's image study to the Image Archive actor and then the modalities execute a storage commitment transaction with the Image Manager to insure that the study was stored properly into the PACS archive before removal from the modalities local storage device. Finally, after reviewing the image study, the radiologist generates a report on the patient diagnosis, completing the SWF integration profile data flow.

DataServer. Although convergence is beginning, the healthcare information industry has resulted in a plethora of medical standards and databases for storing, representing, and accessing patient information. In addition to legacy and proprietary systems, the goal of creating a distributed EMR becomes increasingly complicated in this mixed milieu. Thus, the reality is that there exist many incompatibilities between implementations for a variety of reasons: different versions of a given standard; semantic incongruities; and the development of proprietary formats are common confounders [118]. Rather than advocate a single standard, the open source UCLA DataServer project takes the approach of building atop existing and future clinical systems, while further providing an infrastructure to support large-scale medical research (*e.g.*, population-based studies) [15, 17]. A hybrid P2P architecture is used with data gateways to consolidate access to all patient information (*i.e.*, HIS/RIS/PACS, research databases, etc.) across multiple systems. The system consists of three parts (Fig. 3.10):

1. Local agents. Mediators are responsible for handling local transforms to queryable data sources. DataServer clients use a flexible XML query-by-example syntax to specify the data to retrieve. On receipt, the gateway parses the query elements and constraints, mapping them to an underlying data source and transforming the query into a target source's native format. Information may be contained in a database, a file system, a message protocol, or any other framework that has a queryable interface. The query is then executed, and the results reformatted back into an XML format.
2. Patient record index (PRI). To link patient information from different institutions/ EMRs, a centralized index (akin to that seen in 1st generation P2P) called the *patient record index*, is used per patient to establish a queryable table of available information. As information is discovered/added into a patient's record, the local mediator is responsible for updating the PRI; in this respect, the mediator acts as a servent. The PRI provides time-stamped meta-information on EMR contents, including a brief description and uniform resource locator (URL) to the servent

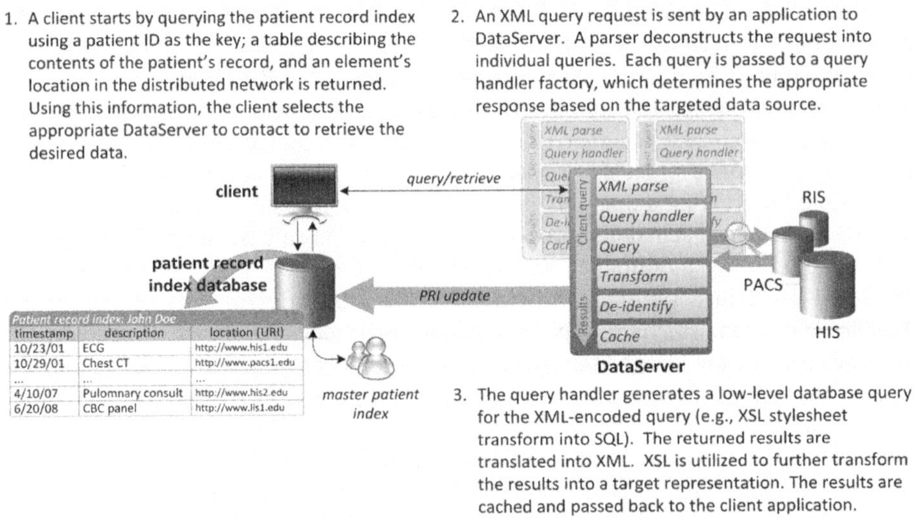

Figure 3.10: Basic DataServer architecture. A group of queryable data sources (within an institution) are made accessible via an agent (*i.e.*, DataServer) that is responsible for mapping queries and results to the correct formats. Automated de-identification and caching techniques are incorporated to offset workflow impact on clinical systems. To find information across multiple sites, the patient record index (PRI) construct is used as a centralized database that containers to pointers to specific EMR elements. The PRI is updated by participant DataServers on discovery of new data.

that can provide the information. An UDDI component was built so that the modular capabilities of a given DataServer implementation could be published and a network of DataServers established. Users looking to find information search the PRI and use the URL to directly access patient information.

3. Automated de-identification. An original objective of DataServer was not only to enable clinical access to information, but also to support medical researchers making use of clinical data. Such usage is tempered by various confidentiality policies set out by institutional review boards (IRBs) and in the United States, the Health Insurance Portability Assurance Act (HIPAA). DataServer includes methods to handle authentication and patient record de-identification. For the latter, a real-time de-identification engine using natural language processing (NLP) was constructed to identify HIPAA-specified fields within all free-text reports [107]; and an *n*-blind masking algorithm was used to replace patient-sensitive information.

In contrast to the more common data warehousing approach, DataServer supports real-time access to underlying clinical data sources; as such, result caching and methods to minimize query impact on clinical information systems were developed.

Visual integration: HL7 Clinical Context Object Workgroup (CCOW). A different viewpoint worth noting, CCOW is a framework for applications to be visually linked [47]. The underlying premise of CCOW is *context management*, wherein a single "information system" (comprised of multiple applications) is presented to the user through a shared, synchronized data environment that supports the user's workflow. For example, the selection of a patient object in one application (*e.g.,* scheduling) will inform other currently running programs that *Patient X* has been selected and their respective graphical interfaces can then be updated (*e.g., Patient X's* last visit record may be shown in an EMR window; his current insurance information presented in another part of the display, etc.). Another rudimentary demonstration of CCOW is a single user login across multiple applications. Thus, central to CCOW is the idea that there exist sharable definitions of common objects between applications (*i.e.,* the patient entity, a user object, etc.). The base architecture consists of three parts: CCOW-compliant applications; a context manager that coordinates information and event exchange between the applications; and mapping agents that handle data synonymy/translation between the applications. Because of the nature of the communication between applications, secure and authenticated communications are a key concern in CCOW; revitalized efforts are thus focused on updating this framework to use XML-based SAML (security assertion markup language) and a service-oriented architecture.

Collaborative Clinical Research: Example Image Repositories

Lastly, in step with the efforts to make clinical data uniformly accessible across distributed systems, many projects are capitalizing upon emergent infrastructure to realize collaborative research at multiple sites. While the idea of multi-site clinical trials is not new, the capacity to electronically collate, validate, and analyze results is achieving new scales of participation and powering larger studies. Demonstrative of this move are efforts under NIH's Roadmap initiatives to "re-engineer" the clinical research enterprise (*e.g.,* National Electronics Clinical Trials and Research (NECTAR) network) [85, 123]. The development of informatics infrastructure permits specialized projects to be launched, bring together multidisciplinary expertise from different sites. For instance, the EU *@neurist* connects clinicians and a gamut of researchers in different fields working on understanding and treating intracranial aneurysms:

separately, each site/investigator would be unable to attain sufficient numbers to generate insight into the disease, but together this cooperative will have a sizeable patient population from which new knowledge can be gleaned. Several groups have examined the issues surrounding the creation of imaging-based clinical research networks and repositories:

- American College of Radiology Imaging Network (ACRIN). With the now crucial role of imaging in cancer screening, diagnosis, and treatment, the NCI-funded imaging-based clinical trials cooperative provides the foundation for multi-institutional studies involving the evaluation of imaging protocols and treatment [45]. ACRIN has developed a data center capable of receiving and archiving DICOM imaging studies alongside associated electronic case report forms and clinical data needed to conduct clinical trial across the United States. The ACRIN data collection process enables normalization and structuring of study variables (*e.g.,* using controlled vocabularies, carefully defined data dictionaries, etc.), along with quality control methods mandatory to rigorous scientific investigations. Expertise in imaging and statistical analysis is also provided by ACRIN.
- Lung Imaging Database Consortium (LIDC). Considerable research has been performed with respect to medical image processing and its application for CAD. Unfortunately, a present barrier to the widespread adoption of these technologies lies in the absence of standardized datasets upon which comparisons of competing algorithms can be objectively evaluated (let alone the need for a "truth" value for gold standard assessment). Another NCI-funded program, LIDC was founded as a group of five academic institutions focused on CAD methods for lung CT, including the creation of an imaging database of standardized thoracic images and quantitative results [5, 77]. Challenges tackled by LIDC include: the creation of methods to ensure unbiased establishment of gold standard findings (*e.g.,* blinded readings, consensus procedures); the development of an extensible database to handle CT images and (future) derived quantitative measures; and the infrastructure to support web-based dissemination of the images.
- National Cancer Image Archive (NCIA). Building from the experiences of ACRIN, LIDC, BIRN, and other imaging-based efforts like caIMAGE, caBIG's *in vivo* imaging workspace has developed several tools for the management and sharing of images. NCIA was released as a web-based image repository linking DICOM imaging, annotations, genomic, and other biomarker (meta)data associated with cancer research [83]. NCIA provides a form-driven search interface to find imaging series matching various criteria (*e.g.,* source clinical trial, anatomy, acquisition parameters) across its centralized database and/or remotely linked collections, thereby federating different image datasets. Image archive resources available include virtual colonoscopy, mammography, and neuroimaging. As with other caBIG initiatives, NCIA is driven by standards and uses controlled vocabularies to describe anatomical/disease categories. The NCIA software suite is a freely downloadable open source package that can be used to create customized standalone systems.

Though the above predominantly draw from the oncology domain, image repositories have been established for other disease areas, including neuro- and musculoskeletal imaging [49, 64]. Markedly, while these major endeavors are opening new venues for image data sharing, the systems are still largely confined to lower-level search parameters (*e.g.,* those found within a DICOM header). Ultimately, as image processing and search algorithms improve, content-based image retrieval (CBIR) methods must be integrated.

References

1. Enabling Grids for E-sciencE (EGEE). http://www.eu-egee.org/. Accessed April 10, 2008.
2. Alaoui A, Levine B, Beauchene P, McAlinden EMA, Hu M, Choi JA, Welsh JCA, Mun SKA (2003) Implementation of a home monitoring network for patients with congestive heart failure. In: Levine B (ed) Information Technology Applications in Biomedicine, 2003. 4th Intl IEEE EMBS Special Topic Conference pp 51-54.
3. Amendolia SR, Estrella F, Hassan W, Hauer T, Manset D, McClatchey R, Rogulin D, Solomonides T (2004) MammoGrid: A service oriented architecture based medical grid application. Lecture Notes in Computer Science, 3251:939-942.
4. Anderson DP (2004) BOINC: A system for public-resource computing and storage. 5th IEEE/ACM International Workshop on Grid Computing, pp 365-372.
5. Armato III SG, McLennan G, McNitt-Gray MF, Meyer CR, Yankelevitz D, Aberle DR, Henschke CI, Hoffman EA, Kazerooni EA, MacMahon H (2004) Lung Image Database Consortium: Developing a resource for the medical imaging research community. Radiology, 232(3):739.
6. Bakker AR (2003) Views on HIS development; recommendations of earlier working conferences compared with present challenges. Int J Med Inform, 69(2-3):91-97.
7. Ball MJ (2003) Hospital information systems: Perspectives on problems and prospects, 1979 and 2002. Int J Med Inform, 69(2-3):83-89.
8. Barrows RC, Jr., Clayton PD (1996) Privacy, confidentiality, and electronic medical records. J Am Med Inform Assoc, 3(2):139-148.
9. Bates DW, Ebell M, Gotlieb E, Zapp J, Mullins HC (2003) A proposal for electronic medical records in U.S. primary care. J Am Med Inform Assoc, 10(1):1-10.
10. Bates DW, Gawande AA (2003) Improving safety with information technology. N Engl J Med, 348(25):2526-2534.
11. Bates DW, Kuperman GJ, Rittenberg E, Teich JM, Fiskio J, Ma'luf N, Onderdonk A, Wybenga D, Winkelman J, Brennan TA, Komaroff AL, Tanasijevic M (1999) A randomized trial of a computer-based intervention to reduce utilization of redundant laboratory tests. Am J Med, 106(2):144-150.
12. Bidgood WD, Jr., Horii SC, Prior FW, Van Syckle DE (1997) Understanding and using DICOM, the data interchange standard for biomedical imaging. J Am Med Inform Assoc, 4(3):199-212.
13. Billingsley KG, Schwartz DL, Lentz S, Vallieres E, Montgomery RB, Schubach W, Penson D, Yueh B, Chansky H, Zink C, Parayno D, Starkebaum G (2002) The development of a telemedical cancer center within the Veterans Affairs Health Care System: a report of preliminary clinical results. Telemed J E Health, 8(1):123-130.
14. Brailer D, Augustinos N, Evans L, Karp S (2003) Moving toward electronic health information exchange: Interim report on the Santa Barbara County data exchange. California Health Care Foundation. http://www.chcf.org/documents/ihealth/SBCCDEInterimReport.pdf. Accessed April 4, 2008.
15. Bui AA, Dionisio JD, Morioka CA, Sinha U, Taira RK, Kangarloo H (2002) DataServer: An infrastructure to support evidence-based radiology. Acad Radiol, 9(6):670-678.
16. Bui AA, Taira RK, Goldman D, Dionisio JD, Aberle DR, El-Saden S, Sayre J, Rice T, Kangarloo H (2004) Effect of an imaging-based streamlined electronic healthcare process on quality and costs. Acad Radiol, 11(1):13-20.

17. Bui AA, Weinger GS, Barretta SJ, Dionisio JD, Kangarloo H (2002) An XML gateway to patient data for medical research applications. Ann N Y Acad Sci, 980:236-246.
18. Chaudhry B, Wang J, Wu S, Maglione M, Mojica W, Roth E, Morton SC, Shekelle PG (2006) Systematic review: Impact of health information technology on quality, efficiency, and costs of medical care. Ann Intern Med, 144(10):742-752.
19. Clarke I, Sandberg O, Wiley B, Hong TW (2001) Freenet: A distributed anonymous information storage and retrieval system. Designing Privacy Enhancing Technologies: International Workshop on Design Issues in Anonymity and Unobservability, Berkeley, CA.
20. Cohen B (2008) The BitTorrent protocol specification. http://www.bittorrent.org/beps-/bep_0003.html. Accessed April 11, 2008.
21. Condor Project (2008) Condor Project Homepage. http://www.cs.wisc.edu-/condor/. Accessed April 9, 2008.
22. Coyle N, Khojainova N, Francavilla JM, Gonzales GR (2002) Audio-visual communication and its use in palliative care. J Pain and Symptom Management, 23(2):171-175.
23. Davidson SM, Heineke J (2007) Toward an effective strategy for the diffusion and use of clinical information systems. J Am Med Inform Assoc, 14(3):361-367.
24. Dolin RH, Alschuler L, Beebe C, Biron PV, Boyer SL, Essin D, Kimber E, Lincoln T, Mattison JE (2001) The HL7 Clinical Document Architecture. J Am Med Inform Assoc, 8(6):552-569.
25. Dolin RH, Alschuler L, Boyer S, Beebe C, Behlen FM, Biron PV, Shabo Shvo A (2006) HL7 Clinical Document Architecture, Release 2. J Am Med Inform Assoc, 13(1):30-39.
26. Doolittle GC, Spaulding AO (2006) Providing access to oncology care for rural patients via telemedicine. J Oncology Practice, 2(5):228.
27. Drake TA, Braun J, Marchevsky A, Kohane IS, Fletcher C, Chueh H, Beckwith B, Berkowicz D, Kuo F, Zeng QT, Balis U, Holzbach A, McMurry A, Gee CE, McDonald CJ, Schadow G, Davis M, Hattab EM, Blevins L, Hook J, Becich M, Crowley RS, Taube SE, Berman J (2007) A system for sharing routine surgical pathology specimens across institutions: the Shared Pathology Informatics Network. Hum Pathol, 38(8):1212-1225.
28. Ebbert TL, Meghea C, Iturbe S, Forman HP, Bhargavan M, Sunshine JH (2007) The state of teleradiology in 2003 and changes since 1999. AJR Am J Roentgenol, 188(2):W103-112.
29. Foster I (2006) Globus Toolkit Version 4: Software for service-oriented systems. IFIP International Conference of Network and Parallel Computing. Springer-Verlag, pp 2-13.
30. Foster I, Iamnitchi A (2003) On death, taxes, and the convergence of peer-to-peer and grid computing. 2nd International Workshop on Peer-to-Peer Systems (IPTPS'03), pp 118–128.
31. Gagliardi F, Jones B, Grey F, Bégin ME, Heikkurinen M (2005) Building an infrastructure for scientific Grid computing: Status and goals of the EGEE project. Philosophical Transactions: Mathematical, Physical and Engineering Sciences, 363(1833):1729-1742.
32. Garg AX, Adhikari NK, McDonald H, Rosas-Arellano MP, Devereaux PJ, Beyene J, Sam J, Haynes RB (2005) Effects of computerized clinical decision support systems on practitioner performance and patient outcomes: A systematic review. JAMA, 293(10):1223-1238.
33. Geissbuhler A, Spahni S, Assimacopoulos A, Raetzo MA, Gobet G (2004) Design of a patient-centered, multi-institutional healthcare information network using peer-to-peer communication in a highly distributed architecture. Medinfo, 11(Pt 2):1048-1052.
34. Gershon-Cohen J, Cooley AG (1950) Telognosis. Radiology, 55(4):582-587.
35. Globus Alliance (2008) Nimbus home page. http://workspace.globus.org/index.-html. Accessed October 15, 2008.
36. Hastings S, Oster S, Langella S, Kurc TM, Pan T, Catalyurek UV, Saltz JH (2005) A grid-based image archival and analysis system. J Am Med Inform Assoc, 12(3):286-295.

37. Haux R (2006) Health information systems - past, present, future. Int J Med Inform, 75(3-4):268-281.
38. Health Level Seven Standards Developing Organization (2008) Health Level 7 main web site. http://www.hl7.org. Accessed May 16, 2008, 2008.
39. Heinzelmann PJ, Williams CM, Lugn NE, Kvedar JC (2005) Clinical outcomes associated with telemedicine/telehealth. Telemed J E Health, 11(3):329-347.
40. Henning M (2006) The rise and fall of CORBA. Queue, 4(5):28-34.
41. Hersh WR, Hickam DH, Severance SM, Dana TL, Pyle Krages K, Helfand M (2006) Diagnosis, access and outcomes: Update of a systematic review of telemedicine services. J Telemed Telecare, 12 Suppl 2:S3-31.
42. Hersh WR, Hickam DH, Severance SM, Dana TL, Pyle Krages K, Helfand M (2006) Telemedicine for the Medicare Population: Update. Agency for Healthcare Research and Quality. US Department of Health and Human Services, Rockville, MD.
43. Hewitt C (2008) ORGs for scalable, robust, privacy-friendly client cloud computing. IEEE Internet Computing, 12(5):96-99.
44. Hillestad R, Bigelow J, Bower A, Girosi F, Meili R, Scoville R, Taylor R (2005) Can electronic medical record systems transform health care? Potential health benefits, savings, and costs. Health Aff (Millwood), 24(5):1103-1117.
45. Hillman BJ (2002) The American College of Radiology Imaging Network (ACRIN): Research educational opportunities for academic radiology. Acad Radiol, 9(5):561-562.
46. Hing ES, Burt CW, Woodwell DA (2007) Electronic medical record use by office-based physicians and their practices: United States, 2006. Adv Data(393):1-7.
47. HL7 CCOW Technical Committee (2008) Clinical Context Object Workgroup (CCOW) web site. http://www.hl7.org/special/committees/visual/index.cfm. Accessed April 22, 2008.
48. Holmquest DL (2007) Another lesson from Santa Barbara. Health Aff (Millwood), 26(5):w592-594.
49. Hsu W, Long LR, Antani S (2007) SPIRS: A framework for content-based image retrieval from large biomedical databases. MedInfo, vol 12, pp 188-192.
50. Huang CH, Konagaya A, Lanza V, Sloot PMA (2008) Introduction to the special section on BioGrid: Biomedical computations on the grid. IEEE Trans Inf Technol Biomed, 12(2):133-137.
51. Huang HK (2004) PACS and imaging informatics: Basic principles and applications. 2nd edition. Wiley-Liss, Hoboken, N.J..
52. Huff SM, Rocha RA, McDonald CJ, De Moor GJ, Fiers T, Bidgood WD, Jr., Forrey AW, Francis WG, Tracy WR, Leavelle D, Stalling F, Griffin B, Maloney P, Leland D, Charles L, Hutchins K, Baenziger J (1998) Development of the Logical Observation Identifier Names and Codes (LOINC) vocabulary. J Am Med Inform Assoc, 5(3):276-292.
53. Hunter DC, Brustrom JE, Goldsmith BJ, Davis LJ, Carlos M, Ashley E, Gardner G, Gaal I (1999) Teleoncology in the Department of Defense: A tale of two systems. Telemed J, 5(3):273-282.
54. Hussein R, Engelmann U, Schroeter A, Meinzer HP (2004) DICOM structured reporting: Part 1. Overview and characteristics. Radiographics, 24(3):891-896.
55. Huston T (2001) Security issues for implementation of e-medical records. Communications of the ACM, 44(9):89-94.
56. Institute of Medicine (1999) To err is human: Building a better health system. The National Academies Press, Washington, DC.
57. Institute of Medicine (2003) Key capabilities of an electronic health record system. The National Academies Press. http://www.nap.edu/catalog/10781.html. Accessed March 3, 2008.

58. Integrating the Healthcare Enterprise (2007) IHE Technical Framework: Integration Profiles. ACC, HIMSS, RSNA. http://www.ihe.net/Technical_Framework/upload/ihe_tf_rev-8.pdf. Accessed October 15, 2008.
59. Jennett PA, Affleck Hall L, Hailey D, Ohinmaa A, Anderson C, Thomas R, Young B, Lorenzetti D, Scott RE (2003) The socio-economic impact of telehealth: A systematic review. J Telemed Telecare, 9(6):311-320.
60. Jha AK, Ferris TG, Donelan K, DesRoches C, Shields A, Rosenbaum S, Blumenthal D (2006) How common are electronic health records in the United States? A summary of the evidence. Health Aff (Millwood), 25(6):w496-507.
61. Jirotka M, Procter ROB, Hartswood M, Slack R, Simpson A, Coopmans C, Hinds C, Voss A (2005) Collaboration and trust in healthcare innovation: The eDiaMoND case study. Computer Supported Cooperative Work (CSCW), 14(4):369-398.
62. Joos D, Chen Q, Jirjis J, Johnson KB (2006) An electronic medical record in primary care: impact on satisfaction, work efficiency and clinic processes. AMIA Annu Symp Proc:394-398.
63. Jovanov E, Milenkovic A, Otto C, de Groen PC (2005) A wireless body area network of intelligent motion sensors for computer assisted physical rehabilitation. Journal of NeuroEngineering and Rehabilitation, 2(6).
64. Keator DB, Grethe JS, Marcus D, Ozyurt B, Gadde S, Murphy S, Pieper SS, Greve D, Notestine R, Bockholt HJ (2008) A national human neuroimaging collaboratory enabled by the Biomedical Informatics Research Network (BIRN). IEEE Trans Inf Technol Biomed, 12(2):162-172.
65. Komatsoulis GA, Warzel DB, Hartel FW, Shanbhag K, Chilukuri R, Fragoso G, Coronado S, Reeves DM, Hadfield JB, Ludet C, Covitz PA (2008) caCORE version 3: Implementation of a model driven, service-oriented architecture for semantic interoperability. J Biomed Inform, 41(1):106-123.
66. Krupinski E, Nypaver M, Poropatich R, Ellis D, Safwat R, Sapci H (2002) Telemedicine/ telehealth: An international perspective. Clinical applications in telemedicine/telehealth. Telemed J E Health, 8(1):13-34.
67. Kuhn KA, Giuse DA (2001) From hospital information systems to health information systems. Problems, challenges, perspectives. Methods Inf Med, 40(4):275-287.
68. Kumar A, Rahman F (2004) System for wireless health monitoring. In: Rahman F (ed) Sensors for Industry Conference, 2004. Proceedings the ISA/IEEE, pp 207-210.
69. Kumar VS, Rutt B, Kurc T, Catalyurek UV, Pan TC, Chow S, Lamont S, Martone M, Saltz JH (2008) Large-scale biomedical image analysis in grid environments. IEEE Trans Inf Technol Biomed, 12(2):154-161.
70. Laure E, Fisher SM, Frohner A, Grandi C, Kunszt P, Krenek A, Mulmo O, Pacini F, Prelz F, White J (2006) Programming the Grid with gLite. Computational Methods in Science and Technology, 12(1):33-45.
71. Lin CC, Chiu MJ, Hsiao CC, Lee RG, Tsai YS (2006) Wireless health care service system for elderly with dementia. IEEE Trans Inf Technol Biomed, 10(4):696-704.
72. Linder JA, Ma J, Bates DW, Middleton B, Stafford RS (2007) Electronic health record use and the quality of ambulatory care in the United States. Arch Intern Med, 167(13):1400-1405.
73. Lukowicz P, Kirstein T, Troster G (2004) Wearable systems for health care applications. Methods Inf Med, 43(3):232-238.
74. Malan D, Fulford-Jones T, Welsh M, Moulton S (2004) CodeBlue: An ad hoc sensor network infrastructure for emergency medical care. Proc MobiSys 2004 Workshop on Applications of Mobile Embedded Systems (WAMES 2004), vol 6, Boston, MA, pp 12-14.

75. Mandl KD, Kohane IS (2008) Tectonic shifts in the health information economy. N Engl J Med, 358(16):1732-1737.
76. McMurry AJ, Gilbert CA, Reis BY, Chueh HC, Kohane IS, Mandl KD (2007) A self-scaling, distributed information architecture for public health, research, and clinical care. J Am Med Inform Assoc, 14(4):527-533.
77. McNitt-Gray MF, Armato SG, Meyer CR, Reeves AP, McLennan G, Pais RC, Freymann J, Brown MS, Engelmann RM, Bland PH (2007) The Lung Image Database Consortium (LIDC) data collection process for nodule detection and annotation. Academic Radiology, 14(12):1464-1474.
78. Miller RH, Miller BS (2007) The Santa Barbara County Care Data Exchange: What happened? Health Aff (Millwood), 26(5):w568-580.
79. Mizushima H, Uchiyama E, Nagata H, Matsuno Y, Sekiguchi R, Ohmatsu H, Hojo F, Shimoda T, Wakao F, Shinkai T, Yamaguchi N, Moriyama N, Kakizoe T, Abe K, Terada M (2001) Japanese experience of telemedicine in oncology. International Journal of Medical Informatics, 61(2-3):207-215.
80. Montagnat J, Jouvenot D, Pera C, Frohner A, Kunszt P, Koblitz B, Santos N, Loomis C (2006) Bridging clinical information systems and grid middleware: A Medical Data Manager. Challenges and Opportunities of Healthgrids: Proceedings of Healthgrid 2006.
81. Montgomery K, Mundt C, Thonier G, Tellier A, Udoh U, Barker V, Ricks R, Giovangrandi L, Davies P, Cagle Y, Swain J, Hines J, Kovacs G (2004) Lifeguard - A personal physiological monitor for extreme environments. Conf Proc IEEE Eng Med Biol Soc, 3:2192-2195.
82. National Cancer Institute Center for Bioinformatics (2007) The caBIG pilot phase report: 2003-2007. https://cabig.nci.nih.gov/overview/pilotreport.pdf. Accessed April 12, 2008.
83. National Cancer Institute Center for Bioinformatics (2008) National Cancer Imaging Archive 3.0 User's Guide. https://imaging.nci.nih.gov/ncia/faces/HTML/help/NCIA_Users_Guide.pdf. Accessed April 28, 2008.
84. National Electric Manufacturer Association (2008) DICOM homepage. http://medical.nema.org. Accessed March 14, 2008.
85. National Institutes of Health (2008) Re-engineering the clinical research enterprise - Overview. http://nihroadmap.nih.gov/clinicalresearch/index.asp. Accessed April 27, 2008.
86. Office for the Advancement of Telehealth (2001) 2001 Telemedicine Report to Congress. US Department of Health and Human Services. http://www.hrsa.gov/telehealth/pubs/report2001.htm. Accessed May 28, 2008.
87. Open Grid Forum (2006) The Open Grid Services Architecture, Version 1.5. http://www.ogf.org/documents/GFD.80.pdf. Accessed April 8, 2007.
88. Oster S, Langella S, Hastings S, Ervin D, Madduri R, Phillips J, Kurc T, Siebenlist F, Covitz P, Shanbhag K, Foster I, Saltz J (2008) caGrid 1.0: An enterprise grid infrastructure for biomedical research. J Am Med Inform Assoc, 15(2):138-149.
89. Pan TC, Gurcan MN, Langella SA, Oster SW, Hastings SL, Sharma A, Rutt BG, Ervin DW, Kurc TM, Siddiqui KM, Saltz JH, Siegel EL (2007) GridCAD: grid-based computer-aided detection system. Radiographics, 27(3):889-897.
90. Piana R (2006) Teleoncology extends access to quality cancer care. Oncology (Williston Park), 20(13):1714.
91. Poissant L, Pereira J, Tamblyn R, Kawasumi Y (2005) The impact of electronic health records on time efficiency of physicians and nurses: a systematic review. J Am Med Inform Assoc, 12(5):505-516.
92. Pryor T, Hripcsak G (1993) The Arden syntax for medical logic modules. J Clinical Monitoring and Computing, 10(4):215-224.

93. Ratnasamy S, Francis P, Handley M, Karp R, Shenker S (2001) A scalable Content-Addressable Network. Computer Communication Review, 31(4):161-172.
94. Ratnasamy S, Stoica I, Shenker S (2002) Routing algorithms for DHTs: Some open questions. Peer-to-Peer Systems, 2429:45-52.
95. Regenstrief Institute (2008) Logical Observation Identifiers Names and Codes (LOINC). http://loinc.org. Accessed May 16, 2008.
96. Ripeanu M (2001) Peer-to-peer architecture case study: Gnutella network. First International Conference on Peer-to-peer Computing, vol 101, pp 99-100.
97. Ritter J (2001) Why Gnutella can't scale. No, really. http://www.darkridge.com/~jpr5-/doc/gnutella.html. Accessed April 5, 2008.
98. Ruland CM, White T, Stevens M, Fanciullo G, Khilani SM (2003) Effects of a computerized system to support shared decision making in symptom management of cancer patients: Preliminary results. J Am Med Inform Assoc, 10(6):573-579.
99. Sarshar N, Boykin PO, Roychowdhury VP (2004) Percolation search in power law networks: Making unstructured peer-to-peer networks scalable. Fourth International Conference on Peer-to-Peer Computing, pp 2-9.
100. Schadow G, Mead CN, Walker DM (2006) The HL7 reference information model under scrutiny. Stud Health Technol Inform, 124:151-156.
101. Shortliffe EH (1998) Health care and the next generation Internet. Annals of Internal Medicine, 129(2):138-140.
102. Sidorov J (2006) It ain't necessarily so: The electronic health record and the unlikely prospect of reducing health care costs. Health Aff (Millwood), 25(4):1079-1085.
103. Simons WW, Mandl KD, Kohane IS (2005) The PING personally controlled electronic medical record system: technical architecture. J Am Med Inform Assoc, 12(1):47-54.
104. Smith B, Ceusters W (2006) HL7 RIM: An incoherent standard. Stud Health Technol Inform, 124:133-138.
105. Stead WW, Kelly BJ, Kolodner RM (2005) Achievable steps toward building a National Health Information infrastructure in the United States. J Am Med Inform Assoc, 12(2):113-120.
106. Stoica I, Morris R, Karger D, Kaashoek MF, Balakrishnan H (2001) Chord: A scalable peer-to-peer lookup service for Internet applications. Computer Communication Review, 31(4):149-160.
107. Taira RK, Bui AA, Kangarloo H (2002) Identification of patient name references within medical documents using semantic selectional restrictions. Proc AMIA Symp:757-761.
108. Thain D, Tannenbaum T, Livny M (2005) Distributed computing in practice: The Condor experience. Concurrency and Computation: Practice & Experience, 17(2):323-356.
109. The Freenet Project (2008) The Free Network Project. http://freenetproject.org. Accessed April 4, 2008.
110. The Globus Alliance (2008) Globus Toolkit Homepage. http://www.globus.org/toolkit. Accessed April 7, 2008.
111. van Halteren A, Bults R, Wac K, Konstantas D, Widya ND, Koprinkov G, Jones V, Herzog R (2004) Mobile patient monitoring: The MobiHealth System. J Info Tech Healthcare, 2(5):365-373.
112. Vannier MW, Summers RM (2003) Sharing images. Radiology, 228(1):23-25.
113. Vossberg M, Tolxdorff T, Krefting D (2008) DICOM image communication in Globus-based medical grids. IEEE Trans Inf Technol Biomed, 12(2):145-153.
114. Wang SJ, Middleton B, Prosser LA, Bardon CG, Spurr CD, Carchidi PJ, Kittler AF, Goldszer RC, Fairchild DG, Sussman AJ, Kuperman GJ, Bates DW (2003) A cost-benefit analysis of electronic medical records in primary care. Am J Med, 114(5):397-403.

115. Wears RL, Berg M (2005) Computer technology and clinical work: Still waiting for Godot. JAMA, 293(10):1261-1263.
116. Weinerman B, den Duyf J, Hughes A, Robertson S (2005) Can subspecialty cancer consultations be delivered to communities using modern technology? A pilot study. Telemed J E Health, 11(5):608-615.
117. Welsh M (2006) CodeBlue: Wireless sensor networks for medical care http://www.eecs.harvard.edu/~mdw/proj/codeblue. Accessed April 18, 2008.
118. Wiederhold G (1993) Intelligent integration of information. ACM SIGMOD. ACM Press, Washington, DC, pp 434-437.
119. Winter A, Brigl B, Wendt T (2003) Modeling hospital information systems. Part 1: The revised three-layer graph-based meta model 3LGM2. Methods Inf Med, 42(5):544-551.
120. Winters JM, Wang Y (2003) Wearable sensors and telerehabilitation. IEEE Eng Med Biol Mag, 22(3):56-65.
121. Wu WH, Bui AA, Batalin MA, Au LK, Binney JD, Kaiser WJ (2008) MEDIC: Medical embedded device for individualized care. Artif Intell Med, 42(2):137-152.
122. Wysocki WM, Komorowski AL, Aapro MS (2005) The new dimension of oncology: Teleoncology ante portas. Critical Reviews in Oncology/Hematology, 53(2):95-100.
123. Zerhouni E (2003) Medicine. The NIH Roadmap. Science, 302(5642):63-72.

Chapter 4
Medical Data Visualization: Toward Integrated Clinical Workstations

ALEX A.T. BUI AND WILLIAM HSU

As our ability to access the abundance of clinical data grows, it is imperative that methods to organize and to visualize this information be in place so as not to overwhelm users: increasingly, users are faced with information overload. Moreover, the manner of presentation is fundamental to how such information is interpreted, and can be the turning point in uncovering new insights and knowledge about a patient or a disease. And of course, medical imaging is itself an inherently visual medium. This chapter presents work related to the visualization of medical data, focusing on issues related to navigation and presentation by drawing upon imaging and other disciplines for examples of display and integration methods. We first cover different visual paradigms that have been developed (*e.g.,* icons, graphs), grouped along dimensions that emphasize the different types of data relationships and workflow. Subsequently, issues related to combining these visualizations are given[1]. As no single graphical user interface (GUI) can accommodate all users and the spectrum of tasks seen in the healthcare environment, the ultimate goal is to create an adaptive graphical interface that integrates clinical information so as to be conducive to a given user's objectives: efforts in this direction are discussed. Throughout, we describe applications that illustrate the many open issues revolving around medical data visualization.

Navigating Clinical Data

Many of the visual components seen in today's electronic medical records (EMRs) are a direct translation of the data's appearance in paper-based charts, albeit adapted to handle user interaction in a computer environment. As the medical record spans a growing plethora of different types of data – nominal, numerical, textual, imaging, and diagrammatic – a mix of these visual elements is often used within a single patient record. Arguably, the result has been a lack of new, persuasive interfaces that support, if not augment, the process of viewing clinical information. Applying ideas in information visualization, creative presentation techniques will help enable users to go beyond traditional patterns of thinking with medical information and the EMR.

The science of information visualization encompasses a broad range of topics, from issues of perception, cognition, human-computer interaction (HCI), and the visual communication of ideas; through to graphical methods for explaining data and promoting knowledge discovery. The general theory and principles of information visualization are well-described in seminal texts by Tufte [169-171], Shneiderman [156], and others. [32] provides some early perspective on information visualization in relation to medicine. Here, we concentrate on graphical representations as they relate to medical information. This section starts with a classification of the basic graphical widgets used to present singular, structured elements of clinical data; these visual

[1] In the model-view-controller (MVC) framework, although the organization of data (*i.e.,* the model) helps drive how information is presented (*i.e.,* the view), we leave that discussion to Chapter 7.

A.A.T. Bui and R.K. Taira (eds.), *Medical Imaging Informatics*,
DOI 10.1007/978-1-4419-0385-3_4, © Springer Science + Business Media, LLC 2010

representations are the mainstay of today's displays. Building from these basic elements, approaches are covered that support the graphical comparison and discovery of relationships between (heterogeneous) data elements.

Elements of the Display

Many taxonomies have been proposed to group visual methods based on data type, structure, data dimensionality, and user task. Notably, [160] describes an object-oriented (OO) categorization of medical data visualization, which we adapt in this section to illustrate increasing levels of graphical abstraction. The lowest level of sophistication involves the sentential presentation of textual and numerical data in a relatively non-interpreted fashion: *lists* and *tables* represent this category of presentation. Next, for quantitative and statistical information, understanding the data involves comparisons and trending: visually, different types of *plots* and *charts* are used to emphasize relative values, with the manner of such displays largely influencing interpretation. Expounding more conceptual relations, *graphs* and *trees* provide a further degree of abstraction. Finally, the top echelon of this hierarchy comprises the visual abstractions brought about through *pictograms*, which collectively aim to be graphical surrogates for real-world entities and concepts seen in the clinical environment.

Lists and tables. Text and numerical data are the predominant component of the patient record. The most familiar method of displaying sequences of related information, *lists* are enumerated or delineated sets of textual and/or numerical items (Fig. 4.1). Typically, the entries in a list are short and concise, presenting a key point or summary that can be quickly read by the user. Examples from clinical practice include an individual's medical problem list; physician worklists (*e.g.,* imaging studies awaiting interpretation); and a patient's set of current medications. Aside from a straightforward display of list items, today's GUIs show lists in a number of different ways, imposing different modes of interaction and selection. For example, combination boxes (combo boxes) enforce selection of a single item, while checkboxes allow for multiple items from a group of related entries. List entries can serve as hyperlinks, allowing a user to access further information. Lists are generally univariate in that a single concept is being communicated per item. *Tables* (also referred to as *grids*) can be seen as extension of lists to present multivariate information, where each row in a table in a single entity, and each column is an attribute of the entity. An archetypal use of tabular views in medicine is the comparison of different lab panel values over a set of dates (Fig. 4.1, right) in flowsheets. Adaptations on tables include colorization and re-orderable matrices.

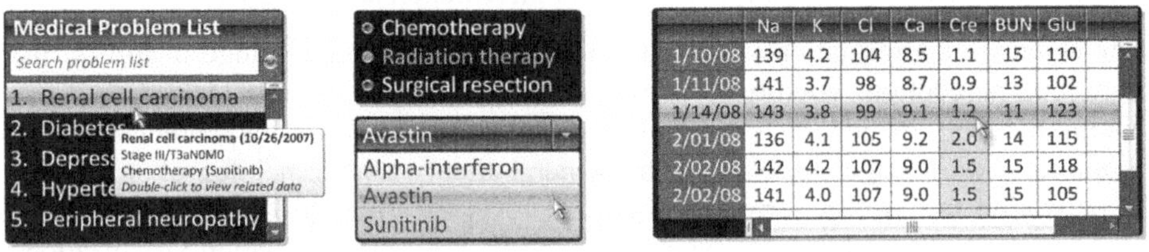

Figure 4.1: Lists and tables are a common graphical component used to present clinical data. From left to right: a patient's medical problem list is shown using a searchable and scrollable list, with tooltips presenting more specific information; radio buttons, checkboxes, and combo boxes are used to provide selection capability from a fixed set of options; and table are often used to present multivariate data for comparison, such as with lab panels.

In the first variant, the range of values for a variable is mapped to a color spectrum so that cells are filled with a color rather than a number. The second variant enables the rows and columns to be sorted or arbitrarily arranged to facilitate pattern discovery. *Heatmaps* use both colorization and re-ordering [50], and are widely used to visualize large quantities of data such as in the analysis of expression data from DNA microarray hybridization experiments.

Plots and charts. Information presented within tables, although precise, fail to foster rapid interpretation of subtle trends, especially over a large number of data points. Thus, the next level of graphical abstraction seen with medical data involves *plots* and *charts* (the terms being used interchangeably), wherein the relative nature of numerical data is contrasted to illustrate changes in values or comparative differences. Data presented in tables can be transformed into a suitable chart to visually accentuate patterns. Elementary graphical charts include:

- Line and scatter plots. A common graphical representation is the 2D *line plot*, wherein one axis (*e.g.*, the *y*-axis) represents the quantitative value of interest (the dependent variable), and the second axis (*e.g.*, the *x*-axis) is the space over which the value is sampled (the independent variable, *e.g.*, time). For instance, an electrocardiogram (ECG) is representative of a line plot, where the amplitude of an electrical signal is charted over time. Likewise, a given laboratory value (*e.g.*, blood glucose) may be plotted to visualize increasing/decreasing trends (Fig. 4.2a). Pediatric growth charts are also indicative of line plots. Care must be given as to how line plots are used to convey information. For instance, dependent on the time scale, interpolating values between subsequent points may be misleading. Consider two blood glucose values taken on 1/1/2007 and 1/31/2008 of 140 and 102 mg/dL, respectively: drawing a line between these two points would suggest that the blood glucose values have decreased over time – but in fact, between these dates the blood glucose may have fluctuated greatly. *Scatter plots* are a generalization of the line plot, often used in research studies to find potential associations/correlations between two variables over a population (*e.g.*, linear relationships, clusters; Fig. 4.2b); again, one variable is explanatory or controlled, and the second variable is the response or observation. Dimensional scaling techniques (*e.g.*, principal component analysis, PCA) can be used to reduce the number of attributes involved, thereby mapping a multivariate visualization problem to 2D where patterns may be more evident. If no association exists between the variables, no discernible visual pattern or trend is seen in the scatter plot.
- Bar charts and histograms. Another well-recognized type of plot is the 2D *bar chart*, where the length of a rectangle is used to proportionally depict the value of a given category (Fig. 4.2c); multiple categories are then compared. Additionally, parallel comparisons between datasets can be visualized in a bar chart, facilitating intra-category comparison. To demonstrate, in a clinical trial for a drug a bar chart may be used to show side effects (*i.e.*, categories) with the percent of individuals affected. Parallel bars may then be placed adjacent to compare these individuals versus a control group (*e.g.*, placebo). *Histograms* are a specific type of statistical bar chart, wherein the categories represent tabulated frequencies of a given value (or values over uniformly divided ranges). An image histogram, which plots the number of pixels with a given intensity value, is representative of this information graphic. For histograms, the choice of discretization can greatly change the understanding of the data. While variations of bar charts exist employing different

graphical techniques (*e.g.,* 3D bar charts, stacked bar charts), the overall complexity of these presentations and a user's ability to correctly interpret the data can outweigh their utility.

- Pie charts. A *pie chart* aims to provide a sense of proportion by dividing a circle into wedges, representing an object and its constituent breakdown. While used frequently, many studies show that pie charts can be harder to correctly interpret [54, 158] and their use is discouraged [171]. One exception to the pie chart paradigm is a variant, the *polar area diagram* [37]: rather than use the angle of a wedge to convey percentage, wedge angles are equal and the radius varies in proportion to the amount. The end effect of a polar area diagram is that the pieces project outwards, making similar quantities easier to compare (Fig. 4.3).
- Radar charts. Less widespread are *radar charts* (also called *circular* or *spider charts*), which compare three or more quantitative variables along multiple axes (Fig. 4.2d). The axes radiate outwards from the center of the plot, along which the data values for each variable are drawn on a shared scale. However, variants of radar charts have been defined to take advantage of shape and area by connecting the plotted points. [75] introduced this concept for clinical labs, with normalized values for laboratory data charted as a shape. Ideally, if the lab values are balanced, the shape will conform to the overall geometry of the radar plot (*e.g.,* for a lab panel with six tests, the overall shape should resemble an equisided hexagon);

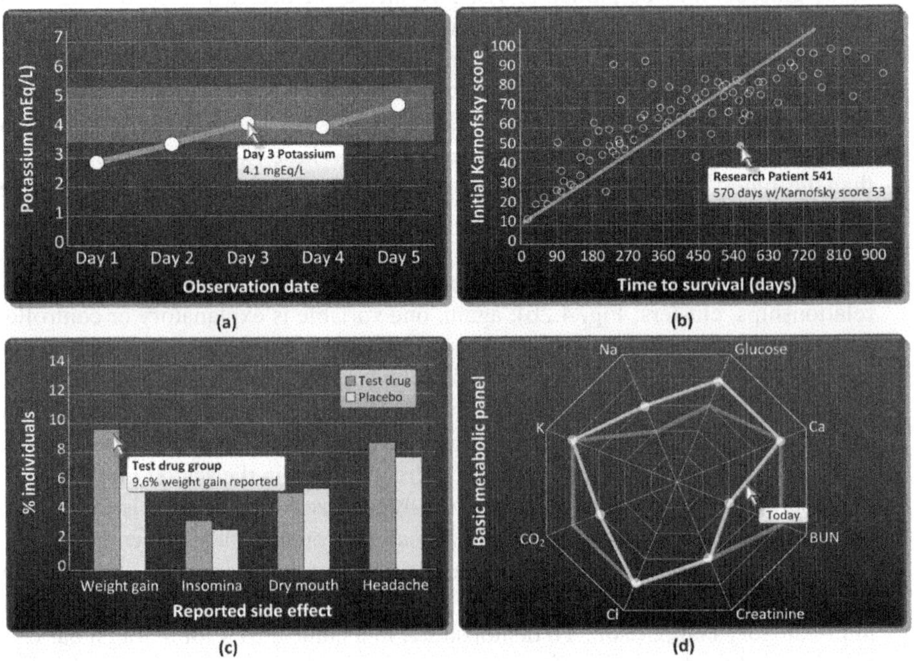

Figure 4.2: Variations of graphical plots and charts, used to present quantitative data. Tooltips can be used to provide specific information on a visualized data point. **(a)** A typical line plot showing a potassium lab over five sequential days. The horizontal axis of a line plot is often a controlled variable, whereas the vertical axis is the observation. **(b)** Scatter plots are a generalization of line plots, comparing two variables to visually ascertain associations. **(c)** Bar charts and histograms are used to compare categorical values. **(d)** Radial charts compare groups of related values for relative proportion.

Figure 4.3: Percentage of labs testing positive for influenza as compiled by the US Centers for Disease Control (CDC), from October 2007 through May 2008 [173]. **(a)** A pie chart of the total number of positive tests per month. **(b)** The same data, represented in a polar area diagram, where each wedge is an eighth of the circle. Although the pie chart allows the reader to establish overall proportions, the polar area diagram provides a better sense for comparison. For instance, it is easier to distinguish subtle differences between February and March based on the small change in radius.

in contrast, skewed labs distort the overall shape, allowing the viewer to quickly identify which axis (*i.e.*, lab) is discrepant and the direction of imbalance (low values gravitating towards the center of the plot, high values being on the edge). Adaptations of the radar graph also use area to compare different observational sets (*e.g.*, two time points): the overlap of an area and trends can be seen.

Unlike lists and tables, where different GUI representations and interactions have been developed, plots are typically displayed as is given users' learned behaviors in their interpretation. Interaction with these graphical elements regularly involves tooltips, filtering, and hyperlinks to the data used to construct the chart.

Graphs and trees. Plots are intended to express numerical data; in contrast, *graphs* and *trees* are designed to demonstrate relations between concepts. In this section, the terms graph and tree refer to the formal constructs defined in computer science, as opposed to more generic pictorial constructs. A graph is a network of objects, comprised of vertices (nodes) and edges, and is said to be *directed* if the edges are arrows defining a path between nodes. A tree is a directed acyclic graph (DAG) in which each node only has one parent.

Apart from their use in evidence-based medical guidelines as *flowcharts* illustrating decision pathways (*e.g.*, eligibility criteria for a clinical trial, study design; Fig. 4.4), graphs are generally not seen in clinical practice. Originally intended for documenting process flow, a well-defined symbol set and visual syntax has evolved for communicating states, actions, and conditional/decision points using flowcharts. The Guideline Interchange Format (GLIF) [19] and other guideline models [130] use flowcharts for the graphical expression of clinical algorithms. Flowcharts have also been used to educate and guide patients [36]. Outside of the clinical arena, conceptual graphs and (probabilistic) graphical models have a longstanding history within medical informatics, being used to represent ontologies and as a part of decision-support frameworks (*e.g.*, Bayesian belief networks (BBNs), hidden Markov Models, Petri nets, etc.). The visualization of these knowledge representations is tackled in Chapter 9.

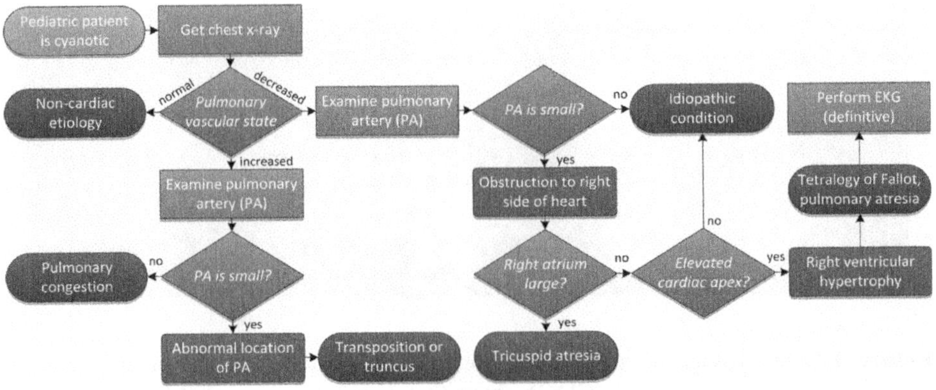

Figure 4.4: An example flowchart for diagnosis of a pediatric patient presenting with cyanosis. Flowcharts are used to illustrate clinical guidelines and decision-support.

Trees are used to illustrate connections between entities where the entire structure of a hierarchy and its encompassing relations are relevant: parent-child relationships (*e.g.*, is-a inheritance); siblings (objects at the same level in the hierarchy); and clusters are visually portrayed, usually with the root of the tree being the most general concept and levels further out (*i.e.*, toward the leaves) becoming more specialized. Information arranged as nested lists are amenable to tree presentations; hence, controlled vocabularies and clinical coding schemes are often shown as trees [17, 152]. A case in point, the ICD-9 (International Classification of Diseases, 9th revision) classification system is organized by increasing specificity of disease (Fig. 4.5a): acute pulmonary heart disease is a broad category (ICD-9 415); while acute cor pulmonale (ICD-9 415.0) and pulmonary embolus and infarction (ICD-9 415.1) are more detailed causes. Other clinical examples of trees include: grouped medical problem lists (*e.g.*, symptoms and diseases by anatomical region); composite lab tests (*e.g.*, a metabolic panel); an imaging study and its constituent series; and structured reports, wherein a document section may consist of multiple sub-parts. *Dendrograms* are a specific type of graphical tree used to envisage related groups; taxonomies and genomic analyses involving hierarchical clustering algorithms are indicative of this graphical element (Fig. 4.5b). Though not seen with clinical data, *phylogenetic* and *ultrametric trees* are noteworthy dendrogram specializations (Fig. 4.5c) used in evolutionary biology to demonstrate speciation, the latter expressing evolutionary time as a function of a branch's length.

The visual elements making up graphs and trees in a GUI are straightforward: textual labels, sometimes coupled with a geometric shape (*e.g.*, a box, ellipse, etc.) or an intersection (*e.g.*, a corner formed by two lines), are used to represent a node and the concept; and lines, arcs, and arrows link the nodes. Trees widgets normally show vertically nested lists with collapsible branches that enable the user to interactively select and view portions of a hierarchy. But a significant challenge comes about in the display of graphs and trees when there are an extensive number of nodes or edges. Visualizing and navigating large and/or highly inter-related datasets is problematic for several reasons: limited (screen) space; difficulty in identify and traversing paths between nodes; and overall readability are recognized issues with graphs. As such, numerous approaches to graph and tree layout have been proposed over the years [43]; [63, 73] provide a thorough overview and a discussion of layout issues. Key concepts for graph and tree presentation are summarized below:

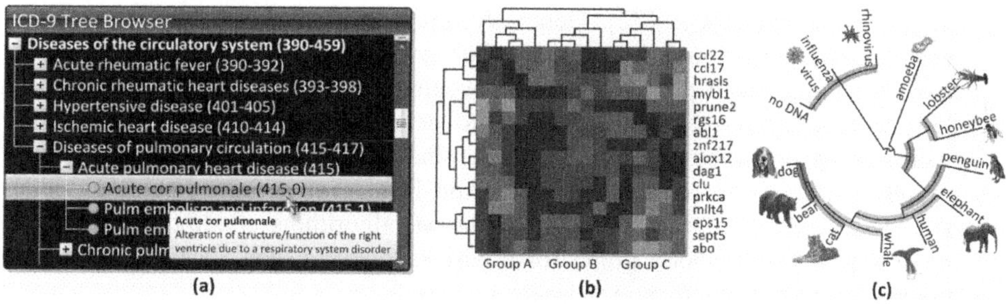

Figure 4.5: Different graphical representations of trees. **(a)** A standard GUI widget shows a tree as a collapsible/expandable, vertical nested list. **(b)** A dendogram is shown along with a heatmap from a gene microarray analysis. The dendrograms on the top and left of the figure show how different genes and test subjects are clustered together. **(c)** A circular phylogenetic tree based on [35], illustrating evolutionary paths.

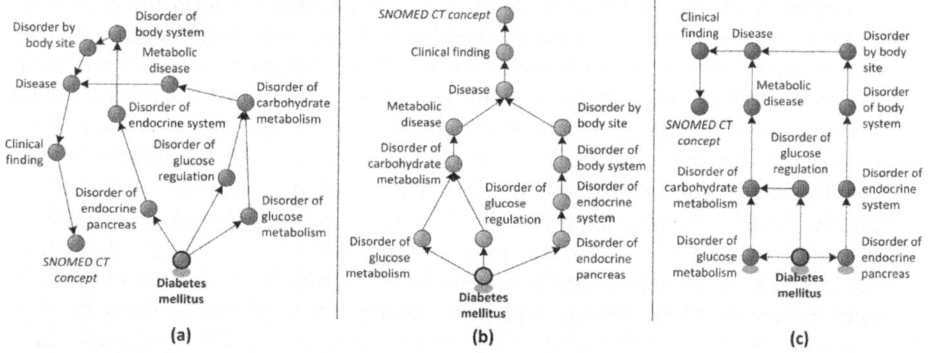

Figure 4.6: Graphs of the *is-a* inheritance hierarchy for the concept *diabetes mellitus* in the SNOMED-CT (Systematized Nomenclature of Medicine, Clinical Terminology) vocabulary. **(a)** A force-directed graph for the concepts, using the number of children terms as the edge weight. Layout in this format does not consider edge crossings, but helps illustrate the relative strength/closeness of relations. **(b)** A top-down layout of the same concepts, better showing the inheritance paths. **(c)** A grid-based layout.

- <u>Graph layout</u>. Several heuristic features have been put forth to capture the visual aesthetics of a graph, including: minimizing the number of edge crossings and the number of bends on a given edge; minimizing edge length and its variance; maximizing the angle between edges sharing a node; and symmetry (*e.g.,* reflective, rotational). Several formal studies have found varying degrees of importance to these features, being largely dependent on the intended usage of the graph and the application domain. Different algorithms for computing graph layout have been suggested, with four broad categories: *force-directed* (spring-embedder) methods; *spectral* layouts; *hierarchical* and *geometric* layouts; and *planarity* methods (Fig. 4.6). Force-directed methods [46] use the strength of a relation between two nodes to compute edge length, with hypothetical "springs" linking each pair. The relative position between all nodes in the graph is thus a function of the overall "tension" in the spring network, attempting to minimize this value. Spectral layout methods [67] use eigenvectors derived from the matrix representation of a graph to compute x-y coordinates (*e.g.,* an adjacency or Laplacian matrix). For directed graphs, hierarchical (top-down) layouts can be used, where each node is assigned to some layer (*e.g.,* via a topological sort) and then rendered such that nodes in the same layer are drawn at the same vertical level [161]. Prefuse provides a library of graph layout techniques [72]. Geometry has been exploited to arrange graphs: orthogonal layouts employ a grid to align nodes and edges; and radial layouts have been applied to place nodes in concentric circles. For instance, DocuBurst [38] uses a radial space-filling layout to delineate relationships among report words using an ontology like WordNet. Finally, for the special case of planar graphs[2], methods for embedding the graph in a plane are used to place nodes [31, 55]. In point of fact, some graph layout algorithms attempt to reduce the node/edge set to a planar graph for a base layout and then successively re-add nodes/edges to complete the rendering.

- Tree layout. In addition to using the techniques for graph layout, specialized methods for trees have been created to address the need for visualizing large numbers of data points. A *treemap* (Fig. 4.7) and its derivatives take a rectangular space and progressively partition the area into smaller (nested) rectangular regions based on the structure of the hierarchy (*e.g.,* number of children) or a value associated with the node: child nodes are thus contained within the parent rectangle [154]. Principal to treemap algorithms are the competing criteria of aspect ratios (to facilitate visual comparison) and predictable locations (to enable discovery). The proportionate sizes of treemap regions can be compared visually. As an example, DBMap [118] uses a treemap to navigate a large collection of biomedical data on neurological diseases such as epilepsy and brain neoplasms. The treemap algorithm is capable of displaying over 4,000 nodes in an average of 200 pixels, allowing for a lot of information to be displayed in a small area. Alternatively, *cone trees* [62] and *disc trees* bring into play a 3D rendering perspective, arranging nodes along a given geometry: a parent node is at the apex or center of the shape, and the children are organized around the base/

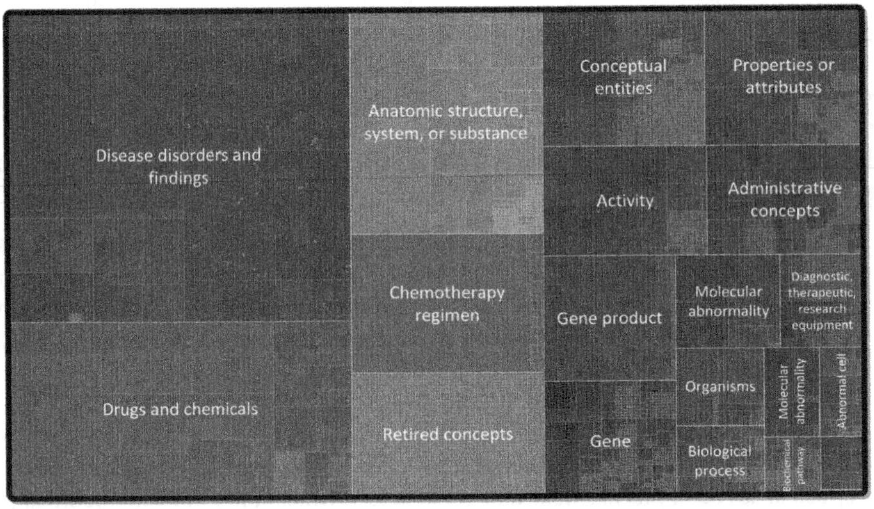

Figure 4.7: A treemap of the April 2008 NCI Thesaurus, constructed of 65,986 terms and parent-child relationships [123], rendered using *prefuse* [72]. Search tools are often provided with treemaps; in this example, nodes with the word *lung* are highlighted.

---

[2] A *planar graph* is a graph that can be drawn on a (2D) plane such that no edges cross. This problem has been well-studied in computer science and graph theory, with many linear-time algorithms for testing if a given graph is planar.

circumference. *Hyperbolic trees* [122] take advantage of a non-Euclidean geometry (*i.e.,* hyperbolic geometry) to position nodes, and have been applied to phylogenetic trees and other bioinformatics data [78, 109]; Fig. 4.8 shows a spanning tree created from the National Cancer Institute (NCI) Thesaurus using this visualization method.

Pictograms. A higher level of visual conceptualization comes about in considering the use of *pictograms* to represent clinical concepts. A pictogram is defined as a graphical symbol that represents a concept or entity. The use of pictograms with medical data can be seen fourfold:

1. Icons. *Icons* are small pictograms, and are a familiar component of modern GUIs representing an action or data object. A classic case of the use of icons in routine clinic review is pain assessment tool using a face scale, a variant of the visual analogue scale (VAS) that uses a spectrum of facial expressions on a Likert-type scale to indicate levels of discomfort and/or pain; the pictograms can be linked to descriptive phrases and the numerical scale (Fig. 4.9). It has been observed that the interpretation of icons can be subject to personal and cultural biases [34, 129], thus making the use of icons across populations complex. In certain cases, the graphic is universally understood [141]; but in domain-specific scenarios, individuals may initially need assistance in understanding the suggested visual cue [92]. Icons can also be derived from an object's content. For instance, TileBar [71] takes in a set of documents and user-specified terms to generate multiple rectangular bars that are divided into smaller, color-coded squares. Each square represents individual terms; properties such as relative document length, query term frequency, and query term distribution are encoded through the bar's visual appearance. TileBar thus provides a quick pictorial representation of document content and relevance to keywords (*e.g.,* disease name).

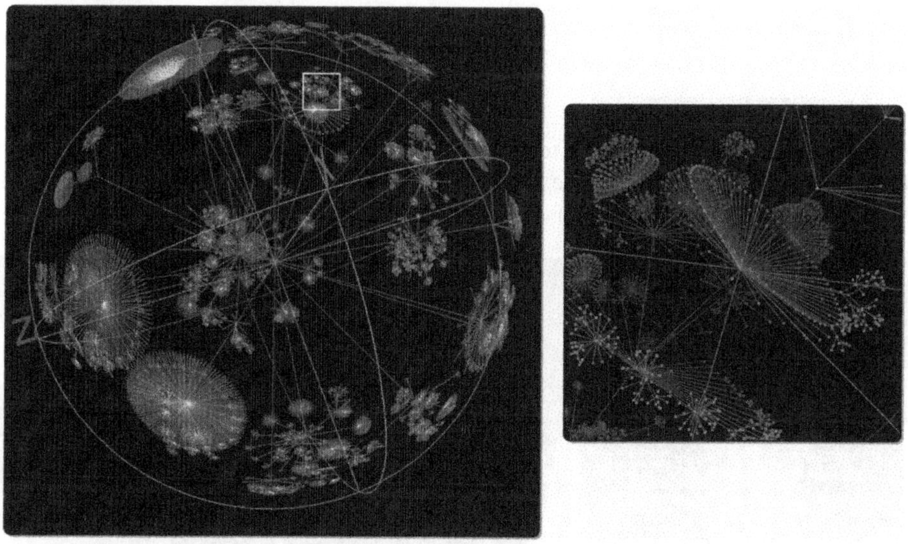

Figure 4.8: 3D hyperbolic tree visualization for the same NCI Thesaurus dataset. As can be seen, distribution patterns are apparent, as there are clear clusters of terms. The right side of the figure shows a zoom-in of the highlighted region on the left. The visualizations were made using the *Walrus* software package [41].

2. Maps. *Maps* are larger pictograms, being mainly concerned with a spatial framework (*e.g.,* an anatomical atlas, such as the human body; Fig. 4.10a-b). For instance, maps are used as surgical drawings to document the planned approach, and the pre- and post-operative state of the region of interest. Whole-body anatomical drawings are also used to quickly demonstrate affected or symptomatic areas. A further discussion of anatomic maps is given subsequently from the perspective of emphasizing spatial relationships. Maps can also used to represent high dimensional data, such as the contents of a clinical report. [104] abstracts a text document into a Kohonen's feature map using visual cues such as dots, clusters, and spatially-related areas to represent the unique concepts (*e.g.,* disease, drugs, chemotherapy), the frequencies of word occurrence in titles and frequency of word co-occurrence respectively.
3. Diagrams. *Diagrams* are illustrated figures that present an abstraction or conceptual metaphor (*e.g.,* a timeline for a therapeutic regimen; a structural depiction of a medical device; a velocity field diagram showing hemodynamic analysis in a blood vessel; Fig. 4.10c). Although a clinical diagram may be anatomically based, the primary difference between a map and a diagram is the intended communication of a spatial vs. non-spatial relationship, respectively.
4. Images. Lastly, medical *images* are the final category, being a physical representation of the real-world based on either light (*e.g.,* optical photography, such as seen with dermatology and post-operative surgical procedures); radiation (*e.g.,* computed tomography, nuclear medicine); or other physical value (*e.g.,* hydrogen nuclei interaction/relaxation, such as under magnetic resonance). The rendering can be a 2D projectional or cross-sectional image, showing spatial relationships (*e.g.,* between a tumor and normal tissue); a 3D reconstruction; or a 4D representation (*e.g.,* an animated 3D visualization showing changes over time).

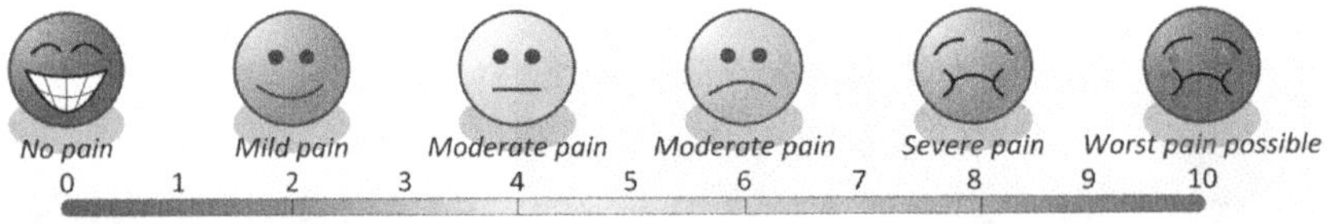

Figure 4.9: A pain assessment tool using a face scale and verbal descriptors. The facial expressions help to provide a visual context for a Likert scale. In this example, color is also used to provide a sense of continuum.

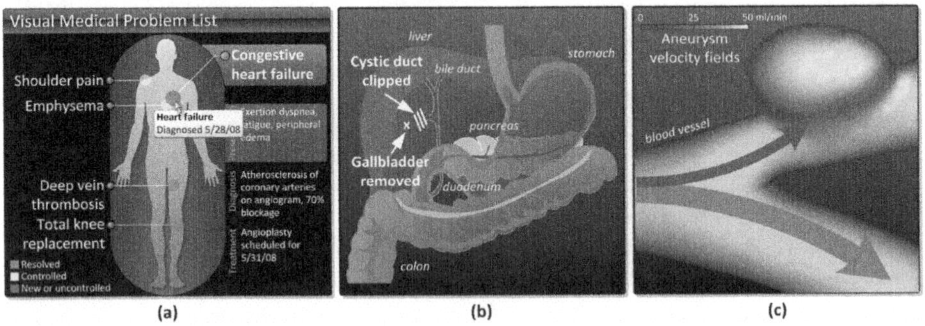

Figure 4.10: **(a)** Example map using a human body as a visual index for medical problems. **(b)** A surgical illustration using an anatomical map to demonstrate post-operative patient state following gallbladder removal. **(c)** Although showing anatomy, the primary goal of this diagram is to show the velocity of blood around an aneurysm region in a 3D surface rendering.

The above categorization of clinical visualizations is only intended to provide insight into some widespread graphical elements used to communicate clinical concepts and data: it is by no means comprehensive. The reader is referred to [63, 70] for a more thorough discussion and examples of general information graphics and visualization.

Visual Metaphors: Emphasizing Different Relationships

The representations described thus far illustrate individual data elements from a medical record. However, clinicians need more sophisticated visualizations that build from these elementary abstractions to uncover insights into disease processes and patient treatments. Indeed, visualization transforms abstract data into a form that amplifies cognition and discovery [27]. As clinical data repositories grow, different visualizations are needed to help identify important trends (*e.g.,* increasing blood pressure) and relationships (*e.g.,* an adverse drug interaction that is causing the increase in blood pressure). This section addresses how various information visualizations have been applied to accentuate associations amongst medical data. Recent innovations have led to novel visual metaphors, tailored to different types of data and tasks. Collectively, these approaches aim: 1) to make explicit the relationships present within the medical record; 2) to present clinical data in more intuitive, easy to understand formats; 3) to magnify any subtle diagnostic, therapeutic, or other management aspects in a patient record that would otherwise be difficult to notice; and 4) to prevent information overload, by simplifying displays and/or facilitating navigation and search. In addition to considering multidimensional visualizations, the remainder of this section loosely adopts a widely cited visual taxonomy [155], further categorized by the three relationships important to understanding medical data: temporal, spatial, and causal.

Temporal Representations

Change is an innate part of a disease process. The capacity to diagnose patients, and to reach conclusions about interventions, comes from an ability to compare two or more time points, noting changes between observed states. Thus, it is natural that temporal relationships are a heavily investigated component of the patient record – and consequently, the visualization of an individual's clinical history. [6] provides a current review of general temporal visualization, with a time-specific taxonomy suggested in [5]. The predominant metaphor with clinical data is the *timeline*. Abstractly, timelines use one axis (usually the horizontal) to express the passage of time; and an orthogonal axis represents the variable of interest and its range. Data elements are plotted on this 2D space to visualize the chronological sequence of events: quantitative values are shown as geometric objects, while more abstract entities (*e.g.,* a document) are depicted as icons. Timelines can be used to represent a mix of both point events (*i.e.,* occurring at a single time point) and interval events (*i.e.,* occurring with distinct start and end points).

As mentioned earlier, lab plots and trending displays are customarily time-based illustrations. [42] first introduced the concept of clinical timelines for non-laboratory data, presenting a framework that allows a user to perform four types of manipulations on a chronology: *slice* (removing events from the start/end of a timeline), *filter* (removing events that do not satisfy some condition), *add* (creating a new event within a timeline), and *overlay* (combining two timelines' events into a single timeline). Two rudimentary operations were defined for visual processing of timelines: *align*, to match up timelines around events; and *scale*, to adjust rendered temporal scales. This initial work is expanded upon in the well-known project, LifeLines [134]: for each medical condition, LifeLines uses a combination of icons, colors, and interactivity to delineate relationships among events. LifeLines2 [177], the successor to the original project, adds a set of operators that allow users to dynamically reorganize the presentation of data based on a certain feature. For instance, all of a patient's events (*e.g.*, physician encounters, high blood pressure) may be aligned based on the proximity to an event of interest (*e.g.*, heart attack). [12] presents other approaches that extend the timeline concept to encode information in visual features: in this work, the blood oxygen saturation is encoded in the height of the timeline while colors indicate values outside normal levels. In general, the issues encountered with timeline visualizations include: 1) how to (visually) identify trends within the data, automatically highlighting patterns, cycles, or progressive changes in values; and 2) how to optimize temporal granularity given a limited amount of display space. [33] also explores the impact of different visualizations (springs, elastic bands, and paint strips) to emphasize the strength of interval-based temporal relationships.

Trending and temporal abstraction. Time series analysis is a well-established area of research, applicable to a wide array of clinical observations to reveal potential patterns of interest. These methods, in addition to temporal abstraction techniques, have been coupled with visualizations to identify abnormal trends in laboratory values [11, 39, 65]. KNAVE II [91, 111, 151] is a well-cited example for pairing temporal analysis with an interactive, exploratory visualization for discerning trends. Graphical queries posed by the user are automatically characterized to extract time span, parameters, values, and context. Results are presented using a series of timelines representing different concepts (*e.g.*, white blood-cell counts). To call attention to such temporal trends and identified temporal abstractions, color schemes have been used: [74] evaluated this method for signifying lab trends, finding that novel color-coded visualization results in faster interpretation relative to traditional tabular views. For period occurrences within serial data (*e.g.*, monthly episodes), *spiral graphs* [179] have been proposed (Fig. 4.11a): timestamps are mapped on a spiral path, with graphical line features (color, texture, thickness) and icons denoting events. The visual

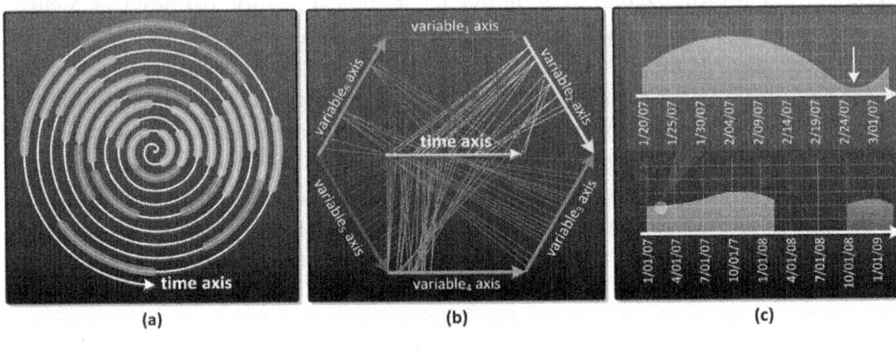

Figure 4.11: **(a)** Example of a spiral graph. The time axis is drawn as a spiral so that periodicity is made visible. In this example, the green and yellow events align and are thus periodic, but no discernible pattern is seen with the event seen in blue. **(b)** An example of the TimeWheel representation. The time axis is shown in the middle; events are then plotted to the edges of the polygon (variable axes). **(c)** Dependent on how values are represented over different scales, trends may be removed. In this case, a dip in the value towards zero (denoted by an arrow in the top plot) is not visible on a wider timescale because of averaging effects (as seen in the bottom plot).

co-occurrence of these features then allows one to quickly identify patterns. An interactive, radial layout of axes for multivariate timestamped data is described in TimeWheel (Fig. 4.11b): the time axis is placed in the center of polygon; and the sides of the polygon represent a variable's axis, with colored lines connecting the time axis to the edge of the polygon. Judging two sequences of events in TimeWheel is performed by arranging two axes horizontal to the time axis but opposite to each other (*i.e.*, above and below the time axis) [166]. VIE-VENT takes a non-graphical approach to highlight trends, applying domain knowledge to classify patterns and determine qualitative descriptors [117]. Time series data can also be transformed into other spaces (*e.g.*, wavelet) to aid pattern discovery [181]. Markedly, the problem of trend visualization and recognizing changes over time is not limited to clinical data; significant work has also been done in other areas, including detecting changes within documents [174], stock markets, and meteorology datasets (*e.g.*, climate and weather patterns).

Temporal granularity. A perennial challenge in timeline visualization is the limited amount of display space along the time axis. Moreover, many time series visualizations are based on a regularly occurring (and thus evenly spaced) data value. Unfortunately, data within the clinical record is unevenly spaced, being recorded at different scales of temporal granularity and frequency (Fig. 4.11c). For instance, an intensive care unit (ICU) may capture information every hour, whereas an outpatient physical exam may only occur annually: selecting too small a time unit will generate overly wide displays and/or significant empty gaps for which no data is present; conversely, choosing too large a time unit may result in lack of (visual) discriminatory space between events and trends diminished. As the longitudinal EMR becomes a reality, users will be faced with longer histories and data, spanning not just a few months or years, but potentially decades of patient information. Several methods can be applied to the granularity problem [9]:

1. Linear scales. The simplest approach, the timeline is presented as a linear scale with all elements shown at a given temporal granularity that enables the entire time series to be displayed. The user can then interact with the timeline, zooming in/out on specific sections, and the temporal resolution is changed on demand to provide more/less information accordingly. Linear time scales can also be coupled with deformation techniques: a user interaction (*e.g.*, mouse over, selection) triggers zoom-in behavior on a portion of the timeline (*e.g.*, 1D fisheye distortion, accordion panels). Linear scales try to preserve a total sense of timeline duration.
2. Non-linear scales. These methods space the time units dependent on an analysis of the data's distribution, thus attempting to optimize the use of space. For example, some algorithms give time spans with more data points wider regions, while shrinking the width of regions with less data. In other situations, where the data sampling follows a known curve (*e.g.*, logarithmic), the time axis can follow a similar scaling. The extreme case of non-linear scaling is the use of event-based indices, where each sequential element is uniformly spaced across the time axis (regardless of the interval between events). A consequence of these non-linear approaches is that the time axis length/scale is no longer representative of the overall duration of the timeline.
3. Re-sampling. Finally, it is possible to resample the dataset to create an evenly spaced distribution, thus taking advantage of linear scaling and existing time series rendering algorithms.

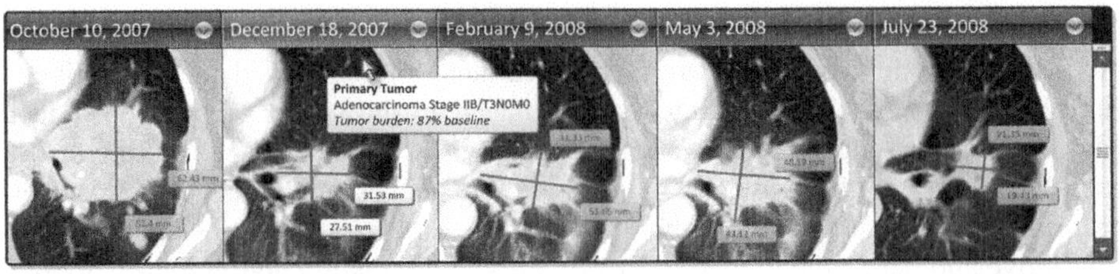

Figure 4.12: Example imaging timeline. Sequential CT studies for a lung cancer patient are illustrated, allowing the viewer to determine how measurements were taken previously, and assess the overall change in tumor size in response to treatment.

In the case of visual scaling and re-sampling, it is important to note that the scheme by which data values are interpolated and/or aggregated into new values may not preserve underlying trends.

Imaging timelines. In addition to the above works, researchers have examined the use of imaging to produce a visual history. Given the central role of imaging in cancer to diagnose and assess response to treatment, a focus of these efforts has been to document the variation in tumor size over time (Fig. 4.12). The Oncology Imaging Timeline (OITL) allows the user to capture sentinel images and bidirectional measurements used in computing changes in tumor burden (*e.g.*, made by a radiologist) in the context of a given chemotherapeutic treatment for lung cancer [2, 23]. Primary tumors and other regions of interest followed by the clinician are aligned in a grid; visual comparison across a row easily details how a given tumor was previously evaluated, and the change in a given tumor. [184] provides a similar framework for lung nodules. Fundamentally, a similar approach can be applied to any disease reliant on imaging-based assessment over time. The TimeLine project, an outgrowth of OITL, provides a problem-centric, chronologic view of medical data; this effort is described in more detail in an ensuing section.

Animation methods. Timelines are typically static in that the graphics do not alter to explicitly illustrate time: each time point is represented simultaneously. Alternatively, animation can be used to expressly demonstrate the changes between states. A clear use of animation is in the rendering of time-based imaging studies, such as with angiography, echocardiograms, dynamic SPECT (single positron emission computed tomography), and other real-time imaging: the progressive change in anatomy or physiology (*e.g.*, a contrast injection) can be visualized in a cine format. Techniques such as using semi-transparent isosurfaces to overlay images from different time points [168] and flow visualization to depict changes over time with a combination of directed arrows and geometric objects [137] are being explored to characterize temporal and spatial changes in a single image.

Spatial Representations

Visualization of real-world spatial features (*e.g.*, size, shape) concerning human anatomy, in both two- and three-dimensions (2D, 3D), are critical to disease understanding. Intrinsically, human anatomy is intertwined with the notion of location and other spatial properties. Imaging, of course, reinforces this view, as descriptions of findings are anatomically based. Terminologies such as SNOMED (Systematized Nomenclature of Medicine) explicitly represent spatial relationships. As a result, the visualization of these spatial relationships is an important aspect of dealing with clinical data.

2D representations of anatomy. Direction, location, size, and distance are finding features that can be effectively shown in a 2D visualization environment. The principal example of the use of 2D is traditional radiographic imaging: projectional (*e.g.*, x-ray) and cross-sectional (*e.g.*, magnetic resonance (MR), computed tomography (CT), ultrasound) are all shown in a two dimensional plane. The Visible Human Project (VHP), run by the National Library of Medicine (NLM), is perhaps the best known 2D imaging dataset, consisting of complete MR, CT, and cryosection images of male and female cadavers [124]. VHP is responsible for several major discoveries, and has spawned several projects, including 3D volume renderings (see below), body atlases, and educational tools. But medical images only show spatial properties – they do not explicitly communicate specific relationships without interpretation or annotation. Rather, initial works in query interfaces for content-based medical image retrieval and spatial databases (*e.g.*, geographic information systems, GIS) are instructive in seeing different approaches to visually express 2D spatial relationships (Fig. 4.13): *query-by-sketch* and *iconic positioning*, where users draw objects of interest (or manipulate iconic shapes) into the desired relative positions [44]; and *iconic spatial primitives*, where different binary spatial relationships (*e.g.*, inside, overlapping, above/below, left/right, etc.) are enumerated for use in a query.

Anatomical maps (Fig. 4.10a) are also common 2D representations for expressing spatial relationships, albeit with less precision than imaging. Michelangelo's Vitruvian Man is a classic example, being an early study of body proportions and symmetry. Today, anatomical illustrations ranging from basic caricatures to precise depictions are used in clinical medicine. Specific domains of medicine use maps (Fig. 4.14): in neurology, Brodmann maps [20] assign regions of the brain to specific functions; pain drawings are used in nursing and other fields to solicit a patient's perception of problem areas; and illustrated atlases of surgical procedures are commonplace.

Anatomical maps are used also as general guides: [88] uses drawings of the human body, termed a *hypermap*, as a visual table of contents to ICD-9 mapped resources available on the Internet; [1] suggests 2D anatomic templates and icons to assist in radiographic reporting whereas [85] maps radiographic findings extracted from a report to a standardized atlas illustration.

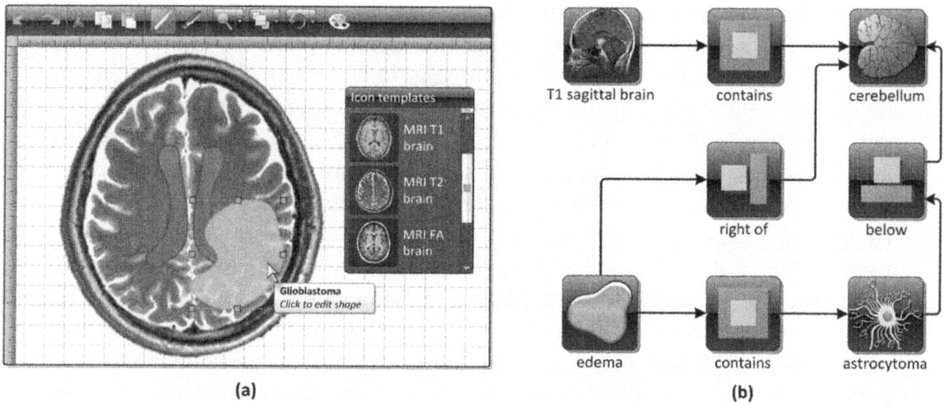

Figure 4.13: Examples of representing 2D spatial relationships. **(a)** Query-by-sketch and iconic positioning permit users to draw and/or directly place and manipulate templates relative to each other, compositing a picture with the desired spatial relations in terms of size, proximity, and shape. **(b)** Iconic spatial primitives are explicit (binary) relationships that can be used to logically describe 2D relationships.

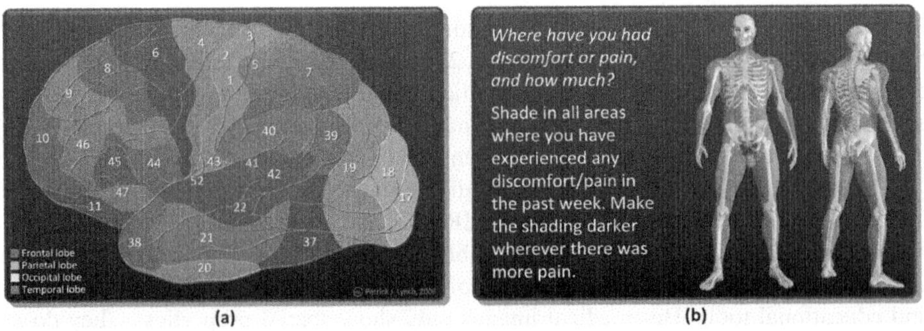

Figure 4.14: 2D anatomical maps demonstrating spatial relationships. **(a)** Brodmann map (*sagittal view of brain adapted from Patrick J. Lynch*). Each numbered region is associated with a brain function. Color overlays are used to further identify the frontal, parietal, occipital, and temporal lobes. **(b)** A pain diagram, asking a patient to draw the location and severity of discomfort (*human figures adapted from Bernhard Ungerer*).

3D representations of anatomy. Advances in computing hardware and data acquisition allow the creation of 3D models of morphology and physical processes: surface and volume renderings are now widespread, and can be created from routine 2D imaging datasets (*e.g.,* via texture-based volume rendering, wire mesh models, surface reconstruction algorithms, etc.). Such visualizations are more intuitive (particularly for non-radiologists) as they match the experience of viewing objects in the real world, and can be made interactive to enable multiple viewpoints. Moreover, non-traditional 2D cross-sectional images can be interpolated from a 3D model (*e.g.,* slices at an angle, rather than the conventional axial/coronal/sagittal views). Three dimensional representations of human anatomy can be divided twofold:

1. Simulation. Numerous works have used 3D graphics to provide virtual experiences that mimic real-world diagnostic and therapeutic procedures. For example, virtual colonoscopy uses a high-resolution CT scan of the abdominal region as the basis for a 3D surface reconstruction of the colon's surface, which can then be examined by a physician for polyps [132]. Similar reconstructions are used for lung volume assessment and bronchial pathways [90]. [128] uses a 3D model to picture entrance/exit wounds and likely damage incurred from stab and bullet wounds. 3D simulation is also popular for surgical planning and training: [148] emulates the reaction of tissue and organs in a virtual dissection; [56] uses a deformable 3D models of the brain to assist image-guided neurosurgery systems with adapting to changes in tissue structure as a result of the surgical procedure; and 3D volume models of intracranial aneurysms are coupled with hemodynamics to understand optimal placement of materials (*e.g.,* coils, clips) and approach [175].
2. Maps and indexing. Like their 2D counterparts, 3D representations are also used as maps. A navigational metaphor similar to hypermaps is used by IBM's Anatomic and Symbolic Mapper Engine to demarcate a patient's past and current medical issues from his patient record using SNOMED codes, showing findings on rotatable 3D human body [80]. Applications in neuroimaging are also demonstrative of 3D spatial representations, with coordinate systems (*e.g.,* Talairach atlas [99, 164]) and probabilistic atlases used to identify structural changes between normal and diseased populations (*e.g.,* depression, Alzheimer's disease) [112].

Although many of the above are 3D spatial representations of anatomy derived from imaging studies, a key drawback is the computational time needed to process these datasets for display. Real-time applications of 3D renderings (*e.g.,* for intra-operative assessment) are rare given this bottleneck. Present research to overcome this issue includes the use of multiple processors and specialized hardware (*e.g.,* graphical processing units, GPUs; field programmable gate arrays, FPGAs) to distribute and to speed up numerical operations involved in (medical) image processing and rendering.

Multidimensional Relationships

From the perspective of dimensionality, temporal relationships can be said to encompass primarily 2D and 3D representations. But as evidenced prior, there are clinical datasets that incorporate more than three or four variables of interest. With the diverse information stemming from multiple (clinical) domains, a need for visualizations to help extract meaningful information from large dimensional datasets has become necessary [57]. [89] groups visual data exploration techniques for multivariate, multidimensional data into three classes: geometric projection, icon-based, and graph-based techniques.

1. Geometric projection. This first group supports users in the task of reducing the dimensionality of a dataset to find meaningful trends. Established linear (*e.g.,* PCA) and non-linear (*e.g.,* Isomap, kernel PCA) dimensional reduction methods, along with factor analysis, can transform a high dimensional feature space into a more tenable number of (combined) variables. These results are traditionally then visualized using scatter plots; but for complex data, such as comparing gene expressions across multiple patients, a matrix of scatter plots [7] can be displayed in an array so that features between scatter plots can be visually correlated. Another technique is *parallel coordinates*, where each dimension is represented by a vertical axis and values for a particular case are linked by lines [81]. [167] applies parallel coordinates to facilitate searching for optimal visualization parameters (*e.g.,* view position, shading coefficients, transfer function) to render a volume. Parameter sets are represented as lines in a parallel coordinate display that connects parameters and the resultant image together, visually linking a unique combination of parameters with its resulting output.
2. Icon-based methods. The second group of methods maps data points to a glyph representation, where values determine the glyph's appearance. For instance, stick figure [131] maps numerical data to be analyzed onto an iconic representation; each data element controls the angles and limb lengths of the stick figure icon. In addition, data points are represented as shapes where each dimension is mapped to a small array of pixels and is given a particular shape and color based on the data values.
3. Hierarchical and graph-based techniques. Lastly, this third category involves subdividing and structuring the underlying data to identify hierarchies and relationships; trees, treemaps, and other graphs are representative of this group.

One example of a visualization tool that utilizes a combination of multidimensional techniques is an exploratory tool, called The Cube [53]. It aims to answer the question, *how does a set of attributes relate to each other from the available data*? The display consists of several planes, one for each attribute that is selected, and allows a clinician to manipulate multivariate data from across a patient population for the purpose of identifying patterns. Each encounter with a patient is represented by a line connecting individual values in the different attribute planes. Several parallel lines, for example, may be interpreted as a group of patients with a similar information profile and potentially identical diagnosis. The researchers demonstrated their system on a dataset of 1,500 examinations, finding that clinical practitioner test subjects preferred the 3D parallel coordinates visualization to other available 2D visualizations.

Causal Relationships

A driving objective of clinical medicine is the determination of underlying etiology: *why* is a patient presenting with a given set of symptoms? In this light, ascertaining the extent of associative and causal relationships between clinical data is essential, as is the visualization of these relations. And because the cause is frequently unclear (for any number of reasons), uncertainty enters into the picture: the degree of belief in a relationship must also be conveyed. A useful set of semantics for visualizing causal relationships is given in [87]: *causal amplification* (x increases the effect of y), *causal dampening* (x negatively impacts the effect of y), *causal multiplicity* (y is the effect of more than one cause); and *causal strength* (z contributes more than x to effect y).

Curiously, few graphical examples of expressing causal relationships and uncertainty exist within the medical domain. Instead, many GUIs depend on the use of directed graphs (*i.e.,* with arrows) to express causation between entities, such as with *Hasse diagrams* (time-space), *fishbone diagrams,* and *causal loops* (Fig. 4.15). In part, these visualizations are based on a linear time representation (*e.g.,* a timeline) to suggest cause-and-effect mechanisms. However, these static, graph-based representations of causality can become cluttered and difficult to understand as additional factors are introduced and/or the number of entities and relationships increases. Hence, several visualizations utilizing animation as a method of specifying causality have been developed. For instance, *growing-squares* [51] assign each process a unique color; when processes interact with one another, their colors mix. Importantly, in a perceptual study of the movement of objects on a screen, Michotte [115] showed how subtle changes in their movement produced large variations in the way test subjects described what they saw. He identified multiple factors (*e.g.,* timing, relative ratio of velocities) that affect a person's ability to perceive causality. [178] uses three metaphors taken from physics (pin-ball, prod, and waves) to establish visual causality vectors that express different types of change; evaluation confirmed that the perception of causality is highly dependent on the temporal synchrony between the cause and the effect.

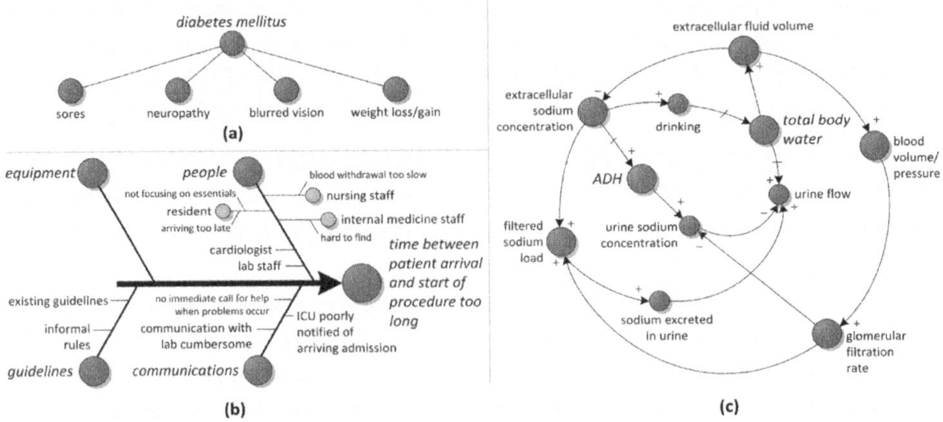

Figure 4.15: Example causal diagrams. **(a)** Hasse diagram for diabetes, showing the disease as the cause, and its symptoms as effects. **(b)** Fishbone diagram, providing a causal breakdown for delays in commencing a patient's procedure (adapted from [18]). **(c)** A causal loop diagram showing the positive/negative feedback on kidney function.

Uncertainty. In non-medical applications, uncertainty has been visualized in various ways, employing dashed lines, arrows, color, overlays, and blurring effects to articulate the degree of belief associated with a value or relationship. For example, *uncertainty maps* have been used to simulate how the introduction of three new ferry routes in San Francisco Bay would affect other vessels in the region [114]. [58] presents studies that measure whether participants are able to visually assess the uncertainty expressed through degraded and blended icons, determining: 1) that participants are able to interpret the meaning associated with such icons' appearances; and 2) that surprisingly, the presence of numeric probabilities expressing uncertainty (in comparison to purely iconic representation) does not provide a statistically significant advantage. Various static (*e.g.*, glyphs) and animation techniques have also been used to specify uncertainty in particle/fluid flow and spatially-related data [48, 106].

The importance of showing uncertainty in medical imaging data is remarked upon in [84], observing the implications of radiographically-guided surgeries and the need for accuracy. One approach to showing errors margins is to combine isosurface and volume rendering methods. Probabilistic animation has been explored as a method to increase diagnostic accuracy in MR angiography [108], finding that uncertainty animation using a probabilistic formulation of the transfer function captures important alternative presentations that would not have been perceived using static, manually tuned renderings. Positional uncertainty can be visually represented using a likelihood volume representation, which assigns visual weight based on certainty: [142] uses this technique for rendering a drug molecule's side chain, generating a 3D volume by overlaying all copies of a drug molecule's possible positions and using increased opacity to denote positions where atoms are more stable.

Navigating Images

The use of visual metaphors expressing spatial and temporal relationships is demonstrable in the context of browsing large image collections and high resolution images. In many CT and MR imaging studies, multiple series and thus potentially hundreds of images are acquired. And in pathology and mammography, the full image resolution often exceeds the viewable screen area. While some of the images/regions contain the pathology of interest, most are not relevant to the patient's diagnosis and treatment. Progressively, the navigation of medical images is time consuming: new ways to overview and to quickly isolating key regions are of rising importance.

Optimizing space. The challenge of reviewing images has been widely looked at in the realm of organizing personal photography. For a small number of images, the traditional approach is to create scaled representations (*i.e.*, thumbnail icons), which can be easily arranged in a grid. However, the popularity of digital cameras has created a proliferation of images such that for many collections it is infeasible to place all images on the screen while still having visually discernible thumbnails (*i.e.*, the scaled images are no longer recognizable). Techniques have therefore been proposed to maximize the use of screen real estate while facilitating image exploration, namely using zoomable interfaces in conjunction with optimized layout schemes. PhotoMesa [15] incorporates a zoomable user interface atop a treemap to help users navigate through large numbers of image previews. The Semantic Image Browser [83] (SIB) characterizes unique features within each image of a collection; related images are then displayed together using a multi-dimensional scaling technique. SIB also generates an image overview of a large collection of images that is interactive, allowing users to learn the contents of a collection, distributions, and relationships at a glance. A comparison of navigating a large image repository using a zoomable interface for browsing large image collections performed versus a traditional 2D grid of thumbnails finds that while user satisfaction is matched for both approaches, the zoomable interface excels in terms of the time required to find an image and overall accuracy in selecting the correct image [40].

Novel 3D metaphors have also been suggested to support navigation of large image collections of natural scenes: [159] applies registration and morphing techniques to an unordered set of photos, providing the ability to explore reconstructed scenes in 3D and to retrieve other images that contain the same object or part of the scene. [125] extends this approach by proposing new techniques for automatically selecting and warping images for display as the user interacts with the scene.

Temporal layout. In maximizing screen space coverage, the process of laying out images may not necessarily take into account additional image features that may serve to index or identify image content. The foremost attribute of a photo compilation is time: many users prefer to see their images organized chronologically [145]. Algorithms for clustering and selecting similar photos by date and other attributes have been proposed by researchers. PhotoTOC [135] introduced the use of representative photos to create an overview, table-of-content summary of photos clustered by data and time. Time Quilt [79] uses a layout that not only conveys temporal order by ordering images along a timeline but also minimizes the white space between photos. The photos are first clustered temporally (*e.g.*, by creation date), laid out into a grid, and wrapped into vertical columns with a predefined maximum height. As the number of thumbnails increase, they become too small to be recognizable; thus instead, a representative photo (chosen as the middle photo of each cluster) is displayed. Alternatively, *cover flow* layouts (Fig. 4.16) and immersive 3D walls (*e.g.*, PicLens) have been used to provide ordered, animated views of images. And as stated before, timeline views have been used to provide immediate assessment of selected regions of interest (ROIs).

Viewing large images. As opposed to the problem of viewing many images, others have examined how to deal with a single image of dimensions extending beyond the available space. Besides using scrollbars to pan the field of view, [133] provides an early survey of single image browsing, identifying several strategies (Fig. 4.17):

- Zoom and replace. The GUI provides a scaled view of the entire image, and an interactive tool that where the user can select a region for zooming in at a higher resolution. On choosing a region, the entire display is replaced with this new view; the reverse operation allows the user to zoom out to a previous view. Variations of this approach include having fixed magnification levels and selectable regions vs. variable magnification and user-drawn selectable regions.

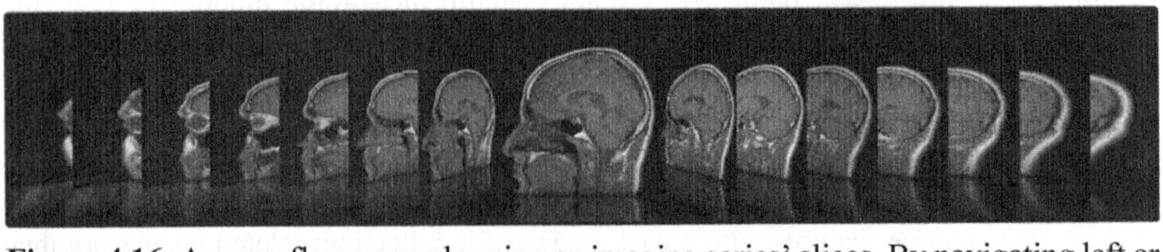

Figure 4.16: A cover flow example using an imaging series' slices. By navigating left or right, the display shifts the images accordingly to maintain the center focal point.

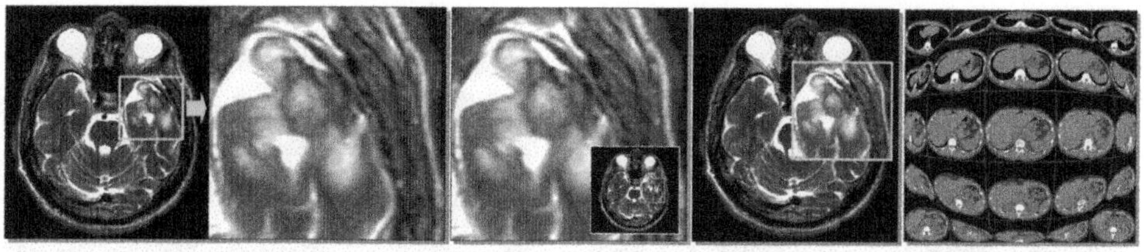

Figure 4.17: Different methods for dealing with magnification of a single part of an image (or image layout).

- Overview-detail. Unlike the zoom and replace paradigm, overview-detail uses a small portion of the screen area as a fixed thumbnail view of the entire image; the remainder of the screen is then used to display a selected region. The thumbnail is overlaid with annotations to indicate the current region zoomed region, thereby maintaining a sense of spatial location overall.
- Magnifying glass metaphors. This category is the reverse of the overview-detail method: the entire image is displayed on the screen (scaled, as necessary), and the user is given an interactive tool that allows for a small portion of the image to be magnified, much like a magnifying glass. The magnified region is overlaid atop the full view, creating a small thumbnail region that moves with mouse (some implementations also fix the location of the thumbnail). In many medical imaging workstations, this tool is referred to as a *magic lens*.
- Fisheye views. Lastly, building from the idea of using a magnifying lens, graphical distortion methods such as fisheye views can be employed: the focal point is magnified at the highest level of resolution, while the immediate region is progressively reduced.

Combinations of these approaches are frequently supported in modern GUIs (*e.g.*, the well-known Adobe Photoshop program provides both zoom and replace with overview-detail for editing graphics). Hierarchical strategies have also been explored: [176] divides larger images of a pathology slide (*e.g.*, 2400 x 3600 pixels) preparations into smaller regions, which are hyperlinked back to a compressed version of the entire high resolution image. Such an approach reduces the time and bandwidth required to view a particular section of the image.

Applications to medical images. Medical image viewers commonly use three modes of image presentation: *tile mode* displays images in a side by side grid, imitating the traditional light box presentation of films; *stack mode* allows users to navigate through images sequentially while only viewing a single image at a given moment; and *cine mode* animates through the entire stack automatically to reproduce real time physiological phenomenon (*e.g.*, cardiac muscle contractions). Tile mode is geared towards optimizing the use of display space, while the stack and cine modes emphasize spatial and temporal relationships. As ongoing improvements in scanner technology enable higher resolution image acquisition, the volume and the degree of detail within routine imaging studies will certainly escalate and it is unclear that the traditional modes of review will remain sufficient.

The solutions for natural images can also be adapted toward the medical domain, with several considerations: 1) the spatial ordering of image series must be preserved; 2) the use of graphical distortion should be limited, as it can unduly deform image

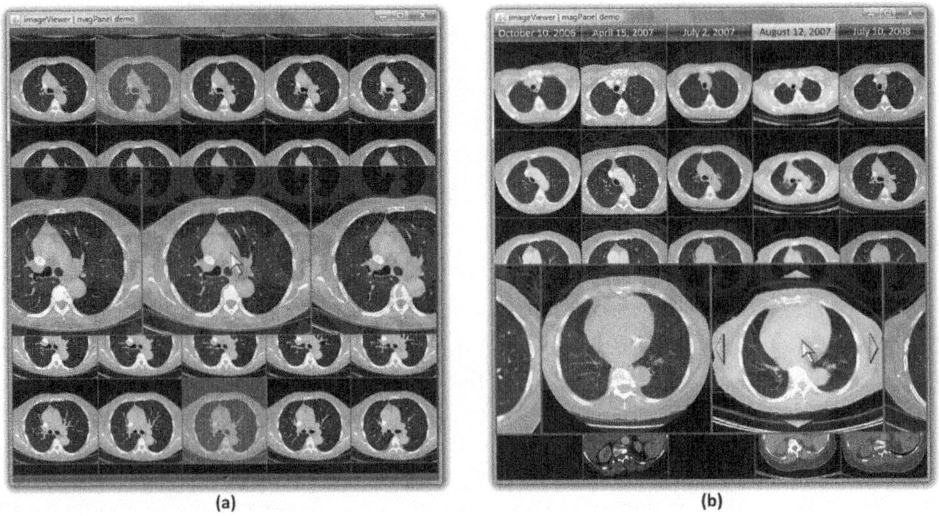

Figure 4.18: (a) A thoracic CT series is shown using thumbnails and row zooming. Spatial ordering is preserved while allowing a large number of images to be shown in a limited amount of space. The user can click to identify key images (shaded in blue). **(b)** Based on key images, past studies are co-registered and slices at the same anatomical level are shown, facilitating temporal comparison. Each column is a series, chronologically ordered to show change. To explore surrounding slices, the GUI permits the user to scroll up/down and to display additional images for further spatial context.

proportions and make visual comparisons difficult; 3) thumbnails keeps pertinent details and are of sufficient size to be useful to the clinician; and 4) methods that work for diverse, contrasted image sets are avoided, given the grayscale nature of radiographic images and the high degree of similarity between slices. As a case in point, one can borrow from the idea of contact sheets used in photography, allowing a user to see a grid of appropriately sized thumbnails (Fig. 4.18). Mousing over a given row in the grid has a "loupe" effect, providing a fully magnified view of those images, and helps to maintain perspective for comparisons. On identifying slices of interest, temporal views can then be generated with other studies automatically aligned to the specified anatomic levels.

New methods of interaction can improve how users view medical images and how they identify spatial and temporal relationships. For example, OsiriX [146] provides a customizable interface for using tailored input devices that simplify how users navigate through large image stacks. [94] describes a way of aligning 2D image views with 3D volume renderings, enabling better navigation in the latter via anatomical ROI selection and automated viewpoint selection. A variety of visual techniques can be employed to help navigate a large dataset of medical images, emphasizing spatial or temporal relationships:

- [126] demonstrates the use of multidimensional visualizations and textual input as a method for selecting and retrieving images from the Visible Human as related to a particular anatomical location.
- *Difference maps*[3] can also be used to highlight (spatial) changes over time: two co-registered and normalized images from different points in time can be subtracted

[3] Also referred to in the literature as *subtraction images* and *difference images*.

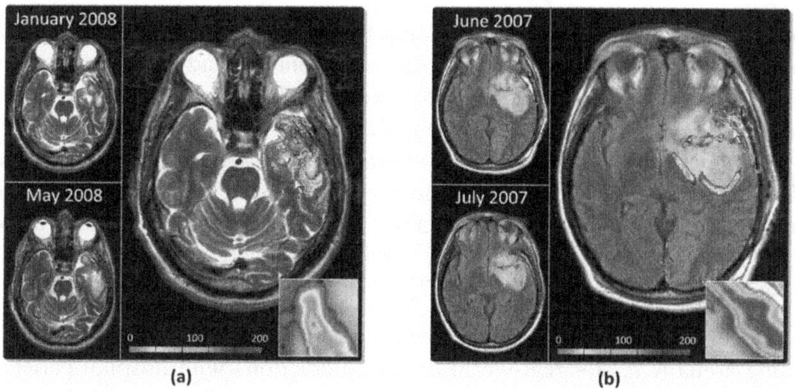

Figure 4.19: Difference maps for a neuro-oncology patient. Adjusting for intensity and geometry, slices at the same anatomical level are compared. **(a)** Two T1 MRIs four months apart are compared, with pixel values "subtracted" and mapped to a color spectrum. The difference map is overlaid on the baseline image. **(b)** Difference maps are useful when subtle changes may not be obvious. Here, two images taken a month apart appear similar, but small alterations on the rim become apparent once mapped.

to find significant alterations; the degree of change is then correlated to a color spectrum and superimposed on the original (shared) image (Fig. 4.19). [139] covers additional techniques for visualizing dynamic behavior within image sets, focusing on perfusion studies.

- Visual highlights of 3D volumetric datasets are also possible, such as through the use of medical illustration techniques (outlining, depth perception, shading) to automatically emphasize anatomical features via perceptual cues [163].

Notably, fundamental work has been done in the perception of radiologic images [95, 98, 113], including distinctions in how radiologists search images in comparison to a lay person and other factors impacting interpretation. Such efforts can provide further insight into methods to improve layout and navigation within medical image data.

Combining Information: Integrating the Medical Data

The first part of this chapter covered different graphical metaphors handling one type of data (*e.g.*, textual documents, images, labs, etc.), but potentially with many features (*i.e.*, multivariate). As any clinician will point out, however, today's process of diagnosis and disease management is multifaceted, drawing upon several types of information to reach a conclusion: the understanding of a disease (and a patient) is usually not predicated upon a single piece of evidence, but rather several observations. We thus address the question of how to coalesce the different graphical metaphors – and their data – into a single, comprehensive display supporting a user's tasks.

Defining Context

Combining information is largely dependent on the end reason for why the data is combined; that is to say, how will the integrated information be used? [10] argues that too often, there is a cognitive leap between the display of information and its use, advocating methods supporting "visual thinking." To bridge this gulf, one strategy is to build applications that are aware of (and hence adapt to) the environment, the user, and his/her goals. The idea of *context-aware* systems (sometimes called *context-sensitive*) was conceived of in ubiquitous computing research in the early 1990s, and has since been adapted in a gamut of fields, including HCI. In theory, by using contextual cues, an application is better able to tailor the display's contents and GUI experience to meet the need(s) of a user. [8] formally defines context: "*Context is any information that can be used to characterize the situation of an entity. An entity is a person, place, or object that is considered relevant to the interaction between a user and an application, including the user an applications themselves.*" In this case, the definition of relevance is dependent on the user's task. The idea of "*focus + context*" is to enable a viewer to locate an object of interest and to view the item in detail, while still maintaining an overview of surrounding information [27].

In healthcare, context-aware approaches are gaining attention not only in the mobile computing arena (*e.g.*, for real-world device interactions with hospital patients given physical proximity, such as based on radio frequency identifier tags, RFID), but also in the design of EMR interfaces [13, 82]. Context-aware GUIs regularly involve user and task models [150]; for the medical environment and the display of patient data, a disease model is also needed to provide a knowledge-base that can further customize content. Below, we briefly focus on user and task modeling in relation to clinical information; a discussion of the disease model is left to Chapter 7.

Defining the User

User modeling is an active area of research in human-computer interaction; [59, 93] provide historic perspectives. The aim of user models is to provide a set of characteristics describing an individual or group of similar users such that a system can query these features to adjust the interface or presented content in support of the user's tasks. The choice of user features varies greatly based on the target application, but can encompass: the demographics of the individual (*e.g.*, geriatric vs. young adult patient; male or female); the types and degree of domain knowledge (*e.g.*, patient vs. physician; urologist vs. pulmonologist); the degree of GUI familiarity (*e.g.*, novice vs. expert); and user preferences. A helpful formulation in categorizing user models considers three questions [143]: 1) is the user model canonical or representative of only one person; 2) does the user explicitly select the values to instantiate his model, or does the system attempt to observe, learn, and abstract the user's features over time; and 3) is the model intended to support short- or long-term usage of the system (*e.g.*, a single session or multiple sessions with the GUI)? Each aspect is elaborated upon further.

- Canonical *vs.* individual. The question of whether an interface is built for entire populations of users (*i.e.*, canonical) or individual users dictates its design: canonical models are generated as part of the implementation of the system and do not change over time while individual models are built and maintained for each new user. In one popular approach, canonical models categorize users into basic groups (*e.g.*, novice, intermediate, expert) called *stereotypes* [143]. Each group has its own unique set of assumptions that guide what interface elements are presented. For instance, novice users may need additional guidance and tooltips that help with familiarization of the interface while expert users may be presented with more functionality to provide quicker access (*e.g.*, keyboard shortcuts). Individual models adapt to the user's preferences over time by learning how the user interacts with the interface. For example, if the user frequently accesses a particular function or needs a particular piece of information, the interface identifies and changes to make the function easier to perform or to automatically display the information. Many adaptive systems take a combined approach where the default settings are based on a canonical model but as the user interacts with the interface, an individual model is generated.

- Explicit *vs.* implicit. Models may also be classified as explicit or implicit. In *explicit models*, information about the user and task is provided manually by the system designer or user. In *implicit models*, information about the user is collected by the system through the course of normal interaction. Explicit models allow users to customize aspects of the user interface such as changing the layout of the display by dragging and dropping elements on the screen or selecting specific preferences or stereotypes from a list of options. Implicit models try to learn the user's preferences by observing and recording the user's system interactions. Straightforward approaches include counting the frequency by which a user accesses a function or restoring the user's last used workspace. More sophisticated machine learning methods can also be used to compute implicit models.
- Short-term *vs.* long-term. Short-term characteristics are often associated with preferences or assumptions about a user that are valid over a single session. For example, a short-term factor would be to use the hospital information system to query for the patient's current list of prescribed medications; meanwhile, another user may query the system for other information such as patient admissions, discharge, and transfer data. Long-term characteristics tend to describe user preferences that do not change across multiple sessions of using the application. The color schemes for an application or the default printer to which documents are outputted are examples of long-term factors.

For the most part, processes for generating user models have centered on explicit (*e.g.*, building stereotypes, measuring user satisfaction [144]) and implicit methods (*e.g.*, collecting data on the user and invoking machine learning techniques to generate structural descriptions about user behavior and interests [61]).

Though user modeling has been incorporated into many applications, two areas are of interest here: customizing GUIs to better suit the user (adaptive interfaces) and tailoring presented information to be relevant to the user's interests (adaptive hypermedia).

1. Adaptive interfaces. *Adaptive interfaces* build a user model based on preferences and goals collected through normal interaction with the user [100]. A general architecture for adaptive systems is presented in [16]. The most basic systems consist of three parts: a model that specifies the components of the system that may be altered; a user model; and a connecting model that defines how the system changes and what it can adapt to. [105] describes an episodic approach to building user models by first observing the interaction between the user and the software application, identifying different episodes from the actions, recognizing user behavior patterns, adaptively helping users according to recognized user plans, and storing this information into a user profile. Lumière [77] uses a Bayesian belief network computed based on the user's past interaction with the system to predict the most likely task that the user is trying to accomplish. The system then presents relevant GUI components to help the user with the identified task. The modeling of interactions and tasks can also be used to inform underlying data models and database views, leading to user-oriented displays [136].

2. Adaptive hypermedia. Hypermedia is the integration of graphics, audio, video, plain text, and hyperlinks. Adaptive hypermedia builds upon traditional hypermedia but incorporates a user model to highlight relevant information, tailoring the presentation of this data based on what is known about the user and task [21]. [107] utilizes stereotypes to customize the type of information, ranked by relevance for a particular user, based on a given diagnosis. The system's user model captures details on the user's familiarity about each medical topic, the system's interface, and unique preferences. For example, ANATOM-TUTOR [14] is a tutoring system to assist medical students with learning about human anatomy; it utilizes a user model to structure the presentation of anatomical knowledge in a method best suited for the users' level of knowledge. Users instantiate the model by completing a questionnaire prior to using the system; this information adjusts the level of difficulty of questions and the depth of explanations provided by the system.

Defining the Task: Incorporating Workflow

Distinct from the user's characteristics and preferences in using a system is a description of the task that the user wants to complete via the system. A *task model* informs the system of the user's intentions. For example, is a physician user reviewing a patient's history through the EMR for the first time, performing a follow-up examination, or documenting an encounter? In each situation, different intellectual and procedural goals are accomplished. A number of task model methodologies have been proposed over the years [103] to structure task requirements from users. Such models can be used to identify possible usability problems with a system; to assess human performance (*e.g.*, time, efficiency); and to design a GUI [102]. Task modeling and its use in GUI development has its provenance in HCI. A survey of task models and methods to elicit such information is beyond the scope of this section; however, the different models share several commonalities (under different semantics):

- Task hierarchies. At the core of these models is a way to describe the end objective of the interaction with the system. Rarely are the specifications of tasks atomic: usually several sub-tasks or steps comprise a single task. Thus hierarchical and object-oriented approaches are taken to organize this information, with higher-level abstractions being aggregates of more elementary goals. An implicit constraint in these frameworks is that a given task cannot be completed without all of its sub-tasks being performed.
- Objects and actions. Objects are the entities that participate in the task, and encompass the user through to the resources required to complete the task (*e.g.*, a database, another individual, the system GUI, etc.). Actions are the basic methods that an object is capable of performing; this concept borrows from the OO paradigm, encapsulating entities with behaviors.
- Roles. A user may change behaviors given different tasks. For instance, a doctor reading a patient medical history may be acting as a physician diagnosing a patient, or may instead be looking at the record as a clinical researcher extracting information. The concept of a role is correlated with that of a user model.
- Operators and events. Although task hierarchies provide compositional rules, they do not impose any temporal ordering on (sub-)tasks. Hence, a task model incorporates some mechanism (*i.e.*, operators) that provides relative sequencing between tasks. To describe these constraints, event descriptions are embedded within the model, specifying milestones and/or conditional dependencies.

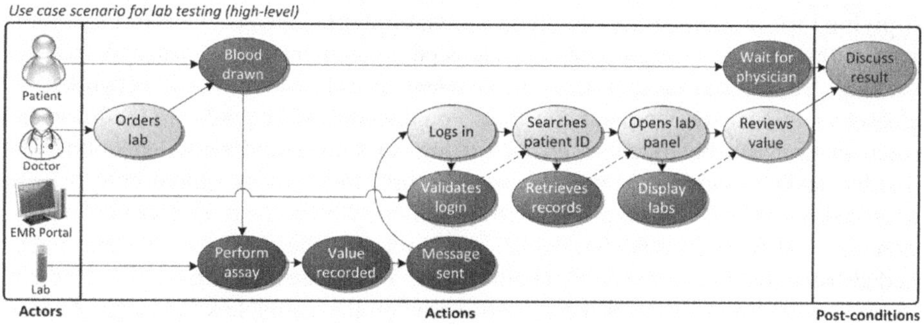

Figure 4.20: A use case scenario. The interaction between different actors (patient, doctor, EMR portal, lab; left side) is illustrated in this high-level perspective of a physician reviewing a lab test. Actions are depicted in labeled ovals, with arrows between the elements denoting dependency.

Another means of stipulating a task and its steps is seen in *use case modeling*, a technique chiefly seen in software engineering. Use cases help define the scope of a system, specifying the components that interact with the system (called *actors*); the expected relationships between components and their functionality; and inherent assumptions by users and/or the system. Each use case provides one or more *scenarios* that convey how the system should interact with users to achieve a specific goal or task. A scenario depicts (Fig. 4.20): the involved actors; the goals of the actors (*i.e.*, why actors interact with the system); the triggering event and pre-conditions for the process (*i.e.*, when the task starts); the main flow of events and possible alternatives (*i.e.*, how the task occurs, with actors and the system's actions stated); and the end state of the system in terms of post-conditions. A UML (Unified Modeling Language) standard exists for use case models, characteristically drawn with users on one side of the diagram and annotated ovals representing actions, moving from left to right; arrows between users and actions show (temporal) dependency. From these different scenarios, an application (and its GUI) can be designed.

Both task and use case models are intimately tied to workflow: if workflow defines a sequence of operations performed by a person or group of individuals, then task models and scenarios are the operations at a finer level of specificity. By observing workflow and tasks, the integration of data components and functions in an interface can be anticipated as part of the users' needs. The ribbon task interface introduced in Microsoft Office 2007 is illustrative: based on usability analysis, tasks are now grouped together into ribbon strips; the ribbon dynamically adapts based on prediction of the user's current activities, adding/removing functions as appropriate. Similar widgets can be used in EMR interfaces and clinical workstations (Fig. 4.21).

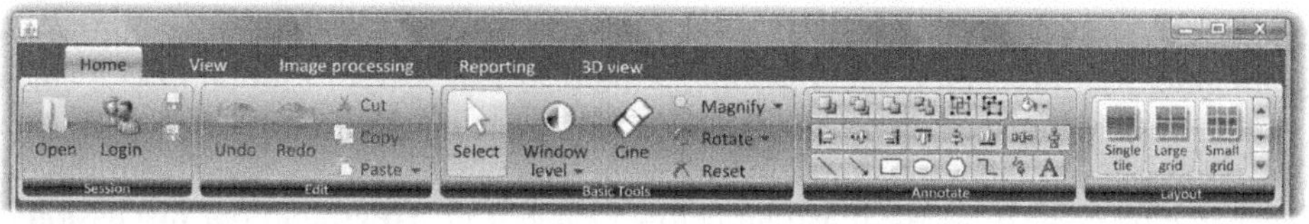

Figure 4.21: Example ribbon interface for an imaging workstation, replacing the standard menus and toolbars. The most common functions for image manipulation, layout, and annotation are grouped together as on the first (home) ribbon strip; other ribbons are made available based on the workflow of the user and his current actions.

Applying task models to the healthcare environment. Much of the effort in defining tasks within medical informatics is relegated to clinical guidelines and treatment planning. Several approaches have been taken in establishing "task networks" that impose conditional and temporal constraints on directed cyclic graphs (*e.g.*, flowcharts), which in turn can be transformed into a logical syntax and reasoned upon [130]. Broadly, tasks in guideline languages can be broken down into three core types, as suggested by the PROforma project [162]: *actions*, which entail a procedure that acts upon the real-world; *enquiries*, which are points in the workflow where input is needed from some resource (*e.g.*, another user, a database); and *decisions*, wherein a choice is made (by the user or the system) based on the available knowledge. Although detailed workflows and tasks have been identified in specific domains (*e.g.*, breast cancer [60]), there is an absence of an ontology (or terminology) for describing generic clinical tasks or research-oriented tasks involving clinical data (*e.g.*, retrospective analysis). Rather, ontological endeavors have focused on guideline structure, depending on existing vocabularies that describe diagnostic/therapeutic procedures for enumerating tasks. Unfortunately, this level of granularity can hinder translation to specific GUI-related activities, which may involve higher levels of abstraction and/or other implied tasks; [97] touches upon this issue of granularity. For example, while searching for a patient's latest lab test is a common activity, most guidelines would specify a specific exam and criteria (*e.g.*, is creatinine abnormal?) – there is a disconnect between the GUI search task, which may be contingent on several other factors (*e.g.*, opening the correct patient's record, accessing the lab values, determining if the assay was performed, etc.), versus the purely cognitive process of assessing the lab value. Moreover, is searching for one type of laboratory value (*i.e.*, creatinine) different than another (*e.g.*, potassium)? The reason for performing a given task influences its interpretation [96]. Arguably, while knowing with specificity which lab a user may need in a given domain can certainly help guide a GUI in a given application, there remains a need to understand the commonality between tasks in order to improve user interface consistency within the EMR and the integration of patient data elements.

Combining Graphical Metaphors

With context driven by the user and task models, it becomes possible to consider how to put clinical data together into a graphical display that suits a given need for an individual. Knowing the user, his intentions, and the sequence of actions to be completed, a program can supply what data is needed to finish a task, choose how best to present this information to communicate ideas, select the layout of this data in an interface beneficial to the user's workflow, and determine the toolset to work with the data.

Creating Integrated Displays

The process of combining data elements into a problem-centric visualization can be decomposed into several steps: *identifying* the data that needs to go into the display; *prioritizing* the selected data; *relating* the data elements; *selecting* the appropriate visual metaphor for the data; and finally *laying out* the visual metaphors. Each stage is shapeable by context, as we illustrate below (Table 4.1). To ground our discussion of visual integration, consider the following data for a patient at high risk for coronary artery disease (CAD): 1) blood tests assessing cholesterol and blood glucose; 2) imaging including ECG, echocardiograms, and cardiac CTs; 3) free-text and structured reports from physicians, including radiologists, cardiologists, and the patient's primary care physician; and 4) the course of treatment during this period, encompassing medications and interventions. Each piece of data can be presented in its own visual metaphor (plots, images, text, treatment descriptions) to provide a viewer with some information, but not the full "story" of the patient's history or current state.

Identifying data. In a medical display, the selection of data, as alluded to earlier, is decided by a disease model that outlines the (clinical) information relevant to the condition's diagnosis and treatment. But by nature such models are usually comprehensive and the scope of information needs to be honed by the user and task models. For instance, in our example of the CAD-risk patient, all of the identified data may be needed by a physician reviewing the history, but the patient himself or his primary care physician (PCP) may only require certain pieces of information. Different knowledge-bases can be used to steer the selection and filtering process: evidence-based medical guidelines can be used for physician users, while educational sources can isolate data useful to a patient. Similarly, task models pinpoint elements of information needed to complete a task; of course, such data are included as requisite components of the display.

	PCP/Internist	**Radiologist**	**Patient**
Task description	Follow-up assessment	Image interpretation	Self-management
Clinical data			
Demographics	•	•	
Medical history	•	•	
Vitals (BP, BMI)	•		
LDL, HDL cholesterol	•		•
Triglyceride lab	•		•
Blood glucose lab	•		•
Electrocardiogram	•		
Imaging (ECG, thoracic)	•	•	
PCP reports	•	•	•
Cardiology reports	•	•	•
Radiology reports	•	•	
Medication history	•	•	•
Prioritization	1. Labs 2. Medical history 3. Medication history 4. Demographics 5. Vitals 6. PCP reports 7. Cardiology reports 8. Radiology reports 9. EKG & imaging	1. Imaging 2. Radiology reports 3. Cardiology reports 4. Demographics 5. Medical history 6. Medication history 7. Labs	1. Labs 2. Medication history 3. PCP reports 4. Cardiology reports
Relationships	Medications → Labs EKG → Cardiology report Imaging → Radiology report	Imaging → Radiology report Medications → Labs	Medications → Labs
Visual metaphors	Labs ⇨Line plot Medical history ⇨List Medication history ⇨Timeline Demographics ⇨List Vitals ⇨Timeline Reports (all) ⇨Icon EKG & imaging ⇨Icon	Imaging ⇨Presentation states Radiology reports ⇨Full text Cardiology reports ⇨Full text Demographics ⇨List Medical history ⇨List Medication history ⇨List Labs ⇨Table	Labs ⇨Line plot Medication history ⇨Timeline PCP reports ⇨Icon Cardiology reports ⇨Icon
Target medium	Clinical workstation	Imaging workstation	Web page

Table 4.1: Example of creating an integrated display for a given medical problem (high risk coronary artery disease patient). The sequence of steps helps to identify the required data, the relative importance of each element, potential relationships between the data, and a means of viewing the information collectively. Each stage is conditioned by the user and task models, providing context to tailor the display.

Prioritizing the data. Simply put, not all data elements are of equal importance in a diagnostic or treatment task. This second step has a practical purpose: by ascertaining what is important, the amount of visual space allocated and prominence given to a graphical component can be gauged. The relative priorities do not necessarily correlate with the order in which data is accessed, but with where users' focus will linger most in the display. Once more, context can help reveal clues as to what a given user or task may deem important. Clinical users of a given category will likely find information from their own specialty of central concern: continuing with the CAD patient, a cardiologist would likely emphasize EKG and cardiac consults, whereas a radiologist may stress the CT images. From task models, processes that are conditional (*i.e.*, decision points) are likely to expose important data points that should be accentuated in the overall display.

Relating the data. The display of clinical information must express a range of spatio-temporal and causal interactions; for instance, is the statin that was prescribed to the CAD patient decreasing his low density lipoprotein (LDL) cholesterol level? Being aware of the interplay between data elements, an interface can better highlight potential relationships. Semantic relations available expressly (*e.g.*, rules encoded for a given medical problem) and indirectly (*e.g.*, structural information derived from a knowledge-base, an underlying data model) can be used to establish these linkages. From these connections, shared visual cues can then be used to demonstrate associations. Conventional cueing approaches include color coding of related data elements; spatial proximity (*e.g.*, overlapping, tooltips); similar line styles, etc.

Selecting the appropriate visual metaphor. For each selected data element, a graphical representation must be chosen, optimizing a user's understanding based on the *raison d'être* for the data in the display and the viewer's ability to comprehend and make use of the visualization. Choosing the appropriate visual metaphor for a given set of data can be thought of as generating a sentence from a graphical language [110]: how do we best communicate the information to a given user? A case in point, lab values in our example are used to ensure that the patient's cholesterol is controlled: a trended lab plot would provide a temporal view, and may be an easier means of educating the novice patient about the need to maintain diet and treatment; whereas a tabular or presentation of only the last lab value may be better for an internist who reviews daily in an in-patient setting. Similarly, a PCP may only be interested in knowing that imaging exams were conducted so that thumbnail icons of the studies

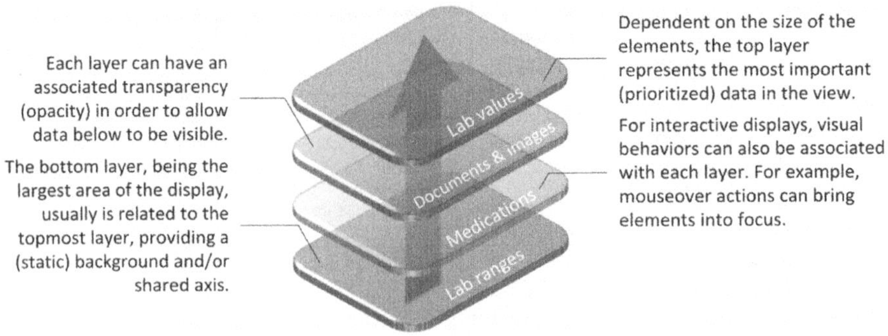

Figure 4.22: The process of layering data elements in the display. Based on the prioritization of the information, layers can be assigned. Opacity and interactive behavior between the layers can then be used to highlight relationships between the data.

may be suitable; but a subspecialty radiologist or cardiologist may need a more detailed graphic. Clearly, the task model also affects the selection of the visual metaphor: if the internist was seeing the patient for the first time, a historic perspective (such as with a trended plot) would be more appropriate than the last lab value. Finally, the anticipated communication medium (*e.g.*, is the display an interactive client, a web page, or a printed report?) will further impact the choice of visual metaphors. For instance, a printed summary sheet affords a higher degree of resolution (relative to a computer screen), but at the expense of interaction [138].

Laying out the data. The last step in integrating the data elements together is to spatially organize the information in the display. Here, workflow from a task model and the relationships between data are helpful in guiding layout, with the intent of creating a visual/focal flow to the presentation and interaction with the data. The layout of the data should take into account possible visual interactions to exploit user search behaviors. Shneiderman's framework of overview, zoom/filter, and details on demand [155] is germane: primary data in the display can be visible in a synopsis state, allowing for selection and augmentation with additional data (*e.g.*, but of lesser priority). Importantly, the ultimate source of information should accessible to the user.

Dependent on the complexity and degree of graphical integration in the display, layout not only comprises arrangement in the *x-y* plane, but also *layering* of the visual metaphors (*i.e.*, *z*-order), allowing for juxtaposition of graphical elements. The extent to which layering of graphical metaphors should occur is debatable. As current EMR interfaces tend to keep data elements separate (*i.e.*, each type of data in its own panel), users may not be familiar with composited graphical displays – however, the design of new displays can encourage exploration and new insights. Expounding upon this idea, consider the visualization of the four data elements for the CAD patient: collating the selected visual metaphors so that a user can view the data together provides integration, but the separate depictions still force the viewer to implicitly relate data elements. The scenario can be improved by creating a single, layered graphical metaphor (based on the original four metaphors), enabling the viewer to overtly see relationships (Fig. 4.22)[4]:

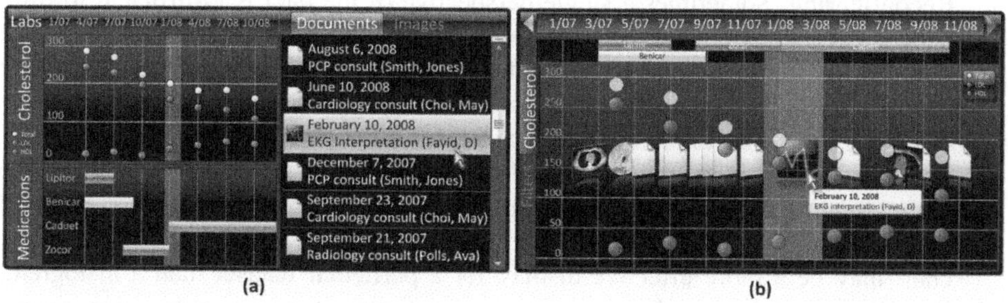

Figure 4.23: Two different integrated displays for a CAD patient, focusing on cholesterol labs, medications, consult reports, and associated imaging. **(a)** A classic GUI assigning each data element to its own panel. The separation of components can make visual correlations harder to appreciate. **(b)** A layered display with the same data elements. The row of document and image icons shifts vertically with the mouse position; and a cover flow technique is used to identify active icons. In this case, it becomes easier to see the correlation between drugs and the patient's decreasing cholesterol.

[4] Of course, *too much* layering of information can also be counterproductive and lead to overly complicated displays and/or data being obscured in the view. The KISS principle (*keep it simple, stupid*) should be followed in any design.

the different data elements can be overlaid using a combination of transparency and interaction to spatially relate information and support details-on-demand behavior. Based on these principles, Fig. 4.23 demonstrates two different interfaces for the CAD patient: Fig. 4.23a lays out separate panels for each data source/type, with visual cues to illustrate temporal correlation; Fig. 4.23b uses layers to aggregate the information with more explicit visual linkage.

Interacting with Data

Thus far, we have described representing medical data: first by using basic display elements and then by combining elements into integrated displays. However, representations are intrinsically *passive* – while they transform the underlying data into a graphical representation, they are typically static and do not provide mechanisms for manipulating the data. This section describes methods that focus on interaction, providing users with active methods to uncover new insights by posing queries to the data and identifying patterns in the results. The variety of interactive methods have been organized into taxonomies (*e.g.*, organized by low-level techniques [45], interactive techniques [172]); in line with the user and task models, the categories in [164] are followed, which categorizes based on a combination of user objectives and the interaction techniques that accomplish them (Fig. 4.24):

1. Selecting. The act of selection uniquely identifies a single data point by highlighting it (*e.g.,* using a different color) so that users may visually track the location of items of interest. Typically, selection occurs as the first step of a series of interaction techniques as a method to identify a subset of data elements that the user is interested in exploring further.
2. Exploring. The amount of information displayed is limited by the screen size and the user's ability to perceive an array of presented information simultaneously. If the amount of data is too much to fit into a single screen, tools are needed to explore the data. Actions such as panning and scrolling allow users to intuitively move data across the screen and configure the display to show data of interest.
3. Reconfiguring. Sometimes a single perspective of the data is insufficient to fully understand any patterns or trends. Reconfiguring the dataset allows users to change how the data is presented by viewing the same data in different arrangements. For instance, in multidimensional scatter plots, new views of the data are generated by changing the attributes presented on the axes.
4. Encoding. While reconfiguring the display allows users to view data using the same visualization but in different arrangements, encoding allows users to transform the representation of a data element from one form to another. For example, a pie chart may be a more effective display for a particular dataset than a histogram. Encoding may also involve reassigning visual attributes (*e.g.*, color, size, shape) to better differentiate clusters of data.
5. Abstracting. Data may be viewed at varying levels of abstraction. A common technique for abstracting data is to allow users to *zoom* between broader and more detailed views. An overview may be used to obtain a general idea of the data; however, users will want to magnify specific regions in the data that is of interest to them to view additional information.
6. Filtering. When displaying large amounts of data simultaneously, users need tools to help identify and focus on the data relevant to their task. Filtering is a technique that allows users to conditionally hide or change the appearance of certain data

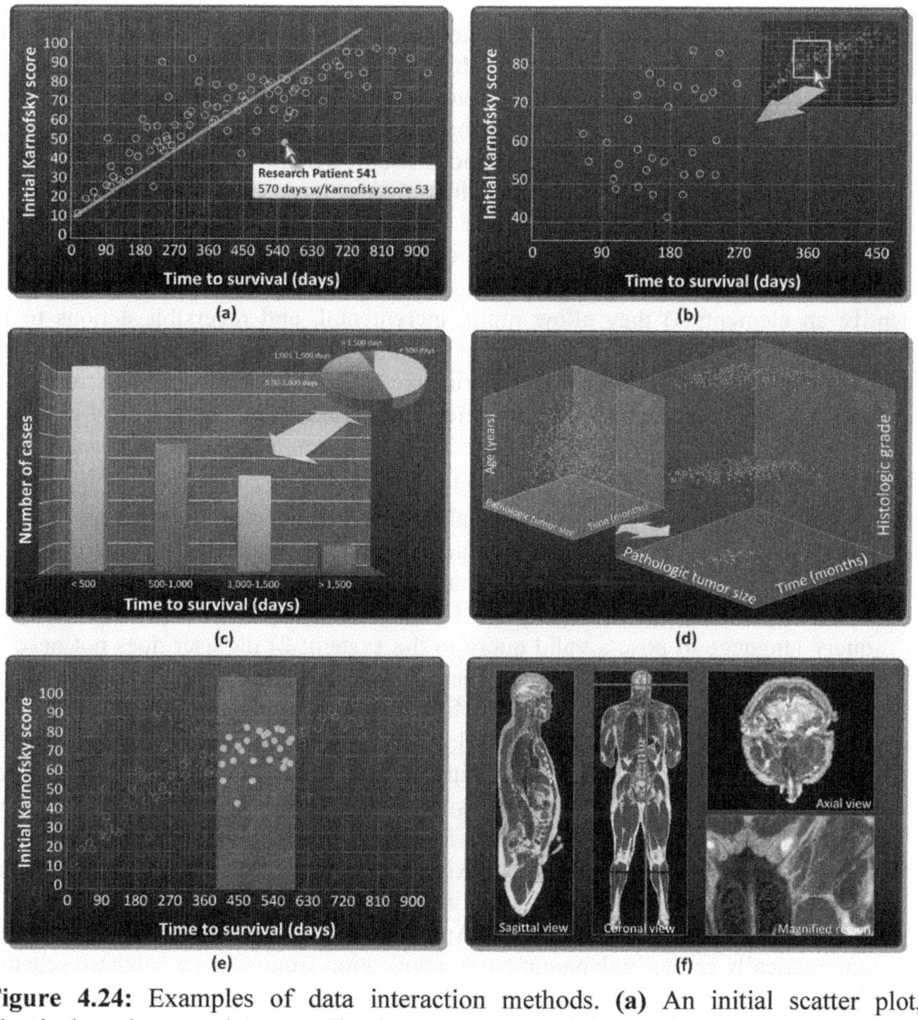

Figure 4.24: Examples of data interaction methods. **(a)** An initial scatter plot, identical to the one shown earlier in Figure 4.2b. *Selection* is demonstrated by the highlighting of the current point in a different color. **(b)** *Exploring* is shown via zooming on a portion of the same dataset. **(c)** *Reconfiguring* entails the use of a different visual metaphor on the same data; here, the pie chart is transformed into a bar chart. **(d)** *Encoding* allows for data reorganization along different features; in this case, by plotting data by histological grade rather than age, clusters become evident. **(e)** *Filtering* allows for the selection/highlighting of data meeting some criteria, such as those patients with a given range of time to survival and Karnofsky score. **(f)** *Connecting* allows for simultaneous views on the same entity to be (visually) linked together.

points that do not fall within specified criteria. If a physician is examining a patient who has hypercholesterolemia, unrelated documents should be filtered (*e.g.*, reports on a broken leg), leaving only a subset of documents pertinent to the treatment of high cholesterol.

7. Connecting. When multiple different visualizations are used to represent the same data, the correspondence between each view may be highlighted by linking them together. For instance, if a user selects a set of data points in one view, all of the views reflect the same selection in their own way. This process is called *brushing*.

Querying frameworks. The aforementioned seven types of interaction are commonly used in combination to provide users with graphical tools for manipulating data and posing queries. Traditionally, users interact with a database by formulating textual queries using machine-understandable languages such as structured query language (SQL), which features a non-intuitive and difficult syntax for non-programmers to learn. To address these issues, querying frameworks have come to support *dynamic queries* and *direct manipulation* as more intuitive interfaces for working with data [127]. These frameworks share several characteristics: 1) they provide graphical representations of real-world objects and actions; 2) they use a pointer to select or identify an element; 3) they allow rapid, incremental, and reversible actions to be performed on the data; and 4) they provide immediate and continuous display of results. In particular, direct manipulation principles have been shown to assist users with navigating large information spaces [4]. Here, we explore two categories of querying frameworks that are used in medicine: direct manipulation and *query-by-example*; [28] provides a survey of querying frameworks in other domains.

- Direct manipulation. Direct manipulation interfaces model how people interact with objects in the real-world by providing users with tools to interact with visual objects that represent the data elements [153]. Several benefits exist for applying direct manipulation to data querying: 1) the user does not need to learn a complex query language to pose a valid query to the system; 2) the user does not need to worry about making syntax errors; and 3) the user obtains immediate feedback about the query and the results [68]. Many applications have been developed using diagrammatic visual querying [68, 69]; a few are mentioned here. ADVIZOR [49] is a commercial system that works with relational data cubes to query aggregated information. Users select data by using either dragging; or a tool that makes predefined (*e.g.,* rectangle, circle) or freeform shapes. The interface allows new selection sets to be related with existing sets by using expressions such as replace, add, and subtract. Changes in a data view automatically propagate across visualizations tied to selection sets. Another system is IVEE/Spotfire [3], which automatically creates a dynamic query application from a given database schema. A collection of common visualizations (*e.g.,* histograms, bar charts, pie charts) is selected based on the attribute data types within the application's data schema. In medicine, systems such as LifeLines and KNAVE-II utilize a combination of interaction techniques to manipulate time series data. Users are able to *select* events (*e.g.,* procedures, medications, laboratory values) of interest at specific time points, *filter* events to show only those occurring within a defined time period, *explore* more recent or older events by dragging a scrollbar, and *abstract* time points by grouping them into semantically related clusters (*e.g.*, events related to a disease). Direct manipulation techniques have also been employed towards studying patient populations: SOVAT [149] facilitates the exploration of large data warehouses to identify common health factors within a community. The system links different information displays together; variables in a table can be dragged into the charting area and plotted. Selecting regions on a geographical map display additional quantitative data specific to those areas.
- Query-by-sketch. This interaction paradigm asks a user to sketch or provide an example object as the basis of a query, finding all objects in the database with similar visual attributes (Fig. 4.25). Sketching provides an intuitive way for users to express the image-like representation of spatial configurations that are in their minds. Early seminal work in this area include [29, 86]. The former presents a

relational query language introduced to simplify the usage and management of image data while the latter defines a set of operations as part of a high-level query language that provides basic methods for querying pictorial databases. MQuery is a visual query language that uses a single set of related query constructs to interact with data stored as time-based streams [44]. These works have laid the foundation allowing future applications to support pictorial querying. One such application that has benefited from their work is GIS, namely because geographic concepts are often vague, imprecise, little understood, and not standardized. [47] presents a spatial query-by-sketch system that automatically translates the spatial layout of query objects into a database-understandable query. In medical images, the visual attributes of abnormalities (*e.g.,* size, shape, location) may be correlated with patient outcome. Query-by-example systems help users search through image databases by these features rather than using a limited set of text keywords such as the header information. Two primary examples of such systems in medicine include ASSERT [157], which indexes high resolution computed tomography lung images based on features calculated in a physician-defined region of interest, and IRMA [101], which is a multilevel, distributed system that supports visual queries on both local (*e.g.,* tumor shape/texture) and global (*e.g.,* overall image intensity, anatomical area) attributes. [147] presents a system that provides a user with tools to query a thoracic imaging database using a combination of template and user-drawn features; an evaluation of this system by radiologists resulted in 98% and 91% recall and precision rates, respectively. Each system provides users with tools to *select* example images or shapes from a collection to form the basis of a query and to manipulate the appearance of the query to match the user's interests. The returned results are then presented using techniques such as zoomable interfaces described earlier in navigating large imaging collections.

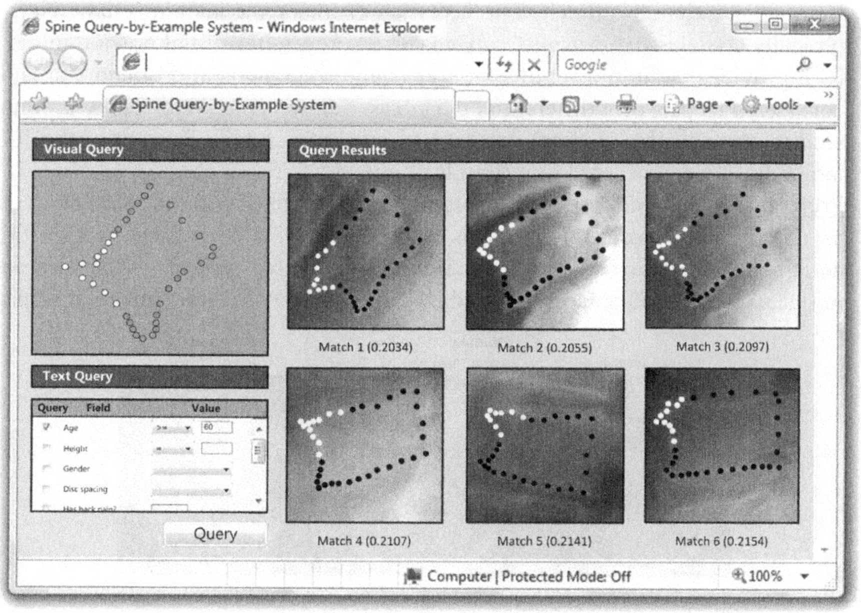

Figure 4.25: Query-by-sketch interface for matching vertebral shapes from a database of spine x-rays. The user draws the desired shape (top left corner) and specifies additional (nominal, numerical) query constraints. The system then matches the drawn shape against its collection and returns the highest ranked images.

Imaging Workflow & Workstations

From an imaging informatics viewpoint, the question arises: how can context and integrated visualization assist the imager? Our observation of imaging workstations over the years is that most use standard, static user interface paradigms for layout (*e.g.*, tile displays) and interaction (*e.g.*, menu bars, tool palettes). Efficiency and functionality, however, are sometimes contradictory goals for a well-designed image workstation. Inspecting the course over which images are ordered, reviewed, and reported upon is informative, as each stage provides hints as to the context for display and interpretation. Taking into account context, tailored presentations for a user and task can be created, proffering new modes of operation and opportunities for efficiencies.

Defining imaging context. Notwithstanding the user and task, imaging context is surely predicated on knowing the reason for the imaging study and the history of the patient: without this information, proper interpretation of the images is unfeasible. Consider the case of a cancer patient with three sequential imaging studies (Fig. 4.26): baseline (pre-treatment), initial post-treatment scan, and a current follow-up study. If interpretation is performed by comparing the current and baseline studies, a conclusion may be reached that a tumor is getting smaller. However, if treatment involved complete resection of the tumor, the interpretation of the current study would change towards progression of the disease. A more subtle example is that of an individual with chronic kidney disease: a finding of osteoporosis and its cause is dependent on whether the patient is on corticosteroids (*i.e.*, metabolic bone disease vs. long-term steroid-induced bone loss). In both cases, without the correct information, the radiologist will reach the wrong conclusion and improper treatment may result. Two commingled aspects of the workflow can be used to improve context:

1. Reason for exam. As intimated in the earlier chapters, the imaging study requisition typically contains the *reason for exam* (RFE). For issues of unknown etiology, the RFE provides the initial set of symptoms and/or the medical problem that is being considered as part of a differential diagnosis. For screening and follow-up imaging studies, the specific medical problem of interest is cited. Thus from the RFE, a target disease can be used to inform the context.
2. Prefetching. In (older) PACS (picture archive and communication system), *prefetching* involves retrieving past imaging studies to a local workstation in anticipation of their use (*e.g.*, based on a schedule of patients). The idea was to mitigate delays in data access, providing radiologists with "on-demand" comparison

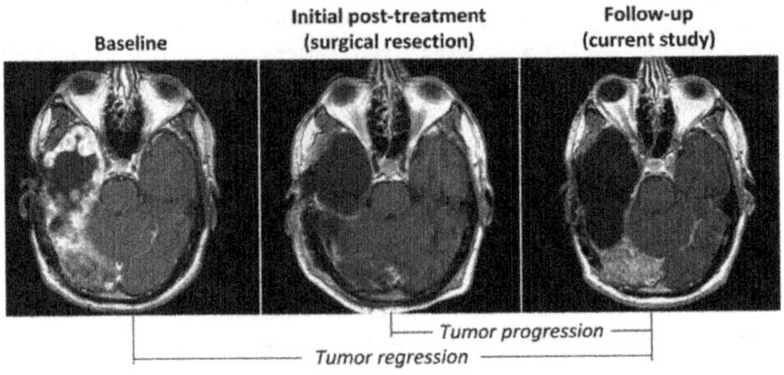

Figure 4.26: Context in image interpretation is critical, as inadequate information can impact patient care. In this example, comparison of the baseline image to the current follow-up may lead a conclusion of tumor regression; however, in knowing that the patient's tumor was removed, the interpretation changes to one of progression.

of a current imaging study to previous images. Although the importance of image prefetching has declined given faster networks and increased storage capacity, the idea of prefetching has evolved in step with PACS integration to other clinical systems, allowing for retrieval of previous reports and labs. So-called *problem-oriented prefetching* aims to retrieve any clinical data relevant to a given medical problem [24, 121], establishing historic context for interpretation.

Aspects of DICOM[5] Supplements 10 and 52 (Modality Worklist Management; General Purpose Worklists) can be used to aid in context definition. For instance, both supplements make use of free-text fields for the reason for the requested procedure (0040,1002) and requested procedure comments (0040,1400). However, the contents of these fields are not standardized (unless an enterprise has established operational policies) and may therefore need processing to extract pertinent medical problem references. Additional information on the targeted disease may be gleaned from the study protocol itself, and may be considered the reverse mapping of the protocol selection algorithms described earlier in this book: often, the image acquisition parameters of the study (*e.g.*, modality, pulse sequences, reconstructions) can be used to help distinguish the condition under investigation.

Viewing the data. Once the current imaging study is acquired and the data needed for interpretation is available, the next step is to present the data. We divide the process into three sub-steps: *data classification and filtering*, in which the retrieved contents of the patient's EMR are categorized and sorted based on priority; *image layout*, where the current imaging study is spatially organized based on the disease and task; and *information integration*, providing non-imaging data alongside the studies. The primary purpose of an integrated imaging display is to not only to facilitate image interpretation, but to enable the radiologist's role as a problem solver within the healthcare process. More detail on each sub-step follows:

1. Data classification and filtering. The first sub-step entails prioritization of the new images and the different data elements identified as part of prefetching [119]. As prefetching generally returns more clinical information than is immediately needed for image interpretation, non-relevant documentation from the EMR are removed from consideration in the integrated display. For instance, images of a patient's herniated disc have no bearing on a fractured distal radius: MR scans of the former can be ignored relative to an x-ray of the latter. Key events in the patient history are uncovered (*e.g.*, interventions, drug history) and image studies are temporally organized around these points in time (*e.g.*, pre-treatment vs. post-treatment). Clusters of data around these events are also created (*e.g.*, admission/discharge reports for an inpatient exam; recent lab work). Many methods can be used to classify and filter the patient data; a discussion is given in Chapter 7.
2. Image layout. From the context, both the disease and patient history (including past imaging exams) are known. The second stage determines how to layout the current set of images given the milieu of previous studies and the user's task. In the analog environment using alternators, a film librarian would manually organize this information for review following (heuristic) rules. With PACS, the equivalent is the *hanging protocol*. At a high level, a hanging protocol is meant to capture a user's preferences in viewing a given imaging study: the availability of prior studies for comparison; the layout and orientation of each imaging series; and

[5] The reader is referred to Chapter 3 for a more in-depth discussion of DICOM.

the visual characteristics of each image panel (*e.g.,* window/level, magnification) are all defined. Additionally, information taking into account the underlying capabilities of the imaging workstation can be used to best accommodate the user's preferences. Supplement 60 of the DICOM (Digital Imaging and Communications in Medicine) standard provides a formal method for the storage and retrieval of hanging protocols. The definition of a DICOM hanging protocol consists of two parts: the *definition module,* which entails a description of the protocol and its intended modality, anatomical region, laterality, procedure type, reason for procedure, and the priors needed for review; and the *environment module,* defining the number of screens and workstation capabilities (screen resolution, color/grayscale bit depth). Thus, a PACS implementing hanging protocols first attempts to match a given current imaging study to one or more hanging protocols, using the definition module for selection and then filtering using the environment module for the current imaging workstation. Based on the selected hanging protocol, the appropriate prior imaging exams are also retrieved. While Supplement 60 supplies the overall structure for hanging protocols, it does not detail how matches should be made, how additional information from other information systems may be used to inform the hanging protocol selection, or how to select amongst priors. Combining patient history, the RFE, and current study, new algorithms for selecting hanging protocols can be developed. Such rules can automatically arrange series for side-by-side study comparison [121] and overlays (*e.g.,* combined PET/CT) as part of the workflow. Notably, layout can also be guided based on past DICOM presentation states (Supplement 33) associated with the patient, identifying past series that were marked as "important" in prior readings. Taking layout design further, the displayed toolset can be tailored (*e.g.,* an adaptive ribbon, see Fig. 4.21) to the hanging protocol and task, and alternatives to standard tile layouts can be used to add visual information generated from image processing (*e.g.,* difference maps).

3. <u>Information integration</u>. Though images are the predominant data element of an imaging workstation, the secondary data obtained from the EMR needs to be made available through the same interface [120]. Even today, imaging workstations fail to integrate data from different clinical sources, and the user is faced with different screens/windows to access radiology and hospital information systems (RIS, HIS: debatably, this separation is counter to workflow. Contemporary methods for data integration can exploit advances in information extraction (*e.g.,* natural language processing, NLP; see Chapter 6) to identify crucial document subsections and values. Techniques for layering information, as described earlier, can be used to

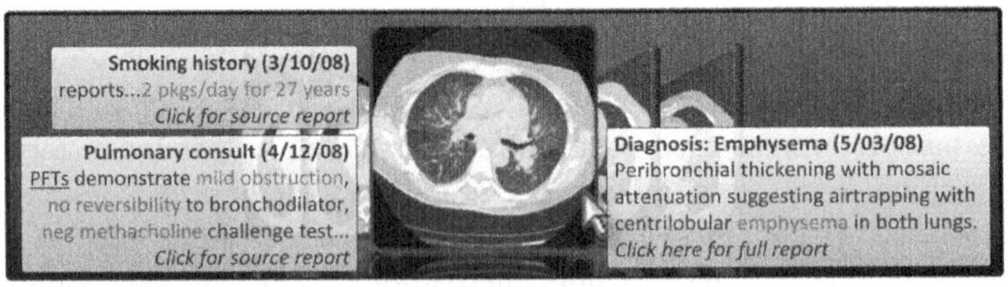

Figure 4.27: Demonstration of text and image integration. Thumbnail icons of past DICOM presentation states are shown in a cover flow widget. Mouseover of a given image presents key text used in establishing the diagnosis of the patient; NLP extracted, color-coded phrases documenting patient history are provided as semi-opaque popup tooltips. Each tooltip allows access to the source report.

link this data with the images directly. By way of illustration, sentinel images can be (automatically) annotated with extracted text (Fig. 4.27), removing the user's need to search for the corresponding report.

Facilitating reporting/feedback. Finally, context and the integrated display provide the means to improve the eventual generation of a report containing the interpretation. Increasingly, structured reports and reporting templates are used in radiology practices, and are usually linked to the study acquisition protocol. Having identified key information in the EMR, fields within these reports can potentially be automatically completed and linked with new sentinel images documenting findings. In effect, a succinct summary can be created to answer the questions set out by the RFE.

Discussion and Applications

As noted by many, a complete review of biomedical information visualization techniques over the decades is well beyond the reach of a single paper or chapter – we have thus attempted to touch upon a few key examples and issues in the area. Indeed, the pursuit of integrated medical displays has been continual since the introduction of computers into healthcare, accelerating first with the uptake of WIMP (windows, icons, menus, and pointing devices) GUIs; and again with HTML-based (hypertext markup language) interfaces. Faced not only with increases in the quantity of information available, but new types of clinical and research data, informaticians are challenged to find systematic means to design and present this data in a usable display. The EMR interface has hence progressed and today can be perceived as a study in collecting the "correct" data elements to tell an evolving story: the patient's medical history. The reader's familiarity with the story and his interest can vary: if the reader is familiar with the narrative, he may only wish to see the latest installment; in some situations, a gestalt sense of key events is sufficient; and in other cases, the reader is intent on examining the nuances of each scene. Moving forward, developments in integrated medical displays will occur on multiple fronts, including:

- Improved understanding of user requirements. The design of medical information displays will surely benefit from a clearer understanding of what drives clinicians' actions. Early efforts investigating the questions primary care physicians ask during routine care have helped establish common patterns of inquiry [52]. In a similar vein, NLM's Clinical Questions collection continues this endeavor to better understand the information needs of clinical users. By taking into account these questions as possible task categories, displays better suited to physicians' intentions and cognitive processes can be constructed.
- Common platforms. As in other informatics endeavors, there is now a push to standardize the components comprising integrated clinical GUIs. For instance, Microsoft's Common User Interface (CUI) [116] is a preliminary compilation of prototype graphical widgets for presenting information related to a range of healthcare processes (*e.g.*, laboratory data, medications, scheduling, imaging). While consensus on the presentation of such data may not be possible given the breadth of the information and user preferences, shared and iterative design will be a valuable tool to reveal potential design issues (*e.g.*, see IBM's Many Eyes website). Alternatively, approaches that sufficiently abstract clinical data visualization to the level of an application programmer interface (API) may provide a basis for open source development, analogous to that seen with the Insight Toolkit (ITK) and Visualization Toolkit for 3D graphics (VTK).

- New designs. Ultimately, new integrated clinical displays must advance past the "one-interface-fits-all" model of source-oriented organization, to views that catalyze exploration and the users' thinking. Any relationships between elements should be made graphically apparent; and as databases of medical knowledge grow, known causal connections can be applied to avail visual design [30]. Moreover, how can we improve what users do today through EMR and clinical workstation interfaces? The evaluation of new designs must also be carefully considered: what is an appropriate baseline for comparison, and what metrics should be considered in terms of usability, efficiency, and (knowledge) discovery?

We conclude with two sample applications that bring together the ideas presented in this chapter. First, the TimeLine project has been an ongoing effort to realize problem-centric visualizations of medical records. Core structures within this project and our experience in creating this integrated display are briefly described. Second, as personal health records become adjunct to the EMR, we describe the development of patient-centric visualizations – a largely different user base from the traditional physician-oriented view.

TimeLine: Problem-centric Visualization

The majority of clinical portals and EMR interfaces presently mirror the underlying modular organization of information around departments and originating data sources.

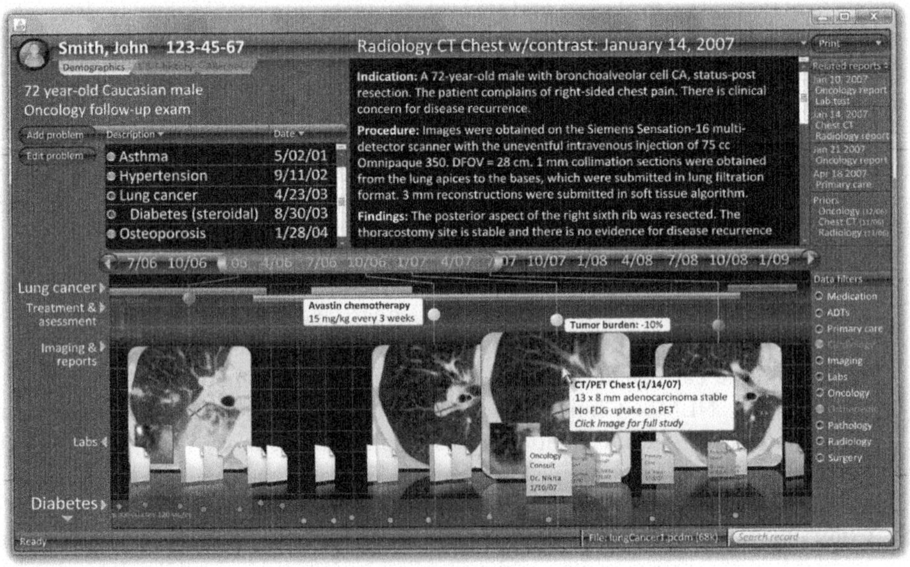

Figure 4.28: The TimeLine interface, organizing information around a patient's medical problem list. The top left of the display contains a chronology of the individual's medical problems. By selecting items from the list, the user adds timelines to the bottom half of the display, presenting icons and summary data for the particular problem. Mousing over the icons, additional information is simultaneously presented for connected data elements. For instance, in this lung cancer example, concurrent chemotherapy, tumor burden, and tumor measurements are displayed when the mouse is over a sentinel image of the tumor. Temporal granularity is controllable by stretching the time bar in the middle of the interface. Newer developments in TimeLine permit the user to merge timelines for visual presentation, and the use of sentinel events to "stretch" horizontal spacing between data.

For example, a patient record view often comprises (tabbed) panels that allow the user to peruse documents from laboratory, radiology/PACS, cardiology, etc. But observations and treatment plans for a given disease span these different databases; hence, finding data can require multiple searches. Largely, this source-oriented information organization does not fit with the mental paradigms of physicians, or how users think about a disease. In contrast, an abundant number of interfaces have been presented in the literature, designed to display information for a single disease entity (*e.g.*, diabetes, cancer, etc.). Although such systems excel at the presentation of the medical problem, the relationship between multiple symptoms and diseases can be lost. A balance between the two extremes of complete EMR access and disease-specific views is needed.

The TimeLine project has been a continuing endeavor to create a problem-centric visualization framework that addresses this need [22, 23]. TimeLine consists of two interlinked parts: *data reorganization*, wherein the EMR is retrieved and automatically sorted around a medical problem list; and a *time-based visualization*, which draws upon a *visualization dictionary* to composite a layered chronology (*i.e.*, timeline) of visual metaphors per medical problem and based on context. The result is a display that permits a user to access the full patient record while simultaneously viewing multiple medical problems in a tailored fashion, all sharing the same time axis (Fig. 4.28). The interface provides an overview to the patient's medical history, allowing drill-down to access specific data elements in detail (*i.e.*, reports, images, etc.).

Data Reorganization

Given access to the clinical patient record, two difficulties became apparent in working toward an implementation of TimeLine. First, as the EMR is adopted, movement toward standards for data representation and content are progressing. Yet the use of standards is often wrought with site- and/or vendor-specific adaptations; and mismatches can occur even between different versions of the same standard. Second, access to this information is still source-oriented: a means of reordering data around different medical problems is necessary. But the concept of a medical "problem" is often mutable based on the clinician's perspective; and a complete dependence on codification schemes (*e.g.*, ICD-9, SNOMED) to provide disease categories is contingent on the proper and consistent use of codes for all data. TimeLine's solutions are twofold: 1) data mapping from XML (eXtensible Markup Language) data representations to an object-oriented representation of clinical data elements; and 2) data classification of the elements into temporal views that correspond to user-defined categories of medical problems. We summarize both processes here; further details are given in [22].

Data mapping. The data mapping process starts with the definition of "core" elements that represent a minimal set of information for a clinical data entity. In essence, for a given data entity, a set of attributes are defined that must be present in order for instantiation. For example, a clinical document usually involves a title, a report body, and an author. In some cases, standards provide a starting point for determining these attributes (*e.g.*, DICOM for medical images; HL7 clinical document architecture (CDA) for reports). However, these standards are comprehensive in their definition, rather than strictly minimal: to manage different sources and representations of the same types of information, TimeLine's tactic is to use the "lowest common denominator" in data representation. For each clinical data entity, TimeLine defines an *element property file* (EPF) that declares an attribute name and data type for the attribute value. The EPF for a clinical data entity is thus the minimal class definition for an object.

Given the EPF, TimeLine uses a separate set of transformation rules to parse XML clinical documents into instances of these object classes.

Data classification. Upon instantiation of a clinical data element from its XML, TimeLine attempts to associate the instance with one or more medical problems. This data classification process uses a pipeline of rule-based and classifier methods:

1. Data model. TimeLine uses a data model that integrates clinical information from the underlying EMR data sources. The data model uses an extended entity-relation framework to represent each clinical data source as a time-stamped stream of elements (see Chapter 7). These temporal sequences are then used to construct logical partitions of the data: subsets of information defined by conditional constraints (*e.g.*, on the elements' attributes) can be specified to create disease-specific views. Consequently, the definition of these views provides a knowledge-base upon which classification of a given object can transpire (*i.e.*, a given data element will appear in all views whose criteria it matches).
2. Classification codes. When available, ICD-9 and SNOMED codes are used as clues to further aid the classification process. For example, rather than use the specific ICD-9 hierarchy, we have developed an anatomic categorization based on ICD-9 codes to assist in medical problem list generation [26]. Notably, TimeLine ignores commonly occurring symptomatic codes (*e.g.*, fever, headache) as they are too non-specific in nature to help in classification.
3. Content-based features. Additional clues are obtained by invoking different techniques, such as NLP, to suggest the primary topic of discussion or concern; or by determining the original reason for a study (*e.g.*, the RFE).

The end result of this classifier pipeline is a tree structure with the following categorical pattern: anatomy → medical problem → data source → clinical data. We begin with broad groupings of the data based on anatomy and systems (*e.g.*, brain, pulmonary, musculoskeletal) given that many (sub-specialty) clinicians concentrate on a given region (*e.g.*, neurologist, pulmonologist, orthopedic surgeon). Subsequently, the hierarchy progresses to a disease entity (*e.g.*, stroke, asthma, osteoarthritis) and then group the data therein based on source (*e.g.*, imaging, reports, labs). This organization of the data serves as the foundation from which visualizations can then be fashioned.

Visualization Dictionary

Given the large number of visualizations that are available to developers when building interfaces for viewing medical records, a challenge lies in determining which representation to use for a given data type and in a particular context. To address this issue, [25] presents a *visualization dictionary*, which guides the choice of visualization based on context, type of medical data, and available visualizations. The dictionary may be thought of conceptually as a three-dimensional space, illustrated in Fig. 4.29a. A point in this space thus determines for a given context (*i.e.*, medical problem, user, task) and data type how to display the information such that it highlights important features in a way relevant to the user and task. Each axis in described in detail.

Data types. In order to match data types to visualizations, each data type needs to be characterized. One method is to re-organize the data into semantically related clusters that are structured hierarchically in a process called *hierarchical data clustering*. For

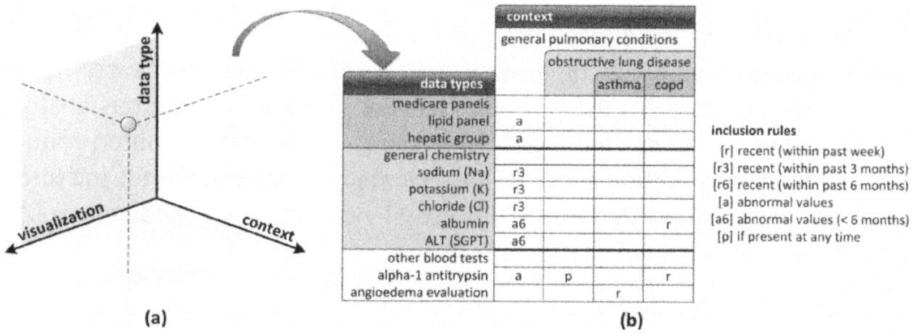

Figure 4.29: TimeLine visualization dictionary. **(a)** The dictionary is conceptualized as three separate axes, defining for a given context and data type an appropriate visual metaphor for rendering in the display. **(b)** The plane defined by the data type and context is a table with inclusion rules. A hierarchy of related diseases and data types are organized in successively more specific physician-defined categorizations; different logical rules for including a given data type are specified as entries in the table.

instance, at the highest level of abstraction, a patient record contains a heterogeneous collection of data: clinical documents, lab results, imaging studies, genetic profiles, and others. Further subclasses are derived by enumerating attributes within each type. For instance, clinical documents come in different forms (*e.g.*, letter, consult, pre-op/post-op), are generated by different departments (*e.g.*, radiology, pathology), and are written by various physicians (*e.g.*, radiologist, oncologist). Knowledge of related clusters may be obtained using a controlled vocabulary such as UMLS. By grouping data elements across multiple hierarchical levels, the dictionary captures the depth and variety of information captured in clinical data. Organizing data elements by clusters allow a visualization to answer questions such as: *given a particular variable and level of abstraction, what are all the related variables that match the defined level?* Higher-level abstractions are used to summarize the patient's data, while lower-level abstractions provide the details. The data type axis is correlated with TimeLine's data classification scheme, and in particular, its temporal data model.

Visualizations. Available visual techniques are also characterized in the dictionary. The process of characterization occurs twofold: first, visualizations are assigned to a classification based on the type of data they are designed to present; and second, attributes and requirements are enumerated. Each visualization type has a set of considerations that determine how effective the visualization is under certain situations. For example, a line plot may not be suited for rendering data streams with low sampling frequency (*e.g.*, lab values with few measurements or large gaps in time between measurements). The properties are stored in the visualization dictionary and recalled in queries that search for the "best" visualization given a set of attributes. For instance, while a bar chart may suffice to represent elementary relationships, for more complex data a configural chart may be better in order to convey multiple relationships at once. In addition, attributes may be used to determine the "next best" visualization if the primary choice is not available (or preferred by the user). By way of illustration, if 2D line plots were not available, a query with the desired properties would then be executed against the visualization dictionary, and the visualization with the closest matching parameters would be returned (*e.g.*, scatter plot, 3D line plot).

Context. Our earlier discussion of context in this chapter is applicable here: a context is driven by a particular requirement to focus on one type of information; all other data is secondary to this principal information. Notably, TimeLine's visualization dictionary uses context in a problem-centric manner (*i.e.*, being disease driven). Though many ontologies exist for disease classification, such works are not focused on the relationship of data for the purposes of visualization, so much as anatomic and etiologic concepts. In fact, too much granularity is in fact counterproductive in organizing clinical information [165]. Thus, the context axis in TimeLine attempts to align multiple disease visualizations with users' perception of medical problems, as defined by the system's underlying data model.

The plane defined by the data type *vs.* context axes comprises a grid. In conjunction with the data classification methods given above, entries in this grid state a type of *data inclusion rule* for a given context, refining the potential view: *always include*, where values for the given data type are always presented; *include based on recent activity* temporally filters information (*e.g.*, within the past six months); *include based on data value* triggers inclusion of a given type of data based on a value (*e.g.*, abnormal laboratory); and *include based on trend* handles the case when the user is interested in sudden value changes (*e.g.*, blood glucose levels are slowly increasing, even if considered normal). These rules may be combined using Boolean logic to create more complex data inclusion behavior (*e.g.*, only include a data element if it is abnormal and within the past three months).

Dictionary representation. The visual dictionary performs three tasks: 1) it indexes the metadata generated about data elements and available visualizations; 2) it matches data elements with the appropriate visualization given contextual information; and 3) it generates a set of instructions that determines the how the information is formatted and rendered in the application. Initially, the application provides a list of data elements to be rendered (*e.g.*, laboratory values) and the context (*e.g.*, patient complains of being constantly tired). Each data element is characterized and hierarchically clustered based on their relationship with the given context (*e.g.*, related laboratory tests are paired together). A query is executed to search for visualizations that are capable of displaying each data element (*e.g.*, time series plot) and are the closest matches to the data's attributes (*e.g.*, able to display the frequency and range of the data). The results are filtered and formatted by the data inclusion rules. For instance, plots with abnormal values are highlighted and displayed prominently while other plots are rendered such that they appear in the background. In addition, if the physician knows that the patient is diabetic, specific lab tests that are relevant to this group of patients (*e.g.*, glucose, insulin, c-peptide) are displayed first.

Patient-centric Visualization

A growing trend in healthcare is the availability and adoption of personal health records (PHRs). A major hurdle towards implementing PHRs has been dealing with data standards and interoperability between systems. But developments addressing these issues, such as the HL7's Continuity of Care Document (CCD), have helped establish interfaces that access federated clinical records (see Chapter 3). Many employers, managed care organizations, and third party vendors are starting to provide patients with online portals that allow them to view and store their medical data. Google Health and Microsoft HealthVault are examples of recent efforts to develop patient-centric web mashups of clinical data by combining information from multiple sources (*e.g.*, providers, drug stores, personal monitoring devices).

PHRs provide new opportunities for communicating information over the web: either through a community of peers who share similar goals and experiences, or with a primary care provider using secure messaging. Peer support communities and forums have been shown to offer individuals emotional as well as informational support in dealing with chronic illnesses. These communities often provide experiential empathy, which is typically beyond the scope of practitioners [76]. [64] studied 780 patients with hypertension and found that individuals actively engaged in their own care, such as communicating regularly with their healthcare provider and viewing/updating their medical records through an online portal, had better outcomes.

Unlike electronic medical records, which are often created and maintained by individual institutions (*e.g.*, hospitals, PCPs), the PHR is initiated and maintained by the patient. The patient has sole control over the people who have the credentials to view this information. As such, the presentation and nature of PHRs is markedly different from that of the EMR:

- User audience. Patients typically do not have the same depth of knowledge as physicians; therefore, patients need to be provided with context when reviewing laboratory results and documents generated during clinical care. In the past, practitioners have expressed concern about giving patients access to documents that are not intended for a lay audience (*e.g.*, are patients interested in reading and contributing to the medical record? Does the record cause confusion, and/or promote/relieve anxiety?) [66]. However, studies that have interviewed patients with access to their own records have shown the potential of improving outcomes and patient experience by actively engaging them in their own care [140, 180]. Indeed, PHRs are in a unique position to educate patients about their conditions if an appropriate interface is developed to allow patients to learn about and explore their data in a guided and understandable way. Unlike physicians who are used to dealing with large quantities of medical data on a regular basis, consideration is needed in designing patient-centric visualizations to prevent information overload by tailoring the type and amount of information presented to a patient.
- Tasks and expectations. Patients store health information in a PHR with a different intent than a practitioner-driven EMR. EMRs are legally binding, thus creating implications for how information is captured. While the PHR may include many of the same types of information – labs, clinical notes, and health histories – people use this information to obtain a greater understanding of their illnesses, manage chronic illnesses, and promote healthy habits such as exercise and dieting.
- Access. While most PHRs are envisioned as part of a website, an emergent class of mobile devices and wearable sensors are now able to provide immediate access to health data and resources while collecting real-time information about patients, such as vitals (see Chapter 3). Visualizations adapted to such devices are needed to provide patients with the ability to explore their own data, integrating it with information from the medical enterprise to provide the lay person with context for interpreting the results.

Presently, patient-centric displays utilize a combination of visualizations that have been described earlier in this chapter. Such GUIs often feature colorful, intuitive interfaces with simple, clear explanations and guided wizards that assist users with the inputting and viewing of data; but largely, the visual design of PHRs is a nascent area. However, it is clear that new strategies must be taken in tailoring the display and methods of interaction with patients. For instance, previous studies have shown that a patient's ability to search for relevant medical information is severely impaired by his lack of domain knowledge. [182] tested an individual's ability to find doctors with certain clinical interests using terms with which the individual is familiar (*e.g.,* cancer specialist versus oncologist). The study concludes that a lay understanding of medical terminology leads to the inability to retrieve desired information because clinical documents are typically encoded using professional terms. One approach has been to develop an automated translation system that supplements documents using medical jargon with consumer-friendly display names that help patients understand the terms [183]. Also, when searching for information, systems can analyze the terms inputted by the patient and automatically suggest synonymous technical terms.

Individuals are becoming increasingly empowered in their own care, and new paradigms for presenting medical information are needed to help patients and physicians alike interpret and act upon the collected information. Unlike physicians, however, patient-centered visualizations need to focus on the goals and expectations of the patient: patient health portals need to tailor their content so that patients can be best informed of their condition to encourage interaction with their practitioner and to enact lifestyle modifications as necessary.

References

1. Abad-Mota S, Kulikowski C, Gong L, Stevenson S, Mezrich R, Tria A (1995) Iconic reporting: A new way of communicating radiologic findings. Annu Symp Comput Appl Med Care, p 915.
2. Aberle DR, Dionisio JD, McNitt-Gray MF, Taira RK, Cardenas AF, Goldin JG, Brown K, Figlin RA, Chu WW (1996) Integrated multimedia timeline of medical images and data for thoracic oncology patients. RadioGraphics, 16(3):669-681.
3. Ahlberg C (1996) Spotfire: An information exploration environment. ACM SIGMOD Record, 25(4):25-29.
4. Ahlberg C, Williamson C, Shneiderman B (1992) Dynamic queries for information exploration: An implementation and evaluation. Proc ACM SIGCHI Conf Human Factors in Computer Systems, pp 619-626.
5. Aigner W, Miksch S, Müller W, Schumann H, Tominski C (2007) Visualizing time-oriented data - A systematic view. Computers & Graphics, 31(3):401-409.
6. Aigner W, Miksch S, Müller W, Schumann H, Tominski C (2008) Visual methods for analyzing time-oriented data. IEEE Trans Vis Comput Graph, 14(1):47-60.
7. Andrews DF (1972) Plots of high dimensional data. Biometrics, 28(1):125-136.
8. Anind KD (2001) Understanding and using context. Personal Ubiquitous Comput, 5(1):4-7.
9. Aris A, Shneiderman B, Plaisant C, Shmueli G, Jank W (2005) Representing unevenly-spaced time series data for visualization and interactive exploration. In: Costabile MF, Paterno F (eds) Proc INTERACT, pp 835-846.
10. Arnheim R (2004) Visual Thinking. 2 edition. University of California Press, Berkeley.
11. Augusto JC (2005) Temporal reasoning for decision support in medicine. Artif Intell Med, 33(1):1-24.
12. Bade R, Schlechtweg S, Miksch S (2004) Connecting time-oriented data and information to a coherent interactive visualization. Proc SIGCHI Conf Human Factors in Computing Systems, pp 105-112.
13. Bardram JE (2004) Applications of context-aware computing in hospital work: Examples and design principles. Proc 2004 ACM Symp Applied Computing. ACM Press, Nicosia, Cyprus.
14. Beaumont IH (1994) User modeling in the interactive anatomy tutoring system ANATOM-TUTOR. User Modeling and User-Adapted Interaction, 4(1):21-45.

15. Bederson BB (2001) PhotoMesa: A zoomable image browser using quantum treemaps and bubblemaps. Proc 14th Annual ACM Symp User Interface Software and Technology, pp 71-80.
16. Benyon D (1993) Adaptive systems: A solution to usability problems. User Modeling and User-Adapted Interaction, 3(1):65-87.
17. Bodenreider O (2002) Experiences in visualizing and navigating biomedical ontologies and knowledge bases. Proc Intelligent Systems for Molecular Biology (ISMB) SIG Meeting (Bio-ontologies).
18. Bonetti PO, Waeckerlin A, Schuepfer G, Frutiger A (2000) Improving time-sensitive processes in the intensive care unit: The example of 'door-to-needle time' in acute myocardial infarction. Int J Qual Health Care, 12(4):311-317.
19. Boxwala AA, Peleg M, Tu S, Ogunyemi O, Zeng QT, Wang D, Patel VL, Greenes RA, Shortliffe EH (2004) GLIF3: A representation format for sharable computer-interpretable clinical practice guidelines. J Biomed Inform, 37(3):147-161.
20. Brodmann K (1909) Vergleichende Lokalisationslehre der Grosshirnrinde: In ihren Prinzipien dargestellt auf Grund des Zellenbaues. JA Barth, Leipzig.
21. Brusilovsky P (1996) Methods and techniques of adaptive hypermedia. User Modeling and User-Adapted Interaction, 6(2):87-129.
22. Bui AA, Aberle DR, Kangarloo H (2007) TimeLine: Visualizing integrated patient records. IEEE Trans Inf Technol Biomed, 11(4):462-473.
23. Bui AA, Aberle DR, McNitt-Gray MF, Cardenas AF, Goldin J (1998) The evolution of an integrated timeline for oncology patient healthcare. Proc AMIA Symp, pp 165-169.
24. Bui AA, McNitt-Gray MF, Goldin JG, Cardenas AF, Aberle DR (2001) Problem-oriented prefetching for an integrated clinical imaging workstation. J Am Med Inform Assoc, 8(3):242-253.
25. Bui AA, Taira RK, Churchill B, Kangarloo H (2002) Integrated visualization of problem-centric urologic patient records. Ann N Y Acad Sci, 980:267-277.
26. Bui AA, Taira RK, El-Saden S, Dordoni A, Aberle DR (2004) Automated medical problem list generation: Towards a patient timeline. Proc MedInfo, vol 107, pp 587-591.
27. Card SK, Mackinlay JD, Shneiderman B (eds) (1999) Readings in Information Visualization: Using Vision to Think. Morgan Kaufmann, Inc. San Francisco, CA.
28. Catarci T, Costabile MF, Levialdi S, Batini C (1997) Visual query systems for databases: A survey. Journal of Visual Languages and Computing, 8(2):215-260.
29. Chang NS, Fu KS (1979) Query-by-pictorial-example. Proc IEEE 3rd Intl Computer Software and Applications Conference (COMPSAC 79), pp 325-330.
30. Chen C (2005) Top 10 unsolved information visualization problems. IEEE Trans Computer Graphics and Applications, 25(4):12-16.
31. Chiba N, Nishizeki T, Abe S, Ozawa T (1985) A linear algorithm for embedding planar graphs using PQ-trees. J Computer and System Sciences, 30(1):54-76.
32. Chitarro L (2001) Information visualization and its application to medicine. Artif Intell Med, 22:81-88.
33. Chittaro L, Combi C (2001) Representation of temporal intervals and relations: Information visualization aspects and their evaluation. Proc 8th Intl Symp Temporal Representation and Reasoning (TIME'01). IEEE Computer Society, pp 13-20.
34. Cho H, Ishida T, Yamashita N, Inaba R, Mori Y, Koda T (2007) Culturally-situated pictogram retrieval. In: Ishida T, Fussell SR, Vossen PTJM (eds) Connecting the Universal to the Specific: Towards the Global Grid (IWIC 2007), pp 211-235.
35. Ciccarelli FD, Doerks T, von Mering C, Creevey CJ, Snel B, Bork P (2006) Toward automatic reconstruction of a highly resolved tree of life. Science, 311(5765):1283-1287.

36. Clayman CB, Curry RH (1992) The American Medical Association Guide to Your Family's Symptoms. 1st updated pbk. edition. Random House, New York.
37. Cohen IB (1984) Florence Nightingale. Scientific American (250):128-137.
38. Collins C (2006) DocuBurst: Document content visualization using language structure. Proc IEEE Symp Information Visualization, Baltimore, MD.
39. Combi C, Shahar Y (1997) Temporal reasoning and temporal data maintenance in medicine: Issues and challenges. Computers in Biology and Medicine, 27(5):353-368.
40. Combs TTA, Bederson BB (1999) Does zooming improve image browsing? Proc 4th ACM Conference on Digital Libraries. ACM, Berkeley, California, United States, pp 130-137.
41. Cooperative Association for Internet Data Analysis (CAIDA) (2008) Walrus visualization tool. http://www.caida.org/tools/visualization/walrus/. Accessed June 6, 2008.
42. Cousins SB, Kahn MG (1991) The visual display of temporal information. Artif Intell Med, 3:341-357.
43. Di Battista G, Eades P, Tamassia R, Tollis IG (1998) Graph Drawing: Algorithms for the Visualization of Graphs. Prentice Hall, Upper Saddle River, NJ, USA.
44. Dionisio JDN, Cardenas AF (1996) MQuery: A visual query language for multimedia, timeline and simulation data. J Visual Languages and Computing, 7(4):377-401.
45. Dix A, Ellis G (1998) Starting simple: Adding value to static visualisation through simple interaction. Proc Working Conf Adv Visual Interfaces, pp 124-134.
46. Eades P (1984) A heuristic for graph drawing. Congressus Numerantium, 42:149-160.
47. Egenhofer MJ (1997) Query processing in spatial-query-by-sketch. J Visual Languages and Computing, 8(4):403-424.
48. Ehlschlaeger CR, Shortridge AM, Goodchild MF (1997) Visualizing spatial data uncertainty using animation. Computers & Geosciences, 23(4):387-395.
49. Eick SG (2000) Visualizing multi-dimensional data. ACM SIGGRAPH Computer Graphics, 34(1):61-67.
50. Eisen MB, Spellman PT, Brown PO, Botstein D (1998) Cluster analysis and display of genome-wide expression patterns. Proc National Academy of Sciences, 95(25):14863-14868.
51. Elmqvist N, Tsigas P (2004) Animated visualization of causal relations through growing 2D geometry. Information Visualization, 3(3):154-172.
52. Ely JW, Osheroff JA, Ebell MH, Bergus GR, Levy BT, Chambliss ML, Evans ER (1999) Analysis of questions asked by family doctors regarding patient care. BMJ, 319(7206):358-361.
53. Falkman G (2001) Information visualisation in clinical odontology: Multidimensional analysis and interactive data exploration. Artif Intell Med, 22(2):133-158.
54. Feldman-Stewart D, Kocovski N, McConnell BA, Brundage MD, Mackillop WJ (2000) Perception of quantitative information for treatment decisions. Med Decis Making, 20(2):228-238.
55. Feng Q (1997) Algorithms for drawing clustered graphs. Department of Computer Science and Software Engineering, PhD Dissertation. University of Newcastle.
56. Ferrant M, Nabavi A, Macq B, Jolesz FA, Kikinis R, Warfield SK (2001) Registration of 3-D intraoperative MR images of the brain using a finite-element biomechanical model. IEEE Trans Med Imaging, 20(12):1384-1397.
57. Ferreira de Oliveira MC, Levkowitz H (2003) From visual data exploration to visual data mining: A survey. IEEE Trans Vis Comput Graph, 9(3):378-394.
58. Finger R, Bisantz AM (2002) Utilizing graphical formats to convey uncertainty in a decision-making task. Theoretical Issues in Ergonomics Science, 3(1):1-25.
59. Fischer G (2001) User modeling in human-computer interaction. User Modeling and User-Adapted Interaction, 11(1):65-86.

60. Fox J, Alabassi A, Patkar V, Rose T, Black E (2006) An ontological approach to modeling tasks and goals. Computers in Biology and Medicine, 36(7-8):837-856.
61. Frias-Martinez E, Magoulas G, Chen S, Macredie R (2006) Automated user modeling for personalized digital libraries. Intl J Information Management, 26(3):234-248.
62. George GR, Jock DM, Stuart KC (1991) Cone trees: Animated 3D visualizations of hierarchical information. Proc SIGCHI Conf Human Factors in Computing Systems. ACM, New Orleans, Louisiana, United States.
63. Görg C, Pohl M, Qeli E, Xu K (2007) Visual representations. In: Kerren A, et al. (eds) Human-centered Visualization Environments, vol 4417. Springer Berlin/Heidelberg, pp 163-200.
64. Green BB, Cook AJ, Ralston JD, Fishman PA, Catz SL, Carlson J, Carrell D, Tyll L, Larson EB, Thompson RS (2008) Effectiveness of home blood pressure monitoring, Web communication, and pharmacist care on hypertension control: A randomized controlled trial. JAMA, 299(24):2857-2867.
65. Haimowitz IJ, Kohane IS (1996) Managing temporal worlds for medical trend diagnosis. Artif Intell Med, 8(3):299-321.
66. Halamka JD, Mandl KD, Tang PC (2008) Early experiences with personal health records. J Am Med Inform Assoc, 15(1):1-7.
67. Hall KM (1970) An r-dimensional quadratic placement algorithm. Management Science, 17:219-229.
68. Hansen MD (2005) An Analysis of the Diagrammatic Visual Data Querying Domain. Computer Science Department, PhD Dissertation. University of California, Santa Cruz.
69. Hansen MD (2005) A survey of systems in the diagrammatic visual data querying domain. University of California, Santa Cruz.
70. Harris RL (1999) Information Graphics: A Comprehensive Illustrated Reference. Oxford University Press, New York.
71. Hearst M (1995) TileBars: Visualization of term distribution information in full text information access. Proc SIGCHI Conf Human Factors in Computing Systems. ACM Press/ Addison-Wesley Publishing Co., Denver, CO, United States.
72. Heer J, Card SK, Landay JA (2005) prefuse: A toolkit for interactive information visualization. Proc SIGCHI Conf Human Factors in Computing Systems, pp 421-430.
73. Herman I, Melançon G, Marshall MS (2000) Graph visualization and navigation in information visualization: A survey. IEEE Trans Vis Comput Graph, 6(1):24-43.
74. Hoeke JO, Bonke B, van Strik R, Gelsema ES (2000) Evaluation of techniques for the presentation of laboratory data: Support of pattern recognition. Methods Inf Med, 39(1):88-92.
75. Hoeke JO, Gelsema ES, Wulkan RW, Leijnse B (1991) Graphical non-linear representation of multi-dimensional laboratory measurements in their clinical context. Methods Inf Med, 30(2):138-144.
76. Hoey LM, Ieropoli SC, White VM, Jefford M (2008) Systematic review of peer-support programs for people with cancer. Patient Educ Couns, 70(3):315-337.
77. Horvitz E, Breese J, Heckerman D, Hovel D, Rommelse K (1998) The Lumiere Project: Bayesian user modeling for inferring the goals and needs of software users. Proc 14th Conf Uncertainty in Artificial Intelligence, pp 256-265.
78. Hughes T, Hyun Y, Liberles D (2004) Visualising very large phylogenetic trees in three dimensional hyperbolic space. BMC Bioinformatics, 5:48.
79. Huynh DF, Drucker SM, Baudisch P, Wong C (2005) Time Quilt: Scaling up zoomable photo browsers for large, unstructured photo collections. Proc ACM SIGCHI Conf Human Factors in Computing Systems:1937-1940.

80. IBM Zurich Research Laboratory (2007) IBM Research unveils 3-D avatar to help doctors visualize patient records and improve care. http://www.zurich.ibm.com/news/07/asme.html. Accessed August 1, 2008.
81. Inselberg A, Dimsdale B (1990) Parallel coordinates: A tool for visualizing multi-dimensional geometry. Proc 1st IEEE Conf Visualization, pp 361-378.
82. Jahnke J, Bychkov Y, Dahlem D, Kawasme L (2004) Context-aware information services for health care. In: Roth-Berghofer T, Schulz S (eds) Proc 1st Intl Workshop on Modeling and Retrieval of Context (MRC).
83. Jing Y, Jianping F, Daniel H, Yuli G, Hangzai L, William R, Matthew W (2006) Semantic Image Browser: Bridging information visualization with automated intelligent image analysis. IEEE Symp Visual Analytics Science And Technology, pp 191-198.
84. Johnson CR, Sanderson AR (2003) A next step: Visualizing errors and uncertainty. IEEE Trans Computer Graphics and Applications, 23(5):6-10.
85. Johnson DB, Taira RK, Zhou W, Goldin JG, Aberle DR (1998) Hyperad: Augmenting and visualizing free text radiology reports. Radiographics, 18(2):507-515.
86. Joseph T, Cardenas AF (1988) PICQUERY: A high level query language for pictorial database management. IEEE Trans Software Engineering, 14(5):630-638.
87. Kadaba NR, Irani PP, Leboe J (2007) Visualizing causal semantics using animations. IEEE Trans Vis Comput Graph, 13(6):1254-1261.
88. Kamel Boulos MN, Roudsari AV, Carso NE (2002) HealthCyberMap: A semantic visual browser of medical Internet resources based on clinical codes and the human body metaphor. Health Info Libr J, 19(4):189-200.
89. Keim DA, Kriegel HP (1996) Visualization techniques for mining large databases: A comparison. IEEE Trans Knowledge and Data Engineering, 8(6):923-938.
90. Kiraly AP, Helferty JP, Hoffman EA, McLennan G, Higgins WE (2004) Three-dimensional path planning for virtual bronchoscopy. IEEE Trans Med Imaging, 23(11):1365-1379.
91. Klimov D, Shahar Y (2005) A framework for intelligent visualization of multiple time-oriented medical records. Proc AMIA Symp, p 405.
92. Knapp P, Raynor DK, Jebar AH, Price SJ (2005) Interpretation of medication pictograms by adults in the UK. Ann Pharmacotherapy, 39(7):1227-1233.
93. Kobsa A (2001) Generic user modeling systems. User Modeling and User-Adapted Interaction, 11(1-2):49-63.
94. Kohlmann P, Bruckner S, Kanitsar A, Gröoller ME (2007) LiveSync: Deformed viewing spheres for knowledge-based navigation. IEEE Trans Visualization and Computer Graphics, 13(6):1544-1551.
95. Krupinski EA, Kundel HL, Judy PF, Nodine CF (1998) The Medical Image Perception Society: Key issues for image perception research. Radiology, 209(3):611-612.
96. Kumar A, Ciccarese P, Smith B, Piazza M (2004) Context-based task ontologies for clinical guidelines. In: Pisanelli DM (ed) Ontologies in Medicine. IOS Press, pp 81-94.
97. Kumar A, Smith B, Pisanelli DM, Gangemi A, Stefanelli M (2004) An ontological framework for the implementation of clinical guidelines in health care organizations. In: Pisanelli DM (ed) Ontologies in Medicine. IOS Press, pp 95-107.
98. Kundel HL (2000) Visual search in medical images. In: Beutel J, Horii SC, Kim Y (eds) Handbook of Medical Imaging, vol 1. SPIE Press, pp 837–858.
99. Lancaster JL, Summerlin JL, Rainey L, Freitas CS, Fox PT (1997) The Talairach Daemon, a database server for Talairach atlas labels. Neuroimage, 5(4):S633.
100. Langley P (1999) User modeling in adaptive interfaces. Proc 7th Intl Conf User Modeling, pp 357-370.

101. Lehmann TM, Guld MO, Thies C, Fischer B, Spitzer K, Keysers D, Ney H, Kohnen M, Schubert H, Wein BB (2004) Content-based image retrieval in medical applications. Methods Inf Med, 43(4):354-361.
102. Lewis C, Rieman J (1993) Task-Centered User Interface Design: A Practical Introduction. University of Colorado, Boulder.
103. Limbourg Q, Vanderdonckt J (2003) Comparing task models for user interface design. In: Diaper D, Stanton N (eds) The Handbook of Task Analysis for Human-Computer Interaction, pp 135-154.
104. Lin X (1992) Visualization for the document space. Proc IEEE Conf Visualization '92, pp 274-281.
105. Liu J, Kuenong C, Hui KK (2003) An adaptive user interface based on personalized learning. IEEE Intelligent Systems, 18(2):52-57.
106. Lodha SK, Pang A, Sheehan RE, Wittenbrink CM, Yagel R, Nielson GM (1996) UFLOW: Visualizing uncertainty in fluid flow. Proc IEEE Conf Visualization, pp 249-254.
107. Luis F-R, Frank M. Shipman, III (2000) Adaptive medical information delivery combining user, task and situation models. Proc 5th Intl Conf Intelligent User Interfaces. ACM, New Orleans, Louisiana, United States, pp 94-97.
108. Lundstrom C, Ljung P, Persson A, Ynnerman A (2007) Uncertainty visualization in medical volume rendering using probabilistic animation. IEEE Trans Vis Comput Graph, 13(6):1648-1655.
109. Lungu M, Xu K (2007) Biomedical information visualization. In: Kerren A, et al. (eds) Human-centered Visualization Environments, vol 4417. Springer Berlin/Heidelberg, pp 311-342.
110. Mackinlay J (1986) Automating the design of graphical presentations of relational information. ACM Trans Graph, 5(2):110-141.
111. Martins SB, Shahar Y, Goren-Bar D, Galperin M, Kaizer H, Basso LV, McNaughton D, Goldstein MK (2008) Evaluation of an architecture for intelligent query and exploration of time-oriented clinical data. Artif Intell Med, 43(1):17-34.
112. Mazziotta J, Toga A, Evans A, Fox P, Lancaster J, Zilles K, Woods R, Paus T, Simpson G, Pike B, Holmes C, Collins L, Thompson P, MacDonald D, Iacoboni M, Schormann T, Amunts K, Palomero-Gallagher N, Geyer S, Parsons L, Narr K, Kabani N, Le Goualher G, Feidler J, Smith K, Boomsma D, Pol HH, Cannon T, Kawashima R, Mazoyer B (2001) A four-dimensional probabilistic atlas of the human brain. J Am Med Inform Assoc, 8(5):401-430.
113. Medical Imaging Perception Society (2008) MIPS main web page. http://www.mips.ws/. Accessed June 26, 2008.
114. Merrick JRW, Dinesh V, Amita S, van Dorp JR, Mazzuchi TA (2003) Propagation of uncertainty in a simulation-based maritime risk assessment model utilizing Bayesian simulation techniques. Proc 2003 Winter Simulation Conf, vol 1, pp 449-455.
115. Michotte A (1963) The Perception of Causality. Methuen, London.
116. Microsoft Health (2008) Common User Interface. http://www.mscui.net/roadmap/road-map.aspx. Accessed May 13, 2008, .
117. Miksch S, Horn W, Popow C, Paky F (1996) Utilizing temporal data abstraction for data validation and therapy planning for artificially ventilated newborn infants. Artif Intell Med, 8(6):543-576.
118. Ming Z, Hong Z, Donny T, Stephen TCW (2003) DBMap: A TreeMap-based framework for data navigation and visualization of brain research registry. In: Huang HK, Ratib OM (eds), vol 5033. SPIE, pp 403-412.
119. Morioka CA, El-Saden S, Duckwiler G, Zou Q, Ying R, Bui A, Johnson D, Kangarloo H (2003) Workflow management of HIS/RIS textual documents with PACS image studies for neuroradiology. AMIA Annu Symp Proc, pp 475-479.

120. Morioka CA, El-Saden S, Pope W, Sayre J, Duckwiler G, Meng F, Bui A, Kangarloo H (2008) A methodology to integrate clinical data for the efficient assessment of brain-tumor patients. Inform Health Soc Care, 33(1):55-68.
121. Morioka CA, Valentino DJ, Duckwiler G, El-Saden S, Sinha U, Bui A, Kangarloo H (2001) Disease specific intelligent pre-fetch and hanging protocol for diagnostic neuroradiology workstations. Proc AMIA Symp, pp 468-472.
122. Munzner T (1998) Exploring large graphs in 3D hyperbolic space. IEEE Trans Computer Graphics and Applications, 18(4):18-23.
123. National Cancer Institute Center for Bioinformatics (2008) NCI Terminology Browser. http://nciterms.nci.nih.gov/NCIBrowser/Dictionary.do. Accessed June 6, 2008.
124. National Institutes of Health (2008) The National Library of Medicine's Visible Human Project. http://www.nlm.nih.gov/research/visible/visible_human.html. Accessed June 18, 2008.
125. Noah S, Rahul G, Steven MS, Richard S (2008) Finding paths through the world's photos. Proc ACM SIGGRAPH Intl Conf Computer Graphics and Interactive Techniques ACM, Los Angeles, California.
126. North C, Korn F (1996) Browsing anatomical image databases: A case study of the Visible Human. Proc SIGCHI Conf Human Factors in Computing Systems.
127. Ogden WC, Brooks SR (1983) Query languages for the casual user: Exploring the middle ground between formal and natural languages. Proc SIGCHI Conf Human Factors in Computing Systems. ACM, Boston, Massachusetts, United States.
128. Ogunyemi OI, Clarke JR, Ash N, Webber BL (2002) Combining geometric and probabilistic reasoning for computer-based penetrating-trauma assessment. J Am Med Inform Assoc, 9(3):273-282.
129. Payne PR, Starren JB (2005) Quantifying visual similarity in clinical iconic graphics. J Am Med Inform Assoc, 12(3):338-345.
130. Peleg M, Tu S, Bury J, Ciccarese P, Fox J, Greenes RA, Hall R, Johnson PD, Jones N, Kumar A, Miksch S, Quaglini S, Seyfang A, Shortliffe EH, Stefanelli M (2003) Comparing computer-interpretable guideline models: A case-study approach. J Am Med Inform Assoc, 10(1):52-68.
131. Pickett RM, Grinstein GG (1988) Iconographic displays for visualizing multidimensional data. Proc IEEE Intl Conf Systems, Man, and Cybernetics.
132. Pickhardt PJ, Choi JR, Hwang I, Butler JA, Puckett ML, Hildebrandt HA, Wong RK, Nugent PA, Mysliwiec PA, Schindler WR (2003) Computed tomographic virtual colonoscopy to screen for colorectal neoplasia in asymptomatic adults. N Engl J Med, 349(23):2191-2200.
133. Plaisant C, Carr D, Shneiderman B (1995) Image-browser taxonomy and guidelines for designers. IEEE Software, 12(2):21-32.
134. Plaisant C, Mushlin R, Snyder A, Li J, Heller D, Shneiderman B (1998) LifeLines: Using visualization to enhance navigation and analysis of patient records. Proc AMIA Symp, pp 76-80.
135. Platt JC, Czerwinski M, Field BA (2003) PhotoTOC: Automatic clustering for browsing personal photographs. Proc 2003 Joint Conf 4th Intl Conf on Information, Communications and Signal Processing and the 4th Pacific Rim Conf on Multimedia, vol 1, pp 6-10.
136. Portoni L, Combi C, Pinciroli F (2002) User-oriented views in health care information systems. IEEE Trans Biomed Eng, 49(12):1387-1398.
137. Post FH, Vrolijk B, Hauser H, Laramee RS, Doleisch H (2003) The state of the art in flow visualisation: Feature extraction and tracking. Comp Graphics Forum, 22(4):775-792.
138. Powsner SM, Tufte ER (1994) Graphical summary of patient status. Lancet, 344(8919):386-389.
139. Preim B, Oeltze S, Mlejnek M, Groller E, Hennemuth A, Behrens S (2009) Survey of the visual exploration and analysis of perfusion data. IEEE Trans Vis Comput Graph, 12(2):1-17.

140. Ralston JD, Revere D, Robins LS, Goldberg HI (2004) Patients' experience with a diabetes support programme based on an interactive electronic medical record: Qualitative study. BMJ, 328(7449):1159-1163.
141. Ramer L, Richardson JL, Cohen MZ, Bedney C, Danley KL, Judge EA (1999) Multimeasure pain assessment in an ethnically diverse group of patients with cancer. J Transcult Nurs, 10(2):94-101.
142. Rheingans P, Joshi S (1999) Visualization of molecules with positional uncertainty. Data Visualization, 99:299-306.
143. Rich E (1979) User modeling via stereotypes. Cognitive Science, 3(4):329-354.
144. Rich E (1983) Users are individuals: Individualizing user models. Intl J Man-Machine Studies, 18(3):199-214.
145. Rodden K, Wood KR (2003) How do people manage their digital photographs? Proc ACM SIGCHI Conf Human Factors in Computing Systems, pp 409-416.
146. Rosset A, Spadola L, Ratib O (2004) OsiriX: An open-source software for navigating in multidimensional DICOM images. J Digital Imaging, 17(3):205-216.
147. Sasso G, Marsiglia HR, Pigatto F, Basilicata A, Gargiulo M, Abate AF, Nappi M, Pulley J, Sasso FS (2005) A visual query-by-example image database for chest CT images: Potential role as a decision and educational support tool for radiologists. J Digital Imaging, 18(1):78-84.
148. Satava RM, Jones SB (1998) Current and future applications of virtual reality for medicine. Proc IEEE, 86(3):484-489.
149. Scotch M, Parmanto B (2006) Development of SOVAT: A numerical-spatial decision support system for community health assessment research. Int J Med Inform, 75(10-11):771-784.
150. Selker T, Burleson W (2000) Context-aware design and interaction in computer systems. IBM Systems Journal, 39(3&4):880-891.
151. Shahar Y, Goren-Bar D, Boaz D, Tahan G (2006) Distributed, intelligent, interactive visualization and exploration of time-oriented clinical data and their abstractions. Artif Intell Med, 38(2):115-135.
152. Sheth N, Cai Q (2003) Visualizing MeSH dataset using radial tree layout. Indiana University. http://iv.slis.indiana.edu/sw/papers/radialtree.pdf. Accessed June 2, 2008.
153. Shneiderman B (1983) Direct manipulation: A step beyond programming languages. Computer, 16(8):57-69.
154. Shneiderman B (1992) Tree visualization with tree-maps: 2D space-filling approach. ACM Trans Graphics, 11(1):92-99.
155. Shneiderman B (1996) The eyes have it: A task by data type taxonomy for information visualizations. Proc IEEE Symp Visual Languages, pp 336-343.
156. Shneiderman B, Plaisant C (2004) Designing the User Interface: Strategies for Effective Human-Computer Interaction. 4th edition. Pearson/Addison Wesley, Boston.
157. Shyu CR, Pavlopoulou C, Kak AC, Brodley CE, Broderick LS (2002) Using human perceptual categories for content-based retrieval from a medical image database. Computer Vision and Image Understanding, 88(3):119-151.
158. Simkin D, Hastie R (1987) An information-processing analysis of graph perception. J Am Statistical Association, 82(398):454-465.
159. Snavely N, Seitz SM, Szeliski R (2006) Photo tourism: Exploring photo collections in 3D. ACM Trans Graphics, 25(3):835-846.
160. Starren J, Johnson SB (2000) An object-oriented taxonomy of medical data presentations. J Am Med Inform Assoc, 7(1):1-20.
161. Sugiyama K, Tagawa S, Toda M (1981) Methods for visual understanding of hierarchical Systems. IEEE Trans Syst Man Cybern, SMC-11(2):109-125.

162. Sutton DR, Fox J (2003) The syntax and semantics of the PROforma guideline modeling language. J Am Med Inform Assoc, 10(5):433-443.
163. Svakhine NA, Ebert DS, Andrews W (2009) Illustration-inspired, depth enhanced volumetric medical visualization. IEEE Trans Vis Comput Graph, 12(2):1-10.
164. Talairach J, Tournoux P (1988) Co-Planar Stereotaxic Atlas of the Human Brain: 3-Dimensional Proportional System: An Approach to Cerebral Imaging. Thieme.
165. Tange HJ, Schouten HC, Kester ADM, Hasman A (1998) The granularity of medical narratives and its effect on the speed and completeness of information retrieval. J Am Med Inform Assoc, 5(6):571-582.
166. Tominski C, Abello J, Schumann H (2004) Axes-based visualizations with radial layouts. Proc ACM Symp Applied Computing, pp 1242-1247.
167. Tory M, Potts S, Moller T (2005) A parallel coordinates style interface for exploratory volume visualization. IEEE Trans Vis Comput Graph, 11(1):71-80.
168. Tory M, Röber N, Möller T, Celler A, Atkins MS (2001) 4D space-time techniques: A medical imaging case study. Proc Conf Visualization '01, pp 473-476.
169. Tufte ER (1990) Envisioning Information. Graphics Press, Cheshire, Conn. (P.O. Box 430, Cheshire 06410).
170. Tufte ER (1997) Visual Explanations: Images and Quantities, Evidence and Narrative. Graphics Press, Cheshire, Conn.
171. Tufte ER (2001) The Visual Display of Quantitative Information. 2nd edition. Graphics Press, Cheshire, Conn.
172. Tweedie L (1997) Characterizing interactive externalizations. Proc SIGCHI Conf Human Factors in Computing Systems, pp 375-382.
173. United States Centers for Disease Control (CDC) and Prevention (2008) Influenza (Flu) Activity. http://www.cdc.gov/flu/weekly/fluactivity.htm. Accessed June 1, 2008, 2008.
174. Viégas FB, Wattenberg M, Dave K (2004) Studying cooperation and conflict between authors with history flow visualizations. Proc SIGCHI Conf Human Factors in Computing Systems, pp 575-582.
175. Villablanca JP, Martin N, Jahan R, Gobin YP, Frazee J, Duckwiler G, Bentson J, Hardart M, Coiteiro D, Sayre J (2000) Volume-rendered helical computerized tomography angiography in the detection and characterization of intracranial aneurysms. J Neurosurg, 93(2):254-264.
176. Wang JZ, Nguyen J, Lo KK, Law C, Regula D (1999) Multiresolution browsing of pathology images using wavelets. Proc AMIA Symp, pp 340-344.
177. Wang TD, Plaisant C, Quinn A, Stanchak R, Murphy S, Shneiderman B (2008) Aligning temporal data by sentinel events: Discovering patterns in electronic health records. Proc SIGCHI Conf Human Factors in Computing Systems. ACM, Florence, Italy.
178. Ware C, Neufeld E, Bartram L (1999) Visualizing causal relations. Proc IEEE Information Visualization, Late Breaking Hot Topics, pp 39-42.
179. Weber M, Alexa M, Muller W (2001) Visualizing time-series on spirals. IEEE Symp Information Visualization, INFOVIS):7-13.
180. Winkelman WJ, Leonard KJ, Rossos PG (2005) Patient-perceived usefulness of online electronic medical records: Employing grounded theory in the development of information and communication technologies for use by patients living with chronic illness. J Am Med Inform Assoc, 12(3):306-314.
181. Woodring J, Shen HW (2009) Multiscale time activity data exploration via temporal clustering visualization spreadsheet. IEEE Trans Vis Comput Graph, 15(1):1-15.
182. Zeng Q, Kogan S, Ash N, Greenes RA (2001) Patient and clinician vocabulary: How different are they? Proc MedInfo, vol 10, pp 399-403.

183. Zeng QT, Tse T (2006) Exploring and developing consumer health vocabularies. J Am Med Inform Assoc, 13(1):24-29.
184. Zhu X, Lee KN, Levin DL, Wong ST, Huang HK, Soo Hoo K, Gamsu G, Webb WR (1996) Temporal image database design for outcome analysis of lung nodule. Comput Med Imaging Graph, 20(4):347-356.

PART III
Documenting Imaging Findings

Wherein methods to analyze information related to medical imaging are described. Daily, we are faced with a growing amount of medical information. Largely, our ability to deal with this abundance of data is quickly being overwhelmed. New computational methods are needed to help understand medical images and text to assist physicians and biomedical researchers with their tasks. The chapters in this section consider the growing libraries of algorithms that comprise standardization and analysis of this data. The first chapter considers image standardization and characterization techniques, leading to the quantitative extraction of image features. Subsequently, we consider the problem of understanding the free-text language often used to describe medical findings (such as in a radiology report or other physician consult): developments in medical natural language processing are covered. Lastly, this section examines data models aimed at organizing this information in order to provide logical access to (clinical) data.

- **Chapter 5 –** Characterizing Imaging Data
- **Chapter 6 –** Natural Language Processing of Medical Reports
- **Chapter 7 –** Organizing Observations: Data Models

Chapter 5
Characterizing Imaging Data

Ricky K. Taira, Juan Eugenio Iglesias, and Neda Jahanshad

Imaging represents a frequent, non-invasive, longitudinal, *in vivo* sampling technique for acquiring objective insight into normal and disease phenomenon. Imaging is increasingly used to document complex patient conditions, for diagnostic purposes as well as for assessment of therapeutic interventions (*e.g.*, drug, surgery, radiation therapy) [81]. Imaging can capture structural, compositional, and functional information across multiple scales of evidence, including manifestations of disease processes at the molecular, genetic, cellular, tissue, and organ level [47]. Imaging allows both global assessment of disease extent as well as the characterization of disease micro-environments. Advances in imaging during the past decade have provided an unparalleled view into the human body; and in all likelihood these advances will continue in the foreseeable future. There has been considerable research directed to developing *imaging biomarkers*, defined as, "...anatomic, physiologic, biochemical, or molecular parameters detectable with imaging methods used to establish the presence or severity of disease which offers the prospect of improved early medical product development and pre-clinical testing" [188]. Yet the full utility of image data is not realized, with prevailing interpretation methods that almost entirely rely on conventional subjective interpretation of images. Quantitative methods to extract the underlying tissue specific parameters that change with pathology will provide a better understanding of pathological processes. The interdisciplinary field of imaging informatics addresses many issues that have prevented the systematic, scientific understanding of radiological evidence and the creation of comprehensive diagnostic models from which the most plausible explanation can be considered for decision making tasks.

In this chapter, we explore issues and approaches directed to understanding the process of extracting information from imaging data. We will cover methods for improving procedural information, improving patient assessment, and creating statistical models of normality and disease. Specifically, we want to ascertain what type of knowledge a medical image represents, and what its constituent elements mean. What do contrast and brightness represent in an image? Why are there different presentations of images even when the patient state has not changed? How do we ground a particular pixel measurement to an originating (biological) process? Understanding of the data generation process will permit more effective top-down and bottom-up processing approaches to image analysis.

What is a Pixel?

We start our discussion of imaging informatics by describing the features of a *pixel*, short for "picture element." As the name suggests, a pixel exists only within the specialized (artificial) world of an image and is a conceptual representation of reality. But what is a pixel? A pixel can be viewed as the lowest level of abstraction of information in a digital image: it serves as the fundamental processing representation for computations involving image understanding and analyses. In computer vision, a pixel is the smallest addressable information unit in an image lattice. In image grammars, a pixel is the minimum unit of image syntax (*i.e.*, the basic building blocks for how an

A.A.T. Bui and R.K. Taira (eds.), *Medical Imaging Informatics*,
DOI 10.1007/978-1-4419-0385-3_5, © Springer Science + Business Media, LLC 2010

image is stitched together from a hierarchical organization of scene parts). Understanding the underlying semantics of what a pixel represents is important if we are to develop richer representations that can be used to improve medical image processing and analysis algorithms (*e.g.*, denoising, segmentation, tissue characterization, etc.).

Representing Space, Time, and Energy

First, a pixel value is some function of 2D space: *pixel value* = $f(x,y)$. In 3D, the basic element of the image is a *voxel*, short for "volume element." The quantity f may be regarded as some physical variable containing information about the object under study (*e.g.*, how the object interacted with x-rays). Pixels (and voxels) are digital representations, being discretized in both its intensity value, f, and its location (x,y,z). As such, a more realistic description of a pixel is that the intensity value is an average over a small neighborhood of points centered about the real point (x,y,z). Because images are composed of values that vary in space, they are often thought of as *fields* – a function that varies with position. In digital systems, the field is partitioned into discretized connected chunks (*i.e.*, pixels/voxels). These chunks typically are square or rectangular, but other types of lattice elements are possible. For example, in crystallography theory, there are five basic types of 2D lattice types (hexagonal, rhombic, square, rectangular, parallelogram shaped pixels) and fourteen types of 3D lattices. Examples of non-orthogonal lattices (*e.g.*, hexagonal) in relation to efficient sampling schemes during image acquisition can be found in [32, 53].

A pixel/voxel value can also be further described as a function of space and time: *voxel value* = $f(x,y,z,t)$. Examples of how a voxel value can change with time include: things flowing into and out of the voxel (*e.g.*, flow studies) arising from physiological changes (*e.g.*, functional magnetic resonance imaging, fMRI); and imaging systems that accumulate counts over time (*e.g.*, nuclear medicine studies, intravenous contrast) (Fig. 5.1).

A pixel/voxel value can be characterized as some function of space-time-energy as well: *voxel value* = $f(x,y,z,t,E)$. We can include "spectral" dependencies for what a particular value of a pixel is for a given image, at a specified location and time. For instance, we often talk about multispectral acquisitions in magnetic resonance (MR), obtaining T1-weighted pre- and post-contrast, T2-weighted, and FLAIR (fluid-attenuated inverse recovery) sequences for a given patient at a given time. In x-ray imaging, energy dependency is something we try to eliminate: for example, if we use an x-ray spectrum from a 150 kVp technique, why should the resulting image appear different from the exact same patient if the technique is changed to 80 kVp? In this case, we try to remove this dependency as the energy source is not directly (intrinsically) related to any physical or biological parameters of interest (*e.g.*, electron density, atomic number) – we would ideally like to have a pixel value that reflects only patient characteristics, not technique parameters.

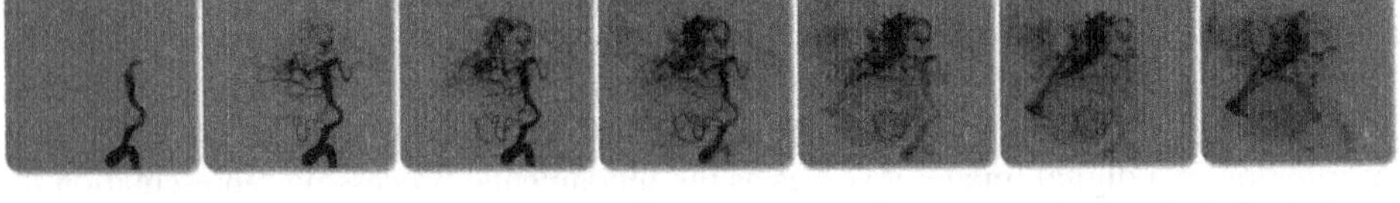

Figure 5.1: Example of a 2D cerebral angiogram. A contrast bolus is injected and images are taken in rapid succession to visualize the vasculature; this process illustrates how a pixel can also be seen as a function of time.

Mathematical Representations of Pixel Values

- Zero-order tensor-scalar fields. At its simplest, a pixel value can be a scalar value, in which case the image is described as a *scalar field.* A scalar value is mathematically a zero-order tensor. A scalar value for a pixel, for example, can be related to some scalar physical property such as electron density, atomic number, or spin-lattice relaxation (Fig. 5.2a). In these cases, the field varies in the intensity of some imaging signal from point to point. The intensity value of a digital image $f(x,y)$ may represent k-tuples of scalar intensity values across several spectral bands (*i.e.,* it can span over an "energy" space). By way of illustration, one can have an imaging study with T1, T2-weighted, and proton density information all spatially registered to a common coordinate system. Scalar images can be shown in grayscale; and color images can be represented visually using a three channel red-green-blue (RGB) representation.
- First-order tensor-vector fields. The next most complicated tensor is an order one tensor, otherwise known as a vector (magnitude and direction)[1]. These often represent some "continuous" magnitude and direction change from point to point (*e.g.,* velocity represents the change in position over time within a field). For example, a vector field may represent a dynamic contrast flow study [106], deformation fields [83], or lung motion during the respiratory cycle [72]. In a 3D spatial representation, a vector field will have three components (magnitude component in each of the axial directions). As a representation, we often use an arrow at a point in space (Fig. 5.2b), with the arrowhead pointing in the direction of the vector and the size of the arrow representing the magnitude of the vector.
- Second-order tensor-tensor fields. A tensor of order 2 is the so-called *inertial matrix* or *tensor* itself. For tensors defined in a 3D space, a 2nd-order tensor has nine components. Tensor fields are associated with images derived from some property of the imaged media that are anisotropic (*i.e.,* directionally dependent, associated with inhomogeneous spatially dependent properties related to a material that is directionally dependent). An example tensor field is the information obtained via diffusion tensor MRI (DTI) (see Chapter 2). The image intensities at each position are attenuated depending on the strength and direction of the magnetic

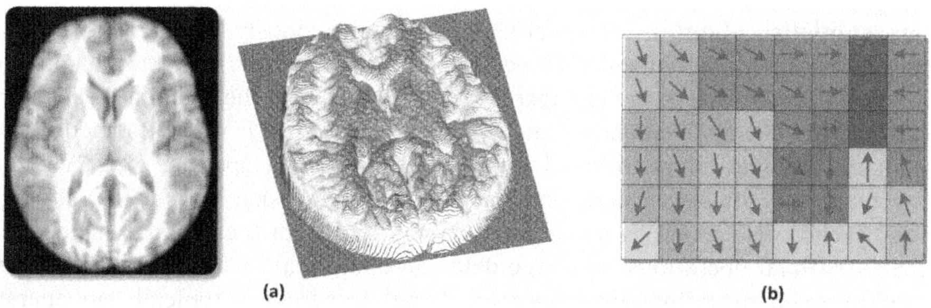

Figure 5.2: (a) An image as a scalar field. The image on the left shows a slice from a brain MR. The right image depicts the gray values in a 3D relief plot, where color and height of pixels in the z-axis are based on the intensity. **(b)** An image vector field. Per pixel, a direction is associated dependent on the magnitude and direction of change. For example, this type of representation is common for deformation fields.

---

[1] Note that here we use the physics definition of a vector and not the one common to computer science.

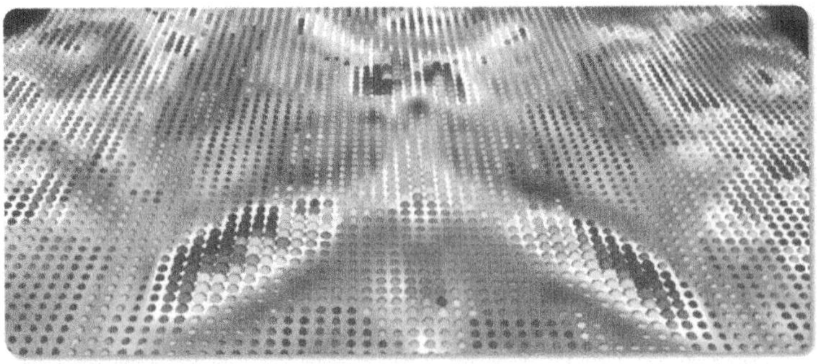

Figure 5.3: Image tensor field. A color-coded diffusion tensor field is overlaid atop a T2-weighted MR image slice.

diffusion gradient as well as the microstructure in which the water molecules diffuse: the more attenuated an image is at a given location, the greater the diffusion in the direction of the gradient. This orientation-dependent contrast is generated by diffusion anisotropy, meaning that the diffusion has directionality. This phenomenon, for example, is useful in determining structures in the brain that can influence the flow of water (*e.g.,* myelinated axons of nerve cells, which are affected by multiple sclerosis). Thus, in DTI, a tensor is used to fully characterize the motion of water in all directions. This tensor is called a *diffusion tensor* (Fig. 5.3), and represents the "normal" and "sheer stresses" to the three faces of a theoretically infinitesimally small cube, relative to an *x*/*y*/*z* reference frame, and is depicted as a 3 x 3 symmetric positive definite matrix:

$$\begin{bmatrix} D_{xx} & D_{xy} & D_{xz} \\ D_{yx} & D_{yy} & D_{yz} \\ D_{zx} & D_{zy} & D_{zz} \end{bmatrix}$$

It is common to represent the diffusion pattern at each image location by an ellipsoid (Fig 5.3) whose principal axes lengths are defined by the eigenvalues of the diagonalized matrix.

Some subtleties about pixel specifications. As previously stated, a pixel is the smallest unit of a digital image. Its precise definition, however, is often left ambiguous in scientific papers due to the fact that it depends upon both the sampling distance and the aperture size employed during image acquisition [65]. It is often assumed in many papers that a pixel refers to equal sampling distance and aperture size. A pixel is characterized by its size, shape, and intensity value. Its size (and shape) should be carefully chosen to retain as much detail as possible from the original analog image signal. Artificial operations on image data can complicate the definition of a pixel. A digital zoom operation (like those on digital cameras) is typically an operation involving pixel replication, doubling the pixel size calibration of the system. A key point concerning image specifications should be made here. The spatial resolution of a grayscale digital image is frequently given in terms of the size of its pixel matrix – quite often, the number of addressable pixels is quoted (*e.g., the resolution is 2048 x 2048 for a 14" x 14" image*). More important, however, is the number of resolvable pixels, which is typically much less (*i.e.,* modulation transfer characteristics).

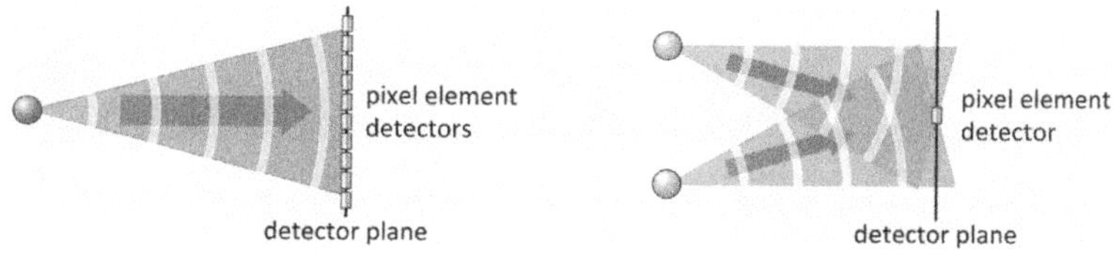

Figure 5.4: On the left, an illustration of a single point in real space generating signals to multiple pixel detectors. The right side shows how multiple points in real space contributing to the intensity of a single pixel value. In general, signals emanating from a point are difficult to localize. In projectional x-ray imaging, a radiographic grid (a type of directional filter) is employed to correct this issue (see Chapter 2).

Physical Correspondence to the Real World

A pixel exists in the imaging world as an addressable value in a mathematical lattice (*e.g.,* a 2D array). To extract meaning, we need to understand which property (or properties) is being sampled from the real-world. What is the correspondence between a location in physical space and the coordinates of pixels in an image matrix? Moreover, what does one dot on an x-ray image mean (*i.e.,* what is the effect that is being observed)? Although these represent seemingly elementary questions, rigorous answers are actually complex: even a radiologist, who looks at thousands of medical images daily, rarely provides adequate answers to these questions. We consider four issues:

1. Signal localization. *Signal localization* refers to how well we can detect an event in real physical space and accurately represent its location in the lattice that constitutes an image. Acquisition methods are imperfect, creating problems in interpreting a pixel value. In imaging, there is never a one-to-one correspondence between a "real-world" point and an imaging pixel. For example, in x-ray imaging, the left side of Fig. 5.4 shows the situation where the signal from a point in the patient's body is spread across many pixels. In contrast, the right side of Fig. 5.4 shows the contribution of two distant real world points toward the value of a single pixel.
2. Sample complexity. The value of a pixel is dependent both on the technique used (*e.g.,* kVp, mA, TE/TR, scan time, etc.) as well as the physical, biological, and pathological state of material within the pixel sample. Models for image signal generation can be quite complex, especially given the number of external and internal factors (physical and biological) involved in signal generation.
3. Sample homogeneity. Pixels are not infinitesimal elements. A pixel ideally is a representation of a relatively homogeneous environment, and we assume that the property we wish to extract from a pixel is also homogenous in the region – but this notion is often incorrect. There can be both physical inhomogeneities (*e.g.,* chemical) and biological inhomogeneities (*e.g.,* different cells, physiological dynamics), so that often a pixel is generated via a mixture of structures and processes (Fig. 5.5). What happens when the calibrated size of a pixel/voxel is larger than the effect one is trying to image? Consider, for instance, the characterization of fiber tracks and the micro-architecture of the brain using relatively large voxels; or trying to detect small micro-calcifications much smaller that a voxel dimension. There can also be pathological non-uniformitics (*e.g.,* tumor cells). These inhomogeneities are often called *partial volume effects*; these types of partial volume artifacts are common at tissue boundaries (*e.g.,* bone and soft tissue) and are related to the effects mentioned in Chapter 2.

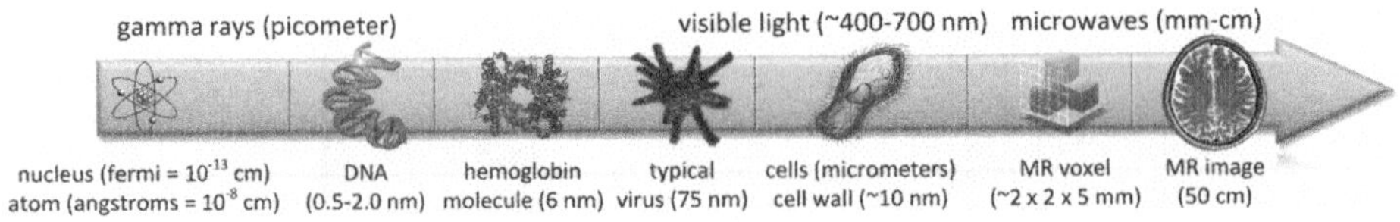

Figure 5.5: The value represented within a single MR image voxel is typically the result of a number of different multi-scale factors.

4. Sample artifacts. Spatial quantization artifacts in the form of aliasing artifacts can result when the spatial sampling frequency of an imaging system is below the Nyquist frequency ($2f_{max}$) criteria, where f_{max} is the maximum spatial frequency of the continuous image signal being digitized [169]. The results of this undersampling during the digitization process are visually disturbing global distortions such as banding, Gibbs ringing, Moiré patterns, streaks, and phase reversals [12].

In summary, pixels are the central level of abstraction in medical imaging. It is critical to understand what information they convey and how various patient factors, technique, and image processing operations impact their state. In the remainder of this chapter, we discuss methods for extracting useful information and knowledge from imaging data – a central objective of medical imaging informatics in order to create clinical applications enhancing the characterization of imaging data and thus aid in diagnostic and prognostic tasks. This goal is divided into issues related to the following four subtasks: 1) the compilation of a scientific-quality database of observations related to normal and disease specific patient phenomena; 2) the identification and instantiation of imaging features important to characterizing and/or discriminating aspects of disease and normal anatomy/physiology; 3) the building of statistical models for integrating diverse evidence; and 4) the application of imaging-based models to assist clinical decision making.

Compiling Scientific-quality Imaging Databases

A significant aspect in understanding both the normal functioning of the human body and a disease is the documentation of observations related to the state and behavior of a phenomenon under investigation (see Chapter 7). The compilation of imaging databases are important given that, thus far, the full potential of imaging data has not been adequately connected to clinical data and outcomes [91]. All efforts should therefore be made to take a disciplined, meticulous approach to image data collection to ensure its availability and quality for patient management and evaluation. Proper data collection entails data that are fully characterized (so results are reproducible) and that the data are amenable to retrieval, processing, distribution, and visualization. In general, there are two fundamentally different approaches to the compilation of imaging databases:

1. Controlled data collection. Preferably, the methods of experimental science (*e.g.*, randomized clinical trials) should be practiced when prospectively gathering image data, detailing and enforcing a study design (*e.g.*, image protocol), patient criteria, documentation requirements (*e.g.*, image quality, language, context), and the resultant analysis of data (*e.g.*, means of interpretation). The experimental approach is investigational from the start in that it is driven by an underlying hypothesis. Data collection is centered on carefully collecting observations driven by the hypothesis describing the effect of interest. The observations are "controlled," striving for the reproducibility of results by accounting for as many variables that

may influence the end outcome and described in a controlled manner (*e.g.*, using ontological references and well-defined concepts and their representations). This high degree of control in the data collection process guarantees that the results are repeatable, and provides the best quality and most conclusive knowledge.

2. Natural data collection. Natural data collection refers to the gathering of imaging studies from a routine (*i.e.*, natural) setting, such as seen in daily clinical practice. This observational approach does not involve intervening or controlling how the data is generated. The advantage of natural data collection is the potentially large number and diversity of cases that can be collected. For example, it is estimated that only 2% of all cancer patients are enrolled in any form of clinical trial [91]. As such, clinical trials are often underpowered and/or make unrealistic assumptions/ generalizations [90]. Instead, natural data collection permits observations to potentially be conducted across a broader range of data. The down side of natural data collection is that such data are messy: in routine practice, images along with the associated metadata and documentation are often highly variable, incomplete, imprecise, and inaccurate. Clinical data is subject to various sampling concerns (precision, randomness, missing data, selection biases) and representational problems (heterogeneous representations, missing context, accuracy of summarizations, source documentation errors, etc.).

An important aspect of medical imaging informatics is thus the development of methods and tools to assist in the standardization of data associated with clinical studies. Standardization is important for data to be meaningful, and is vital in the comparison of results (*e.g.*, between subjects from studies conducted at different institutions; over time for the same individual to assess response to treatment). From the perspective of research, the pooling of standardized data from multiple institutions is critical to increasing the statistical power needed for uncovering significant correlations between imaging findings and other clinical variables (see Chapter 3 for a discussion of image sharing efforts). Below, we consider three image conditioning tasks as fundamental to image standardization: 1) improving signal characterization, encompassing the semantic characterization of pixel/voxel values and their calibration; 2) noise reduction and artifact correction; and 3) improving positional characterization, addressing issues of spatial location within an image volume.

Improving Pixel Characterization

The normalization of an imaging signal is a crucial conditioning step for improving the use of pixel values in characterizing normal and disease states. To illustrate this point, Fig. 5.6 shows different histogram profiles derived from ten different MR brain studies. If one were to try to create a univariate pixel intensity model to classify brain tissue based on a large sample of patients, the result would be a distribution with large variance for each tissue type. Individual study histograms can vary even for the same patient and sequence type. The intensity of MR images varies dependent on acquisition parameters (*e.g.*, echo time, repetition time), making it difficult to compare MR studies. Ideally, methods are needed to standardize images – removing technique-dependent variables affecting pixel values – to enable accurate measurements and evaluations between studies. And though the above example is based on MR, similar situations apply to other imaging modalities. Dependent on the modality, a variety of approaches can be applied to improve signal characterization and hence its normalization, including methods pre- and post-acquisition to calibrate and correct pixel values.

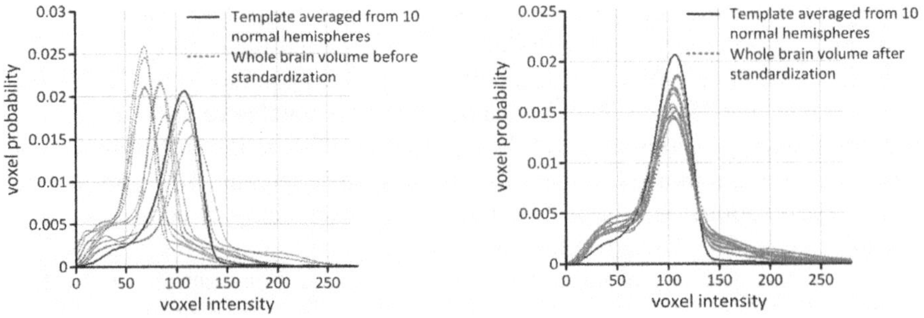

Figure 5.6: Signal normalization via histogram matching approach. The leftmost plot shows the result of acquiring several different normal subjects brain studies using the same MR pulse sequence; as evidenced, there is a large amount of inter-subject variation. Histogram matching techniques can be used to establish a mapping that better standardizes the voxel intensities. The result is shown in the rightmost plot.

Pre-acquisition: Standardizing Imaging Protocols

Perhaps the most straightforward approach to standardize the image acquisition process, the standardization of imaging protocols is an important part of image data collection for clinical trials research and routine longitudinal patient assessment (*e.g.,* for chronic disease, such as cancer). The benefit of protocol standardization is that it allows physicians to develop a subjective mental calibration of brightness and contrast levels seen on images; clinicians establish mental models of the limitations of what they are seeing on images (*e.g.,* resolution, inherent noise, etc.) and minimize interpretation errors. The use of standardized imaging protocols is important to the performance of image analysis algorithms, such as for object segmentation [161]. The disadvantage of standardizing imaging protocols lies in the difficulty of enforcing the acquisition method across time and different institutions. While a group of individuals may agree to acquire images using the same protocol, such an approach requires consensus, is static (*i.e.,* it does not incorporate new machine capabilities or techniques), and often employs protocols that are clinically impractical (*e.g.,* lengthy acquisition times).

Post-acquisition: Pixel Value Calibration and Mapping

Several approaches have been developed over the years to handle image standardization once the study is obtained, including: normalization with respect to reference materials; physics-based models that provide quantitative metrics per pixel independent of acquisition parameters; and data-driven corrections within an image.

Histogram matching. A simplistic approach to signal normalization is to force the histogram of images derived from a certain study class or mode (*e.g.,* T1-weighted brain MR studies) to conform to a standard parameterized shape. This process is known as *histogram matching* [63, 92, 150]. For example, [50] addresses issues related to pathology within brain MR images by first aligning imaging studies to a normal atlas, automatically detecting the hemisphere with pathology, and then using the contralateral side to derive an intensity histogram to remap values to standard image intensity distributions. Histogram matching can be performed easily on any class of images whose members demonstrate similar shaped histograms, and the results are generally acceptable to physicians. But while histogram matching in post-acquisition provides for intensity standardization, it does not provide contrast standardization as a custom lookup table is calculated for each imaging study.

Partial calibration: Hounsfield units in CT. The Hounsfield scale used in computed tomography (CT) is a quantitative scale of *radiodensity*. Radiodensity is a property related to the relative transparency of a material to x-ray radiation, and is an attempt to normalize the brightness and contrast of a CT image relative to that of water. Specifically, the Hounsfield scale is a measure of the attenuating properties relative to distilled water at standard pressure and temperature:

$$1000 \times \frac{\mu(x,y) - \mu_{water}}{\mu_{water}}$$

The Hounsfield scale is linear and is defined using two points: 1000 Hounsfield units (HU) for the attenuation of dry air and 0 HU for pure water at 25°C. The polyenergetic output spectrum of CT scanners can vary widely from scanner to scanner. As such, the same tissue will not produce in general the same CT number (*i.e.,* Hounsfield units) if scanned with different machines because of differences in the x-ray beam spectrum (kVp and filtration) and system calibration procedures (equalizing detector responses). Therefore the Hounsfield scale will differ between scanners and even with different energies on the same scanner; thus CT numbers cannot be directly compared amongst machines [18, 132, 133]. The ability of a material to attenuate x-rays is related in a complex way to the electron density and atomic number of the material as well as extrinsically to the energy spectrum of the incident x-ray photons. The Hounsfield value of a homogeneous material can vary from 1-2% [33]. Beam hardening can cause the Hounsfield value of a given material to be location dependent. Scanner specific parameters such as photon energy, the scan diameter, and the matrix size may affect the CT number [177].

Physics-based models. Physics-based models of medical imaging signal generation can be used to map pixel values generated from technique-dependent factors (*e.g.,* echo time, flip angle, transducer frequency, etc.) to calibrated quantitative physical values independent of the acquisition technique. This calibration of imaging signals can effectively transition current imaging studies from simply being "contrast" maps of a patient to maps related to the intrinsic physical properties of tissue. For example, MR images can have a wide range of intensities depending on a number of factors, such as magnetic field strength/quality, pulse sequence, and coil design. Even comparison of images acquired with the same contrast (*e.g.,* T1) is limited, as pixel intensities will depend on the specifics of the acquisition method. In multiple sclerosis, for instance, hypointense regions in T1-weighted images have been correlated with clinical symptoms [30]. But hypointensity is a subjective criterion of the image interpreter and the degree of hypointensity is dependent on the amount of T1-weighting in the image itself. Hence, methods providing more objective quantitative metrics can serve to provide a better foundation upon which image interpretation can occur.

For many imaging modalities (*e.g.,* MR [184], ultrasound [97]), precise physics-based models are theoretically known regarding signal generation with respect to technique parameters. By way of illustration, for MR, equations relating the technique factors of echo time (TE) and repetition time (TR) to generated signals are well characterized:

$$signal \propto \rho_H \cdot f(v)(1 - e^{-\mathrm{TR}/\mathrm{T1}})e^{-\mathrm{TE}/\mathrm{T2}}$$

By acquiring multiple sequences with different values of TR and TE, approximate solutions for true ρ_{H} (proton density), T1, and T2 at each voxel location can be computed using an array of computational methods including non-linear least square fitting (for T1 and T2-weighted sequences); multivariate linear regression (for DTI); non-negative least squares; and other statistical methods [184, 186, 187]. Pixel-level calculations of T1 values for MR sequences can be performed on the magnitude images as described by [145] using Powell's method (a direction set method for searching for a minimum of a function; for example, see [59]); and T1 derivations from echo planar images have been demonstrated [30]. Several sequences have been proposed to estimate MR parameters from imaging sequences, including work using fast and ultra-fast techniques [30, 135, 176, 219]. The advantage of physics-based models to transform acquired clinical signals to meaningful physical properties is that lengthy calibration procedures are not needed: technique dependencies in the imaging signal can be eliminated. Moreover, compared to non-calibrated pixel intensities, calibrated MR signals of true T1 and T2 values reduce the variability of MR signal correlation to different types of tissue structures, making possible more accurate statistical characterization. The disadvantage of physics-based models lies in the difficulty of actually estimating analytic solutions from routine protocols that operate under real-world clinical constraints (*e.g.*, short scan times, reasonable signal-to-noise ratios (SNR), sufficient spatial resolution and anatomic coverage). Additionally, the models are often simplified and ignore secondary signal generating sub-phenomena (*e.g.*, eddy currents, effect of non-uniformities). From routine studies, the precision with which the physical parameters can be estimated is quite variable. Protocols designed to improve the estimation within a tolerable level of certainty and with clinically realizable pulse sequences have been designed [45, 84, 130, 176, 219], usually involving adjustments to the scan time, slice thickness, number of volume acquisitions, and/or sampling methods [153]. However, still these quantitative imaging protocols can affect patient scan time, making them untenable for routine clinical application.

Signal calibration using basis materials. Calibration of imaging signals can also be performed via methods that rely on expressing imaging signals in terms of equivalent thicknesses of materials with known physical properties. For example, quantitative results in x-ray imaging (*i.e.*, projectional radiography or CT) can be obtained using this approach. Quantitative CT is commonly used, for example, to assess bone mineral density for evaluation of osteoporosis [139].

Dual energy radiography is a method developed to remove the energy dependence in the final displayed image. This approach requires spatially aligned image acquisitions taken at two different effective x-ray energies: a high energy acquisition and a low energy acquisition. The high and low energy acquisitions allow the computation of *atomic number equivalent* and *electron density* equivalent images. We briefly summarize the theory and implementation here. In conventional radiography, the primary signal image is a representation of (x,y,μ) space, where (x,y) is the location of a pixel and μ is the measure x-ray attenuation. The linear attenuation coefficient μ is itself a function of several variables:

$$\mu = f(E, Z, \rho, \frac{N_\alpha Z}{A})$$

where ρ is the mass density, N_a is Avogadro's number, and Z is the atomic number. In the diagnostic energy range, the attenuation coefficient is the sum of the individual attenuation coefficients due to the Compton interactions and the photoelectric effect: $\mu_{total} = \mu_{p.e.} + \mu_{Compton}$. Mathematically, one can decompose the total attenuation coefficient into a linear combination of *energy basis functions*: $\mu_{total}(x,y,z,E) = a_1(x,y,z)f_1(E) + a_2(x,y,z)f_2(E)$. Here, the energy basis functions, $f_1(E)$ and $f_2(E)$, represent the two major energy dependent processes, namely the photoelectric effect and the Compton effect. The energy dependencies of the interaction cross-sections for both photoelectric and Compton interactions have been well-studied [6]. For a polyenergetic x-ray source, we can calculate the percent of primary beam with spectrum, $P(E)$, that passes through a heterogeneous absorber of thickness, t, as follows:

$$\frac{I(x,y)}{I_0(x,y)} = \int_0^{E_{\max}} P(E) e^{-\int_o^t \mu_{total}(x,y,z,E)dz} dE$$

Replacing the linear attenuation coefficient with its energy basis form we have:

$$\frac{I(x,y)}{I_0(x,y)} = \int_0^{E_{\max}} P(E) e^{-\int_o^t [a_1(x,y,z)f_1(E)+a_2(x,y,z)f_2(E)]dz} dE$$

Representing the results of the line integrals (*i.e.*, 3D to 2D projection collapse) as A_1 and A_2 results in:

$$A_1(x,y) = \int_0^t a_1(x,y,z)dz,\ A_2(x,y) = \int_0^t a_2(x,y,z)dz$$

$$\frac{I(x,y)}{I_0(x,y)} = \int_0^{E_{\max}} P(E) e^{-[A_1(x,y)f_1(E)+A_2(x,y)f_2(E)]dz} dE$$

When integration is performed on the above equation, the result is purely a function of $A_1(x,y)$ and $A_2(x,y)$. The integral can therefore be approximated by a 2nd-order power series in $A_1(x,y)$ and $A_2(x,y)$, where the coefficients are energy dependent:

$$\frac{I(x,y)}{I_0(x,y)} = b_0 + b_1 A_1(x,y) + b_2 A_2(x,y) + b_3 A_1(x,y) A_2(x,y) + b_4 A_1^2(x,y) + b_5 A_2^2(x,y)$$

In dual energy radiographic imaging, two registered exposures of the patient, a high-energy exposure (I_H) and a low-energy exposure (I_L) are obtained:

$$\frac{I_H(x,y)}{I_0(x,y)} = b_0 + b_1 A_1(x,y) + b_2 A_2(x,y) + b_3 A_1(x,y) A_2(x,y) + b_4 A_1^2(x,y) + b_5 A_2^2(x,y)$$

$$\frac{I_L(x,y)}{I_0(x,y)} = c_0 + c_1 A_1(x,y) + c_2 A_2(x,y) + c_3 A_1(x,y) A_2(x,y) + c_4 A_1^2(x,y) + c_5 A_2^2(x,y)$$

To determine the value of the constants b's and c's, we perform what is known as a *calibration procedure* of some basis materials. The procedure is referred to as a calibration as we determine the coefficients for two known materials and describe all other materials as a linear combination of these chosen substances. For the dual energy problem, we would like to express the material properties of an unknown tissue as a linear combination of basis (*i.e.*, pure) materials that have a well-known response to x-rays. Two basis materials that are easy to create, of high purity, and that straddle the range of almost all effective atomic numbers seen in the human body are methyl methacrylate plastic (Pl, $Z_{avg} = 6.56$) and aluminum (Al, $Z_{avg} = 13.0$). The material calibration procedure is performed as summarized:

- Obtain several steps (*e.g.*, 20) of the two basis materials (*i.e.*, plastic and aluminum). Each step is a few millimeters thick.

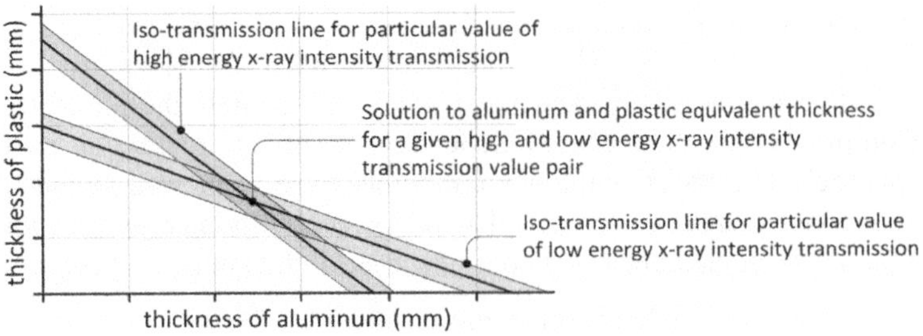

Figure 5.7: Iso-transmission lines for high and low energy material calibration data. The lower line corresponds to the possible combinations of aluminum and plastic thicknesses that would result in a particular optical density seen on a low energy scan. The higher line corresponds to possible combinations of aluminum and plastic thicknesses that would result in a particular optical density seen on a high energy scan. The intersection of these lines gives the estimated Al and Pl equivalent thicknesses for an observed high and low energy optical density value pair. *Figure adapted from* [28].

- Take both a high energy and a low energy radiograph of all possible combinations of Al and Pl steps. Compile 2D tables showing the optical intensity values as a function of the combination of the number of Al and Pl steps imaged. A high energy and a low energy calibration table (optical intensity vs. thickness of Al and thickness of Pl) are hence determined.

From the calibration tables, one can identify *iso-transmission curves*, which give the locus of points (*i.e.*, combinations of plastic steps and aluminum steps) that result in the same intensity transmission value (Fig. 5.7). One can draw a family of iso-transmission curves from the calibration data. Once the calibration procedure is completed, the tables can be used to generate aluminum equivalent (which is approximately equivalent to bone) and plastic equivalent (which is approximately equivalent to soft tissue) images for a given patient study. The steps are outlined below:

1. Take a low energy and high energy image of the patient. The imaging geometry (focus to film distance, focus to patient distance, field of view, etc.) must be exactly the same for the two exposures. Given this assumption and ignoring scatter, the differences seen in the optical densities between the high- and low-energy radiographs of the object are strictly a function of the characteristics of the x-rays (*i.e.*, energy spectrum). Differences will be manifested in the percent of Compton versus photoelectric interactions.
2. For each pixel in the high and low energy images, obtain the intensity value.
3. To determine the equivalent plastic and aluminum thicknesses corresponding to a pixel location (x,y), perform the following sequence of operations. First, find the iso-transmission line on the high energy calibration table corresponding to the pixel intensity on the high energy patient image. Second, find the iso-transmission line for the low energy pixel intensity. The intersection of the two iso-transmission lines gives the equivalent thickness of Al and Pl for the combination of the low and high energy pixel value at a given coordinate (x,y).

Once an equivalent plastic and aluminum thickness for each pixel value is determined, a variety of synthesized images can be visualized. For example, one can visualize only the aluminum (bone) or plastic (soft tissue) component of the image. It is also possible to estimate an effective Z and electron density for each pixel, as these are known for the aluminum and plastic steps used to construct the calibration table. One can then synthesize bone subtracted or soft tissue subtracted images.

Dealing with Image Noise

Noise is a basic part of any imaging system. The numeric intensity of an imaging pixel reflects components of the desirable signal (*e.g.,* transmission of primary photons, true T1, etc.) and an undesirable noise component. Noisy images are particularly problematic when the primary signal effect is small and there are practical limitations related to pixel size, patient radiation exposure, and/or scan time. For example, diffusion tensor images suffer from poor SNR due to some of these factors, and must balance the practicality of repeated scans versus the quality of the image and noise. Thus, in addition to standardization/calibration of an image signal, noise reduction should be considered. A reduction of image noise in the raw data can translate into two benefits: 1) better accuracy and reproducibility of the imaging data; and 2) faster scanning times due to reduced averaging requirements (*e.g.,* in DTI MR studies).

In imaging, a variety of noise sources exist that are random stochastic processes. Examples of such noise include: x-ray quantum mottle noise due to statistical fluctuations in the number of photons emitted from a source per unit time/area and due to the random nature of x-ray attenuation and detection processes; film and screen grain noise due to non-uniform distribution and density of grains; light conversion noise in radiographic screens; speckle noise in ultrasound; amplification noise in analog electronics; and thermal noise due to atomic collisions. Additionally, there is noise associated with quantization effects in converting real-world imaging signals to discrete pixel values, which is referred to as *bit noise* or *quantization noise* [165]. Dealing specifically with x-ray and light photons, which are quantum mechanical entities, there is inherent uncertainty in position, as well as time, energy, and momentum. Indeed, radiographs are often noise-limited as low noise often translates into a high radiation dose to the patient. At the contrast levels obtained in radiographs, our ability to discern objects of interest is more often limited by the presence of noise rather than, for example, limitations in spatial resolution. In many cases, the quality of imaging data, especially if quantitative analysis is desired, can often be more easily improved through the reduction of noise (*i.e., denoising*), rather than by increasing the signal.

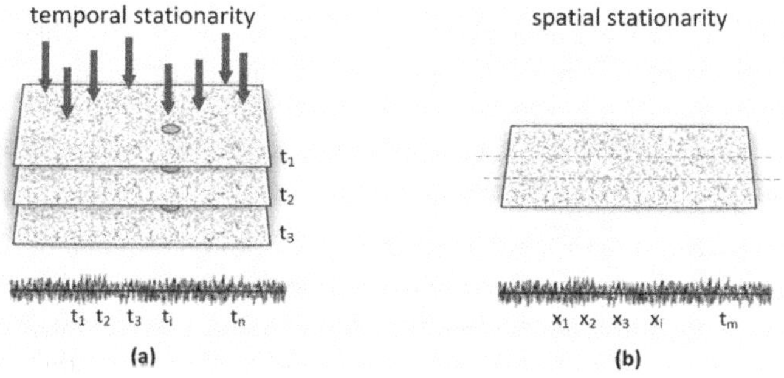

Figure 5.8: The ergodic hypothesis states that the statistics associated with fluctuations in time match the statistics for fluctuations over space.

Characterizing Noise

The characteristics of noise are different depending upon the underlying generating process. The first step in designing a denoising algorithm, then, is to characterize the imaging characteristics of the noise generating process for the particular imaging modality of interest. Once the noise is characterized, we can design a filter to eliminate it. Often, noise is characterized in terms of an additive model, $v(x,y) = u(x,y) + n(x,y)$ where v is the observed image, u is the true noiseless image, and n is the pure noise image. Alternatively, noise can be modeled as a multiplicative factor when there is coherent interference, as in the modeling of speckle noise in ultrasound imaging [233]. Notably, noise in magnetic resonance imaging is not additive as the noise is signal dependent, making separation of signal and noise components a difficult task [149].

Noise is modeled as the result of a random fluctuation of a stochastic signal generating process. Generally, we work with random signals that obey a simplifying constraint, known as *statistical stationarity*. The underlying physical process that gives rise to stationary random signals is described by statistics that are not time dependent. For a spatially statistical stationary process, the statistics are the same over all areas and are not influenced by a shift of origin. Yet in practice, this stationarity assumption for noise is unwarranted. For example, speckle noise in ultrasound is non-stationary and changes according to ultrasound attenuation.

Noise is the result of an unpredictable event, most often due to natural laws. But even though quantum fluctuations produce random signals that are responsible for noise, they also have a great deal of associated regularity and future values can be estimated in a statistical probability sense. In principle, we have two ways of studying the statistical fluctuation of a system:

1. We can assume that we know the function over a large area from which we can then determine probability distribution functions of density fluctuations from a single realization.
2. We can assume that we have an ensemble of similar functions from which we can determine probability distribution functions by an examination of all members of the ensemble.

An ensemble of density distributions can be obtained as follows. Suppose that the density distribution of a uniform exposed and developed area of radiographic film is denoted by $D(x,y;\ sample_1)$, and that a second sample that has been exposed and developed in an identical way, represented as $D(x,y;\ sample_2)$. By repeating this procedure many times we can generate an *ensemble* of density distributions $D(x,y;\ sample_i)$. This ensemble defines a random process, and each member of the ensemble is called a *realization* of the process. The random process is said to be *ergodic* if the coordinate and ensemble statistical averages are the same (Fig. 5.8). Radiographic images are often assumed to have this property.

First order statistics. If the film density $D(x,y)$ at a point (x,y) on a uniformly exposed film is measured by means of a high-resolution scanning device (*e.g.*, a micro-densitometer or laser film scanner) as a function of x and y over a large area of the film (defined by the limits $-X \leq x \leq X$ and $-Y \leq y \leq Y$), we can crudely summarize the information content within this uniformly exposed film using first order (descriptive) statistics (*i.e.*, via its mean and standard deviation, σ). The variance of the noise can be obtained as an empirical measurement or formally computed when the noise model and parameters are known:

$$\overline{D} = \frac{1}{2X2Y} \int_{-X}^{X} \int_{-Y}^{Y} D(x,y)\,dxdy$$

$$\sigma^2 = \frac{1}{2X2Y} \int_{-X}^{X} \int_{-Y}^{Y} [D(x,y) - \overline{D}]^2\,dxdy = \frac{1}{2X2Y} \int_{-X}^{X} \int_{-Y}^{Y} [\Delta D(x,y)]^2\,dxdy$$

Measuring the signal-to-noise ratio usually requires that the noise be measured separately, in the absence of signal. The magnitude of the noise often depends on the level of the signal as in photon quantum noise. Classic examples of formal noise models in imaging due to various physical processes include:

- Poisson probability distribution. X-ray photon absorption is a random process that obeys Poisson statistics. In nuclear medicine, detected photon counting also follows Poisson statistics.
- Rayleigh distribution. The speckle noise seen in ultrasound images follows this distribution.
- Flat spectral distribution. White noise is described by a flat power spectral density. Speckle patterns from film scanning systems can result in this form of noise [40]. Film grain noise is also often modeled as white noise.
- Rice distribution. Rician noise seen is demonstrated in complex value images (*e.g.*, DTI data, MR sequences) that are characterized by real and imaginary components. Unlike the normal distribution, the probability density function is not symmetric about the true signal value, especially at low signal values where it tends to a Rayleigh distribution [71, 183]. At high signal intensities, the Rician distribution tends toward a Gaussian distribution.
- Gaussian/normal distribution. Noise is often characterized by a Gaussian probability distribution as it is often generated by the combined action of many microscopic processes. Whenever noise is generated out of a large sum of very small contributions, the central limit theorem tells us that the probability distribution $P(N)$, which quantifies the probability to find a certain noise amplitude N, indeed has a Gaussian shape. Additionally, the sum of different noises tends to approach a Gaussian distribution.

Here, we consider noise in the context of x-ray imaging systems, though similar principles apply to other modalities. Empirical measurements of noise should proceed with caution. Note that in general the degree of variability of a line profile trace across a uniformly exposed imaging field will depend upon the exposure technique used (*i.e.*, *exposure fluence*, the amount of x-ray quanta per unit area) and the size of the aperture A, used for sampling the raw imaging signal. Intuitively, the standard deviation of a noise distribution decrease as the area of the measuring aperture increases. For film noise, the following relation holds approximately true [178]:

$$\sigma_A \propto \frac{1}{\sqrt{A}}$$

In general, the noise level in imaging systems is connected to the modulation transfer function (MTF) characteristics of the image acquisition chain [31, 170]. For a linear system, the MTF describes the degradation of the imaging signal with respect to spatial frequency [38].

In [70], an approach for creating a noise model is described that relies on the measurement of the relationship between the image intensity, I, and the noise variance: $\sigma_N^2 = f(I, \alpha_1, \alpha_2, \alpha_3, \ldots)$ where α_i are determined by the image acquisition protocol. This relationship is applied to every pixel in the image. The relationship between the intensity and noise variance is determined by first creating a noise image generated as the difference between the original image and a smoothed version. A mask is used to avoid edge pixels in the analysis. A histogram of the local mean intensity, I, is generated from the unmasked pixels. For each histogram bin, the mean intensity, the noise standard deviation, and the associated error are then estimated.

Second order statistics. In systems with multiple noise sources, where the noise amplitude can be written as $N(t) = N_1(t) + N_2(t) + \ldots + N_k(t)$, the key question is whether the noise sources are independent. If yes, the variance of the total noise is just the sum of the individual noise variances. For example, as film granulation, screen mottle, and quantum mottle are independent random phenomena, the total density standard deviation of the density fluctuation from one area to another can be expressed as the square root of the sum of the squares of the components:

$$\sigma_{total}(D) = \sqrt{\sigma^2_{grain}(D) + \sigma^2_{screen}(D) + \sigma^2_{quantum}(D)}$$

In many cases, noise values at different pixels are modeled as being independent and identically distributed, and hence uncorrelated. However, this assumption is not always justified. The problem with first order noise measures is that they are global measures that contain no spatial information. For example, in Fig. 5.9, the noise traces for the two samples have the exact same first order statistics (*i.e.*, the histograms result in the same mean and standard deviation) – but the noise profiles are vastly different. First order statistics do not tell us whether there are correlations between the densities measured at different points. For instance, in the case of an x-ray screen-film combination, even though the absorption of x-ray and light photons is a random process, there does exist some spatial correlation between the energy absorbed due to the spread of the light from the phosphor screen material. In other words, the density fluctuations are dependent upon the system MTF.

The second order (or joint) probability density function of a stationary ergodic process gives the probability that at (x,y) the process lies in the range D_1 to $D_1 + dD_1$, and at $(x + \alpha, y + \beta)$ the process lies in the range D_2 to $D_2 + dD_2$. To understand how one pixel in an image is "correlated" to another pixel in the image, we use second order statistics because there is a need to understand how noise (a random process) is correlated over different distances. The extent of such correlation can be specified by means of the *autocorrelation function* (ACF). The ACF is defined by the first joint moment of D and its shifted version, which for a stationary ergodic process may be written in terms of film density fluctuations as follows:

$$ACF(\alpha, \beta) = \lim_{X,Y \to \infty} \frac{1}{2X2Y} \int_{-X}^{X} \Delta D(x, y) \int_{-Y}^{Y} \Delta D(x + \alpha, y + \beta) dx dy$$

where $\Delta D(x + \alpha, y + \beta)$ is the density fluctuation (*i.e.*, noise) measured at a point displaced from the point (x,y) by a distance α along the x-axis and a distance β along the y-axis (Fig. 5.10). Over long distances, the autocorrelation function tends to zero

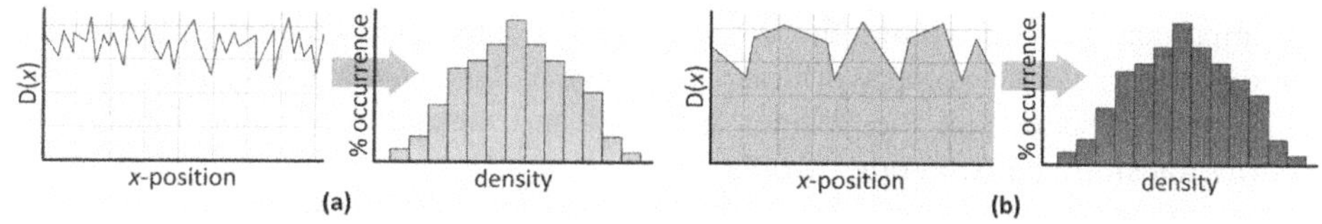

Figure 5.9: Shortcomings of first-order statistics to describe random noise. Both noise profiles in **(a)** and **(b)**, though widely different in nature, result in exactly the same first-order statistics.

as two pixels that are far apart are typically uncorrelated. One point of note is that the scale value of the autocorrelation function is simply equal to the measured variance: $ACF(0,0) = \sigma^2$. As the autocorrelation function includes the variance, it completely specifies the first and second order statistics of the measured density fluctuations, provided the process is Gaussian. Furthermore, it can be shown that for a Gaussian process all the higher order probability density functions and their moments can be specified in terms of the autocorrelation function, and so this function defines the whole random process. The autocorrelation function provides a more complete description of the noise character in a radiographic film than σ alone.

In many applications, it is more useful to specify the spatial frequency content of the noise, which can be done by calculating what is known as the *noise power spectrum* (NPS) or *Wiener spectrum* (WS) of the measured density fluctuations. The Wiener spectrum of the fluctuations of a stationary ergodic process is defined by:

$$WS(u,v) = \lim_{X,Y\to\infty}\left(\frac{1}{2X2Y}\left[\int_{-X}^{X}\int_{-Y}^{Y}\Delta D(x,y)e^{-j2\pi(ux+vy)}dxdy\right]^2\right)$$

In other words, the Wiener spectrum is the ensemble average of the Fourier transform squared of the density fluctuations. The WS is measured by finding the variance at each frequency of an ensemble of images containing no signal (other than perhaps a constant DC component). As the ensemble of signal-less images has Fourier transform (FT) components with zero mean at all frequencies (except perhaps at the DC value), the variance in amplitude is just $<|FT(u,v)|^2>$ and hence the formula for the Wiener spectrum. WS represents the energy spectral density of the noise. That is, it describes the variance in amplitude of each frequency component of the noise generated by a system. The Wiener spectrum is two-dimensional, and when integrated over a frequency space gives a value equal to the root mean square (RMS) variance of image values.

The Wiener spectrum has the units of variance per frequency bin (typically mm^2 for 2D images as $(1/mm^{-1})^2 = mm^2$). Note that the Wiener spectrum is an absolute noise measurement as opposed to an MTF which is dimensionless.

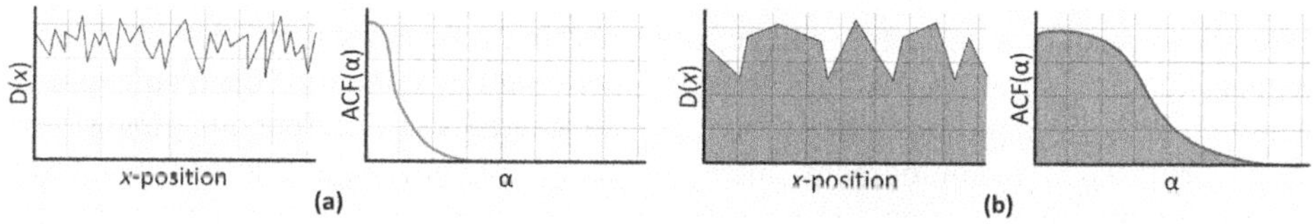

Figure 5.10: Autocorrelation function (ACF) of a typical radiographic flood field. The function tends to zero for large pixel distances (*i.e.*, two pixels far apart tend to be uncorrelated). The two areas of high correlated verse relative uncorrelated are shown.

An important theorem relating the WS and the autocorrelation function is found from the digital signal processing world and is known as the *Wiener-Khintchine Theorem*. This theorem states that the Wiener spectrum and the autocorrelation function are Fourier transform pairs. In other words:

$$WS(u,v) = \int_{-X}^{X}\int_{-Y}^{Y} ACF(x,y)e^{-j2\pi(ux+vy)}dxdy$$

$$ACF(x,y) = \int_{-\infty}^{\infty}\int_{-\infty}^{\infty} WS(u,v)e^{j2\pi(ux+vy)}dudv$$

The energy spectral density of the noise (WS) is the Fourier transform of its autocorrelation sequence. This observation is a very important result: it means that the autocorrelation sequence of a signal and its energy spectral density contain the same information about the signal. As neither of these functions contains any phase information, it is impossible to reconstruct the signal uniquely from the autocorrelation function or the energy density spectrum.

Noise Reduction

Statistical models that characterize the noise generation process within an imaging system can assist noise reduction algorithms. A variety of approaches can be applied to reduce noise in a medical image, and include: 1) improving image acquisition and reconstruction protocols; 2) purely data-driven denoising algorithms; and 3) knowledge-based methods based on *a priori* (*e.g.*, physics or statistical) noise models.

Effective noise reduction methods can be performed during acquisition of the imaging study. For example, noise reduction filters can be used in CT [102]. Tube current modulation in CT can lead to noise reduction, which in turn can be used to achieve patient dose reduction without any loss of image quality. Time averaging of image acquisitions is a common approach in MR studies and angiography to reduce the effects of noise on image quality. Increase in measurement time generally leads to improved SNR because signals are coherent (*i.e.*, they have a stable amplitude, frequency, and phase) while noise is incoherent (fluctuating amplitude and phase) and tends to average out over time. *Ensemble averaging* exploits the distinction that noise is not the same from one measurement of the signal to the next, whereas the genuine signal is at least partially reproducible between time points. When the signal can be measured more than once, by measuring the signal repeatedly and as fast as practical, then adding up all the measurements point-by-point will produce an ensemble average that can improve the signal-to-noise ratio of the image by a factor proportional to the square root of the number of averages. Ensemble averaging is one of the most powerful methods for improving signals when it can be practically applied.

Post-processing noise reduction algorithms input the acquired noisy image data, v, and attempt to recover a noise free image, u. Generally, post-processing methods lead to image signal distortions as well. A trade-off between the amount of denoising and the amount of artifact generation needs to be considered for different imaging studies (*e.g.*, artifacts in mammography may not be as tolerant as a fluoroscopy study). Denoising algorithms often see no difference between small details and noise, and therefore they remove both. Artifacts from denoising algorithms include ringing, blurring, staircase effects, and checkerboard effects.

Gaussian smoothing. Noise reduction through spatial smoothing depends on the assumption that true signal amplitudes change smoothly across the image while noise is seen as rapid random changes in amplitude from point to point within the image. In smoothing, the data points of a signal are modified so that individual points that are higher than the immediately adjacent points (presumably because of noise) are reduced, and points that are lower than the adjacent points are increased. This procedure naturally leads to a smoother signal. As long as the true underlying signal is actually smooth, then the true signal will not be distorted much by the smoothing process and the noise will be reduced. In *Gaussian smoothing*, a Gaussian kernel with standard deviation σ_g is convolved with the image, relying on the fact that the neighborhood involved is large enough so that noise is reduced by averaging. The standard deviation of the noise in this case is reduced proportional to the inverse of the Gaussian kernel size. However, the trade-off in this simple approach is a loss of resolution (*i.e.,* image blurring). The Gaussian convolution works well in areas of constant regions, but performs poorly in areas near edges or areas that contain texture. Gaussian smoothing is often effective in reducing noise that is problematic for subsequent image analysis algorithms, such as segmentation, which are adversely effected by the presence of too many local minima/maxima and inflection points in the data.

Neighborhood filters. Gaussian filters perform denoising by local averaging of pixels. The restored pixel value is obtained as a weighted average where the weight of each pixel depends on the distance to the restored pixel. In effect, the approach applies a low pass filter to the image, creating an image that is blurred. *Neighborhood noise filters* explore the idea of denoising by averaging groups of "similar" nearby pixels: only those neighboring pixels within a certain threshold to the center pixel are used to compute the average. It is assumed that inside a homogeneous region, the gray level values fluctuate because of noise. Strategies include computing an arithmetic mean or Gaussian mean for a target pixel based on qualified neighboring homogeneous pixels. The *Yarosloavsky neighborhood filter* is an example [224]. Many variants of the filters exist. The *Nagao-Matsuyama filter* calculates the variance of nine sub-windows centered around the pixel to be restored. The window with the lowest variance is used for computing the average value to be assigned to the restored pixel. Thus, the filter finds the most homogeneous local region centered around the central pixel and uses this local region to assign the average value. The idea is extended in *non-local* (NL) *means algorithms* that redefine the "neighborhood" of a pixel to include any set of pixels anywhere in the image such that a window around that pixel looks like the window around the central pixel to be modified [52]. This method has been used and experimentally tuned specifically for modeling Rician noise in MR images [124].

Anisotropic filtering. To avoid blurring near edges, Gaussian convolution can be restricted to parts of the image for which the gradient of the image is near zero [49, 64]. *Anisotropic filtering algorithms* time evolve the image under a smoothing partial differential equation, with the diffusion coefficient designed to detect edges [156, 183]. In this manner, noise can be removed without significantly blurring the edges of the image.

Total variation minimization. The total variation (TV) method for noise reduction was invented by [172]. The original image, u, is assumed to have a simple geometric description: the image is smooth (has regularity) within the objects of the image and has jumps across object boundaries. The total variation semi-norm for scalar valued data u is defined as:

$$TV_{\Omega}(u) = \int_{\Omega} |\nabla u| \, dx$$

where ∇u is the gradient of the image, u. Given a noisy image $v(\mathbf{x})$, the TV algorithm recovers the original image $u(\mathbf{x})$ as the solution of the constrained minimization problem. In [172], the TV semi-norm with an added L2 fidelity norm constraint is minimized:

$$\min_u \left\{ G(u,v) = TV(u) + \frac{\lambda}{2} \|u - v\|_2^2 \right\}$$

Note that we can write the functional G more abstractly as $G = R + {}^{\lambda}\!/_{2}F$, where R is a TV regularization functional and F is a fidelity functional. The regularization term is a geometric functional measuring the smoothness of the estimated solution. The fidelity term is a least squares measure of fitness of the estimated solution compared to the observed data. The solution to the minimization problem provides a resulting image that offers a good combination of noise removal and feature preservation. This framework leads to the Euler-Lagrange equation:

$$\partial_u G = -\nabla \cdot \left(\frac{\nabla u}{|\nabla u|} \right) + \lambda (u - f)$$

We can find a minimum by searching for a steady state of:

$$\frac{\partial u}{\partial t} = -\partial_u G$$

or by directly attacking the zero of the Euler-Lagrange equation (*i.e.*, $-\partial_u G = 0$) with a fixed-point iteration [9]. The Lagrange multiplier, λ, controls the tradeoff between the regularity and fidelity terms. Edges tend to be preserved with this method, but some details and texture can be over-smoothed if λ is too small. Examples of various implementations of TV noise reduction algorithms for medical images can be found in [27, 119, 129, 136].

Frequency domain noise filters. Frequency domain filters exploit the assumption that noise can be isolated or predominate within certain spectral bands. Noise in an image generally has a higher spatial frequency spectrum than the signal image component because of its spatial de-correlatedness. Noise thus tends to have a broader frequency spectrum. Hence basic low-pass spatial filtering can be effective for noise smoothing.

In general, frequency domain filters are applied independently to every transform coefficient and the solution is estimated by the inverse transform of the new coefficients. If the noise characteristics for an imaging chain can be estimated *a priori* in the frequency domain, then a Wiener deconvolution process can be applied to filter out the noise spectrum. The Wiener formulation filters the frequency components of the image spectrum in a manner that reflects the amount of noise at that frequency: the higher the noise level at a particular frequency, the more the frequency component is attenuated [218]. The general Wiener filter assumes that the noise is spatially stationary and is an optimal filter in that it minimizes the difference between the desired output image signal and actual noisy signal in the least squares sense. *Adaptive Wiener filters* [108, 157] use a spatially varying model of the noise parameters, allowing the filter to adapt to areas of the image that may contain objects of rich frequency content for which less blurring is appropriate. Local adaptive filters in the transform domain compute a local spectrum centered on a region of a central point that is modified and an inverse transform is used to compute the new value for the central pixel.

Frequency domain filters that use a Fourier basis set suffer from the fact that local frequency filtering can cause global spurious periodic patterns that can be visually disturbing. To avoid this effect, a basis function that takes into account more local features is often preferred, such as wavelet basis functions. *Wavelet coefficient thresholding* [104] assumes that the small coefficients in a wavelet transform are more likely to be due to noise and that the large coefficients are more important signal features. These thresholding methods have two main concerns: 1) the choice of the threshold is often done in an *ad hoc* manner; and 2) the specific distributions of the signal and noise may not be well matched at different scales. As such, noise reduction can be based on various criteria:

- Wavelet coefficient shrinkage based on scale and space consistency. This approach is motivated by edge localization and distinguishes noise from meaningful data using the correlation of wavelet coefficients between consecutive scales [48, 175]. A spatially selective noise filter based on the spatial correlation of the wavelet transform at several adjacent scales can be developed. A high correlation between scales is used to infer that there is a significant feature at the position that should pass through the filter.
- Probabilistic methods. [122, 168] demonstrate estimation of the original signal from wavelet coefficients using probabilistic approaches.

Examples of various wavelet-based denoising approaches can also be found in [113, 122, 174]. Wavelet-based methods have the advantage of being fast and introduce no ringing artifacts. However, these techniques may not preserve edges well. On the other hand, TV methods are extremely efficient in preserving edges, but sometimes lead to a loss of texture and staircase artifacts. Hybrid approaches that combine both approaches to obtain denoising without the artifacts introduced by either algorithm have thus been developed [109]. Fig. 5.11 shows the original and denoised images using a combined TV and wavelet approach (Daubechies 4 basis functions with BayesShrink wavelet coefficient attenuation thresholding) for fractional anisotropy MR images. The reduction in noise in the raw and calculated images is visually apparent and standard deviation of image signal measured in homogenous tissue (average of four regions of interest) decreased from 11.9 to 8.3 after denoising. Furthermore, no significant blurring or loss of texture is seen in the denoised images.

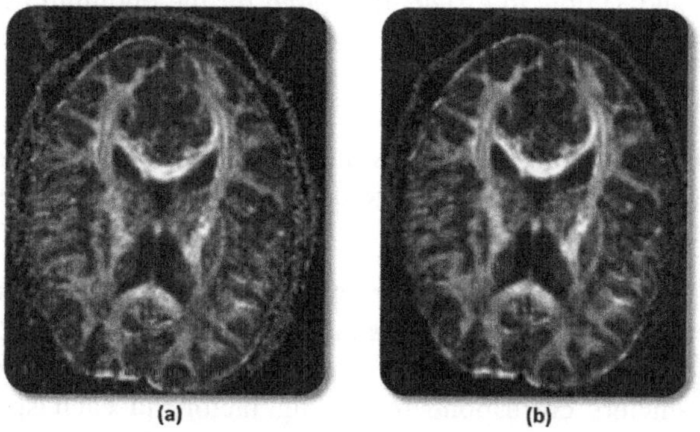

Figure 5.11: **(a)** Original image and **(b)** denoised fractional anisotropy MR image using TV and wavelet methods with BayesShrink (*courtesy of TinMan Lee*).

Registration: Improving Pixel Positional Characterization

Associated with each pixel (or voxel) of an image is a spatial coordinate (x,y). In this section, we discuss methods for improving the meaning of positional information within a clinical imaging study. In particular, methods of image registration are discussed here, while the important application of atlas creation is discussed later in the chapter. The problem of *image registration* can be stated as follows: given a reference image set of a certain view from a patient, co-align a different image set of the same view to the given reference images. Registration of imaging datasets is useful in analyzing various combinations of datasets:

- Same patient, different study type, same time. Applications in this category involve a single patient who, within a short timeframe, has acquired imaging data from a given anatomic region using multiple modalities. Registration of the imaging datasets can help to provide multiple independent pieces of evidence for a single spatial location. For example, PET (positron emission tomography) and MRI information can provide a view of the patient that combines functional and anatomical information.
- Same patient, same study type, different time. A fundamental task of radiologists is to monitor changes in image appearance and to compare these finding to some baseline (*e.g.*, a prior study). For instance, in therapeutic assessment, physicians will analyze a pre- versus post-interventional scan. Registration of the involved datasets can allow physicians to visualize *difference* or *distortion maps* of two registered datasets for the patient (see Chapter 4), highlighting anatomical and/or physiologic changes on follow-up.
- Different patients, same disease. Probabilistic atlases compiled from patients belonging to healthy and/or diseased states are being developed to investigate statistical trends related to anatomic morphology and imaging signal levels.

Image registration is often described formally as an optimization problem:

$$\underset{T}{\operatorname{argmax}}\ similarity\left(image_1(\vec{r}), image_2\left[T(\vec{r})\right]\right)$$

where r is the position vector. This problem statement involves finding the transformation that when applied to $image_2$ results in an image that is maximally similar to some specified reference $image_1$. Thus, there are three essential elements of a registration algorithm: 1) what transformations to apply; 2) how to represent similarity between images; and 3) how to efficiently search for the optimum transformation that maximizes similarity.

Transformations

The transformation algorithm in the registration process defines a mapping function to between the source and destination image data sets. The transformation functions can be roughly classified as follows:

- Parametric vs. non-parametric. Parametric methods have a mapping function characterized by a low number of parameters, usually much less than the number of pixels in the target image. Examples of parametric transformations include: scaling (parameters correspond to scaling factors in each spatial dimension); transforms based on basis function coefficients [4]; thin-plate splines [10]; and the use of a linear combination of polynomial terms [220].

- Global vs. local transformation models. Global models apply one set of mapping parameters to the entire source image dataset. Transformation parameters for local methods can depend on the location of the pixel in the source image dataset.
- Linear vs. nonlinear methods. Linear methods are represented by transforms that preserve the operations of vector addition and scalar multiplication (*i.e.*, $f(x + y) = f(x) + f(y)$; $f(cx) = cf(x)$). An example of a linear transformation is an affine transform.

Registration of two imaging datasets into spatial alignment typically follows a two-step process: first, global linear registration for coarse alignment; and second, non-linear registration to account for local variations in the morphology represented within the datasets. Note that registration algorithms can be facilitated by using extrinsic markers (*e.g.*, foreign objects) in a defined configuration (fiducial) placed in the imaging field (*e.g.*, the use of a stereotactic frame within neuroimaging procedures).

Global linear registration. Spatial normalization through linear alignment of images is typically the initial step in registering a given scan to a target dataset. *Linear registration* is a rigid body transformation that attempts to translate, rotate, scale, and shear the images in an attempt to maximize the correspondence of voxel values between the source and target datasets (*i.e.*, affine transforms are determined in this step). Linear registration results in the imaging volumes being in the same general stereotaxic space, and groups of voxels roughly represent similar anatomy.

Linear registration is relatively fast and computationally straightforward to implement, and can be performed in 2D or 3D space. In 3D, up to twelve parameters can be applied to control each voxel level transformation: three translations, three rotations, three scaling, and three shearing variables handle transforms in each of the three axes. Recall from basic linear algebra the following transforms:

- Translations. Translations move every pixel in the image a constant distance in a given direction. This involves sliding the image in any direction along a vector in space. The following describes the 2D formulation of a translation of t_x and t_y in the horizontal and vertical directions, respectively:

$$\begin{bmatrix} \hat{x} \\ \hat{y} \end{bmatrix} = \begin{bmatrix} 1 & 0 \\ 0 & 1 \end{bmatrix} \begin{bmatrix} x \\ y \end{bmatrix} + \begin{bmatrix} t_x \\ t_y \end{bmatrix}$$

- Scaling. Scaling defines the resizing (*i.e.*, expansion/compression) of an image, and is expressed as follows, where s is the scaling factor:

$$\begin{bmatrix} \hat{x} \\ \hat{y} \end{bmatrix} = \begin{bmatrix} s & 0 \\ 0 & s \end{bmatrix} \begin{bmatrix} x \\ y \end{bmatrix}$$

- Rotations. Rotations describe the motion of an image about a fixed point. The following describes the 2D formulation of a rotation of angle θ counterclockwise about the origin:

$$\begin{bmatrix} \hat{x} \\ \hat{y} \end{bmatrix} = \begin{bmatrix} \cos\theta & -\sin\theta \\ \sin\theta & \cos\theta \end{bmatrix} \begin{bmatrix} x \\ y \end{bmatrix}$$

- Shearing. Shearing fixes all points on one axis and shifts all points along the other axes by a given distance or slope, $^1\!/_k$. Although shearing may appear to distort the original image more than other transforms, it does not change the grid on which the image lies, as areas are still carried onto equal areas. The following describes the 2D formulation of a horizontal shearing by a factor of k:

$$\begin{bmatrix} \hat{x} \\ \hat{y} \end{bmatrix} = \begin{bmatrix} 1 & k \\ 0 & 1 \end{bmatrix} \begin{bmatrix} x \\ y \end{bmatrix}$$

Any combination of these parameters can be applied to perform a single linear registration. The general form of an affine transform in 2D is:

$$\begin{bmatrix} \hat{x} \\ \hat{y} \end{bmatrix} = \begin{bmatrix} a & b \\ c & d \end{bmatrix} \begin{bmatrix} x \\ y \end{bmatrix} + \begin{bmatrix} t_x \\ t_y \end{bmatrix}$$

where the matrix can be seen as a product of the matrices involving rotation, scaling, and shearing. Linear registration methods can be sufficient for aligning imaging datasets, especially when both sets belong to the same patient and are at similar points in time (*i.e.,* consecutive image acquisitions, when there is little change in patient disease state). However, researchers and physicians often desire to delve further into images and observe group-wise similarities/differences in detail. For this task, nonlinear registration methods are required.

Nonlinear registration. While linear image registration permits comparison of images in studies related to the same patient, more often than not, further processing is required to ensure a more precise 1-to-1 anatomic correspondence of voxels. Nonlinear methods are required due to the fact that biological variability among patient anatomy is relatively unconstrained. In particular, this degree of precise registration is required for aligning each study to a common coordinate system, which can then provide a framework for statistical comparison of data in an automated fashion. In such situations, quantifications of exact size differences, volume change, or shape alterations of structures are often of interest. For instance, a neuroscientist examining differences in the hippocampus of Alzheimer's patients would like to calculate the exact changes in the hippocampal shape of patients (vs. normal subjects). Nonlinear methods are thus frequently used to compute deformations: to register data volumes from different subjects; to describe changes that occur because of a patient's conditions, such as due to natural growth, surgery, or disease progression; and to correct for geometric distortions during acquisitions from various imaging systems.

Nonlinear registration involves distorting the 2D or 3D grid on which the image lies. Through these registration methods, pixels or voxels of the subject image are expanded or contracted in various directions along the coordinate system to match the desired region of the target image volume. This deformation from one image to the other is visualized through vector fields created at each voxel to denote the direction and magnitude of change for that particular voxel. A deformation grid is also able to capture the degree to which the size and shape of each voxel was altered so that the subject and target images are well matched. For example, in the case of finding cancerous nodules in the lung, one can set limits, or thresholds, on the extent of deformation allowed by the registration algorithm. If the registration algorithm detects a possible nodule in a location where the voxels in the original image had to be greatly expanded to match the template, the actual nodule found may be too small to be deemed cancerous or to be of immediate concern.

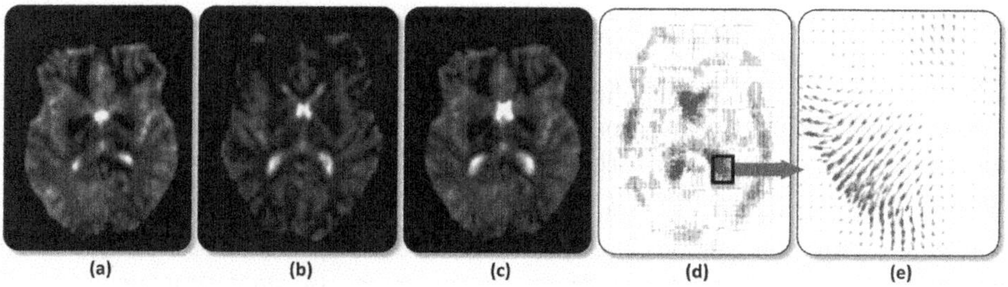

Figure 5.12: Example of the registration process. **(a)** Subject image. **(b)** Target image. **(c)** The result of mapping the subject to target. **(d)** The deformation vector fields obtained from the mapping from the subject to target is shown. **(e)** The greater focus on vector fields around the ventricles shows enlargement. It is evident that the target ventricles are larger, hence the deformation fields show vectors moving outwards with respect to the ventricles to enlarge this region in the mapping.

Mathematical formulations for nonlinear registration methods are largely beyond the scope of this chapter; however, references such as [11, 25, 26, 54] provide examples of some techniques and the practicality of their applications in the medical arena. Common approaches to nonlinear registration include polynomial warping algorithms [221], Bayesian formulations [5], and physical models such as optical flow [44]. The product of these algorithms is a deformation field that encodes the correspondence between all points in the source and target data volumes. An example of an optical flow-based algorithm [73, 185, 201, 215] in which a deformation field is computed iteratively is shown below:

$$v_{n+1}(x) = G_\sigma \otimes \left\{ v_n + \frac{T(x) - S(x)}{\|\nabla T(x)\|^2 + \alpha^2 (T(x) - S(x))^2} \cdot \nabla S(x) \right\}$$

where $v_{n+1}(x)$ is the correction vector field at iteration $n + 1$, G_σ is a Gaussian filter with variance σ^2, $\otimes$ denotes a convolution, α is a parameter controlling the magnitude of deformation allowed, and T and S are the target and source images, respectively. The Gaussian filter regularizes the deformation field, using the fact that adjacent pixels will deform to the same extent. Based on the level of image noise, the Gaussian filter size is chosen (*e.g.*, a 7 x 7 or 9 x 9 kernel with standard deviations of 3 and 4 respectively). An example of a result of a nonlinear registration with resulting deformation field map is shown in Fig. 5.12.

Similarity Metrics

Images are aligned to the desired template by systematically altering the transformation matrix, using a cost function to compute a surrogate metric that represents a minimized difference between the source and target reference volumes [220, 221]. Various cost functions or *similarity measures* exist such as mutual information, correlation ratio, *cross correlations*, *sum of squares intensity differences*, and *ratio image uniformity* [77, 86]. The cost function used depends on the task at hand. Cost functions such as mutual information and correlation ratio are often used to align images obtained from mixed modalities such as CT and PET, or CT and MRI; while cross correlation, sum of squares intensity, and ratio image uniformity are mostly used with image sets of the same modality. Minimizing cost functions are typically sensitive to contributions by voxels that do not have matching voxels in the second dataset (*i.e.*, outliers; for instance, a resected mass). As a result, non-overlapping volumes result in a large value for the cost function. Because of this problem, automatic registration methods often fail in the presence of gross pathology that significantly alters the appearance of medical images. An automated method for outlier identification is thus necessary to remove this problem; for instance, least trimmed squares optimization can be implemented to reject such outliers as described in [95, 171, 195].

Nonlinear registration methods attempt to minimize the difference between the details present in the subject and target datasets using *image warping* strategies. As with linear registration, nonlinear methods are based on a global cost function and a minimization strategy that searches for the optimal parameters for the transformation model in a rapid and reproducible manner. There have been many proposed cost functions associated with nonlinear registration. In general, these cost functions depend upon the features that best characterize the structures within the imaging data. By way of illustration, consider a researcher interested in examining structures that are identifiable through intensity patterns, region curvature, shape/surface features, or specific landmarks prevalent in a class of images of interest. These features can be detected in the source and target datasets (see below) and can be used as the driving force for the nonlinear registration algorithm to obtain maximal similarity between these specific features.

Optimal search strategies. Various search strategies must be applied to registration algorithms to identify the transformation that maximizes the similarity between the source and target image datasets. If the transform operations can be performed quickly and new proposals are systematic, then sometimes an exhaustive search can be performed – for example, if only translations are needed. As the transformations become more complex and/or similarity metrics computationally taxing, efficient non-exhaustive search methods must be used. Most optimization methods apply a multi-resolution search scheme in which registration is first run at a coarse resolution, then progressively refined at higher resolution. The application of Gauss-Newton numerical methods for minimizing the sum of squared differences and Levenberg-Marquardt optimization methods are described in [76, 137]. Gradient descent and conjugate gradient descent methods and deterministic annealing approaches are described in [14, 126]. Partial differential equation methods that include optical flow models and level sets are described in [210]. Finally, a particle swarm search strategy approach for optimization of medical registration is presented in [213].

Preprocessing

If intensity measures are to be compared, then it is important to ensure that all the images being mapped have approximately the same intensity range as the structures within the image. To account for variations in imaging parameters and image acquisition methods, intensity normalization should be performed before proceeding to intensity-based registration methods (see above). As discussed earlier, one general approach to intensity normalization is histogram normalization and equalization: obtaining a histogram of all the intensity values within each image and optimizing the spread such that all images have approximately the same intensity distribution for distinguishable structures. For example, in T1-weighted MR images of the brain, the gray matter, white matter, and the cerebral spinal fluid (CSF) have clearly distinguishable intensity ranges; the CSF is generally more distinguishable on a histogram plot from the gray

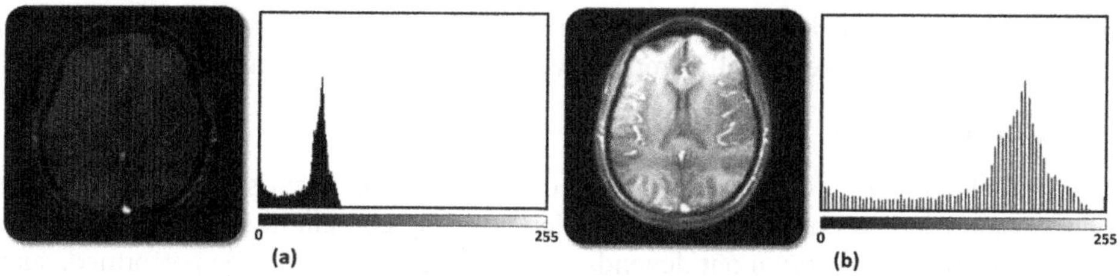

Figure 5.13: (a) The original image with its corresponding histogram. **(b)** The same image and new histogram after an equalization algorithm is applied.

and white matter than gray and white matter are from each other. If this image were to be mapped to a template before normalization, the nonlinear registration method would focus on the boundaries separating CSF from the gray matter and the brain from the background. Less emphasis would be placed on mappings of the gray and white matter regions. Through histogram normalization of intensity values, one can increase the intensity range of the white matter so that the three tissue types are equally spaced and distributed along the entire intensity range. Not only does this allow a more accurate mapping of an image to a template, but when performed for all images within the set, reduces bias in the registration due to differences in intensity ranges for tissue types as well as ensures similar mappings of all images. Fig. 5.13 demonstrates the effect of histogram equalization.

User Interaction

Most nonlinear registration methods require user input to specify parameters, such as the degree of stretching and elasticity, to be allowed during the registration process. For example, high elasticity may be required to obtain the optimal mapping between corresponding anatomical structures (*e.g.,* hippocampus) in two image sets; however, if the elasticity is too high, other portions of the image may be distorted when mapped to the target. Some nonlinear registration methods are automated so that they rely only on global properties of the images for comparisons, such as intensity values across the images. Other algorithms require some user input to define landmarks or features to be matched across images. In these landmark-based registration methods, multiple single-voxel landmarks are placed on both the target and the image to be mapped. The registration algorithm's primary focus then is to match the landmarked regions between the images, and subsequently match surrounding voxels according to a nonlinear optimization procedure. This method of *landmarking*, relying on expert input to delineate the structures, can be particularly useful when attempting to map exact shapes or regions that may not be well-defined through intensity differences alone. In this case, placing a few landmarks or points around the region of interest helps to guide the registration process and prevents shape distortion due to surrounding structures of similar intensity. An example of such landmark-based registration is presented in [100].

Comparison of Methods

There have been several studies comparing registration methods for various tasks. [37] looked at four spatial normalization techniques and found that the ideal choice of registration is of greater importance when examining higher resolution images. More recently, [107] compared 14 nonlinear deformation algorithms after linear alignment and ranked the algorithms based on various measures. Although much of today's medical image processing research focuses on finding optimal methods for mapping one image to another, accurately and consistently extracting information and knowledge from images is still an important task. Once images are aligned in the same space and anatomical structures of interest are found, comparative statistical analysis can be performed.

Imaging Features

After signal calibration, noise reduction, and registration, the next step is to summarize the relevant information in the images, which typically contain highly redundant data. The information to be drawn out depends on the subsequent task to be performed, such as segmentation, indexing or classification. This overarching step is known as *feature extraction.* In certain situations, it is possible for an expert to define appropriate features based on his/her knowledge of the image modality, the anatomical structures to be analyzed, and a disease. For example, if the goal were to classify a patch of a CT volume as liver or non-liver, one could think of calculating the difference between the average pixel value in the patch and 50, which is the typical Hounsfield value for the liver. However, when the problem becomes more complicated (*e.g.,* detecting a nodule in the breast from an ultrasound image), assigning efficient feature also becomes harder. In this case, the classical approach is to extract a high number of apparently valid features, and then to uncover a subset that provides better results in subsequent processing. This approach is the well-known problem of *feature selection.* The problem with this approach is that the dimensionality of the data grows very quickly, requiring impractically large amounts of training data for parameter estimation (*i.e.,* the curse of dimensionality [93]), which results from the fact that the dependence of the number of parameters in a model with respect to the number of dimensions is typically exponential. The purpose of *dimensionality reduction techniques* is to overcome this problem. In this section, we will first explore some of the more popular choices for features in medical imaging; a detailed discussion of general image feature extraction can be found in [148]. Then, feature selection and dimensionality reduction strategies will be discussed.

Appearance-based Image Features

Some features are computed directly from the image pixels (voxels, if the data is 3D). The simplest possible feature uses only the intensity/value of the pixel under study; in medical imaging, the efficacy of this approach is limited, working only for rudimentary problems. More sophisticated features take into account not only the pixel under consideration but its neighbors as well. For example, one set of features results from applying *linear filters,* which generate a linear combination of the pixels in the neighborhood of each pixel (Fig. 5.14a). This calculation is relatively fast, provided that the size of the neighborhood is not too large. Largely, appearance-based image features can be categorized as follows:

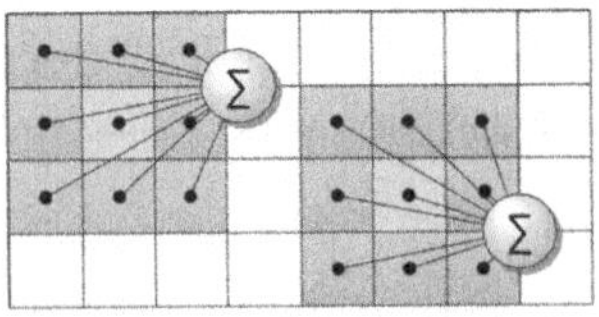

(a)

-1	0	1
-1	0	1
-1	0	1

(b)

-1	0	1
-2	0	2
-1	0	1

(c)

1	1	1
1	-8	1
1	1	1

(d)

Figure 5.14: (a) Linear filtering in two dimensions. The middle, lighter squares mark pixels of interest. **(b)** 2D Prewitt operator in horizontal direction; **(c)** 2D Sobel operator in the horizontal direction; and **(d)** 2D Laplacian operator. The Prewitt and Sobel filters for the vertical direction can easily be obtained by transposing the mask.

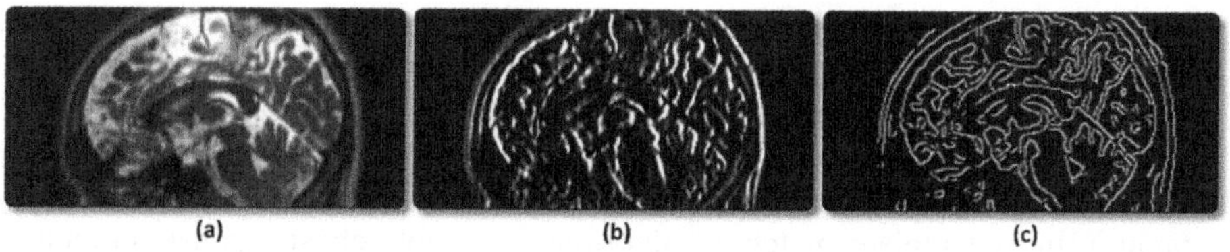

Figure 5.15: **(a)** An interpolated sagittal slice of a normal brain MR image is shown (3T, T2-weighted). **(b)** The result of applying a 2D horizontal Sobel edge operator is illustrated. **(c)** The output of a Canny edge detector is depicted.

- Color. An obvious method of describing an image region is to use its color distribution; in medical images, this translates into the grayscale distribution of the pixels. In addition to being used to normalize signal intensity (as described earlier), histograms are also used to describe objects when associated with an area/volume of an imaging study. Notably, color features are largely invariant to spatial transforms, but are still affected by acquisition even given normalization techniques.
- Edge detection. *Edge detectors* aim to find discontinuities in images, which usually correspond to salient points. An object boundary is defined as one or more contiguous paths between pixels/voxels, defining the spatial extent of the object. Identified edges are predicated upon a change in the gradient (contrast) between pixels, corresponding in the image to a change in material, depth, illumination (*e.g.*, lighting in natural images), or surface orientation. Such change is therefore computed based on identifying differences between adjacent pixels. Edges are frequently classified into three types: 1) a step edge (akin to a step function); 2) a ramp edge (where the change is gradual); and 3) a peak edge (where the change occurs quickly like a step function and reverts). The basic variants of edge detectors, such as the Sobel [190], Prewitt [166], and Laplacian [169] filters rely on different discretizations of the derivative and Laplacian operators (Fig. 5.14b-d). Sobel and Prewitt detectors are based on first-order derivatives, whereas the Laplacian is based on a second-order differentiation. A more complicated and very often used approach is the Canny edge detector [19] (Fig. 5.15c), a multistage algorithm that includes linear filtering, but also non-maximum suppression and intensity thresholding. An overview of edge detection methods is given in [231]. Unto themselves, edge detectors only provide low-level information on boundaries; additional insight can be found in examining corners and curvature. For instance, corners are found when there is a sharp change in the direction of an edge; whereas curvature examines the rate of directional change within an edge. Masks can be specifically constructed to detect corners and ridges. Applying edge detection methods directly to medical images is problematic in the presence of noise and blurring artifacts (*e.g.*, patient motion, physiologic movement); moreover, many diseases do not necessarily appear with clear boundaries (*e.g.*, edematous regions, stroke, tumor infiltration).
- Template matching and filtering. Edges are often considered low-level features. A higher-level feature extraction process, *shape* or *template matching* [67], is another technique based on filtering, using a mask that mimics the appearance of an object being searched for. For example, prototypical examples of a liver-like shape can be applied to an abdominal CT to approximate the anatomical region of the organ (with the assumption that its gross shape is relatively normal). Note that template matching therefore entails: 1) a means of representing candidate

shapes/regions within an image; and 2) a similarity metric to compare the candidate shapes against the template. *Hough transforms* (HT) can be used to find regular geometric shapes (lines, circles, ellipses); and *generalized Hough transforms* (GHT) has been developed to locate arbitrary shapes with unknown location, scale, and rotation/orientation. But apart from GHT (which is relatively computationally taxing), most template matching algorithms are not robust against rotation or scaling, making it very difficult to directly use for medical images acquired in routine clinical environments. This issue is one of the reasons that have made *Gabor filters* [56] popular: the input image is fed to a bank of filters that pick up the energy contents in different directions and frequency bands. The (continuous) expression for a Gabor kernel is:

$$g(x,y)=\exp\left(-\frac{x'^2+\gamma^2 y'^2}{2\sigma^2}\right)\cos\left(2\pi\frac{x'}{\lambda}+\psi\right)$$

where $x' = x\cos\theta + y\sin\theta$ and $y' = -x\sin\theta + y\cos\theta$. The parameter θ is the orientation of the filter; λ is the wavelength at which the filter is tuned; ψ is the phase shift; σ^{-1} is the spatial frequency bandwidth; and γ is the aspect ratio, which determines the width of the covered angle. The filter bank is constructed by assigning different values to the parameters (Fig. 5.16a). Applications in the medical imaging literature include [57, 58, 110].

- Multi-scale and scale invariant methods. Multi-scale techniques apply operations across several different scaled versions of an image, looking to extract visual features globally and locally. One increasingly common strategy is to compute Gaussian derivatives at different scales [114, 192, 209, 222]. The image is first smoothed with a Gaussian kernel, which selects the scale of interest by blurring details whose size is below the scale under consideration. Then, derivatives of different orders are extracted using the corresponding discretized operators. The idea behind this type of feature extraction is that a function can be locally approximated around a given point by its Taylor expansion, which is given by the value of the function and its derivatives at that point. As both blurring and differentiating are linear operators, they can be combined into a single filter. A typical filter bank is shown in Fig 5.16b. The main complication that multi-scale approaches introduce is that of scale selection (*i.e.,* what is the appropriate scale to enhance a certain structure in the image for further analysis?) The *scale-invariant feature transform* (SIFT) [118, 146, 181] attempts to addresses this issue by finding local extrema in the extended image-scale space. This operation cannot be represented by a linear filter. The result is a set of candidates, called keypoints, which is refined by discarding points with low-contrast and points on edges. The core SIFT algorithm comprises four stages: 1) scale-space extrema detection, in which the image is convolved with Gaussian filters at different scales, computing the difference of Gaussians (DoG) from across these images, thereby defining keypoints at the maxima/minima of the DoG; 2) keypoint localization, which positions the key-points within the image space and removes low-contrast keypoints; 3) orientation assignment of the keypoints to provide rotation invariance; and 4) generation of a keypoint descriptor that consists of ~128 features in 2D. SIFT has been found to be moderately robust in terms of illumination, noise, and minor shifts in viewpoint/angle and has been used for object recognition within (natural) images [117].

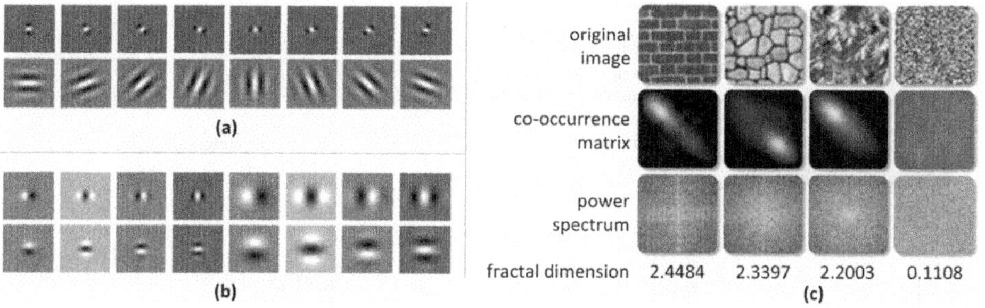

Figure 5.16: **(a)** Example of Gabor filter bank. The first row corresponds to $\lambda = 8$ pix, $\gamma = 1.0$, $\psi = 0$, $\sigma = 5$ pix. and different values for θ, uniformly sweeping the interval $[0, \pi]$. The second row corresponds to $\lambda = 6$ pix, $\gamma = 2.0$, $\psi = 0$, $\sigma = 2.5$ pix. and the same values for θ. **(b)** Example of scale space filter bank. The first row corresponds to horizontal derivatives of orders one through four at two different scales ($\sigma = 2$, $\sigma = 4$ pix.). The second row displays the corresponding vertical derivatives. **(c)** Four different textures of decreasing complexity are shown, along with their six-neighborhood co-occurrence matrices, power spectrum, and computed fractal dimension.

- Texture methods. Finally, features based on texture analysis are sometimes used in medical imaging. Texture can be defined as a function of the spatial variation in pixel intensities. *Statistical texture analysis* often involves calculating statistics on the co-occurrence matrices $C_{\bar{d}}$ [79, 158, 207], which are defined as follows: $C_{\bar{d}}(i,j)$ stores the number of occurrences of the image values i and j at positions $\bar{p}_1, \bar{p}_2$ such that $\bar{p}_1 - \bar{p}_2 = \bar{d}$. In essence, co-occurrence matrices compute a matrix based on a given displacement, $\bar{d} = (\Delta x, \Delta y)$. Measures on the matrix include energy, entropy, contrast, homogeneity, and correlation measures. Another statistical approach is to use spectral features computed from the Fourier power spectrum of the image, for instance, the fractal dimension [15, 23, 123] (Fig. 5.16b). The power spectrum is closely related to the autocorrelation function of the image, which captures the repetitive patterns in the data. *Geometric texture analysis* aims for decomposing the texture of the image into a set of simple texture elements, for example salient points and edges. The primitive texture elements are composited together to create more sophisticated textures. For example, Law's micro-textures are based on convolution kernels. The distribution of these elements can then be studied with statistical methods [159, 206]. Finally, model-based texture analysis is based on fitting the observed image patch to a predefined model. Once the model parameters have been adjusted, they can not only be used as features, but also utilized to synthesize texture [163, 194].

Shape-based Image Features

Some commonly employed image features are derived from the shape of a previously segmented object. The output of an image segmentation algorithm typically is a binary mask or a list of boundary pixels; this demarcated region can serve as the basis for computing further information about the shape. Several types of shape descriptors have been proposed in the literature; below, we focus on descriptors for 2D images, but most of them can be easily generalized to 3D (see also Chapter 7 for a discussion of spatial representations; and [225] also provides a review of shape-based metrics):

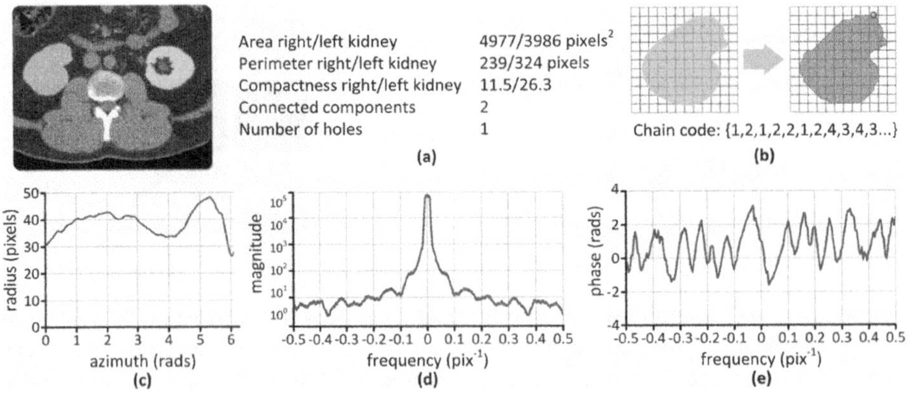

Figure 5.17: An axial CT slice of the kidney is shown, with the left and right kidneys segmented. A stone in the left kidney creates a hole in the topology. **(a)** Elementary geometric shape descriptors are computed. **(b)** The right kidney contour is approximated with straight line segments, generating a 8-direction chain code in a clockwise direction. The circle marks the start point. **(c)** The signature for the right kidney boundary is given, plotting the edge in terms of polar coordinates. **(d-e)** The Fourier descriptors for the same boundary are plotted in terms of magnitude and phase.

- Geometry. The most elementary quantitative shape descriptors are based on the geometry or statistics of the shape. Simple geometric details include area and perimeter. Other features include derivative metrics, such as compactness ($C = P^2/A$, where P is the perimeter and A is the area); and topological descriptors, such as the number of holes or the number of connected components (Fig. 5.17a). Geometry-based features can be divided twofold: 1) contour-based, in which the calculation only depends on knowing the shape boundary; and 2) region-based, wherein the boundary and the internal pixels must be known. Examples of the former include perimeter, compactness, eccentricity, and the minimum bounding box. Examples of the latter include area, shape moments (*e.g.*, centroid), and convex hull.
- Alternate shape representations. The shape boundary can be recast into a different representation, mapping it to a new domain from which features can be calculated. *Chain codes* [39, 147] (Fig. 5.17b) approximate the boundary by a set of straight-line segments, which usually have a fixed length and a discrete set of orientations, and then code the direction of each segment with a numbering system. The *signature* and the *Fourier descriptors* [55, 179] (Fig. 5.17c-e) are 1D representations of the boundary. The signature is a description in polar coordinates. The Fourier descriptors of a contour are the discrete Fourier transform of the sequence, $s_k = x_k + iy_k$, where (x_k, y_k) are the coordinates of the points along the boundary. Fourier descriptors have the advantage of well-known, simple equivalents for affine transforms (rotation, scaling, translation, shearing) in the frequency domain, allowing for direct comparisons of image data in the frequency domain (and providing a measure of positional invariance, as a change in the position of an object only results in a phase change) [67].
- Skeletonization. Another common representation of a shape is its *skeleton*. The skeleton can be defined as the set of points in the object for which the closest point in the boundary is ambiguous (Fig. 5.18). Intuitively, it is a thin version of the shape that summarizes its geometry and topology. Skeletons are very popular

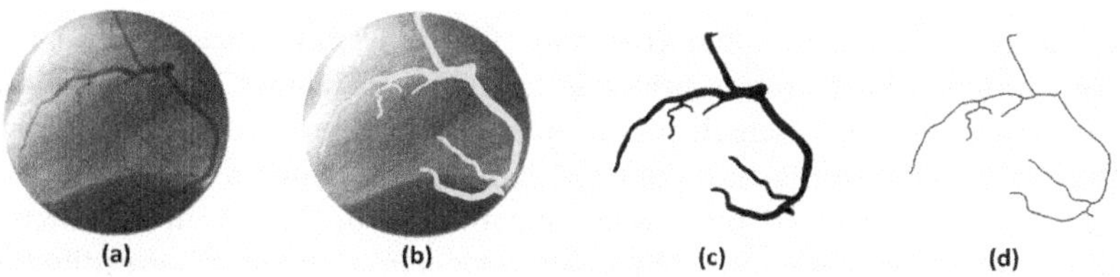

Figure 5.18: **(a)** Example of cardiac angiography. **(b)** The blood vessel is identified and segmented based on the flow of contrast. **(c)** A binarized version of the segmented blood vessel is created. **(d)** The skeletonized version of the segmented region represents the area as a single line.

in medical imaging for representing linear and branching structures, such as bronchi, blood vessels, or the colon [80, 182, 214].

- Point distribution models. Lastly, *point distribution models* [35] have become widespread in medical imaging to represent shapes given the (comparatively) low variance in the shape of organs across individuals. A point distribution model can be seen as a geometric average across several different representative shapes, providing (statistical) information on the degree of spatial and regional appearance variations within a sampled population. Central to a point distribution model is the identification of landmark points that are used to identify the shape boundary [202]. The generation of a point distribution model follows three basic steps: 1) a data vector is built for each shape by stacking the x and y coordinates of N_i landmarks into a single column vector.; 2) the training shapes are aligned using the Procrustes method [69], and the mean shape calculated; and 3) principal component analysis (PCA) is applied to reduce the dimensionality from N_i to N_c. Each shape can now be approximated by a low-dimensional vector **b**[2] of N_c principal components that are uncorrelated: PCA results in the determination of eigenvectors that form an orthogonal basis set from which the original vector can be reconstructed as a linear combination. If the variability of the landmark position is low in the training data, then it is possible to make $N_c << N_i$ without losing too much information in the process. Hence, it is possible to sum up a shape in N_c features. Point distribution models are also utilized very frequently in image segmentation by iteratively deforming the base model to fit a shape detected in an image. This technique is commonly referred to as an *active shape model* (ASMs) [34, 51, 209]. *Active appearance models* (AAMs) extend the ASM framework to provide pixel value (intensity) information within the region.

Feature Selection

As already mentioned, in many applications it is difficult to design an efficient set of features. In such cases, it is common to calculate a very high number of plausible features, such as the output from Gabor filters or Gaussian derivatives at different scales, and then find the subset that provides better results in the subsequent processing. As an example, we will henceforth assume that the features will be fed to a classifier. To select the optimal set of features, the available data is divided into at least two groups: one for training and one for testing. To evaluate the performance of a set of features, the classifier is trained using the values of features from the training data, and then the classification error on the test data is recorded. The optimal set of features is

---

[2] We follow the standard notation of using bold characters to symbolize sets or vectors of variables.

the one that leads to the minimal error rate. Other performance criteria can also be used. For example, the minimax principle approach to feature selection involves the use of entropy measures to search for an optimal combination of features [230]. First, given the set of all features and statistics about these features, a distribution fusing these features is constructed by maximizing the entropy over all distributions that result in the observed statistics. Next, feature selection occurs by greedily searching among all plausible sets of feature statistics generated in the first step that set whose maximum entropy distribution has the minimum entropy.

Evaluating all possible combinations of features is often impractical. Thus, *feature selection* techniques [41] aim to speed up the process by choosing a subset of sub-optimal features. A number of approaches have been proposed over the years for feature selection; one way of categorizing these methods differentiates between *feature ranking* techniques, where each considered feature is evaluated independently by some metric and the top *n* features are chosen; and *subset selection*, where a search is performed for across different subsets of features that are considered simultaneously and scored relative to some metric. An example of the former includes the well-known stepwise regression method (popular in statistics; see Chapter 10). Examples of the latter include greedy algorithms to grow/shrink the feature set and/or evaluate candidate subsets:

- Forward and backward selection. Basic forward/backward selection add/remove features based on minimization of an error criterion. In *forward feature selection*, the pool of selected features is initially empty and, at each step, the feature that leads to the lowest error rate is added. The algorithm terminates when the error rate increases instead of decreasing after adding a new feature, or when the pre-established number of features is reached. In *backward feature selection*, the pool is initially populated with all the features, which are subsequently dropped in a similar manner until termination conditions are met.
- Hybrid forward/backward methods. Better results can be achieved with slightly less greedy approaches, in which steps in both directions (add/drop) are permitted. In *plus l – take away r* ($l > r$) [193], l steps forward are followed by r steps backwards iteratively. Starting with an empty selected set, the algorithm terminates when a predefined number of features is reached. Conversely, *take away l – plus r* works the same way but in the other direction, starting with a pool containing all the defined features. *Floating strategies* [167] allow for steps in both directions (forward/backwards), choosing based on whatever action leads to the smallest error rate per iteration. These heuristics provide results almost as good as an exhaustive search, but can also be rather slow, because the number of selected features tends to oscillate before reaching the terminating limit. In [167], a discussion and quantitative comparison of performance for feature selection strategies is provided.

Feature selection algorithms often employ some information criterion to evaluate the utility of including the feature in the final selection (*i.e.*, does the feature provide any new information?). Popular choices include the Akaike information criterion (AIC) and the Bayesian information criterion (BIC). [75, 115] provides a full handling of the issues surrounding feature selection.

Aggregating Features: Dimensionality Reduction

Another means of dealing with a large number of features is to combine the features together; in some cases, individual feature spaces overlap or are related, so that a smaller number of features can be used to represent the entirety of the feature space with minimal information loss (if any). In effect, an (almost) invertible mapping is created between a small subset of the features to the larger group of features. This strategy underlies the use of *dimensionality reduction* techniques, which aim to find the subspace that contains most of the information present in an initial feature space. Feature selection can be seen as a particular case in which the subspace is built by extracting some of the components of the feature vectors in the original space. Largely, dimensionality reduction techniques can be divided into two groups: linear and nonlinear. There are many variants of these strategies, which are an ongoing topic of research; the reader is referred to [20, 60, 208] for additional details and reviews.

Linear dimensionality reduction. Examples of linear dimensional reduction methods are the well-known *principal component analysis* (PCA) [99] and *linear discriminant analysis* (LDA) [131]. To demonstrate, we consider the former. In PCA, the aim is to approximate d-dimensional data points as $\mathbf{x} \approx \bar{\mathbf{x}} + \mathbf{Pb}$, where $\bar{\mathbf{x}}$ is the average of the data, $\mathbf{b}$ is the vector of n principal components (of lower dimensionality than $\mathbf{x}$), and $\mathbf{P}$ is an orthogonal matrix (by approximate, the new coordinates are a linear function of the old coordinates). If the matrix $\mathbf{P}$ is known, the vector $\mathbf{b}$ that best describes (in a least-squares sense) a data point, $\mathbf{x}$, can be calculated by the projection, $\mathbf{b} \approx \mathbf{P}^t(\mathbf{x} - \bar{\mathbf{x}})$. It can be shown that, if the columns of the matrix $\mathbf{P}$ are the n eigenvectors corresponding to the n largest eigenvalues of the sample covariance matrix of the dataset, $\mathbf{P}$ is optimal among all $d \times n$ matrices in that it minimizes the approximation error:

$$C = \sum_{j=1}^{N_{samples}} \left\| \mathbf{x} - (\bar{\mathbf{x}} + \mathbf{Pb}) \right\|^2 = \sum_{j=1}^{N_{samples}} \left\| (\mathbf{I} - \mathbf{PP}^t)(\mathbf{x} - \bar{\mathbf{x}}) \right\|^2$$

Moreover, it can be shown that, under Gaussian assumptions, the principal components are independent from one another and follow a Gaussian distribution, $b_i \sim N(0,(\lambda_i)^{1/2})$, where λ_i is the ith largest eigenvalue. The number of components to keep is defined by the user, depending on the required precision for the approximation. The typical strategy is to plot the function: $e(n) = \sum_{j=1}^{n}\lambda_j / \sum_{k=1}^{d}\lambda_k$, which represents the proportion of the total variance preserved by the principal components, finding the smallest n such that $e(n) > 0.95$. The meaning of the principal components can be illustrated by plotting the modes of variation, which are calculated as $\mathbf{m}_i(t) = \bar{\mathbf{x}} + t((\lambda_i)^{1/2}\mathbf{v}_i$, where $\mathbf{v}_i$ is the eigenvector corresponding to the ith largest eigenvalue, λ_i. By letting t vary across an interval of plausible values for the normalized Gaussian distribution (such as on the intervals [-3, 3] or [-2, 2]) and plotting the resulting shape, the effect of each principal component on the shape is shown.

PCA is widely used for dimensionality reduction because it is the optimal approximation in the least-squares sense. But if the goal is not to approximate the data as well as possible, but to separate it into classes, there are better approaches. *Linear discriminant analysis* (LDA) [131] is one such method. For two classes, and under Gaussian assumptions, it can be shown that the direction, $\mathbf{w}$, which yields optimal separation between the two classes when the data points are projected is $\mathbf{w} = (\sum_1 + \sum_2)^{-1}(\boldsymbol{\mu}_1 - \boldsymbol{\mu}_2)$, where $\boldsymbol{\mu}_l$ and $\sum_l$ are the means and covariance matrix of the two groups. The algorithm can be easily generalized for multiple classes. PCA and LDA are compared with a simple example in Fig. 5.19.

Nonlinear dimensionality reduction. PCA effectively computes a set of orthogonal bases such that a linear weighting of the bases can be used to represent the feature

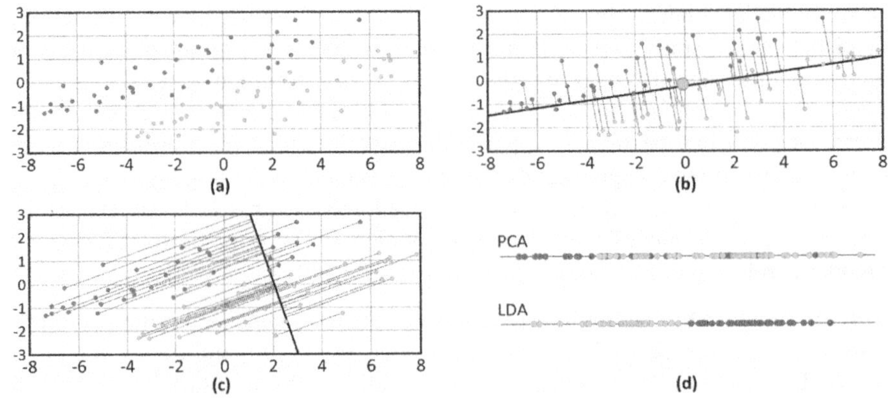

Figure 5.19: **(a)** Data points are shown corresponding to two different classes of features. **(b)** PCA of the data points are projected onto the first principal component (solid black line). The mean is represented by the large dot in the center. **(c)** LDA of the data points are projected onto the direction that yields the biggest separation. **(d)** Projections for both PCA and LDA algorithms are shown; in this case, LDA does a better job of class discrimination.

space. PCA assumes that the data follow a multivariate Gaussian distribution. In contrast, nonlinear dimensional reduction techniques are capable of modeling data in low-dimensional manifolds of different geometries (*e.g.,* spirals) and use nonlinear techniques to best approximate the data. One way of achieving this mapping is a *kernel trick*, wherein the observations are mapped into an augmented, higher-dimensionality space prior to performing a linear dimensionality reduction. Representative methods in this class of techniques include kernel PCA, multidimensional scaling, and Isomap [200]. For example, kernel PCA extends the original PCA framework by performing the mapping in a Hilbert space. Isomap uses the concept of geodesic distances in a weighted graph; for instance, the sum of edge weights along the shortest path between two points is regularly used (*e.g.,* as opposed to a linear Euclidean distance).

Considerations when using dimensional reduction. As it is frequently not possible to know *a priori* which technique is optimal for performing dimensional reduction, one method to choose between the different algorithms is to select based on the minimization of generalization errors in representing the dataset. Lastly, it is important to note that dimensional reduction techniques allow one to choose how many dimensions will be used to approximate the original feature space (with the tradeoff being the degree of approximation error in the representation). A given feature space and dataset has an *intrinsic dimensionality* (ID), which is the minimum number of variables (features) that are needed to represent the data. Ideally, the chosen number of dimensions should not be below the ID value. [212, 223] cite well-known procedures for estimating ID (*e.g.,* [62]) while [16, 17] describes newer ID estimation techniques. Arguably, reliably estimating ID is an open challenge, as many such methods are heuristic. [125] suggests error analysis over different iterations of increasing dimension to arrive at an appropriate estimate of intrinsic dimensionality.

Imaging Atlases and Group-wise Image Analysis

The range of steps described thus far – pixel value standardization, denoising, registration, and feature extraction – are all important steps that that permit imaging studies from a population of patients to be analyzed and compared within a common anatomic coordinate system. *Anatomic imaging atlases* provide an important representational framework for characterizing the spatial variability of physical features associated with a group of individuals aggregated from across imaging studies. In this section, we describe the need for standardized anatomical atlases and templates, including a brief discussion of how atlases are created; and how atlases can be used to extract both individual and group-wise features via image normalization and registration.

The Need for Atlases

Radiologists analyze countless imaging studies on a daily basis, yet no two images will ever be the same. This fact is due not only to the differences among individuals, but a number of factors including the details of the image acquisition, the patient's orientation and position within the scanner, physiologic changes, etc. How then are radiologists able to accurately diagnose patients solely on the basis of the images they encounter? Moreover, how can radiologists assess the extent of a disease by visual inspection? While skill, perfected through years of practice and familiarity with medical images, helps answer such questions, it is not a complete answer. Rather, a radiologist's abilities are not only based upon experience, but more fundamentally upon *norms* that have been established for populations. Norms are extracted from the analysis and comparison of imaging data compiled for various populations. Thus, norms define an approximate expectation for a feature for a particular group and provide a baseline for determining the relative normality of a patient's presentation. From this perspective, a standardized way of determining and representing norms is needed – a function that is fulfilled by medical imaging atlases.

Philosophers and scientists have been fascinated for centuries by the human body, seeking to illuminate its many complexities and mysteries. General anatomical atlases, in a sense, have always been a part of civilization, with early atlases appearing among the ancient Egyptians, Persians, Chinese, and Europeans [143]. Such atlases have been a critical tool for learning about the human body, providing approximate locations and the relative size of anatomical components. With today's advances in medicine, a new generation of imaging-based atlases is being generated, which provide an unparalleled degree of detail and incorporate a wider range of information (*e.g.*, histologic, radiologic, disease-specific, micro- and macro-scale features, etc.).

Creating Atlases

In a way, anatomical atlases can be likened to the more familiar geographical atlases in that a standard for visualization is set; furthermore, details on relative locations are provided through a given coordinate system, and labels are specified for each set of coordinates. But unlike geographical atlases, anatomical atlases are not necessarily definitive in nature [128]. While atlases of the Earth can be created through satellite images and precise measurements taken of a single object or area of interest, medical imaging atlases are designed primarily based on the perception of what entails a study population. Whether this population is the entire human race, the disease-free portion of a population of a certain age group, or patients with a certain type of cancer, it is often infeasible to obtain information on the makeup of every member of the target group. Atlases are thus created from images of a particular subset of the desired population. Frequently, this "subset" involves taking one or multiple high resolution images of only one person – a designated prototype of the cohort of interest – to act as a standard for the group. This subject's image set is then considered to be the template to which all other members of the population are referenced and compared. For example, the atlases from the National Library of Medicine (NLM) Visible Human Project are each derived from a single individual and are regularly used to teach human anatomy [144]. However, it has been demonstrated that using a single subject as a template is not an ideal way to represent a population [46, 205]. Depending on the task at hand, an atlas based upon a single subject may be able to provide sufficient information, for instance describing the relative location of anatomical regions of interest with great detail. Yet this level of detail cannot take into account the variability of the location, shape, size, or even its existence within the broader population.

As such, if a single subject atlas is inadequate, then the atlas requires input from a larger subset of the population. If a sufficiently large group exists, it may be assumed to represent a wide range of the population pool and generalizations may be made with respect to the entire population's (anatomical) properties. Atlases created from larger populations are often statistically determined (*e.g.,* through averaging). The variability in anatomy within populations results in fewer fine details present in the atlas, as size and shape may be lost or distorted. Such distortion is compounded by the fact that individual images are subject to acquisition bias and general error, even if taken from the same person.

Methods of combining group information to obtain a precise and efficient atlas, which aim to optimize the desired contribution from the images of each group member, involve image registration and normalization (see above). For example, if one wanted to create an atlas of the brain from a group of 20 subjects to delineate particular structures of interest, the following rudimentary steps can be taken:

1. First, intensity normalization of all subject images is conducted so that the intensity ranges are all similar.
2. Next, linearly align all images (*i.e.,* perform affine registration) so that each study is in the same general, global space.
3. Average all the images to create an average template, referred to as Template 1.
4. Perform an optimized form of nonlinear registration, mapping all linearly aligned subjects to Template 1. This process will ensure that the anatomical information of each subject will be mapped to a space shared by all of the subjects.
5. Average the resulting images from the nonlinear registration to create a new template, referred to as Template 2. Template 2 will better preserve the finer details of anatomy rather than merely averaging the linearly aligned images.
6. Label the desired structures of interest in Template 2; this step should be done using expert guidance (*e.g.,* with an individual with knowledge of the field; or an accepted anatomical text). In point of fact, application requirements should drive the level of anatomic granularity expected from the labeled atlas. The level of detail that can be described by an imaging-based atlas, however, also depends on the resolution and tissue discrimination abilities of the imaging modality used to create the atlas (*e.g.,* MR, CT, PET, etc).
7. The labeled Template 2 now serves as an atlas with segmented, labeled regions of interest (ROIs).

Notably, this method results in the ROIs being outlined in all of the subjects without the labor intensive task of manually labeling each set of images individually. Instead, one image is labeled, and the labels are then applied to all images: each subject's deformation field determined as part of the nonlinear registration can be inverted and applied to the atlas, resulting in the labeling of each subject's images, delineating the regions of interest per subject study. Another advantage of this approach is the relative ease of adding new subjects to the sample: new individuals do not need to be individually labeled, but instead one can merely apply the labeled regions of the atlas to each new image, assuming perfect registration.

Using Atlases

Virtually all fields of medicine can benefit from the use of imaging-based atlases. As remarked upon earlier, The Visible Human project [144] is an effort put forth by the National Institutes of Health (NIH) to develop complete, extremely detailed full body atlases using information from cryosection, CT, and MR images from one male and one female normal cadaver. When a specific organ is of interest, atlases of that particular anatomy are instead used. For example, atlases of the heart have been created [116] to assist in segmenting out ROIs such as the left ventricle of the heart (a common location for myocardial diseases). Finer structures, such as coronary arteries, can be segmented to aid in the diagnosis and treatment of arteriosclerosis. Atlases of the lung are also of vital importance for establishing the range of normal values in quantitative measures of lung structure and function. This requirement is particularly acute for 3D images of the lungs taken with CT, as finer pulmonary structures may be difficult to match across imaging volumes of populations [111]. Due to the relatively static nature of the brain, the most extensive work to date on atlases have been applied to the field of neurology (Fig. 5.20). One of the first standardized atlases was developed by Talairach and Tournoux [198], where the entire brain anatomy is addressed using Cartesian coordinates. This coordinate system has been used extensively in functional neuroimaging applications where entire images are mapped onto the Talairach coordinate system [61]. [204] outlines several atlases used for structural and functional neuroanatomy studies, describing advantages and possible limitations of each; they suggest possible advantages to combining the many available atlases for maximizing the anatomical information extractable at given locations.

In general, atlases obtained through imaging have a variety of uses for students, physicians, and researchers alike. Important applications of imaging-based atlases include:

- Teaching. Atlases created from imaging information can be critical tools in teaching normal and abnormal anatomy – when used as references, they allow students to learn the relative size, shape, and position of a desired part or region of the body. Probabilistic atlases can detail patterns of variability in both anatomic structure and function [128].

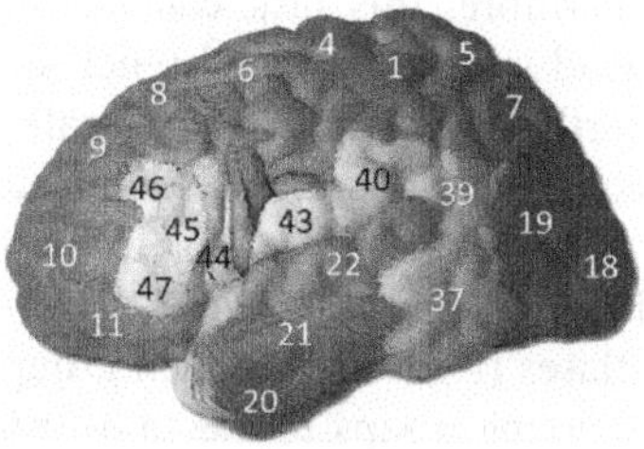

Figure 5.20: A 3D labeled atlas of the human brain. This atlas segments the brain into various color coded regions of interest, each with a corresponding anatomical label. *Image courtesy of Dr. P. Thompson at the Laboratory of NeuroImaging at UCLA.*

- Clinical decision-making. Atlases can be used to guide the decision-making process of physicians in determining the relative (ab)normality of a particular anatomical structure. Atlases are useful diagnostic tools, offering physicians a method of confirming a diagnosis when general symptoms are an inadequate basis. Thus, atlases may improve diagnosis and reduce medical complications. Consider, for example, a person admitted for evaluation who exhibits symptoms including involuntary movements – a characteristic of both Parkinson's and Huntington's disease. To differentiate between these two degenerative diseases, a physician can compare brain scans taken from the patient to various imaging atlases, including ones representative of Parkinson's and Huntington's disease. As anatomical variations within the brains of these two disorders are not identical, the clinician is able to confirm the presence of one disease over the other, if there are substantial similarities between the patient's image set and one of the comparative atlases. An accurate diagnosis may then be made and treatment begun.
- Assessment of change. Atlases can also be used to track changes associated with the progression of a disease. In such cases, a physician will compare images of a single person at various points in time. The physician may compare present images with those taken in the past, in which case the previous image will serve as a template from which differences can be measured. It is also possible to map the entire set of images to a normal, disease-free atlas of the same region, or conversely, one of a fully progressed disease state. These comparisons can help to determine the rates of change in the anatomy or the degree of normality/disease within a patient.
- Disease characterization and segmentation. Biomedical research makes extensive use of atlases, especially to quantitatively compare multiple images from groups of individuals. This process is important in the study of disease and disease progression. What differentiates a certain disease group from the normal population? What other diseases follow similar patterns or involve correlated anatomical abnormalities? What biomarkers should clinicians look for in diagnostic imaging tests to confirm a diagnosis? Do certain medications or treatments alter, slow the rate of change, or perhaps entirely reverse the abnormalities shown in images? Questions like these can be answered and explored by using labeled atlases to help segment imaging studies into various regions of interest that are known to be affected by a disease (or are yet to be discovered as potential biomarkers). Studies that purport to link anatomical abnormalities to a disease, however, require the evaluation of large populations of both affected and unaffected groups. Manually delineating the ROIs in each subjects can be an extremely time consuming task; for example, it can take several hours to accurately segment the hippocampus in one person alone. Furthermore, this task can have high intra- and inter-labeler variability, which can lead to study inconsistencies, so that manually segmented regions must then be further confirmed as accurate by an expert before further analysis. In the presence of a single, already delineated atlas, the ROI delineation task can be made far simpler if the labels from the atlas are (automatically) mapped onto each individual's images. Note that in some situations, it may be useful to use multiple atlases (*e.g.*, one for each group under study), to reduce bias and errors in mapping several groups to the same atlas. This tactic is particularly prudent when there are well-known and well-defined anatomical variations within the populations under study.

- Pathology detection. A probabilistic atlas characterizing a diseased population can be utilized to automatically detect pathology in new patient scans [203]. The next section discusses issues of disease characterization and pathology detection further.

Morphometry

In medical research, it is desirable to identify differences in anatomy that occur due to particular diseases. To effectively study these variations and make sound conclusions, researchers should use large groups of patients and controls to determine statistically significant differences in anatomy, whether it is variations in volume, position, or shape. When mapping an anatomical region to a template (*i.e.*, atlas), one can measure the amount of change required to map a given subsection/landmark of the patient group to the atlas and compare it to the control group without disease. In some sense, one can imagine warping one atlas, representing a disease population, to another normative atlas, in order to determine the degree of change either globally or within a given anatomical ROI. For example, consider mapping an MR brain scan to an atlas, and assessing changes in the hippocampus relative to a normal population. Differences in mapping can, for example, be associated with the hippocampal curvature at a particular axis of the hippocampus: the degree of curvature may signify changes that can possibly be used for early detection of disease or the efficacy of a treatment. Current research focuses on three main methods of group-wise statistical comparisons, all of which are extremely prevalent in the field of neuroscience; we briefly describe these three approaches below.

Voxel-based morphometry. Voxel-based methods of analysis compare scalar values at each corresponding voxel (or pixel in 2D) in groups of images. *Voxel-based morphometry* (VBM) is a method of obtaining statistical parametric maps of volumetric differences. Once images have been standardized (intensity normalization, linear registration), this type of analysis can be applied by summing up voxel counts of the intensity range of interest in a particular location, corresponding to the approximate volume in that region. Volumetric comparisons may then be made between populations. For this type of volumetric analysis, it would be counterproductive to perform nonlinear registration to deform the grid on which the image lies because the voxels will not be of uniform size (thereby making the summing operation invalid). VBM is a very popular approach and has been used in numerous neuroimaging studies of gray matter volume to quantify differences between populations [7, 68, 127].

Markedly, voxel-based statistical analysis is not limited to solely accounting for volumetric differences between regions or to linearly-aligned images. For example, in DTI, images are collected using multiple directional gradients to determine the relative diffusion rate of water in each direction. The directional diffusion information collected is often combined and represented as a scalar quantity depicting the degree of diffusion anisotropy at each voxel, as seen in common fractional anisotropy (FA) images (see Chapter 2). When acquiring these diffusion images, at least one set of images are taken with no diffusion gradients applied; this image set is identical to obtaining a T2-weighted MR scan. As atlases of T2-weighted MR scans are available and can also be constructed easily, linear and nonlinear registration can be performed on these T2-weighted images for each subject. The resulting transformation matrices from the linear registration along with the deformation fields obtained from the nonlinear registration can then be applied to the FA maps created from diffusion imaging. Once complete, the scalar anisotropy maps are then in the same stereotaxic space as the registered

T2 images, and each voxel is representative of approximately the same anatomical region throughout all subject images. Voxel-based statistical analysis can then be performed to determine relative differences in anisotropy across populations.

When performing statistical tests on the numerous voxels making up images, one must account for the error rate in multiple comparisons: performing tests at every voxel results in a vast amount of independent statistical tests; and the number of tests is proportional to the number of false positives found. False positives, in this case, are voxels that are deemed significantly different between the two groups. In other words, the more independent tests that are run between large imaging volumes, the more voxels are labeled as significantly different when in fact this might not be the case. Several methods have been proposed to account for multiple comparisons, all of which aim to reduce the effect of statistical errors on significance. The use of methods such as Bonferroni correction, Gaussian random field theory [5], non-parametric permutation tests, and false discovery rate [8] are thus commonplace.

Deformation-based morphometry. As mentioned previously, nonlinear registration results in vector fields describing the deformation encountered at each voxel in the image. These deformation fields describe the spatial transformations of voxels required to map different images to the same template. Using the deformation fields obtained from a nonlinear transformation, one can assess the degree to which certain regions differ in one image with respect to a set of comparison images. This form of analysis can also be used to statistically compare differences in volume for certain ROIs. *Deformation-based morphometry* (DBM) can also use the deformation fields to identify differences in the relative position of structures within an image volume.

Generally, a single multivariate test is performed on the deformations using a few descriptive parameters extracted from the deformation fields using, for instance, single value decomposition (SVD). The deformation field may be decomposed into components that describe global position, orientation, size, and shape, of which only shape may be of interest to a researcher. Hotelling's T^2 statistic can be used for simple comparisons between two groups of subjects: $t^2 = n(\mathbf{x} - \mu)\mathbf{S}^{-1}(\mathbf{x} - \mu)$, where n is the number of points, $\mathbf{x}$ is the vector of elements, μ is the mean, and $\mathbf{S}$ is the sample covariance matrix. Equivalently, a multivariate analysis of the covariance can be performed using Wilk's λ statistic on more complicated group experimental designs (see Chapter 10). In fact, atlases that are created from the imaging data of many subjects need to be created *after* all images are registered to the same stereotaxic space – only then can their anatomical information be combined to create the desired atlas. Atlases containing data from many sources can be created by simply averaging all images once in the same space, or through other methods such as a *minimal deformation template* (MDT). MDT creation involves nonlinearly mapping several subjects to a single subject, designated to be the template. The deformation fields obtained from these mappings are then combined and used to deform the template to better represent the entire group of subjects under study.

Tensor-based morphometry. As with DBM, *tensor-based morphometry* (TBM) also takes advantage of the deformation fields obtained from nonlinear registration. Though DBM is able to access changes in the global shape of the anatomical structure being imaged, TBM is further able to identify differences in the local shape of structures within each imaging voxel. In fact, the objective of TBM is to localize regions of shape differences among groups of images using the deformation fields to quantify the mappings of a point in a template, $T(x_1, x_2, x_3)$, to the same point in a subject's image, $S(y_1, y_2, y_3)$. The Jacobian matrix of the deformations is calculated by partial derivatives to depict information regarding the local stretching, shearing, and rotation of the deformation necessary to map the subject image to the template. The calculation of the Jacobian at each point is based on the following:

$$J = \begin{bmatrix} \frac{\partial y_1}{\partial x_1} & \frac{\partial y_1}{\partial x_2} & \frac{\partial y_1}{\partial x_3} \\ \frac{\partial y_2}{\partial x_1} & \frac{\partial y_2}{\partial x_2} & \frac{\partial y_2}{\partial x_3} \\ \frac{\partial y_3}{\partial x_1} & \frac{\partial y_3}{\partial x_2} & \frac{\partial y_3}{\partial x_3} \end{bmatrix}$$

Often, the determinant of this Jacobian matrix is taken to obtain a scalar quantity reflecting volumetric differences in the size of each voxel. This quantity is often referred to as simply the *Jacobian*. Assuming the deformation field mapped the subject to the template, and not *vice versa* (template to subject) the following data can be inferred at the voxel level: $J > 1$ represents growth of the region; $J < 1$ represents shrinkage of the region; and $J = 1$ represents no change in the volume of the voxel. When calculating the determinant of the Jacobian, it is possible to obtain a value of 1; therefore, one would assume that there is no change in the volume or position of a particular voxel. However, this result does not necessarily indicate that the shape of the voxel remained the same. More powerful forms of TBM take this change in length into account, along with changes in area and amount of shear in each voxel, using multivariate statistics on measures derived from the Jacobian matrix [13].

Discussion

In this chapter, we have discussed the importance of characterizing imaging data from the standpoint of improving its interpretation and for standardizing its representation to facilitate the comparison and sharing of both clinical and research study results. The standardization process includes both signal and positional level calibrations. An overview of the importance of anatomical atlases along with some of their potential uses in medicine was then presented. Methods for conducting group-wise comparisons pertaining to groups of images that have been registered have also been described.

The information presented in this chapter is but a brief survey of a field that has a long history and an enormous body of work in developing practical medical image understanding systems. However, important issues remain: how to model the interaction of features associated with images; how to compile data and training sets; how to learn the parameters for a given model; and how to utilize the model for inferring targeted user interpretations from the imaging data are all still unanswered questions. These issues motivate efforts to develop robust systems that can automatically extract quantitative observations (*e.g.*, size, shape, texture, intensity distributions) from both anatomical structures and pathological processes that manifest as imaging findings (*e.g.*, a tumor or mass). These quantitative observations can then be used to develop practical phenotypic models to better characterize the human condition (*i.e.*, normal or diseased states). Additionally, the characterization of imaging data must prospectively be made in the context of the entire clinical context of the patient. Spatial-temporal

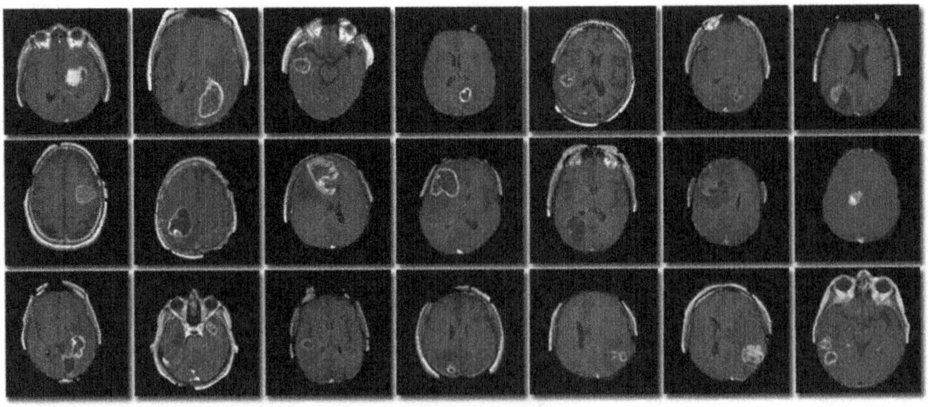

Figure 5.21: Axial T1-post contrast images of the brain showing the highly diverse imaging presentation of tumors in glioblastoma patients.

imaging patterns are likely degenerate with respect to many diseases, thus higher-level characterization of imaging data should be made in the context of other patient information (*e.g.*, demographic, social, laboratory, histopathology, surgical, and other observational and assessment data).

The implementation of systems that attempt to characterize patient data for decision support must be integrated into a larger development system that includes automated clinical data collection, structuring, modeling, and individualized patient model instantiation. This end-to-end development plan requires the consorted efforts of clinical staff, clinicians, informaticians, statisticians/mathematicians, and computer scientists.

The realization of clinical image analysis systems, however, includes a number of well-documented challenges associated with imaging data that make practical implementation difficult [164]. For example, Fig. 5.21 shows a sample collection of brain tumor images, as seen on T1-weighted, contrast-enhanced MR, demonstrating the high degree of variability in presentation. Brain tumors may be of any size, may have a variety of shapes, may appear at any location, and may appear with different image intensities [43]. Some tumors also deform neighboring structures and appear together with edema, which changes the intensity properties of nearby regions. Edema is especially hard to identify, with some knowledge of tissue flow resistance needed for accurate characterization. Given this difficulty, medical image analysis is a major area of focus within several disciplines, including applied mathematics, statistics, computer science, electrical engineering, cognitive science, neuroscience, medical informatics, and biomedical physics. Comprehensive toolkits are available to promote the application and development of medical image analysis methods (*e.g*, the Insight Segmentation and Registration Toolkit, ITK).

Towards Medical Image Analysis

While a complete discussion of medical image analysis methods is ultimately beyond the scope of this text, some major points of note are summarized below.

Mathematical Foundations

Various mathematical formalisms are used in imaging analysis systems to integrate variables, rules, constraints, transformations, and search strategies into a single unified representation. Energy models that can address both local and global characteristics of image states are among the most powerful. There are two important mathematical foundations that are common in medical image analysis: 1) Bayesian statistical formulations; and 2) partial differential equation (PDE) methods.

Bayesian statistical methods. The Bayesian framework leverages the fact that medical image data have a number of statistical regularities that can be exploited in certain image processing operations (*i.e.,* medical images are statistically redundant). Examples of statistical regularities include: the statistics related to the correlation of intensity values of neighboring pixels; the regularities associated with the strength and smoothness of anatomic boundaries; and the regularity of textures associated with homogeneous tissue regions. Additionally, one can employ the regularity of views in medical images in which anatomical sites and regions appear with preferred orientations. These regularities can be modeled statistically within a Bayesian model in which prior expectations of these regularities can be formally represented by a probability distribution. Probabilistic distributions over structured representations, such as graphs, grammars, predicate logic statements, and schemas, are being used to model medical imaging data. The Bayesian inferencing approach is applied by searching for a *maximum a posteriori* (MAP) estimate of the most likely interpretation of the imaging data for a given set of observed imaging features (see Chapter 9). The MAP estimate is found by methods such as expectation-maximization and Markov Chain Monte Carlo (MCMC) algorithms.

Partial differential equation methods. PDE-based frameworks are the second approach for building applications of medical imaging processing systems. The increasing amount of available computational power has made it possible to numerically solve PDEs of relatively large size with high precision. Finite element analysis, a numerical technique for finding PDE solutions, was first proposed in the 1940s and, along with its variants, has become a standard tool in science and engineering. Image processing and analysis algorithms involve solving an initial value problem for some PDE. The PDE solution can be either the image itself at different stages of modification (*e.g.,* de-noising or deblurring) or a "filtered" view of the image such as a closed curve delineating object boundaries. The medical image community has applied PDEs to imaging problems including restoration, segmentation, and registration:

- Image restoration (denoising). PDE-based image restoration focuses on optimizing a functional of the corrupted image. This functional typically consists of a data fidelity term and a regularization term. For instance, in the popular Rudin-Osher-Fatemi (ROF) model [172], the goal is to minimize:

$$F(r(x)) = \int_{\Omega} [(x) - r(x)]^2 \, dx + \lambda \int_{\Omega} |\nabla r(x)| \, dx$$

  where $i(x)$ is the original, noisy image; and $r(x)$ is the restored version. The stationary point of a functional can be calculated by solving the associated Euler-Lagrange equations, which are a system of PDEs. Other approaches involving PDEs are based on nonlinear diffusion [21, 156]; and [173, 217] summarize the most usual methods in the literature.
- Image segmentation. In image segmentation problems, PDEs typically arise from the Euler-Lagrange equations of a functional to optimize. This functional no longer depends on a restored image, $i(x)$, but on a discrete labeling, $l(x)$, instead. Three of the most prominent methods are active contours [103], and the Mumford-Shah [140] and Chan-Vese [22] techniques. Another popular approach that requires solving PDEs are *level set methods* [152], which have the advantage of handling

changes of topology within a region in a seamless manner. A good (though slightly outdated) survey of these and other methods can be found in [138].

- Image registration. PDE-based methods have been proposed for registration using level sets [154, 211] and viscous fluid models. Surveys of image registration methods, including PDE-based algorithms, can be found in [120, 232].

Image Modeling

Building atop these mathematical formalisms, image modeling for medical image analysis continues to be driven by the investigation of new types of image features. The current trend in the field has been to use a larger number of semantically richer, higher-order features specific to targeted entities, including: signal-level features; multi-spectral features; edge primitives (edgelets); curve primitives (curvelets); local texture features (textons); kernel basis dictionaries (*e.g.,* Gabor filters, wavelet); responses to parameterized object templates (bones, heart, ventricles, tumors); and summarization features related to regional statistics (*e.g.,* variance). Apart from the relatively low-level nature of image features, researchers are also investigating how to aggregate image features together into meaningful models, capturing contextual interactions and hierarchical relationships.

Modeling the contextual interaction of imaging features. Graphical models (see Chapter 8) are commonly used in statistical-based methods to model a joint distribution by decomposing the distribution into a smaller number of factors, thus simulating the interaction of feature variables. For images, these models can be either *descriptive* or *generative* [74, 228], and can be used to capture: localized correlations between pixel features (*e.g.,* texture fields using Markov random fields); intermediate-level correlations (*e.g.,* sparse coding methods [151]); and/or higher-order topological correlations (*e.g.,* context-free grammars). Generative models can be formulated using a stochastic grammar, enabling one to model the medical image in terms of a basic vocabulary of patterns corresponding to normal or abnormal structures (*e.g.,* anatomy, tumor, edema) [229].

Hierarchical compositionality and graphical models. Related to graphical models is the concept of hierarchical composition. Work on *compositional hierarchies* in which imaging objects/scenes are represented as a hierarchy of connected parts is an emerging trend [98, 226, 227]. These hierarchies allow for improved probabilistic expectation models for compositional regularities and can hence be used for better identification of objects and their constituents. Medical image descriptions are seen as the result of a complex hierarchy of features, some directly referring to the physical world (*e.g.,* T1 spin-spin relaxation), some referring to patterns within the imaging world (*e.g.,* high signal intensity region), and others referring to abstract objects summarizing lower-level regularities (*e.g.,* "mass"). Description of increasingly complex patterns can be represented via this mode of hierarchical feature composition using probabilistic graphs and grammars [24, 191].

Linking Images to Additional Knowledge

In many applications, better results are often achieved by applying domain-specific knowledge. In medicine, this knowledge comes from a bevy of sources; and several efforts in medical image analysis are looking to link image features and findings with this knowledge to enable a range of quantification tasks and to answer prognostic questions. For example, an important area of basic research is investigating how

MR imaging features	Classification task
Mass effect, cyst formation, necrosis on MR	Grading of supratentorial gliomas [42]
Heterogeneity of contrast enhancement, edema, mass effect, cyst formation, necrosis, flow void	Grading of gliomas [3, 160]
Contrast enhancement, volume of peritumoral edema	Grading of gliomas [87, 141]
Tumor capsule, vascular supply, calcification	Degree of malignancy [112, 216]
Degree of necrosis, edema, presence of tumor cysts	Patient survival times [105, 121]
Intensity of tumoral mass	Patient survival times [78]
Non-contrast enhancing tumor, edema, satellites and multi-focality	Patient survival times [162]
Indistinct border on T1 images and mixed signal intensity on T1 and T2	Genetic allelic loss of 1p and 19q, evidence of incomplete surgical resection [96, 134, 199]
T2/T1 volume ratio, border sharpness on T2	Molecular GBM subtypes [1]
ADC value	Ki-67 labeling index (negative correlation) [85]

Table 5.1: Features relating various MR imaging findings to aspects of glioblastoma multiforme nature and outcomes.

features and patterns seen in medical images are linked to biophysical parameters in order to establish *in vivo* disease biomarkers. As a point in case, several studies have investigated the correlation of imaging features to biological parameters associated with brain tumor microenvironments and patient survival times [3, 42, 66, 87, 101, 112, 121, 134, 141]; Table 5.1 summarizes a few of these studies. It has been observed that histologically similar tumors often demonstrate highly distinct imaging profiles on MRI. Recently, several studies have attempted to correlate imaging findings with molecular markers, but no consistent associations have emerged, and many of the imaging features that characterize tumors currently lack biological or molecular correlates [47]. Image appearance models of MRI data conditioned on gene-expression states are being researched to improve discrimination and characterization of disease based on imaging patterns [82]. In a slightly different light, a number of spatio-temporal growth models have been developed to describe tumor cell density [94, 196, 197]. [196] considered the effect of chemotherapeutic agents in heterogeneous tissue where drug delivery may vary with vascular density; [94] extended this model by incorporating T1 and DTI MR atlases derived from normal subjects to estimate the anisotropic diffusion for glial cells throughout the brain. Although these examples draw upon only brain tumors, similar efforts to link imaging with additional knowledge and models are seen for other cancers and diseases.

Biomechanical models. Biomechanical models that simulate the complex shape and motion of the heart have been developed using a variety of approaches including spectral-based methods [142], physics-based elastic models [88], and bending/stretching thin-plate models [2, 180]. These models, when registered and fitted to image data (*e.g.*, tagged MRIs) can be used to characterize cardiac function, identifying useful clinical parameters that give insight into disease processes [155]. The idea is that medical imaging data from a specific patient can provide shape, landmark, and orientation information for various anatomic structures. This information can then be incorporated into a dynamic model via the specification of initial conditions and constraints on which the model should adhere to (*e.g.*, shape, size, location and orientation constraints). The model can then simulate biomechanical behavior based on the individual's presentation.

Physiologic models. Likewise, the joining of theoretical *in silico* models of physiologic processes with imaging data derived from patient studies can potentially provide a power approach to garner insight into an individual's disease, allowing a physician to simulate the course of a disease and/or predict response to a given intervention. The Physiome and Virtual Physiological Human Projects are substantial efforts related to building multi-scale models of human physiology [29, 89]. Spatial-temporal information (*e.g.*, geometry, motion, perfusion) derived from imaging studies (*e.g.*, MR, CT, PET) for a given patient can be used to estimate a few parameters of a theoretic model, which can then drive estimations of parameters associated with lower level phenomena (*e.g.*, molecular, cellular, electrical) that can be used to provide mechanistic insights into a patient's disease. For instance, detailed multi-scale physiological models of the heart are available that include membrane-bound channels, myocytes, and protein interactions [36, 189]. Information derived from MR patient studies are used to generate information related to fibroblast and collagen microstructure, which are then used to compute tensors related to electrical conductivity and mechanical stiffness. Simulations can then be conducted to assess the overall normality or deficiencies of the heart.

References

1. Aghi M, Gaviani P, Henson JW, Batchelor TT, Louis DN, Barker FG, 2nd (2005) Magnetic resonance imaging characteristics predict epidermal growth factor receptor amplification status in glioblastoma. Clin Cancer Res, 11(24 Pt 1):8600-8605.
2. Amini AA, Duncan JS (1991) Pointwise tracking of left-ventricular motion in 3D. Proc IEEE Workshop on Visual Motion, pp 294-299.
3. Asari S, Makabe T, Katayama S, Itoh T, Tsuchida S, Ohmoto T (1994) Assessment of the pathological grade of astrocytic gliomas using an MRI score. Neuroradiology, 36(4):308-310.
4. Ashburner J, Friston KJ (1999) Nonlinear spatial normalization using basis functions. Hum Brain Mapp, 7(4):254-266.
5. Ashburner J, Neelin P, Collins DL, Evans A, Friston K (1997) Incorporating prior knowledge into image registration. Neuroimage, 6(4):344-352.
6. Attix FH (1986) Introduction to Radiological Physics and Radiation Dosimetry. John Wiley & Sons Inc, New York, NY.
7. Baron JC, Chételat G, Desgranges B, Perchey G, Landeau B, de la Sayette V, Eustache F (2001) In vivo mapping of gray matter loss with voxel-based morphometry in mild Alzheimer's disease. Neuroimage, 14(2):298-309.
8. Benjamini Y, Hochberg Y (1995) Controlling the false discovery rate: A practical and powerful approach to multiple testing. J Royal Statistical Society Series B - Methodological:289-300.
9. Blomgren P, Chan TF (1998) Color TV: Total variation methods for restoration of vector-valued images. IEEE Trans Image Process, 7(3):304-309.
10. Bookstein FL (1989) Principal warps - Thin-plate splines and the decomposition of deformations. IEEE Trans Pattern Analysis and Machine Intelligence, 11(6):567-585.
11. Bro-Nielsen M, Gramkow C (1996) Fast fluid registration of medical images. Proc Visualization in Biomedical Computing: 4th Intl Conf. Springer-Verlag, pp 267-276.
12. Brooks RD, Glover GH, Talbert AJ, Eisner RL, DiBianca FA (1979) Aliasing: a source of streaks in computed tomograms. J Comput Assist Tomogr, 3(4):511-518.
13. Brun C, Lepore N, Pennec X, Lee AD, Barysheva M, Madsen SK, Avedissian C, Chou YY, de Zubicaray GI, McMahon K, Wright M, Toga AW, Thompson PM (2009) Mapping the regional influence of genetics on brain structure variability - A tensor-based morphometry study. Neuroimage.

14. Cachier P, Pennec X, Inria SA (2000) 3D non-rigid registration by gradient descent on a Gaussian-windowed similarity measure using convolutions. Proc IEEE Workshop on Mathematical Methods in Biomedical Image Analysis, pp 182-189.
15. Caldwell CB, Stapleton SJ, Holdsworth DW, Jong RA, Weiser WJ, Cooke G, Yaffe MJ (1990) Characterisation of mammographic parenchymal pattern by fractal dimension. Phys Med Biol, 35(2):235-247.
16. Camastra F (2003) Data dimensionality estimation methods: A survey. Pattern Recognition, 36:2945-2954.
17. Camastra F, Vinciarelli A (2002) Estimating the intrinsic dimension of data with a fractal-based approach. IEEE Trans Pattern Analysis and Machine Intelligence, 24(10):1404-1407.
18. Cann CE (1988) Quantitative CT for determination of bone mineral density: A review. Radiology, 166(2):509-522.
19. Canny J (1986) A computational approach to edge-detection. IEEE Trans Pattern Analysis and Machine Intelligence, 8(6):679-698.
20. Carrerira-Perpinan MA (1997) A review of dimension reduction techniques (Technical Report). Dept Computer Science, University of Sheffield. www.dcs.shef.ac.uk/intranet/re-search/resmes/CS9609.pdf. Accessed February 5, 2009.
21. Catte F, Lions PL, Morel JM, Coll T (1992) Image selective smoothing and edge-detection by nonlinear diffusion. SIAM J Numerical Analysis, 29(1):182-193.
22. Chan TF, Vese LA (2001) Active contours without edges. IEEE Trans Image Process, 10(2):266-277.
23. Chen CC, Daponte JS, Fox MD (1989) Fractal feature analysis and classification in medical imaging. IEEE Trans Med Imaging, 8(2):133-142.
24. Chen Y, Zhu L, Lin C, Yuille A, Zhang H (2008) Rapid Inference on a novel AND/OR graph for object detection, segmentation, and parsing. In: Platt J, Koller D, Singer Y, Roweis S (eds) Advances in Neural Information Processing Systems. MIT Press, Cambridge, pp 289-296.
25. Chou YY, Leporé N, Avedissian C, Madsen SK, Parikshak N, Hua X, Shaw LM, Trojanowski JQ, Weiner MW, Toga AW (2009) Mapping correlations between ventricular expansion and CSF amyloid and tau biomarkers in 240 subjects with Alzheimer's disease, mild cognitive impairment and elderly controls. Neuroimage, 46(2):394-410.
26. Christensen GE, Rabbitt RD, Miller MI (1996) Deformable templates using large deformation kinematics. IEEE Trans Image Process, 5(10):1435-1447.
27. Christiansen O, Lee TM, Lie J, Sinha U, Chan TF (2007) Total variation regularization of matrix-valued images. Int J Biomed Imaging, 2007:27432.
28. Chuang KS, Huang HK (1987) A fast dual-energy computational method using iso-transmission lines and table lookup. Medical Physics, 14(2):186-192.
29. Clapworthy G, Viceconti M, Coveney PV, Kohl P (2008) The virtual physiological human: Building a framework for computational biomedicine (editorial). Philos Transact A Math Phys Eng Sci, 366(1878):2975-2978.
30. Clare S, Jezzard P (2001) Fast T1 mapping using multislice EPI. Brain, 1504(2):76.
31. Cleare HM, Splettstosser HR, Seemann HE (1962) An experimental study of the mottle produced by x-ray intensifying screens. Am J Roentgenol Radium Ther Nucl Med, 88:168-174.
32. Condat L, Van De Ville D, Blu T (2005) Hexagonal versus orthogonal lattices: A new comparison using approximation theory. IEEE Intl Conf Image Processing, vol 3, pp 1116-1119.

33. Constantinou C, Harrington JC, DeWerd LA (1992) An electron density calibration phantom for CT-based treatment planning computers. Med Phys, 19(2):325-327.
34. Cootes TF, Hill A, Taylor CJ, Haslam J (1994) Use of active shape models for locating structure in medical images. Image and Vision Computing, 12(6):355-365.
35. Cootes TF, Taylor CJ, Cooper DH, Graham J (1995) Active shape models - Their training and application. Computer Vision and Image Understanding, 61(1):38-59.
36. Crampin EJ, Halstead M, Hunter P, Nielsen P, Noble D, Smith N, Tawhai M (2004) Computational physiology and the Physiome Project. Exp Physiol, 89(1):1-26.
37. Crivello F, Schormann T, Tzourio-Mazoyer N, Roland PE, Zilles K, Mazoyer BM (2002) Comparison of spatial normalization procedures and their impact on functional maps. Human Brain Mapping, 16(4):228-250.
38. Curry TS, Dowdey JE, Murry RC, Christensen EE (1990) Christensen's physics of diagnostic radiology. 4th edition. Lea & Febiger, Philadelphia, PA.
39. Dai XL, Khorram S (1999) A feature-based image registration algorithm using improved chain-code representation combined with invariant moments. IEEE Trans Geoscience and Remote Sensing, 37(5):2351-2362.
40. Dainty JC, Shaw R (1974) Image Science: Principles, Analysis and Evaluation of Photographic-type Imaging Processes. Academic Press, London, UK.
41. Dash M, Liu H (1997) Feature selection for classification. Intelligent Data Analysis, 1(3):131-156.
42. Dean BL, Drayer BP, Bird CR, Flom RA, Hodak JA, Coons SW, Carey RG (1990) Gliomas: Classification with MR imaging. Radiology, 174(2):411-415.
43. DeAngelis LM (2001) Brain tumors. N Engl J Med, 344(2):114-123.
44. Dengler J, Schmidt M (1988) The Dynamic Pyramid - A model for motion analysis with controlled continuity. Int J Pattern Recog Artif Intell, 2(2):275-286.
45. Deoni SC, Rutt BK, Peters TM (2003) Rapid combined T1 and T2 mapping using gradient recalled acquisition in the steady state. Magn Reson Med, 49(3):515-526.
46. Devlin JT, Poldrack RA (2007) In praise of tedious anatomy. Neuroimage, 37(4):1033-1041.
47. Diehn M, Nardini C, Wang DS, McGovern S, Jayaraman M, Liang Y, Aldape K, Cha S, Kuo MD (2008) Identification of noninvasive imaging surrogates for brain tumor gene-expression modules. Proc Natl Acad Sci USA, 105(13):5213-5218.
48. Donoho DL, Johnstone IM (1994) Ideal spatial adaptation by wavelet shrinkage. Biometrika, 81(3):425-455.
49. Du J, Fain SB, Gu T, Grist TM, Mistretta CA (2004) Noise reduction in MR angiography with nonlinear anisotropic filtering. J Magn Reson Imaging, 19(5):632-639.
50. Dube S (2009) An Automated System for Quantitative Hierarchical Image Analysis of Malignant Gliomas: Developing Robust Techniques for Integrated Segmentation/Classification and Prognosis of Glioblastoma Multiforme. Department of Biomedical Engineering, PhD Dissertation. UCLA.
51. Duta N, Sonka M (1998) Segmentation and interpretation of MR brain images: An improved active shape model. IEEE Trans Med Imaging, 17(6):1049-1062.
52. Efros A, Leung T (1999) Texture synthesis by non-parametric sampling. Proc IEEE Intl Conf Computer Vision, vol 2, pp 1033-1038.
53. Ehrhardt JC (1990) MR data acquisition and reconstruction using efficient sampling schemes. IEEE Trans Med Imaging, 9(3):305-309.
54. El-Baz A, Gimel'farb G, Falk R, Abo El-Ghar M (2009) Automatic analysis of 3D low dose CT images for early diagnosis of lung cancer. Pattern Recognition, 42(6):1041-1051.

55. El-Naqa I, Yang Y, Galatsanos NP, Nishikawa RM, Wernick MN (2004) A similarity learning approach to content-based image retrieval: application to digital mammography. IEEE Trans Med Imaging, 23(10):1233-1244.
56. Faugeras O (1978) Texture analysis and classification using a human visual model. Proc IEEE Intl Conf Pattern Recognition, pp 549-552.
57. Ferrari RJ, Rangayyan RM, Desautels JE, Borges RA, Frere AF (2004) Automatic identification of the pectoral muscle in mammograms. IEEE Trans Med Imaging, 23(2):232-245.
58. Ferrari RJ, Rangayyan RM, Desautels JE, Frere AF (2001) Analysis of asymmetry in mammograms via directional filtering with Gabor wavelets. IEEE Trans Med Imaging, 20(9):953-964.
59. Flannery B, Press W, Teukolsky S, Vetterling W (1992) Numerical Recipes in C. 2nd edition. Press Syndicate of the University of Cambridge, New York.
60. Fodor IK (2002) A survey of dimension reduction techniques. Lawrence Livermore National Laboratory.
61. Fox PT (1995) Spatial normalization origins: Objectives, applications, and alternatives. Human Brain Mapping, 3(3):161-164.
62. Fukunaga K, Olsen DR (1971) An algorithm for finding intrinsic dimensionality of data. IEEE Trans Computers, 20(2):176-183.
63. Ge Y, Udupa JK, Nyul LG, Wei L, Grossman RI (2000) Numerical tissue characterization in MS via standardization of the MR image intensity scale. J Magn Reson Imaging, 12(5):715-721.
64. Gerig G, Kubler O, Kikinis R, Jolesz FA (1992) Nonlinear anisotropic filtering of MRI data. IEEE Trans Med Imaging, 11(2):221-232.
65. Giger ML, Doi K (1984) Investigation of basic imaging properties in digital radiography. I. Modulation transfer function. Med Phys, 11(3):287-295.
66. Gillies RJ, Raghunand N, Karczmar GS, Bhujwalla ZM (2002) MRI of the tumor microenvironment. J Magn Reson Imaging, 16(4):430-450.
67. Gonzalez RC, Woods RE (2008) Digital Image Processing. 3rd edition. Prentice Hall, Upper Saddle River.
68. Good CD, Johnsrude IS, Ashburner J, Henson RN, Friston KJ, Frackowiak RS (2001) A voxel-based morphometric study of ageing in 465 normal adult human brains. Neuroimage, 14(1 Pt 1):21-36.
69. Goodall C (1991) Procrustes methods in the statistical analysis of shape. J Royal Statistical Society Series B - Methodological, 53(2):285-339.
70. Gravel P, Beaudoin G, De Guise JA (2004) A method for modeling noise in medical images. IEEE Trans Med Imaging, 23(10):1221-1232.
71. Gudbjartsson H, Patz S (1995) The Rician distribution of noisy MRI data. Magn Reson Med, 34(6):910-914.
72. Guerrero T, Zhang G, Huang TC, Lin KP (2004) Intrathoracic tumour motion estimation from CT imaging using the 3D optical flow method. Phys Med Biol, 49(17):4147-4161.
73. Guimond A, Roche A, Ayache N, Meunier J (1999) Multimodal brain warping using the demons algorithm and adaptative intensity corrections. INRIA.
74. Guo CE, Zhu SC, Wu YN (2003) Modeling visual patterns by integrating descriptive and generative methods. Intl J Computer Vision, 53(1):5-29.
75. Guyon I, Elisseff A (2003) An introduction to variable and feature selection. J Mach Learn Res, 3:1157-1182.
76. Haber E, Modersitzki J (2006) A multilevel method for image registration. SIAM J Scientific Computing, 27(5):1594-1607.

77. Hajnal JV, Hawkes DJ, Hill DLG (2001) Medical Image Registration. CRC Press, Boca Raton, FL.
78. Hammoud MA, Sawaya R, Shi W, Thall PF, Leeds NE (1996) Prognostic significance of preoperative MRI scans in glioblastoma multiforme. J Neurooncol, 27(1):65-73.
79. Haralick RM, Shanmuga K, Dinstein I (1973) Textural features for image classification. IEEE Trans Systems Man and Cybernetics, Smc3(6):610-621.
80. Hasegawa M, Nasuhara Y, Onodera Y, Makita H, Nagai K, Fuke S, Ito Y, Betsuyaku T, Nishimura M (2006) Airflow limitation and airway dimensions in chronic obstructive pulmonary disease. Am J Respir Crit Care Med, 173(12):1309-1315.
81. Haux R, Kulikowski C (eds) (2002) Medical Imaging Informatics: Yearbook of Medical Informatics 2002. Schattauer Germany.
82. Hegi ME, Diserens AC, Gorlia T, Hamou MF, de Tribolet N, Weller M, Kros JM, Hainfellner JA, Mason W, Mariani L, Bromberg JE, Hau P, Mirimanoff RO, Cairncross JG, Janzer RC, Stupp R (2005) MGMT gene silencing and benefit from temozolomide in glioblastoma. N Engl J Med, 352(10):997-1003.
83. Hellier P, Barillot C, Memin E, Perez P (2001) Hierarchical estimation of a dense deformation field for 3-D robust registration. IEEE Trans Med Imaging, 20(5):388-402.
84. Henderson E, McKinnon G, Lee TY, Rutt BK (1999) A fast 3D look-locker method for volumetric T1 mapping. Magn Reson Imaging, 17(8):1163-1171.
85. Higano S, Yun X, Kumabe T, Watanabe M, Mugikura S, Umetsu A, Sato A, Yamada T, Takahashi S (2006) Malignant astrocytic tumors: Clinical importance of apparent diffusion coefficient in prediction of grade and prognosis. Radiology, 241(3):839-846.
86. Hill DL, Batchelor PG, Holden M, Hawkes DJ (2001) Medical image registration. Phys Med Biol, 46(3):R1-45.
87. Holodny AI, Nusbaum AO, Festa S, Pronin IN, Lee HJ, Kalnin AJ (1999) Correlation between the degree of contrast enhancement and the volume of peritumoral edema in meningiomas and malignant gliomas. Neuroradiology, 41(11):820-825.
88. Huang WC, Goldgof DB (1993) Adaptive-size meshes for rigid and nonrigid shape-analysis and synthesis. IEEE Trans Pattern Analysis and Machine Intelligence, 15(6):611-616.
89. Hunter PJ, Borg TK (2003) Integration from proteins to organs: The Physiome Project. Nat Rev Mol Cell Biol, 4(3):237-243.
90. Ioannidis JPA (2005) Why most published research findings are false. PLoS Medicine, 2(8):696-701.
91. Jaffe C (2005) caBIG Imaging Workspace. National Institute of Health. http: //cabig.nci.nih.gov/workspaces-/Imaging/Meetings/FacetoFace/December_2005. Accessed June 8, 2009.
92. Jäger F, Hornegger J (2009) Nonrigid registration of joint histograms for intensity standardization in magnetic resonance imaging. IEEE Trans Med Imaging, 28(1):137-150.
93. Jain AK, Duin RPW, Mao JC (2000) Statistical pattern recognition: A review. IEEE Trans Pattern Analysis and Machine Intelligence, 22(1):4-37.
94. Jbabdi S, Mandonnet E, Duffau H, Capelle L, Swanson KR, Pelegrini-Issac M, Guillevin R, Benali H (2005) Simulation of anisotropic growth of low-grade gliomas using diffusion tensor imaging. Magn Reson Med, 54(3):616-624.
95. Jenkinson M, Smith S (2001) A global optimisation method for robust affine registration of brain images. Medical Image Analysis, 5(2):143-156.
96. Jenkinson MD, du Plessis DG, Smith TS, Joyce KA, Warnke PC, Walker C (2006) Histological growth patterns and genotype in oligodendroglial tumours: Correlation with MRI features. Brain, 129(Pt 7):1884-1891.

97. Jensen JA (1991) A model for the propagation and scattering of ultrasound in tissue. J Acoust Soc Am, 89(1):182-190.
98. Jin Y, Geman S (2006) Context and hierarchy in a probabilistic image model. Proc IEEE Conf Computer Vision and Pattern Recognition, vol 2, pp 2145-2152.
99. Jolliffe IT (2002) Principal Component Analysis. Springer, New York, NY.
100. Joshi SC, Miller MI (2000) Landmark matching via large deformation diffeomorphisms. Ieee Transactions on Image Processing, 9(8):1357-1370.
101. Just M, Thelen M (1988) Tissue characterization with T1, T2, and proton density values: Results in 160 patients with brain tumors. Radiology, 169(3):779-785.
102. Kalra MK, Wittram C, Maher MM, Sharma A, Avinash GB, Karau K, Toth TL, Halpern E, Saini S, Shepard JA (2003) Can noise reduction filters improve low-radiation-dose chest CT images? Pilot study. Radiology, 228(1):257-264.
103. Kass M, Witkin A, erzopoulos D (1988) Snakes: Active contour models. Intl J Comp Vision, 1(4):321–331.
104. Kaur L, Gupta S, Chauhan R (2002) Image denoising using wavelet thresholding. Proc Indian Conf Computer Vision, Graphics and Image Processing. Citeseer.
105. Keles GE, Anderson B, Berger MS (1999) The effect of extent of resection on time to tumor progression and survival in patients with glioblastoma multiforme of the cerebral hemisphere. Surg Neurol, 52(4):371-379.
106. Kim WY, Danias PG, Stuber M, Flamm SD, Plein S, Nagel E, Langerak SE, Weber OM, Pedersen EM, Schmidt M, Botnar RM, Manning WJ (2001) Coronary magnetic resonance angiography for the detection of coronary stenoses. N Engl J Med, 345(26):1863-1869.
107. Klein A, Andersson J, Ardekani BA, Ashburner J, Avants B, Chiang MC, Christensen GE, Collins DL, Gee J, Hellier P, Song JH, Jenkinson M, Lepage C, Rueckert D, Thompson P, Vercauteren T, Woods RP, Mann JJ, Parsey RV (2009) Evaluation of 14 nonlinear deformation algorithms applied to human brain MRI registration. Neuroimage, 46(3):786-802.
108. Lee JS (1980) Digital image-enhancement and noise filtering by use of local statistics. IEEE Trans Pattern Analysis and Machine Intelligence, 2(2):165-168.
109. Lee TM, Sinha U (2006) Denoising diffusion tensor images: Preprocessing for automated detection of subtle diffusion tensor abnormalities between populations. Proc SPIE on Medical Image Processing, vol 6144, p 61446O.
110. Lee WL, Chen YC, Hsieh KS (2003) Ultrasonic liver tissues classification by fractal feature vector based on M-band wavelet transform. IEEE Trans Med Imaging, 22(3):382-392.
111. Li BJ, Christensen GE, Hoffman EA, McLennan G, Reinhardt JM (2003) Establishing a normative atlas of the human lung: Intersubject warping and registration of volumetric CT images. Academic Radiology, 10(3):255-265.
112. Li GZ, Yang J, Ye CZ, Geng DY (2006) Degree prediction of malignancy in brain glioma using support vector machines. Comput Biol Med, 36(3):313-325.
113. Lin JW, Sciacca RR, Chou RL, Laine AF, Bergmann SR (2001) Quantification of myocardial perfusion in human subjects using Rb-82 and wavelet-based noise reduction. J Nuclear Medicine, 42(2):201-208.
114. Lindeberg T (1994) Scale-space Theory in Computer Vision. Kluwer Academic, Boston, MA.
115. Liu H, Motoda H (eds) (2007) Computational Methods of Feature Selection. Chapman & Hall/CRC Boca Raton, FL.
116. Lorenz C, von Berg J (2006) A comprehensive shape model of the heart. Medical Image Analysis, 10(4):657-670.

117. Lowe DG (1999) Object recognition from local scale-invariant features. Proc 7th IEEE Intl Conf Computer Vision, vol 2, pp 1150-1157.
118. Lowe DG (2004) Distinctive image features from scale-invariant keypoints. Intl J Computer Vision, 60(2):91-110.
119. Lysaker M, Lundervold A, Tai XC (2003) Noise removal using fourth-order partial differential equation with application to medical magnetic resonance images in space and time. IEEE Trans Image Process, 12(12):1579-1590.
120. Maintz JB, Viergever MA (1998) A survey of medical image registration. Med Image Anal, 2(1):1-36.
121. Maldaun MV, Suki D, Lang FF, Prabhu S, Shi W, Fuller GN, Wildrick DM, Sawaya R (2004) Cystic glioblastoma multiforme: Survival outcomes in 22 cases. J Neurosurg, 100(1):61-67.
122. Malfait M, Roose D (1997) Wavelet-based image denoising using a Markov random field a priori model. IEEE Trans Image Process, 6(4):549-565.
123. Mandelbrot B (1982) The Fractal Geometry of Nature. W.H. Freeman, New York, NY.
124. Manjon JV, Carbonell-Caballero J, Lull JJ, Garcia-Marti G, Marti-Bonmati L, Robles M (2008) MRI denoising using non-local means. Medical Image Analysis, 12(4):514-523.
125. Martin S, Backer A (2005) Estimating manifold dimension by inversion error. Proc ACM Symp Applied Computing, pp 22-26.
126. Matsopoulos GK, Mouravliansky NA, Delibasis KK, Nikita KS (1999) Automatic retinal image registration scheme using global optimization techniques. IEEE Trans Inf Technol Biomed, 3(1):47-60.
127. May A, Ashburner J, Büchel C, McGonigle DJ, Friston KJ, Frackowiak RSJ, Goadsby PJ (1999) Correlation between structural and functional changes in brain in an idiopathic headache syndrome. Nature Medicine, 5(7):836-838.
128. Mazziotta JC, Toga AW, Evans A, Fox P, Lancaster J (1995) A probabilistic atlas of the human brain: Theory and rationale for its development. The International Consortium for Brain Mapping (ICBM). Neuroimage, 2(2):89-101.
129. McGraw T, Vemuri BC, Chen Y, Rao M, Mareci T (2004) DT-MRI denoising and neuronal fiber tracking. Medical Image Analysis, 8(2):95-111.
130. McKenzie CA, Chen Z, Drost DJ, Prato FS (1999) Fast acquisition of quantitative T2 maps. Magn Reson Med, 41(1):208-212.
131. McLachlan GJ (2004) Discriminant Analysis and Statistical Pattern Recognition. Wiley-Interscience, New York, NY.
132. McNitt-Gray MF (2002) AAPM/RSNA physics tutorial for residents: Topics in CT - Radiation dose in CT1. Radiographics, 22(6):1541-1553.
133. McNitt-Gray MF, Cagnon CH, Solberg TD, Chetty I (1999) Radiation dose in spiral CT: The relative effects of collimation and pitch. Medical Physics, 26(3):409-414.
134. Megyesi JF, Kachur E, Lee DH, Zlatescu MC, Betensky RA, Forsyth PA, Okada Y, Sasaki H, Mizoguchi M, Louis DN, Cairncross JG (2004) Imaging correlates of molecular signatures in oligodendrogliomas. Clin Cancer Res, 10(13):4303-4306.
135. Melhem ER, Gotwald TF, Itoh R, Zinreich SJ, Moser HW (2001) T2 relaxation measurements in X-linked adrenoleukodystrophy performed using dual-echo fast fluid-attenuated inversion recovery MR imaging. AJNR Am J Neuroradiol, 22(4):773-776.
136. Michailovich OV, Tannenbaum A (2006) Despeckling of medical ultrasound images. IEEE Trans Ultrasonics Ferroelectrics and Frequency Control, 53(1):64-78.
137. Modersitzki J (2004) Numerical Methods for Image Registration. Oxford University Press, New York, NY.

138. Morel J-M, Solimini S (1995) Variational Methods in Image Segmentation: With Seven Image Processing Experiments. Birkhäuser, Boston, MA.
139. Muller A, Ruegsegger E, Ruegsegger P (1989) Peripheral QCT: A low-risk procedure to identify women predisposed to osteoporosis. Phys Med Biol, 34(6):741-749.
140. Mumford D, Shah J (1989) Optimal approximations by piecewise smooth functions and associated variational-problems. Comm Pure and Applied Mathematics, 42(5):577-685.
141. Nakano T, Asano K, Miura H, Itoh S, Suzuki S (2002) Meningiomas with brain edema: Radiological characteristics on MRI and review of the literature. Clin Imaging, 26(4):243-249.
142. Nastar C, Ayache N (1996) Frequency-based nonrigid motion analysis: Application to four dimensional medical images. IEEE Trans Pattern Analysis and Machine Intelligence, 18(11):1067-1079.
143. National Library of Medicine (NLM) (2009) Historical Anatomies on the Web. http://www.nlm.nih.gov/exhibition/historicalanatomies/browse.html. Accessed June 3, 2009.
144. National Library of Medicine (NLM) (2009) The Visible Human Project. http: //www.nlm.nih.gov/research/visible/visible_human.html. Accessed June 3, 2009.
145. Nekolla S, Gneiting T, Syha J, Deichmann R, Haase A (1992) T1 maps by k-space reduced snapshot-FLASH MRI. J Comput Assist Tomogr, 16(2):327-332.
146. Ni D, Qul Y, Yang X, Chui YP, Wong TT, Ho SS, Heng PA (2008) Volumetric ultrasound panorama based on 3D SIFT. Proc 11th Intl Conf Medical Image Computing and Computer-Assisted Intervention, Part II, vol 11, pp 52-60.
147. Niemeijer M, van Ginneken B, Staal J, Suttorp-Schulten MS, Abramoff MD (2005) Automatic detection of red lesions in digital color fundus photographs. IEEE Trans Med Imaging, 24(5):584-592.
148. Nixon M, Aguado AS (2008) Feature Extraction & Image Processing. 2nd edition. Academic Press, Oxford, UK.
149. Nowak RD (1999) Wavelet-based Rician noise removal for magnetic resonance imaging. IEEE Transactions on Image Processing, 8(10):1408-1419.
150. Nyul LG, Udupa JK, Zhang X (2000) New variants of a method of MRI scale standardization. IEEE Trans Med Imaging, 19(2):143-150.
151. Olshausen BA, Field DJ (1997) Sparse coding with an overcomplete basis set: A strategy employed by V1? Vision Res, 37(23):3311-3325.
152. Osher S, Fedkiw RP (2003) Level Set Methods and Dynamic Implicit Surfaces. Springer, New York, NY.
153. Papadakis NG, Murrills CD, Hall LD, Huang CL, Adrian Carpenter T (2000) Minimal gradient encoding for robust estimation of diffusion anisotropy. Magn Reson Imaging, 18(6):671-679.
154. Paragios N (2003) A level set approach for shape-driven segmentation and tracking of the left ventricle. IEEE Trans Med Imaging, 22(6):773-776.
155. Park J, Metaxas D, Young AA, Axel L (1996) Deformable models with parameter functions for cardiac motion analysis from tagged MRI data. IEEE Trans Med Imaging, 15(3):278-289.
156. Perona P, Malik J (1990) Scale-space and edge-detection using anisotropic diffusion. IEEE Trans Pattern Analysis and Machine Intelligence, 12(7):629-639.
157. Peters B, Meyer-Ebrecht D, Lehmann T, Schmitt W (1996) System analysis of x-ray-sensitive CCDs and adaptive restoration of intraoral radiographs. Proc SPIE on Medical Imaging, vol 2710, pp 450-461.

158. Petrick N, Chan HP, Wei D, Sahiner B, Helvie MA, Adler DD (1996) Automated detection of breast masses on mammograms using adaptive contrast enhancement and texture classification. Med Phys, 23(10):1685-1696.
159. Petroudi S, Kadir T, Brady M (2003) Automatic classification of mammographic parenchymal patterns: A statistical approach. Proc 25th Annual Intl Conf IEEE Engineering in Medicine and Biology Society (EMBS), vol 1, pp 798-801.
160. Pierallini A, Bonamini M, Bozzao A, Pantano P, Stefano DD, Ferone E, Raguso M, Bosman C, Bozzao L (1997) Supratentorial diffuse astrocytic tumours: Proposal of an MRI classification. Eur Radiol, 7(3):395-399.
161. Pizer SM, Fritsch DS, Yushkevich PA, Johnson VE, Chaney EL (1999) Segmentation, registration, and measurement of shape variation via image object shape. IEEE Trans Med Imaging, 18(10):851-865.
162. Pope WB, Sayre J, Perlina A, Villablanca JP, Mischel PS, Cloughesy TF (2005) MR imaging correlates of survival in patients with high-grade gliomas. AJNR Am J Neuroradiol, 26(10):2466-2474.
163. Portilla J, Simoncelli EP (2000) A parametric texture model based on joint statistics of complex wavelet coefficients. Intl J Computer Vision, 40(1):49-71.
164. Prastawa M, Bullitt E, Ho S, Gerig G (2004) A brain tumor segmentation framework based on outlier detection. Med Image Anal, 8(3):275-283.
165. Pratt WK (1991) Digital Image Processing. 2nd edition. Wiley, New York, NY.
166. Prewitt JMS (1970) Object enhancement and extraction. In: Lipkin BS, Rosenfeld A (eds) Picture Processing and Psychopictorics. Academic Press, New York, NY, pp 75-149.
167. Pudil P, Novovicova J, Kittler J (1994) Floating search methods in feature-selection. Pattern Recognition Letters, 15(11):1119-1125.
168. Romberg JK, Choi H, Baraniuk RG (1999) Shift-invariant denoising using wavelet-domain hidden Markov trees. Proc 33rd Asilomar Conf Signals, Systems, and Computers, vol 2, pp 1277-1281.
169. Rosenfeld A, Kak AC (1982) Digital Picture Processing. 2nd edition. Academic Press, New York, NY.
170. Rossmann K (1963) Spatial fluctuations of x-ray quanta and the recording of radiographic mottle. AJR Am J Roentgenol, 90:863–869.
171. Rousseeuw PJ, Leroy AM (1987) Robust Regression and Outlier Detection. Wiley, New York, NY.
172. Rudin LI, Osher S, Fatemi E (1992) Nonlinear total variation based noise removal algorithms. Physica D, 60(1-4):259-268.
173. Sapiro G (2001) Geometric Partial Differential Equations and Image Analysis. Cambridge University Press, Cambridge, UK.
174. Scharcanski J, Jung CR (2006) Denoising and enhancing digital mammographic images for visual screening. Comput Med Imaging Graph, 30(4):243-254.
175. Scharcanski J, Jung CR, Clarke RT (2002) Adaptive image denoising using scale and space consistency. IEEE Trans Image Process, 11(9):1092-1101.
176. Scheffler K, Hennig J (2001) T1 quantification with inversion recovery TrueFISP. Magn Reson Med, 45(4):720-723.
177. Schneider U, Pedroni E, Lomax A (1996) The calibration of CT Hounsfield units for radiotherapy treatment planning. Phys Med Biol, 41(1):111-124.
178. Shaw A, Moores BM (1985) Noise transfer in screen-film subtraction radiography. Phys Med Biol, 30(3):229-238.
179. Shen L, Rangayyan RM, Desautels JL (1994) Application of shape analysis to mammographic calcifications. IEEE Trans Med Imaging, 13(2):263-274.

180. Shi P, Amini AA, Robinson G, Sinusas A, Constable CT, Duncan J (1994) Shape-based 4D left ventricular myocardial function analysis. Proc IEEE Workshop on Biomedical Image Analysis, pp 88-97.
181. Shi Y, Qi F, Xue Z, Chen L, Ito K, Matsuo H, Shen D (2008) Segmenting lung fields in serial chest radiographs using both population-based and patient-specific shape statistics. IEEE Trans Med Imaging, 27(4):481-494.
182. Siddiqi K, Bouix S, Tannenbaum A, Zucker SW (2002) Hamilton-Jacobi skeletons. Intl J Computer Vision, 48(3):215-231.
183. Sijbers J, den Dekker AJ, Scheunders P, Van Dyck D (1998) Maximum-likelihood estimation of Rician distribution parameters. IEEE Trans Med Imaging, 17(3):357-361.
184. Sinha S, Sinha U, Kangarloo H, Huang HK (1992) A PACS-based interactive teaching module for radiologic sciences. AJR Am J Roentgenol, 159(1):199-205.
185. Sinha U, Ardekani S (2004) Parametric brain MR atlases: Standardization for imaging informatics. Stud Health Technol Inform, 107(Pt 2):1374-1378.
186. Sinha U, El-Saden S, Duckwiler G, Thompson L, Ardekani S, Kangarloo H (2003) A customizable MR brain imaging atlas of structure and function for decision support. Proc AMIA Annu Symp, pp 604-608.
187. Sinha U, Yao L (2002) In vivo diffusion tensor imaging of human calf muscle. J Magn Reson Imaging, 15(1):87-95.
188. Smith JJ, Sorensen AG, Thrall JH (2003) Biomarkers in imaging: realizing radiology's future. Radiology, 227(3):633-638.
189. Smith NP, Crampin EJ, Niederer SA, Bassingthwaighte JB, Beard DA (2007) Computational biology of cardiac myocytes: Proposed standards for the physiome. J Exp Biol, 210(Pt 9):1576-1583.
190. Sobel I, Feldman G (1968) A 3x3 isotropic gradient operator for image processing. Stanford Artificial Project (Presentation). Stanford University, Palo Alto, CA.
191. Srinivasan P, Shi J (2007) Bottom-up recognition and parsing of the human body. IEEE Conf Computer Vision and Pattern Recognition (CVPR '07), Minneapolis, MN, pp 1-8.
192. Staal J, Abramoff MD, Niemeijer M, Viergever MA, van Ginneken B (2004) Ridge-based vessel segmentation in color images of the retina. IEEE Trans Med Imaging, 23(4):501-509.
193. Stearns SD (1976) On selecting features for pattern classifiers. Proc 3rd Int Joint Conf Pattern Recognition, pp 71–75.
194. Stegmann MB, Ersboll BK, Larsen R (2003) FAME - A flexible appearance modeling environment. IEEE Trans Med Imaging, 22(10):1319-1331.
195. Studholme C, Hill DLG, Hawkes DJ (1997) Automated three-dimensional registration of magnetic resonance and positron emission tomography brain images by multiresolution optimization of voxel similarity measures. Medical Physics, 24(1):25-35.
196. Swanson KR, Alvord EC, Jr., Murray JD (2002) Quantifying efficacy of chemotherapy of brain tumors with homogeneous and heterogeneous drug delivery. Acta Biotheor, 50(4):223-237.
197. Swanson KR, Alvord EC, Murray JD (2004) Dynamics of a model for brain tumors reveals a small window for therapeutic intervention. Discrete and Continuous Dynamical Systems - Series B, 4(1):289-295.
198. Talairach J, Tournoux P (1988) Co-planar Stereotaxic Atlas of the Human Brain: 3-dimensional Proportional System - An Approach to Cerebral Imaging. Thieme, New York, NY.
199. Talos IF, Zou KH, Ohno-Machado L, Bhagwat JG, Kikinis R, Black PM, Jolesz FA (2006) Supratentorial low-grade glioma resectability: Statistical predictive analysis based on anatomic MR features and tumor characteristics. Radiology, 239(2):506-513.

200. Tenenbaum JB, da Silva V, Landford JC (2000) A global framework for nonlinear dimensionality reduction. Science, 29:2319-2321.
201. Thirion J (1998) Image matching as a diffusion process: an analogy with Maxwell's demons. Medical Image Analysis, 2(3):243-260.
202. Thodberg HH (2003) Minimum description length shape and appearance models. Proc Information Processing in Medical Imaging (IPMI), Ambleside, UK, pp 51-62.
203. Thompson PM, MacDonald D, Mega MS, Holmes CJ, Evans AC, Toga AW (1997) Detection and mapping of abnormal brain structure with a probabilistic atlas of cortical surfaces. J Comput Assist Tomogr, 21(4):567-581.
204. Toga AW, Thompson PM (2007) What is where and why it is important. Neuroimage, 37(4):1045-1049.
205. Toga AW, Thompson PM, Mori S, Amunts K, Zilles K (2006) Towards multimodal atlases of the human brain. Nature Reviews Neuroscience, 7(12):952-966.
206. Tuceryan M, Jain AK (1990) Texture segmentation using Voronoi polygons. IEEE Trans Pattern Analysis and Machine Intelligence, 12(2):211-216.
207. Valckx FM, Thijssen JM (1997) Characterization of echographic image texture by cooccurrence matrix parameters. Ultrasound Med Biol, 23(4):559-571.
208. van der Maaten LJP, Postma EO, van den Jerik HJ (2007) Dimensionality reduction: A comparative review. Maastricht University. http://tsam-fich.wdfiles.com/local-files/apuntes/TPAMI_Paper.pdf. Accessed February 5, 2009.
209. Van Ginneken B, Frangi AF, Staal JJ, ter Haar Romeny BM, Viergever MA (2002) Active shape model segmentation with optimal features. IEEE Trans Med Imaging, 21(8):924-933.
210. Vemuri BC, Ye J, Chen Y, Leonard CM (2003) Image registration via level-set motion: Applications to atlas-based segmentation. Medical Image Analysis, 7(1):1-20.
211. Vemuri BC, Ye J, Chen Y, Leonard CM (2003) Image registration via level-set motion: Applications to atlas-based segmentation. Med Image Anal, 7(1):1-20.
212. Verveer PJ, Duin RPW (1995) An evaluation of intrinsic dimensionality estimators. IEEE Trans Pattern Analysis and Machine Intelligence, 17(1):81-86.
213. Wachowiak MP, Smolikova R, Zheng YF, Zurada JM, Elmaghraby AS (2004) An approach to multimodal biomedical image registration utilizing particle swarm optimization. IEEE Trans Evolutionary Computation, 8(3):289-301.
214. Wan M, Liang ZR, Ke Q, Hong LC, Bitter I, Kaufman A (2002) Automatic centerline extraction for virtual colonoscopy. IEEE Trans Med Imaging, 21(12):1450-1460.
215. Wang H, Dong L, O'Daniel J, Mohan R, Garden AS, Ang KK, Kuban DA, Bonnen M, Chang JY, Cheung R (2005) Validation of an accelerated 'demons' algorithm for deformable image registration in radiation therapy. Phys Med Biol, 50(12):2887-2905.
216. Wang X, Yang J, Jensen R, Liu X (2006) Rough set feature selection and rule induction for prediction of malignancy degree in brain glioma. Comput Methods Programs Biomed, 83(2):147-156.
217. Weickert J (1998) Anisotropic Diffusion in Image Processing. Teubner-Verlag, Stuttgart, Germany.
218. Wiener N (1949) Extrapolation, Interpolation, and Smoothing of Stationary Time Series, with Engineering Applications. MIT Technology Press, Cambridge, MA.
219. Woermann FG, Steiner H, Barker GJ, Bartlett PA, Elger CE, Duncan JS, Symms MR (2001) A fast FLAIR dual-echo technique for hippocampal T2 relaxometry: First experiences in patients with temporal lobe epilepsy. J Magn Reson Imaging, 13(4):547-552

220. Woods RP, Grafton ST, Holmes CJ, Cherry SR, Mazziotta JC (1998) Automated image registration: I. General methods and intrasubject, intramodality validation. J Comput Assist Tomogr, 22(1):139-152.
221. Woods RP, Grafton ST, Watson JD, Sicotte NL, Mazziotta JC (1998) Automated image registration: II. Intersubject validation of linear and nonlinear models. J Comput Assist Tomogr, 22(1):153-165.
222. Worsley KJ, Marrett S, Neelin P, Evans AC (1996) Searching scale space for activation in PET images. Human Brain Mapping, 4(1):74-90.
223. Yang L (2004) Distance-preserving projection of high-dimensional data for nonlinear dimensionality reduction. IEEE Trans Pattern Analysis and Machine Intelligence, 26(9):1243-1246.
224. Yaroslavsky L (1985) Digital Picture Processing. Springer-Verlag, New York, NY.
225. Zhang D, Lu G (2004) Review of shape representation and description techniques. Pattern Recognition, 37(1):1-19.
226. Zhu L, Chen Y, Yuille A (2007) Unsupervised learning of a probabilistic grammar for object detection and parsing. Advances in Neural Information Processing Systems, 19:1617-1625.
227. Zhu L, Yuille A (2006) A hierarchical compositional system for rapid object detection. Advances in Neural Information Processing Systems, 18:1633-1640.
228. Zhu SC (2003) Statistical modeling and conceptualization of visual patterns. IEEE Trans Pattern Analysis and Machine Intelligence, 25(6):691-712.
229. Zhu SC, Mumford D (2006) A Stochastic Grammar of Images. Now Publishers, Hanover, MA.
230. Zhu SC, Wu YN, Mumford D (1997) Minimax entropy principle and its application to texture modeling. Neural Computation, 9(8).
231. Ziou D, Tabbone S (1998) Edge detection techniques - An overview. Intl J Pattern Recognition and Image Analysis, 8:537-559.
232. Zitova B, Flusser J (2003) Image registration methods: A survey. Image and Vision Computing, 21(11):977-1000.
233. Zong X, Laine AF, Geiser EA (1998) Speckle reduction and contrast enhancement of echocardiograms via multiscale nonlinear processing. IEEE Trans Med Imaging, 17(4):532-540.

Chapter 6
Natural Language Processing of Medical Reports

RICKY K. TAIRA

A significant amount of information regarding the observations, assessments, and recommendations related to a patient's case is documented within free-text medical reports. The ability to structure and standardize clinical patient data has been a grand goal of medical informatics since the inception of the field – especially if this structuring can be (automatically) achieved at the patient bedside and within the *modus operandi* of current medical practice. A computational infrastructure that transforms the process of clinical data collection from an uncontrolled to highly controlled operation (*i.e.*, precise, completely specified, standard representation) can facilitate medical knowledge acquisition and its application to improve healthcare. *Medical natural language processing* (NLP) systems attempt to interpret free-text to facilitate a clinical, research, or teaching task. An NLP system translates a source language (*e.g.*, free-text) to a target surrogate, computer-understandable representation (*e.g.*, first-order logic), which in turn can support the operations of a driving application. NLP is really then a transformation from a representational form that is not very useful from the perspective of a computer (a sequence of characters) to a form that is useful (a logic-based representation of the text meaning). In general, the accuracy and speed of translation is heavily dependent on the end application. This chapter presents work related to natural language processing of clinical reports, covering issues related to representation, computation, and evaluation. We first summarize a number of typical clinical applications. We then present a high-level formalization of the medical NLP problem in order to provide structure as to how various aspects of NLP fit and complement one another. Examples of approaches that target various forms of representations and degrees of potential accuracy are discussed. Individual NLP subtasks are subsequently discussed. We conclude this chapter with evaluation methods and a discussion of the directions expected in the processing of clinical medical reports. Throughout, we describe applications illustrating the many open issues revolving around medical natural language processing.

An Introduction to Medical NLP

NLP systems for diagnostic reports (*e.g.*, radiology and pathology) are popular due to the large volume of procedures performed, the high information content, and the grammatical style (a large portion of the text is written in a very direct, declarative manner). Medical NLP systems for clinical documents have the following desirable

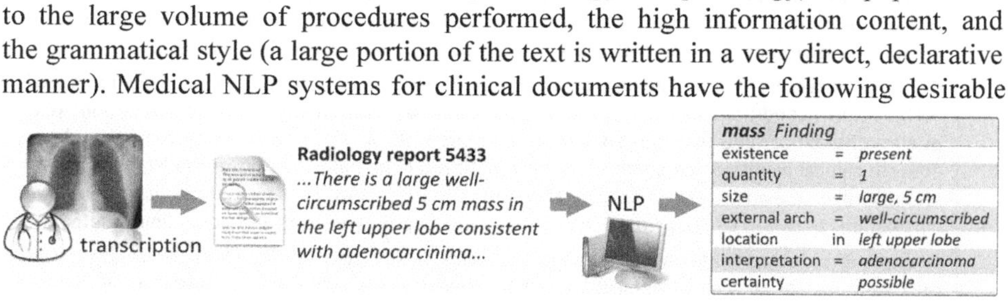

Figure 6.1: A typical goal of a medical natural language processing system is the transformation of free-text documentation to a structured computer representation that includes both concept and grammatical normalization.

A.A.T. Bui and R.K. Taira (eds.), *Medical Imaging Informatics*,
DOI 10.1007/978-1-4419-0385-3_6, © Springer Science + Business Media, LLC 2010

features: 1) they do not require alteration in the reporting methods by physicians; 2) the final representation of the targeted information is in a form that a computer can operate upon; and 3) the process can be automated within a clinical environment. Fig. 6.1 shows a schematic of a typical approach toward the analysis of an imaging study report. The procedure is summarized as follows:

- A radiologist reviews current and comparison images for a given procedure.
- The radiologist dictates the findings, which are stored as audio signals.
- A transcription service (either manual or automated, such as speech-to-text dictation systems) transforms the audio dictations into a text report. The report may (or may not) have additional markups or meta-information (*e.g.*, hypertext markup language (HTML) or eXtensible Markup Language (XML) syntax to structure and/or format the content).
- The input to the NLP system is the free-text report describing the results of a medical procedure or observation.
- The output of the NLP system is a set of structured semantic frames containing a formal representation of relevant information stated within the report.

	Application Description
Report/procedure level	**Prediction of next word to facilitate manual transcription.** [54] developed a radiology report entry system with automatic phrase completion driven by a trigram language model.
	Transcription spelling correction. Transcriptions of medical reports can contain typing errors that may compromise quality of patient care. NLP systems based on both semantic and syntactic local context have been used to improve on generic spell checker algorithms [134]
	Language models for automated transcription by speech recognition. Language models can improve the transformation of speech signals to intended words [19].
	Procedure/disease coding. Billing systems code medical procedures and reason for procedure using typically, CPT-4 and ICD-9 codes respectively. Accurate coding is important for insurance billing purposes and for estimating costs on a disease basis. Accurate coding can also be used for literature retrieval and information mining. Examples of NLP-based coding systems can be found in [9, 37, 72].
	Language translation. Teleradiology services are now being conducted internationally. There is thus a need for translating medical reports between languages (*e.g.*, English to Spanish). A number of companies offer these services on a word-charge basis. Translation services using NLP methods are currently being explored [52].
	De-identification. De-identification of a patient's personal data from medical records is a protective legal requirement imposed before medical documents can be used for research purposes or transferred to other healthcare providers (*e.g.*, teachers, students, teleconsultation). This de-identification process is tedious if performed manually, and is known to be quite faulty in direct search and replace strategies [4, 132, 135, 147].
	Report summarization for patients (translation to layman's language). Reports often contain obscure terminology and abbreviations. A patient who understands more clearly the information within their own report will be better able to communicate with their physician [26].
	Outcome assessment and flagging of alarm conditions. NLP systems can be used to identify various alarm conditions in medical reports; for example, identification of alarm conditions (*e.g.*, pneumothorax) at time of emergency room discharge [98, 137]. [8] reviews systems that detect adverse clinical events using a variety of information technologies. [122] applies NLP methods to identify adverse events related to central venous catheters. [46] uses NLP methods to search for correlations between language used in prostate cancer surgery reports and poor clinical outcomes.
	Document indexing. Indexing of studies to UMLS terms also allows characterization of the document by way of MeSH (Medical Subject Headings) terms. When this linkage is done, one can associate relevant literature to a given patient study [14].

Table 6.1: Examples of medical NLP applications operating at the report/procedural level of the healthcare process. Efforts include applications that identify key concepts and/or map to a specified target representation/language.

For example, if the text report included the sentence, "*There is a large well-circumscribed 5cm mass in the left upper lobe consistent with adenocarcinoma,*" then the desired output of the system is a structured medical finding frame. The head of this frame would contain the name of the finding, mass. The slots corresponding to this frame would encompass the certainty, quantity, size, external architecture, location, and interpretation. For the most part, applications for medical NLP systems are seen to operate at the level of the report, at the level of the patient record, and at the level of an entire collection of medical documents (Tables 6.1, 6.2):

1. Report-level applications. Medical NLP applications at this level operate within the context of a single report, often to identify targeted concepts and codify to a given representation (*e.g.,* a controlled vocabulary, another language). For example, the National Library of Medicine's (NLM) *MetaMap* system [3, 71, 156] automatically identifies UMLS phrases (Unified Medical Language System) phrases within free-text reports. MetaMap is often used in information retrieval (IR) applications to aid in processing queries such as, "*Find example cases in which the patient histology is* X *and the radiological finding is* Y." The representation of this text is transformed from free-text to a vector of UMLS concepts.
2. Patient-level applications. Reconstruction of a patient's medical state from clinical reports is a key objective of many medical NLP systems. If this task can be accomplished, then context-dependent decision support modules can help physicians interpret patient findings and assist in interpretation, prognostic

	Application Description
Patient level	**Medical problem list generation.** What are the current and historical problems (*e.g.,* episodes) for a given patient as documented within their medical record? [29] developed a system for identifying non-negated findings in medical text. Automatic problem list generators have been developed by several academic medical informatics research groups [108, 109].
	Clinical trials recruitment. NLP systems have been used to analyze clinical reports for conditions and eligibility criteria for potential recruitment of patients in targeted clinical trials [67, 94].
	Patient state timeline generation. [148] developed an application that used NLP methods to assist in the identification of findings, and descriptions related to time, location, existence, and causal connections to other findings/pathologic processes.
	Teaching files. NLP systems can be used to automatically identify important concepts related to a specialty or disease for the purpose of creating teaching files [84].
	Identification of patients with targeted disease. NLP systems can provide indexing services that improve information retrieval for the purpose of epidemiology studies, outcomes assessment studies, patient trials recruitment, teaching files, etc. The MedLEE system at Columbia University has been used for these purposes.
Population level	**Disease characterization: Mining relational knowledge from the literature.** This task often involves extraction of cause and effect relations described in journal articles, such as with genetic pathways or drug interactions. For instance, this ability is important for drug companies looking to identify likely gene-protein-molecular pathway mappings associated with a disease process. Statistical analysis of co-occurrence relationship among radiological findings has been studied in [24, 75].
	Bio-surveillance. Early detection of epidemic conditions can be realized by monitoring conditions described in emergency department reports [31, 64, 71].
	Ontology development. What are the terms, concepts, and relations described within documents from a given domain? NLP systems can assist in identifying the proper granularity and diversity of terms mentioned within a target corpus of reports [68, 92].

Table 6.2: Continuation of examples of medical NLP applications, grouped around the patient and population levels.

predictions, and procedural recommendations. Importantly, the task involves a comprehensive representation of findings, problems, and interventions grounded along the dimensions of time, space, existence, and causality (genesis).

3. Population-level applications. Lastly, a growing set of applications look to uncover potential causal chains/associations or large-scale trends across a large corpora of documents, facilitating knowledge discovery. For instance, efforts to extract reported side effects from published literature in order to suggest drug interactions or complications typify this group of applications.

Assessment of Application Requirements

The start of any medical NLP effort must begin with an understanding of the intended use of the system. The requirements to be specified by a medical natural language understanding task can be summarized fourfold: 1) the definition of the domain of clinical documents to be processed; 2) the identification of descriptions of clinical information to be extracted; 3) the definition of a final target format representation; and 4) the estimation of the expected recall and precision performance needed. Notably, performance expectations describe how tolerant the NLP analysis can be with respect to false positive/negative errors; for some clinical applications, there may be little tolerance for errors. Moreover, such expectations will drive the number of training samples, degree of feature richness, and required domain knowledge.

The specification of the domain and task conditions governs a number of subsequent design decisions. For example, once a final surrogate representation for the free-text is defined, then the types of computations that can be performed on this representation become evident. The final representation also guides how we may define relevant subtasks to the NLP process. By way of illustration, language includes a hierarchy of structures (*e.g.,* words, phrases, sentences, concepts, propositions, objects, phenomenon, events, topics, etc.). Once the representation and relationships between these entities are settled on, generative/descriptive/discriminative models for each of these elements can then be addressed.

Depending on its comprehensiveness, a target representation may or may not be able to express the essential information intended by the communicator of the sentence. So an important aspect of medical NLP is the development of a sufficiently rich target representation model. Circumscribing the scope of sanctioned interpretations is part of the domain modeling problem; and of creating application-specific ontologies and semantic models [62]. Specific to radiology, early work in this area was performed by the Canon Group to develop a canonical logic-based representation based on the NLP analysis of radiology reports [60]. More recently, the American College of Radiology Imaging Network (ACRIN) has been developing specifications for reporting radiographic findings within clinical trial studies [10]. HL7 established the Templates Special Interest Group Workgroup with the mission of creating and promoting the development and management of document templates based on the Reference Information Model (RIM) [50]. And the RadLex Project is a working group of the RSNA (Radiological Society of North America) working to complement knowledge sources such as UMLS with concepts from the medical imaging domain [129]. See Chapter 7 for a discussion of various types of data models that are applicable to NLP systems (*e.g.,* phenomenon-centric data model).

Overview of the Medical NLP Problem

Medical NLP systems are typically driven by specific application goals, which as demonstrated, can be quite diverse. Although there are currently no agreed-upon "integrated" models for performing medical language processing, we attempt to provide some perspective on problem representation and the array of computational approaches that have been developed in this field. We hope to motivate insights as to how components and knowledge sources can be shared and integrated toward the goal of large-scale, sophisticated and accurate NLP systems that can be widely deployed.

At its core, language is a means of communication. The free-text within a medical report attempts to communicate findings, assumptions, uncertainties, conclusions, and recommendations made by a reporting physician to other clinicians involved in a patient case. The ideal NLP system takes as input a free-text report and outputs the most probable computer-understandable semantic interpretation of the text. Thus, an NLP engine attempts to maximize the following probability: *P*(semantic ontologic interpretation | input text). Within this formalization, we view medical NLP as a huge classification problem: given any text that can be generated by a medical reporting system, we wish to map this text to one of any possible sanctioned interpretations for the text. From this perspective, medical language processing implies the development of mathematical models to represent language phenomena (*e.g.*, words, meaning, syntactic constructions) and the study of transformations that map (well-formed) texts from a medical domain into computer understandable representations that preserve meaning. We remark here that for any medical NLP system, it is recommended that a well-designed tagging interface be created for each major classification task as an aid to creating training data. Such a tool is independent of the design of the classifier used and is needed for both development and evaluation. Furthermore, the quality and quantity of training data are of utmost importance. If examples are erroneous, noise is introduced into the NLP classifiers; and a sufficient number of training examples is needed to fully represent the spectrum of patterns (feature space) seen for a particular domain. Ensuing discussion in this chapter will illustrate these considerations.

Fig. 6.2 illustrates this high-level description of the medical NLP problem. The lowest level of the hierarchy in the diagram shows the observables of the system, which are the sequence of characters that comprise the inputted free-text. Toward the top of Fig. 6.2 is the final representation stored by the computer, which acts as a surrogate for the text report and from which target applications operate upon. For instance, the surrogate representation could be a vector space model of the document, a list of keywords, a list of UMLS concepts, a set of semantic frames, or a number of conceptual graphs. Often, medical NLP systems generate hidden intermediates that assist in the generation of the final representation (*e.g.*, part-of-speech tag sequences, syntactic parse

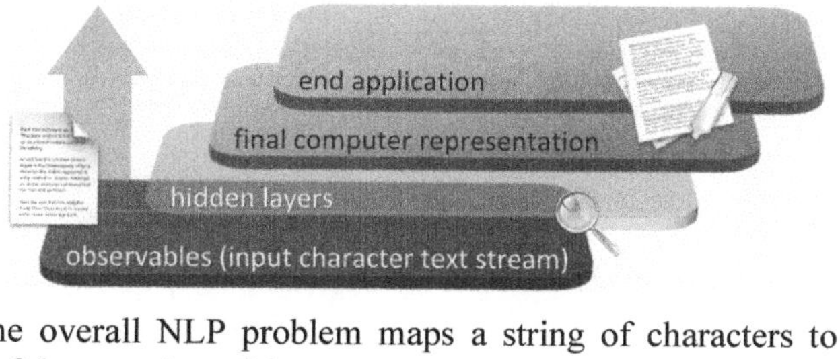

Figure 6.2: The overall NLP problem maps a string of characters to a conceptual representation of the meaning of the text.

trees). These hidden layers of representation used in medical NLP systems can be chosen in several ways depending upon system requirements and design choices. We now review various sub-problems and design issues related to medical NLP below.

Medical NLP System Components & Tasks

Identifying Document Structure: Structural Analysis

A common first step for a medical natural language processor is the identification of all topical and grammatical structural boundaries explicit within a medical report. In this step, we seek a representation for the larger structural abstractions that compose a medical document. The problem definition includes isolating all sections/subsections within the document; all paragraphs within sections; and all sentences within paragraphs (Fig. 6.3). We define the task of identifying the boundaries of all of these units as the *structural analysis* of a medical report. The identification of the structural boundaries of a medical document is useful for the following reasons:

- All natural language processing systems utilize sentences as the basic input unit.
- Different sections of a document can have different language models. Hence, we can improve the understanding of a document using NLP if we include the section type as part of the context for expectation models of the information content. Markedly, we want to identify the section boundaries of a report that correspond to various standard ways of presenting topically coherent text.
- Indexing in medical document retrieval systems can be improved by knowing which parts of the report are relevant for specific types of queries (*e.g.*, findings, conclusions, procedure).
- Report visualization and formatting can be enhanced if section breaks are known (especially for long documents or large corpora). Often, specific users are interested only in a subset of a report. Consider for example a clinician, who may wish only to see the diagnostic conclusion of a radiology report, whereas an administrator may only be interested in the study description and reason for request. Accordingly, an interface can be designed to accommodate rapid review of reports by separating these sections clearly.
- An automated coding system can benefit by knowing which sections of a report contain subjective versus objective patient descriptions (*e.g.*, chief complaint vs. image findings).

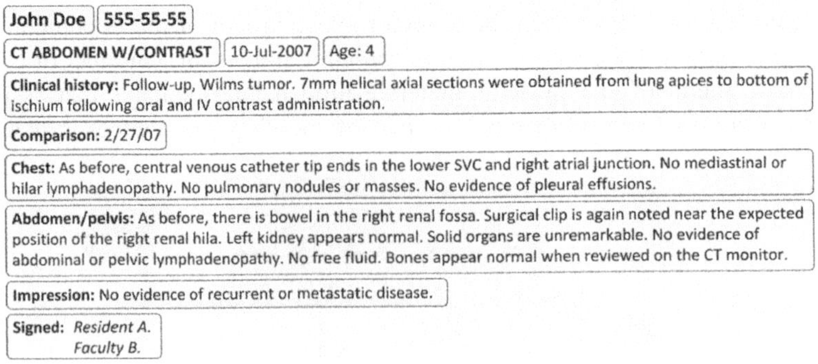
John Doe | 555-55-55

CT ABDOMEN W/CONTRAST | 10-Jul-2007 | Age: 4

Clinical history: Follow-up, Wilms tumor. 7mm helical axial sections were obtained from lung apices to bottom of ischium following oral and IV contrast administration.

Comparison: 2/27/07

Chest: As before, central venous catheter tip ends in the lower SVC and right atrial junction. No mediastinal or hilar lymphadenopathy. No pulmonary nodules or masses. No evidence of pleural effusions.

Abdomen/pelvis: As before, there is bowel in the right renal fossa. Surgical clip is again noted near the expected position of the right renal hila. Left kidney appears normal. Solid organs are unremarkable. No evidence of abdominal or pelvic lymphadenopathy. No free fluid. Bones appear normal when reviewed on the CT monitor.

Impression: No evidence of recurrent or metastatic disease.

Signed: *Resident A.*
Faculty B.

Figure 6.3: Structural analysis involves demarcation of section, sentence, and word boundaries, as shown here.

Section Boundary Detection and Classification

Presently, there is no standardized structure for medical reports, although HL7 is attempting to address this problem with the specification of the XML-based Clinical Document Architecture (CDA; see Chapter 3). However, if the input document has been formatted using a form-based or hierarchical model (*e.g.*, an XML document template definition, DTD), these can also be specified as input to a section boundary detector. The output of structural analysis is a data structure that encodes the hierarchical decomposition of the document into sections, paragraphs, and sentences. For instance, the output structure of a report can be represented using XML tags for report, section, sectionLabel, sectionType, subsection, subsection label, and paragraph.

While seemingly a trivial task for human readers, in many medical record systems section boundaries are still problematic for NLP systems [36, 69], especially when formatting has not yet been applied. Like many NLP problems, formalization of the section boundary detection task entails defining a classification problem. In this case, we treat the problem as a discriminative task, in which we simply represent a document as a linear sequence of symbols (*e.g.*, characters, words) and attempt to discriminate at each character position the state of some hidden variable, y, which can take on one of the following six states: $y \in \{y_0,\ldots,y_5\}$: y_0 = start of section; y_1 = end of section label; y_2 = inside section body; y_3 = inside section label; y_4 = end of section; y_5 = in between sections.

The design of a section boundary classifier can be highly variable. The observable features used to provide evidence/context to discriminate between the possible states for each character need to be determined. These can include both features to rule in a classification state, or conversely, to rule out. Integration and weighting of the rules can assume many forms, as discussed extensively in the pattern recognition literature [51, 81, 149, 159]. Posited as a classifier problem, training data is needed to reflect how an ideal section boundary detector would behave under various input conditions. Fig. 6.4 shows an implementation of a section boundary tagging interface. Users manually specify the locations of the start of each section, start of the section header label (if present); end of the label; and the end of section positions within a text. Each section is also manually assigned to a section class (*e.g.*, Patient History, Findings, Conclusions, Reason for Study, etc.). Each type of medical document (*e.g.*, radiology, pathology, discharge summary) will likely have a domain-dependent set of section types.

As an example of section boundary detection in medical reports, [36] describes a two-pass algorithm that employs a mixture of classification approaches (rule-based and

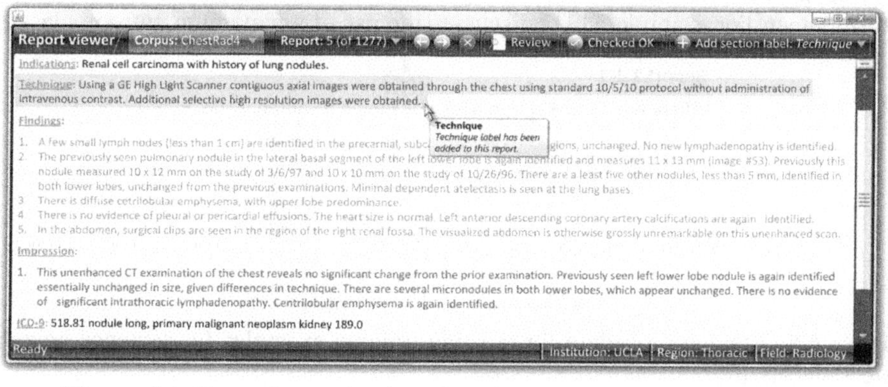

Figure 6.4: Example tagging interface for section boundary detection.

statistical methods) and a search for both labeled sections and sections without header labels. First, a high-precision rule-based algorithm (*i.e.*, rules that are ~100% always true) is employed to detect obvious starts to new sections using a knowledge-base of commonly employed heading labels (*e.g.*, Findings, History, Impressions) and linguistic cues (*e.g.*, colons, all capitals). For example, a specific colon analyzer classifies phrases that use a colon into categories of time (*e.g.*, 7:45 PM), a numeric ratio (*e.g.*, 1:2 mixture), a list, or title to eliminate some false positive boundary instances. Second, the algorithm handles the detection of section boundaries that do not have predictable markers (*i.e.*, no heading labels) using a probabilistic classifier based on an expectation model for the document structure. To this end, the classifier maintains knowledge of: 1) the distribution of the number of words associated with targeted section heading (*e.g.*, Procedure, Findings, History); 2) the statistics on the order of appearance of each section type within various classes of medical reports (*e.g.*, radiology, discharge, surgical); and 3) the types of communications expressed within these sections (e.g., a Conclusion will describe medical findings and their possible etiologies). Descriptive statistics (mean and standard deviation) and *t*-tests are then used to make decisions about section boundary locations.

Sentence Boundary Detection

Once sections have been identified and classified for a medical document, the next step is to identify sentence boundaries. The definition of a sentence is actually not simple: [78, 126] defines a sentence as the largest stretch of language forming a syntactic construction. Usually, sentence boundaries are demarcated by a period, exclamation, or question mark. But in medical text, one often encounters telegraphic writing styles in which sentence punctuation is ignored. For example, consider the following statement for which sentence boundaries are unclear:

> The two largest of these are: (1). In the caudate, a confluent mass measuring 4 cm and (2) In the medial segment of the left lobe of the liver, a large mass measuring 7.5 cm.

As with section boundary detection, a common strategy applied to sentence boundary detection is to frame this task in terms of a discriminative model for a sentence. This approach translates into classifying potential end-of-sentence (EOS) markers (periods, colons, question marks, carriage returns) as being true or false. Fig. 6.5 shows this tagging procedure for creating training data in which the position of EOS markers are manually indicated from text randomly sampled from a target document pool. [110, 167] discusses some issues with tagging sentence boundaries. Once a training set is developed, then various classifier designs and features can be defined for model development. For example, [131] discusses the use of a maximum entropy classifier

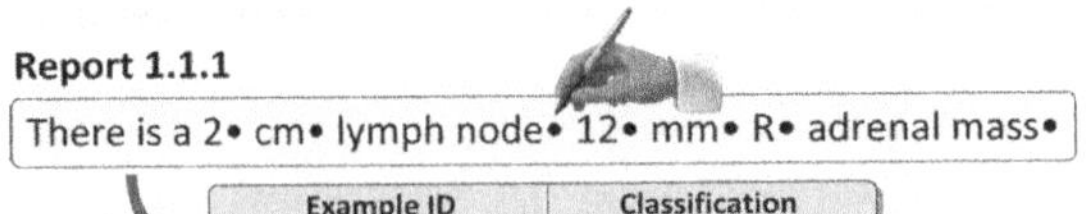

Example ID		Classification	
Report ID	*Byte Offset*	*Truth*	*Local Context*
1.1.1	13	false	(2.)
1.1.1	17	false	(cm.)
1.1.1	29	true	(node.)
1.1.1	34	false	(12.)
1.1.1	38	false	(mm.)
1.1.1	41	false	(R.)
1.1.1	55	true	(mass.)

Figure 6.5: Tagging procedure for creating training examples for sentence boundary detection classifier.

for general text. [35] demonstrates the need to train sentence boundary classifiers on a target domain by comparing the performance using a system trained on medical reports versus a generic classifier found in a common Java library. [77, 106] reports on a similar maximum entropy classifier for radiology reports.

Tokenization

The most elementary structural analysis of text involves transforming the text from a character sequence to a sequence of basic units, typically what we think of as words. This initial phase of medical NLP is called *character stream tokenization*. Tokenization performs an initial surface level grouping of characters to words. In English, this is typically a straightforward task; yet in other languages such as Chinese, there is an absence of recognizable delimiters. Most NLP systems use a word tokenization algorithm that simply searches for whitespace (or sometimes dashes and slashes) within a sequence of characters to define word boundaries – there are no whitespaces within a word. This designation is the *orthographic definition* of a word. The orthographic definition of a word is unambiguous. However, what we really need in language processing is to isolate the "functional" boundaries of the words in a text. The *functional definition* of a word reflects how an NLP system will strategize making semantic sense for a given segment of text. Different strategies for word-level tokenization will in point of fact lead an NLP system to process a given input text in different ways due to the functional "atomic units" of the text being different [158]. As such, some NLP implementations expect a specific type of tokenization, and there are no agreed upon definitions for this task. Below we present some issues regarding this matter:

- Multi-word words. There are some multi-word phrases that act as if they are single words from the perspective of syntax and/or semantics. The most obvious English examples are phrasal verbs that consist of a simple verb plus a participle (*e.g.*, put up with, take off). These are examples of *collocations* (*e.g.*, by means of, computed tomography, vena cava, flare up, blood pressure, in terms of), which also include idiomatic phrases (*e.g.*, throw up, follow up, break out, to stand on one's own feet). A collocation is defined as a sequence of two or more consecutive words that has characteristics of a syntactic and semantic unit, and whose exact and unambiguous meaning or connotation cannot be derived directly from the meaning or connotation of its components. A collocation has two characteristics: 1) *non-compositionality*, in that the meaning of a collocation is not a straightforward composition of the meanings of its parts. (*e.g.*, coming around); and 2) *non-substitutability*, in that we cannot substitute near-synonyms for the components of a collocation (*e.g.*, flared up ≠ fired up). Collocations, including idiomatic phrases, are typically recognized using some dictionary compilations of these expressions. [125] further reports on collocation examples found in medical and nursing text reports; but currently, there are no publically available collocation resources for the medical NLP community. Automated discovery methods to assist in the compilation of such a dictionary of common collocations in a given domain are discussed in [99].
- Single word multi-words. Some orthographically defined words can also be characterized as functionally multiple words, including contractions and compound words that consist of two or more stems joined together (*e.g.*, cardiothoracic). The study of word decomposition is typically discussed under the heading of morphological analysis (see below).
- Abbreviations and acronyms. Abbreviations and acronyms are identified by most NLP systems as a word token. These abbreviations need to be properly expanded

and interpreted. Many times, collocations and idiomatic phrases have common abbreviations that are dependent upon the domain (*e.g.,* CT, CABG, AP, F/U). [13] discusses the various types of abbreviations in pathology reports and [18] elucidates the possible dangers when abbreviations are misinterpreted in medical pathology notes (*e.g.,* IBD as inflammatory bowel disease or irritable bowel disease; DOA as date of admission or dead on arrival; LLL meaning left lower lid, left lower lip, left lower lobe or left lower lung; NC as no change or noncontributory; PE being pulmonary effusion, pulmonary edema, pulmonary embolectomy, or pulmonary embolism). [166] describes a method of resolving abbreviations to their full forms using a bank of pattern-matching rules. Drug dosage expressions can be especially difficult for NLP processors given the combined problems of dealing with periods, spaces, abbreviations, and mixed expressions (*e.g.,* p.o./p.r. q.2h. p.r.n., p.o. q.a.m, q.Saturday; p.o. q.6h (RTC); p.o. b.i.d).

- Symbol expressions. There is a wide range of special symbol expressions present in medical text whose functions can be classified as tokens in regard to the meaning of a given text. Table 6.3 shows some common types of special symbol word units found in medical text. Identifying common types of special symbols that have well formed localized syntax (*e.g.,* dates, patient identification numbers, telephone numbers, and zip codes) is often done using regular expression operators or finite state machines [88]. These methods have the advantage of being fast and easy to implement.

What is a word? The question of what constitutes a "word" is often discussed. But what is a word in regards to a medical NLP system that has some very specific target tasks? *Morphological analysis* provides one method for analyzing a word, being the study of the way words are built up from smaller meaning units. In this linguistic view, a word is composed of *morphemes,* which are the smallest meaningful unit in the grammar of a language. There are two types of morphemes, *stems* and *affixes*. For instance, the word arms is actually a composite of two morphemes: arm + s. Affixes can be prefixes or suffixes. A word with two (or more) stems is called a *compound word.* Analyzing words into their linguistic components is a comparatively uncomplicated task. Common analysis tools include rule-based transformations and finite-state transducers. There are two classes of morphological rules: 1) *inflectional rules*, which relate different word forms of the same lexeme (*e.g.,* lung and lungs); and 2) *word formation rules*, which relate two different lexemes and include derivations in which a word's final lexeme categorization is different than its constituent (*e.g.,* the word independent is a new lexeme item derived from the word dependent that is a member of the lexeme depend) and compounding, in which words are formed by combining complete word forms (*e.g.,* musculoskeletal). Morphological analysis is a common operation in medical IR systems [142], as with vector space models, stemming operations can significantly reduce the dimensionality of the search space. Morphological composition rules can be used for analyzing unknown words [15], and therefore effectively extend the lexical coverage of medical dictionaries [97, 141].

Semantic Type	Example	Semantic Type	Example
Therapeutic protocols	Tarceva OS1774	Date	2006.07.02
Measurement: *Volume*	3 cm x 4 cm x 2.0 mm	Chemical symbols	PO_2
Measurement: *Blood pressure*	125/70	Genetic descriptions	P53
Medical code: *TNM cancer staging*	T1N2M0		

Table 6.3: Common special symbol tokens in medical natural language processing.

Word features. Different NLP tasks will create different surrogate representations for words. For example, an IR system may represent a word by either its entire surface form or its root stem form. In syntactic parsing, it is common to map a word to its part-of-speech (POS) tag (called *pre-terminals*). All subsequent parsing operations make decisions based on these POS tags, and typically not the word itself. Word features are often used by NLP tasks in place of the surface form of the word itself. Thus, the design of NLP-specific lexicons is common, with the advantage of greatly reducing the number of "symbols" that a computational NLP system needs to deal with. For instance, in the domain of thoracic radiology, it is estimated that there are approximately 6,000 unique words (99% coverage) [7]. By contrast, MedPost, a biomedical text POS tagger, contains a tagset size of approximately 60 symbols [145]; and there are about 36 labels defined within the Penn Treebank POS tag set [100]. The features for a word can span many aspects. Not all aspects apply to all words, so subclassing of words is performed – which is one reason why different systems use different subclasses of symbols. Some examples of features are shown in Table 6.4. Features for the surface form of words are maintained in lexicons. Care must be taken in understanding the features used within each lexicon.

Feature	Description	Comments	Task
Character string	The character string that constitutes the word.	Character level *n*-gram models can help model the surface appearance of words. Capitalization can be an important sub-feature.	Is this a proper noun? Is this the start of a new sentence?
Tense	Refers to a time in relation to the moment of utterance.	Future, past, present.	Temporal modeling of observational/event-related information
Gender	Gender constraints.	Male, female, unspecified.	Co-reference resolution of his/her pronouns
Plurality/ number	Number of entities described or involved.	Single, pleural, unspecified (*e.g.,* lesion, lesions).	Subject verb agreement
Part-of-speech tag	Assigned at the word-level within context of a sentence.	Elementary word classes from a structural point of view. (*e.g.,* kidney: anatomy.organ).	Parsing
Semantic label	Assigned at the lexeme level within the context of a sentence.	Elementary word classes from a semantic modeling point of view.	Frame building
Aspect (verbs)	Expresses a temporal view of the event .	Example values include progressive (now) and cessative (terminating). He *is taking* expresses progressive aspect (now).	Temporal analysis
Transitivity/ valency	The number of objects a verb requires in a given instance. *Valency* refers to the capacity of a verb to take a specific number and type of arguments.	Univalent: (X coughs); (X die). Divalent: (X observes Y); (X eat Y) Trivalent: (X gives Y, Z); (X put Y, Z) Admitted (who, where, when, why, under care of)	Link disambiguation, semantic analysis

Table 6.4: Typical types of word features. Categories for individual word class features are highly variable and task dependent.

Defining Word Sequences

Thus far, we have discussed a medical report as having a basic structural organization that includes sections, paragraphs, sentences and words. Indeed, models of medical documents are commonly created based solely on this level of modeling. As an example, medical IR systems commonly create document models based solely on the frequency of individual words. These are known as *bag-of-word representations*.

In NLP analysis, we often think of starting with the observables of the problem, which is a sequence of the surface form of the M words of the sentence: $(w_1, w_2, w_3,\ldots, w_M)$. At the sentence level, the simplest model we can create uses the surface appearance of the sequence of words: the text is just a spatially-ordered sequential set of words. This model ignores any relational associations between words other than their spatial relations (*i.e.,* word distance within the sentence). As this type of model attempts only to describe the sequence of words, it can help to answer the question: how can we predict the occurrence of a word given that we have some knowledge of prior words in a sentence/document?

Such *sequence models* in medicine have been developed for applications such as: 1) automatic spell correction, using bigrams to rank possible alternatives for a misspelled word; 2) medical speech recognition, using language models in conjunction with acoustic models to improve performance [169]; and 3) automatic phrase completion to increase the accuracy and efficiency of medical transcription (*e.g.,* when typing long words that are often misspelled, like *coccidioidiomycosis*) [54]. In each of these applications, the goal was to identify the most likely occurrence of words given the presence of other words. In effect, a language model is maintained for the probability of all possible sequences within a domain, $P(W) = P(w_1, w_2, w_3,\ldots, w_M)$. This sequence can be modeled using a Markov chain that estimates the *a priori* likelihood of seeing a given word sequence [33, 88]. Briefly, in a Markov chain, the sequence probability is first expanded using the chain rule for probabilities:

$$\begin{aligned} P(W) &= P(w_1, w_2, \ldots, w_M) \\ &= P(w_1 \mid w_2, \ldots, w_M) \times P(w_2 \mid w_3, \ldots, w_M) \times \ldots \times P(w_{M-1} \mid w_M) \times P(w_M) \\ &= P(w_1) \prod_{i=2}^{M} P(w_i \mid w_1, \ldots, w_{i-1}) \end{aligned}$$

So the probability of the next word in a sequence depends on the history of words that have come before it. With this factorization, the complexity of the model grows exponentially with the length of the history.

For a more practical and parsimonious model, only some aspects of the history are used to affect the probability of the next word. One way to achieve this is to apply a mapping, H, that divides the space of histories into a manageable number of equivalence classes; we can then estimate the history as follows: $P(w_i \mid w_1,\ldots, w_{i-1}) \approx P(w_i \mid H(w_1,\ldots, w_{i-1}))$. Markov chain models, also called *n-gram models*, establish the following mapping: $H(w_1,\ldots, w_{i-1}) \triangleq w_{i-n+1},\ldots,w_{i-1}$, where all histories up to the n-1 preceding words are assumed to be equivalent. For example, a bigram model of the sentence conditions the probability of a word based on the preceding word within a sentence:

$$P(W) \approx P(w_1)\prod_{i=2}^{M} P(w_i \mid w_{i-1})$$

To improve the estimation, we can use a trigram model that uses the previous two words for context:

$$P(W) \approx P(w_1)P(w_2 \mid w_1)\prod_{i=3}^{M} P(w_i \mid w_{i-2}, w_{i-1})$$

In practice, *n*-gram models (trigrams and higher) are common and fairly straightforward to implement given a large collection of representative text. Estimates of *n*-gram probabilities are made from both empirical count frequencies and various smoothing and interpolation algorithms to improve low-count estimates from the training corpus and for assignment of finite probabilities to words not seen within a training corpus. Design decisions include the order of the Markov chain model, the expected reliability of an estimate, and the generality of the model to related medical domains. An example of a medical application that utilizes a word sequence language model is *ForeWord* [54]. This system employs a hidden Markov model (HMM) to predict the full word or phrase being typed during medical transcription. A trigram model was trained on over 36,000 radiology reports. A Katz backoff procedure with Witten-Bell discounting [88] was used to obtain an improved estimate of the sequence distribution (accounting for unseen words in training, smoothing the distribution) over the empirical frequency estimation. Overall, ForeWord reduced the number of keystrokes required for an average report by a factor of 3.3x.

Pre-terminals: Mapping word sequences to word feature sequences. The Markov chain model described above provides a description of the surface appearance of the sequence of words in a sentence. But in many NLP subtasks, we represent a word sequence instead as a sequence of word features (or a vector sequence of word features) to reduce the overall dimensionality of the NLP problem. For example, if the vocabulary of a typical medical domain consists of 10,000 words and an average sentence length is 20 words, then the maximum number of possible 20-word sequences from this vocabulary is $10{,}000^{20}$. Replacing the word space with a surrogate can reduce the scale of the problem: for instance, using part-of-speech tags, the 10,000 word space can be

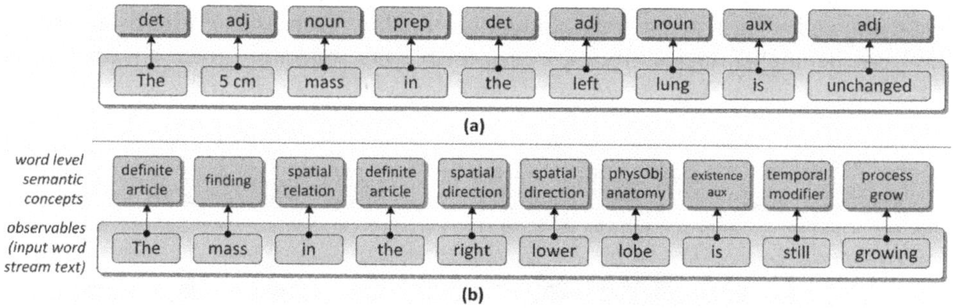

Figure 6.6: Mapping words to **(a)** part-of-speech tags and **(b)** semantic concepts. POS tagging involves the input of an observable sequence of words. The task is to determine what the corresponding (hidden) sequence of POS tags is.

transformed into a smaller number of POS tags (*e.g.,* 50), thereby reducing the overall complexity (*i.e.,* $50^{20} << 10{,}000^{20}$). POS tags are often used when purely phrasal structural analysis is the goal (Fig. 6.6). Semantic word-level tags are also common in medical NLP, though they can number in the hundreds [17].

An important step in many NLP applications is the tagging of individual words with semantic and part-of-speech tags. We can pose the assignment of such word features as a classification task: given a word in the context of a sentence, what is its most likely word feature (*e.g.,* POS tag)? This task is often referred to as the *hidden label problem* in NLP. Note that there are two types of sequence labeling problems: 1) *raw labeling*, where each element is given a single tag (*e.g.,* POS); and 2) *joint segmentation and labeling*, in which words are aggregated and assigned a single tag (*e.g.,* identification of noun phrases, or property nouns). For now, we deal with the former and one-to-one mappings, in which the hidden labels can be referred to as a type of pre-terminal. The general approach for raw sequence labeling in language is shown in Fig. 6.7. The task definition is shown on the right side of Fig. 6.7, where an input word sequence (A, mass, is, seen) is mapped to its appropriate POS label sequence (determiner, noun, auxiliary verb, verb). This task is facilitated by a number of knowledge sources, and this knowledge is operationalized in some optimization algorithm.

Apart from POS tagging [41] and the assignment of semantic classes, two other important medical NLP subtasks that require this type of classification include word sense disambiguation [140] and the general problem of feature assignment to unknown words. Each of these subtasks can be formalized as a general sequence labeling problem, which has been widely discussed in the literature on machine learning and natural language processing [119].

The classic approach to the hidden labeling problem is to implement an HMM [48, 53, 66, 82, 128]. The HMM requires estimation of two types of distributions. First, the transition probability distribution is defined by *P*(tag | previous *n* tags), which formulates a Markov chain (typically 1st order) over the hidden variable. This Markov chain provides the probability of seeing some hidden state for a word *i* in a sentence (*e.g.,* noun) given the previous hidden state for word *i*-1 (*e.g.,* determiner). Second, the emission probability distribution is defined as *P*(word | tag). This subsequent distribution estimates, given a hidden state (*e.g.,* determiner), the probability of seeing a given word (*e.g.,* "the"). The probabilities can be learned empirically from a set of training examples and the use of statistical smoothing algorithms [88]. Once these probabilities are estimated, then a sequence optimization method such as the Viterbi algorithm [59] can be applied to identify the best possible tag sequence, given an input word

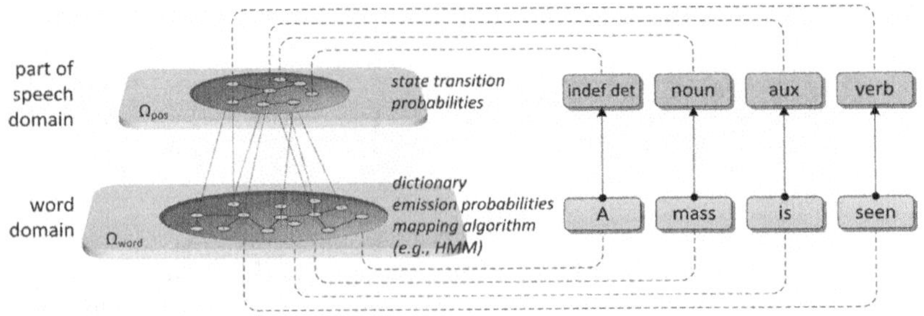

Figure 6.7: Typical approach for assigning features to individual words from a given input sentence.

sequence. The Viterbi algorithm uses a greedy approach that relies on local evidence and assumes that the maximum probability of a path is composed of maximum probability segments. Thus, the Viterbi algorithm does not take into account some long range evidence and can lead to results that visit unlikely/forbidden states.

Some considerations for implementing a medical pre-terminal tagger include:

- What are the possible labels for a word? At the word level, there is no consensus in medical NLP regarding what features should be used or their corresponding values. The set of pre-terminal labels chosen is critical in an NLP application as it must capture the necessary information contained by the words in order to perform higher-level NLP tasks (*e.g.*, parsing). For example, context-free rules relating how sentences are generated are predicated on the chosen pre-terminal labels. Markedly, POS tags are known to have limitations in providing enough information to solve problems associated with prepositional phrase attachment. [144] points out the importance of the tagset in tagging biomedical text. In general, both semantic and part-of-speech features are necessary. Example POS tagsets include those from the Penn Treebank (~36 labels), MedPost (~60 labels), UCREL CLAWS7 (~135 labels), and the minimalist tagset used in the UMLS SPECIALIST lexicon. Given the wide range of granularity definable for the semantics of a word, lexical semantic class tagsets are often application-specific. Arguably, the specification of the lexical semantics requisite for medical NLP is still an open issue [86]. A high-level specification of semantic types in medical language is described in the UMLS Semantic Network [95]; however, qualifiers and relations are not well developed.
- What order Markov chain model should be used? Most implementations use 1st-order models for the transition probabilities. 2nd-order HMMs integrate a longer range of evidence at the cost of requiring a greater number of training examples and better smoothing algorithms. An example of a 2nd-order POS tagger can be found in [153].
- Are there any available training sets? The medical informatics community has begun efforts to compile POS training examples for clinical text. Anonymization is a concern for these public resources. The Mayo Clinic has produced a manually-tagged POS corpus comprising clinical notes [120], while [123] describes the efforts in creating a tagged POS corpus of pediatric clinical notes.
- Can a POS tagger trained in a different domain work in a medical domain? Typically, general POS taggers do not perform well largely due to the number of unknown words (symbols, abbreviations, etc.) seen in medical text. [112] describes an approach using morphological knowledge of medical terms to build a domain-specific lexicon containing POS tags and probabilities for each word. [96] used heuristic sample selection methods to greatly reduce (by 84%) the number of manually tagged sentences required to satisfactorily retrain a maximum entropy POS tagger.

Although HMMs are traditionally used for pre-terminal tagging, other models have been implemented in response to some of the deficiencies of HMMs. One weakness of using HMMs is that the emission probabilities are of the form $P(\text{word} \mid \text{tag})$, whereas it is actually more informative to model the reverse conditional probability, $P(\text{tag} \mid \text{word})$. This probability is potentially more powerful as one can include features related to the specific morphology of a word, which often provides strong evidence regarding its POS and/or semantic class (*e.g.*, Latin root stems, capitalizations, suffixes such as -ology, -ectomy, -itis). Additionally, surrounding words can be used to predict the tag corresponding to the word of focus. POS taggers that directly compute $P(\text{tag} \mid \text{word})$ are commonly implemented using maximum entropy (MaxEnt) classifiers [12, 96, 130]. Adaptation of maximum entropy models to include linear sequence optimization of the tagged labels can be implemented using MaxEnt Markov models (MEMMs) [101, 164], which tend to perform better than HMMs due to a richer feature set. In addition to concerns about the emission probabilities in HMMs, there is also the *label bias problem* that affects the training of the transition probability $P(\text{tag} \mid \text{previous tags})$. The label bias problem arises as we train our model on correct previous tags. The conditional probability of a tag at position *t* uses the feature of the previous tag at position *t*-1: during training we always have the correct tag at *t*-1; but during actual execution on unseen data, these tags are exactly what we are trying to predict. To account for this concern, models based on conditional random field formulations are used [56, 91, 157]. Finally, a common variation in classifier design is to replace MaxEnt models with maximum margin models (*e.g.*, a support vector machine, SVM), giving rise to max-margin Markov networks [151].

Word sense disambiguation. An important task related to pre-terminal feature mapping is the disambiguation (resolution of ambiguities) of homographic words – that is, a word with more than one sublanguage meaning – within the context of a sentence in a report. These ambiguities are also known as *word sense ambiguities*. *Word sense disambiguation* (WSD) is an NLP task that identifies the intended meaning of word that can have many definitions [116]. Consider the following statements:

"The patient had a *round* of chemotherapy," vs. "The mass is *round*."

"There is an oozing *discharge*," vs. "The patient's *discharge* papers from the ICU were lost yesterday."

"The patient experienced pain at the level of the *T1* spine," vs. "A *T1* MR scan was performed yesterday."

"The patient's *back* was in pain," vs. "The pain was in the *back* of the neck."

"There is a *density* in the left lower lobe," vs. "There has been an increase in the size and *density* of the left lower lobe mass," and, "The film *density* was poor in the region of the mediastinum."

The main hindrance to improving word sense disambiguation algorithms in clinical NLP is a lack of training data. Presently, the Mayo Clinic corpus is one of the few available resources for manually tagged word senses for medical NLP developers [140, 163]. [140] uses UMLS semantic classes as the target set of categories for representing the different senses for a word. One challenge to creating such training data is that taggers struggle to identify which words are ambiguous; even after being found, there remain difficulties in stating all possible meanings for a word. Furthermore, for each ambiguous word, one would like to collect a number of examples in different use contexts. WSD training sets are often created for specific sub-domains (*e.g.*, thoracic radiology) in which the number of senses of a word may be much lower than as seen in larger, more general domains. Many word sense disambiguation algorithms thus rely on semi-supervised learning methods that utilize both labeled and unlabeled data [1, 16, 165].

A cornerstone concept for the features used to classify the sense of an ambiguous word comes from the British linguist, J.R. Firth, who coined the phrase, *you shall know a word by the company it keeps* [57]. The principle behind this idea is that a set of words occurring together in context will determine the appropriate senses for one another even though each individual word may be ambiguous. Compared to POS taggers, WSD often requires longer range context to resolve ambiguities. This word association construct can also be applied to include not only how words are used within free-text reports, but how words are related within external lexical knowledge sources such as MeSH and UMLS (*i.e.*, defining conceptual company). [79] show how utilizing such associations in external knowledge sources can be extended to include Journal Descriptor Indexing assignments with words in a training set of PubMed citations. [121] explored the use of similarity measures and relatedness in neighboring words based on paths in a conceptual network. A general discussion of WSD features and algorithms can be found in [116]; and a discussion of medical specific features is given in [140].

Related to WSD is the task of acronym expansion where an ambiguous acronym is to be assigned the correct expansion. Supervised machine learning approaches are common but require a substantial number of manually labeled samples to generate a robust model. [87] described a system that combines limited training data, context from unlabeled text, and ontological knowledge from sources such as WordNet to construct a classifier for identifying the correct expansions of an abbreviation, as well as resolving word sense ambiguities. [120] describes the use of surrounding word context as harvested from the Internet via a Google API to develop expansion models for clinical acronyms.

Spell checking. Spelling errors are common in clinical text reports. [73] identified that 4.9% of unique word tokens mined from a collection of over 200,000 reports had spelling errors. A separate analysis conducted by the Veteran's Administration (VA) reported an estimated spelling error rate of 4% as seen in a large collection of clinical documents [58]. The subtasks for spelling correction include: 1) the detection of the misspelled word; 2) the generation of a list of candidate correct spellings; and 3) the ranking of candidates. The NLM Specialist software tools include spelling correction tools (*e.g., GSpell*) that generate a list of suggested spelling corrections facilitated by a list of commonly misspelled words in medicine and an *n*-gram model to search for words that have common beginning and ending character sequences. Candidate ranking of alternate words is done by computing an edit distance from the candidate word to the unknown word. [156] built a spell checker that used UMLS Specialist Lexicon as a primary source for candidate generation and WordNet [55, 111] as a secondary source. Various character-level operations are commonly applied to generate possible candidate words from the source word (*e.g.*, transposition, deletion, insertion, and substitution); lexical sources (UMLS and WordNet) are then used to sanction those that are deemed appropriate candidates. Matching to lexical sources can include exact matches, and those that sound similar.

Named Entity Recognition and De-identification

De-identification of a patient's personal data from medical records is a protective legal requirement imposed before medical documents can be used for research purposes or transferred to other healthcare providers not directly involved in a patient's care (*e.g.*, teachers, students, tele-consultations). Unfortunately, this de-identification process is time-consuming and tedious if performed manually and is impractical for large-scale research (*e.g.*, population-based studies). A review of various methods for name matching can be found in [11]. Unfortunately, the process is also known to be quite faulty in direct search and replace strategies [147]; Table 6.5 shows examples of reasons for the failure of direct matches. Researchers (and other authorized personnel) who require access to large corpora of confidential medical documents need methods to de-identify these records, as specified by various organizations and regulatory standards set up to protect patient privacy (*e.g.,* Health Insurance Portability and Accountability Act, HIPAA; institutional review boards (IRBs), Federal Policy for the Protection of Human Subjects). HIPAA guidelines define "de-identified" patient data, specifying 18 distinct types of references that should be removed to ensure patient confidentiality. These data items include: patient name, medical record number, age, gender, ethnicity, healthcare provider names, relative names, institution names, address, telephone numbers, fax numbers, e-mail addresses, social security number, license number, vehicle identifiers, URL/IP addresses, full face photos, and service dates. A complete specification can be found in [27]. Until methods to automatically – and accurately – replace and/or mask these patient identifiers are developed, healthcare researchers wanting to use the wealth of data now contained in the electronic medical record (EMR) will continue to be burdened with the responsibility of de-identifying data. The problem of locating proper names in free-text has been a long-standing challenge of the DARPA-sponsored Message Understanding Conferences (MUC) [34] and a topic of interest at various conferences sponsored by the Association for Computational Linguistics (ACL). The topic has also been part of a competition run at the first Workshop on Challenges in Natural Language Processing for Clinical Data [159].

Specific implementations related to medicine can be found in [69, 117, 135, 147, 149]. A publicly available gold standard de-identified corpus of medical records is available and described in [117].

Foundational works in this area of named entity recognition include [40, 103]. [40] approached the problem in the context of formal linguistics theory, looking at the natural language sub-grammar of named entities, including discussion of: the morphological analysis of name components; the analysis of internal syntactic structure of names; and the semantics associated with words that preserve their meaning when associated with a named entity and words that are collocations of an atomic phrase. [103] enumerates various practical issues, emphasizing the importance of internal evidence (*i.e.,* local contexts, such as *Mrs.*, *PhD*), and external evidence (*i.e.,* longer-range contexts, like verbal attachments) for accurate name recognition. The evidence for supporting a given candidate can be obtained at four different levels:

1. Knowledge sources. Compilations of names, drugs, institutions, and geographic locations (*e.g.,* gazetteers) can be use to determine whether a given candidate matches any listings within these knowledge sources.
2. Local level. Words immediately surrounding the candidate (*e.g.,* Baby Boy Jones, John Jones, MD) can suggest a given classification; [104] refers to this as

Category	Example	Category	Example
Apostrophes	John *vs.* John's	**Missed double**	Michele *vs.* Michelle
Substrings	John *vs.* Johnny	**Missed hyphen**	Worthy-Smith *vs.* Worthy Smith
Abbreviations	John Q. *vs.* John Quincy	**Repeated sequences**	Jeremias *vs.* Jerimimias
Modifications	Bob *vs.* Robert	**Sequencing errors**	Ricky *vs.* Rikcy
Acoustical	Gayle *vs.* Gail	**Typing errors**	Kenneth *vs.* Kenhneth

Table 6.5: Different categories of common reasons for name mismatches.

"internal evidence." Features of this type are encoded as *n*-gram token sequences, where the token can be the word, the word's syntactic class, the word's semantic class, or a mixture of the three. Important types of local word features include capitalization, presence of non-alphabetic characters (except hyphens and/or apostrophes), non-presence of a vowel (or letter *y*), unusual internal capitalization for class type (*e.g.,* Di, Le, and Mc are common in people names but not HBsAG). Additionally, statistical evidence can be used to determine what list of words occurs more frequently as lowercase rather than uppercase [44]. For example, although the token "*The*" frequently starts a sentence, over a corpus of reports, the word "*the*" is most often lower case.

3. Sentence-level features. It may help to look at associated verbs, adjectives, and/or complements to disambiguate the classification of a named entity. [149] used semantic selectional restrictions to hypothesize strong associations between some classes of words (*e.g.,* admitted) and the semantic constraints on concepts that can fill their thematic roles (*e.g.,* patient names). [104] called these types of constraints as "external evidence" about a named entity. Semantic selectional restriction rules have previously been used mostly with verbs and the types of words that can fulfill their argument slots. By way of illustration, the verb underwent strongly suggests that the head slot is filled by a patient reference. Other example verb forms with strong associations to patients include: vomited, administered, discharged, and returned. But as noted by [88], verbs are not the only types of words that can impose selectional restrictions: within medical documents, certain adjectives (*e.g.,* 3 year old, male, Asian) can also impose these strong associations. Linking these words grammatically to their corresponding related heads can provide strong contextual evidence for patient name identification. The semantic selectional restrictions are also used to disambiguate candidates that are actual names *vs.* candidates that represent a medical procedure, condition, device, or location. Table 6.6 shows twelve common logical relations that can be parsed at the sentence level.
4. Document and patient record level features. *Co-reference resolution* involves finding words that seem to refer to the same entity. [32] reported an improvement in performance of a named entity recognizer from 92.6% to 97.0% using a co-reference model. Thus, if the phrase, "*Johnny Smith*" is classified as a patient identifier, then other references within the same document are likely also to be tagged with the same class. Co-reference uses the initial set of guesses made by a semantic interpreter to make decisions about candidate instances, the idea being that if one or more patient name instances within a report can be reliably identified then the identified entity can be used as evidence to co-reference more

Logical Relation	Example	Logical Relation	Example
Patient gender	John is a 5yo male	**Patient relative**	John's mother
Patient age	John is a 3 year old	**Patient care provider**	John followed by Dr. Smith
Patient ethnicity	John is Caucasian	**Patient birth**	John born by C-section
Patient procedure	John received therapy	**Patient knowness**	John is well-known to this service
Patient health	John developed a fever	**Patient activity**	John went to school today
Patient status	John responded well		

Table 6.6: Logical relations used in patient name identification algorithm.

difficult cases. As such, co-reference information can provide mutual constraints for classifying patient name instances. The identification of logical relations can then be thought of as a way to build a set of reliable guesses for patient name references. Such guesses can then be used to identify all instances of these guesses within the document: if one phrase is tagged with a particular class, then it is likely that all such phrases should also be tagged with the same class. However, there are rare but real cases in which this is clearly not the case (*e.g.,* Dr. Martin examined Martin today); and care must be taken in case of partial name matches between, for example, the patient and a relative. Co-reference evidence is especially important in identifying instances where a patient name candidate has no logical relation context modeled at the sentence level (*e.g.,* We can't wait to see Johnny start to have a fun time with his toys again). Possible co-reference features for a candidate word include the number of string-level matches with instances marked positive/negative at the sentence level and sentence-level classification (addressing the problem of Dr. Martin examined Martin). In addition to performing co-reference resolution on the patient, it is helpful to know who are the other named characters within the report, including relatives, healthcare workers (*e.g.,* nurses, dieticians, physicians, physical therapists), and others (*e.g.,* teachers, lawyers). Knowledge of these non-patient names can be used as negative evidence in discriminating patient name references from non-references.

Concept Coding: Ontological Mapping

The identification of phrases that map to concepts within a target ontology has been a challenge for medical NLP systems [3, 64, 115, 134, 170]. This mapping is important because the representation changes from character strings to an ontological concept with specific meaning. Concept coding has many uses. At an elementary level, we attempt to identify phrases found in free-text reports to concepts found in ontologies such as UMLS or SNOMED-CT (Systematized Nomenclature of Medicine, Clinical Terms). These concepts can then be used as index items for IR systems. From a clinical perspective, concept coding can be used to summarize a patient's clinical status, where ICD (International Classification of Diseases) codes are assigned based on guidelines. The difficulty of this particular task is exemplified by the fact that clinical document coding was the 2007 Medical NLP Challenge for the Computational Medicine Center [45].

The general problem of coding a real-world textual expression to a controlled vocabulary is that corresponding exact string matches occur rarely. [102] studied this problem in attempting to index a set of Medline documents to UMLS terms, finding that only 17% of document terms could be identified. Even after applying operations such as stemming, synonym analysis, and some grammar normalization, published performance metrics are only on the order of 70-80% [3]. Arguably, current efforts in building controlled vocabularies (*e.g.,* UMLS, SNOMED, etc.) are designed for humans as the intended users rather than machines. Fundamentally, the difficulties lie in the

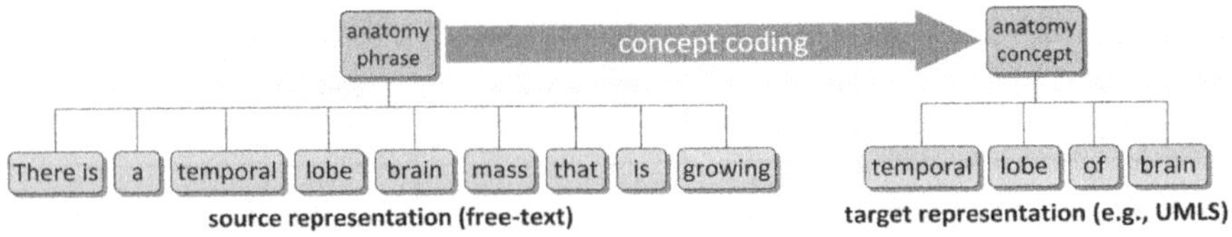

Figure 6.8: Example of semantic phrase chunking and subsequent coding.

unavailability of a canonical representation for medical concepts, both at the target representation level (*e.g.,* UMLS) and at the source document level (free-text report phrase). Marvin Minsky, a pioneer in artificial intelligence, aptly stated the problem: *no machine can learn to recognize X unless it possesses, at least potentially, some scheme for representing X* [113]. This observation emphasizes the idea that knowledge representation is an important part of any NLP system.

The MetaMap Approach

MetaMap is a tool developed by the US National Library of Medicine that codes noun phrases uncovered in free-text to UMLS concepts [3]. The processing steps include parsing, variant generation, candidate retrieval, candidate evaluation, and mapping construction. We use these steps to outline various mapping strategies below:

- Parsing. This step refers to identification of phrases within the free-text that have a mapping to UMLS. The simplest approach is to use a left-to-right marching algorithm, such as the Aho-Corasic algorithm that starts at the beginning of the text string [2] and searches for the longest string match in a given target vocabulary. Most medical applications, however, map only noun phrases so that a more efficient approach is to identify noun phrases within a shallow syntactic parser. [6] describes a system that specifically parses complex anatomy phrasal chunks.
- Variant generation. This step inputs a parsed text phrase and outputs a list of alternative expressions based on synonyms, acronyms, abbreviations, inflection, and derivational variants [49]. A derivational score based on the types of transformations applied by the generator is computed. Filters can be applied to eliminate variants with a given part-of-speech tag (*e.g.,* determiners, conjunctions, prepositions, etc.) from this list and/or to ignore capitalizations. Note that some systems normalize the words in a phrase to their root stems using a *lemmatiser*.
- Candidate retrieval. This third step retrieves all UMLS Metathesaurus strings containing at least one of the variants determined in the prior step. Various indexing schemes can be applied to speed up this retrieval step. For example, the IndexFinder system maps UMLS concepts to four indexing data structures: 1) a hash table mapping a word to a unique word ID (in UMLS there are more than 431,000 distinct words); 2) an inverted index that maps word IDs to a list of phrase identifiers; 3) an array that maps phrase IDs to UMLS concept IDs; and 4) an array indicating the upper bound for a given phrase. This information is used in the mapping algorithm to determine if input words contain the complete phrase [170].
- Candidate evaluation. The evaluation step provides a score to each candidate, indicating how well the input phrase maps to the candidate ontological concept. IR techniques that use a vector space model over individual word tokens are a common approach to candidate evaluation [134, 170]. The candidate and source phrase can be compared by using similarity measures such as the cosine coefficient, Dice coefficient, or Jaccard coefficient [93, 138]. [5] extends this concept to include parser dependency links (*i.e.,* head-modifier pairs) into a vector space model for candidate evaluation [146]. The dependency link model is intended to capture semantic uniformity across a variety of surface forms and attempts to overcome errors seen in long phrases and/or phrases with conjunctions made by bag-of-word approaches. The core MetaMap algorithm also uses syntactic information related to *centrality* (the involvement of the syntactic head) in evaluation, as well as measures of lexical variation, coverage, and cohesiveness.

- Mapping construction. This final step considers all possible final configurations of mappings for a given input text. At the phrasal level, difficulties can arise for various reasons. For instance, conjunctions are often difficult to analyze. [115] mentions the mapping of the phrase spleen rupture and normal stomach to the concept stomach rupture as a possible spurious mapping. The distribution of modifiers to coordinating constituents should also be handled in this last step (*e.g.*, the posterior right third and fourth ribs). Clinical text also often contains shortened descriptions that are underspecified if seen at the phrasal level, but clear from the report context. For example, left apex within a chest x-ray report should be interpreted by the system as the apical aspect of the left upper lobe of the lung. Additionally, in clinical free-text, we often see extraneous terms not related to a given concept (*e.g.*, loop of unopacified bowel; on the right side, the femur is; lower pole of the transplant kidney); as such, some specification as to whether to treat these types of phrases as a single chunk needs to be agreed upon by users.

Data Mining and Lookup-Table Caches

The coding problem can also be tackled using corpus-based approaches in which targeted mapping tables are manually created based on mined phrases from a large set of reports from a given target domain [64]. The collection of words and phrases for a given phrasal type from actual reports ensures that a coding system works well at a practical level and that most of the string representations for multi-word concepts (*e.g.*, anatomy phrases, findings, etc.) are included within the knowledge-base. The result will be a knowledge source that provides reliable results, matches the expected granularity of detail, and will provide direct hits for a very high percentage of coding query inputs.

Phrasal Chunking

Many applications require the identification of key semantic phrases within the text, which are then used to index a given medical document. *Phrasal chunking* is the process of identifying logically coherent non-overlapping sequences of words within a sentence. Generally, a mapping from word tokens and/or their pre-terminal tags is made for a phrasal group. This grouping can be either structurally defined (*e.g.*, noun phrase) [77] or semantically defined (e.g., anatomic phrase) [6]. This task is also closely related to the problem of named entity recognition. Examples of semantic phrasal types seen in medical reports and an example of each type are listed in Table 6.7. The

Phrase Type	Example	Phrase Type	Example
Anatomy phrase	right upper lobe of lung	**Finding phrase**	Wedge compression fracture, focus of increased density
Spatial relation	is located just posterior to	**Image slice**	Cuts 5 to 10
Anatomy-perturbation	fullness of the right upper pole collecting system	**Causal relation**	(mass) is consistent with that of (a tumor)
Existential relation	There is no sign of	**Physical object**	size of the tumor
Drug and dose phrase	Zosyn 4.5 g IV x 14 days; Bactrim DS one tab p.o. twice a week	**Measurement observation**	Increased to 5 cm in size since last exam
Physical object handle	(small) amount of debris (seen)	**Temporal event relation**	(improved) in comparison to (10/25/02)

Table 6.7: Examples of semantic phrasal chunks in medical reports.

semantic phrasal chunking problem in medicine extends to complex expressions such as, the superior aspect of the mid pole of the right kidney, as well as compounds (e.g., the left upper lobe and the right upper lobe). Note that semantic phrasal chunking is a slight variation of the definition of syntactic phrase chunks (semantic vs. syntactic) defined as part of a shared task in the Conference on Natural Language Learning [155]. The phrasal chunking task is limited to determining only the external boundaries of targeted semantic phrases and not the internal syntactic structure.

One straightforward method of phrasal chunking is to compile a comprehensive listing of phrases related to the target semantic phrasal type. This approach is an effective means of pattern recognition within clinical text for phrasal types that have a limited number of instances and/or are of limited word length. For example, the ConText application [30] uses a phrasal micro-glossary describing existence status. Coding the grammar of targeted phrases using regular expression matchers and/or finite state machines complements the use of a phrasal glossary. For phrasal types that have a large number of instances (*e.g.*, anatomy phrases, patient names, institution names, etc.) with high-order complexity, there are various approaches. The abstraction of the phrasal chunking problem into a classification/hidden sequence optimization problem is typical given the computational machinery and software availability for this general class of problems. The task is abstracted as follows: given a phrase type (*e.g.*, anatomy phrase), the goal of the classifier is to tag each word in the sentence with one of the following five outcomes: begin (B), end (E), inside (I), single (S), or outside (O) (BEISO) [90]. The definition for each outcome is given in Table 6.8. Other tagging schemes variants have also been used (*e.g.*, begin, in, out, whole; yes, no; begin, in, out; etc.), not surprisingly affecting classifier performance. For example, in the phrase, "*A chest mass in the right upper lobe is seen*," the markup for the anatomy description phrase is as follows:

A	chest	mass	in	the	right	upper	lobe	is	seen
O	S	O	O	O	B	I	E	O	O

The phrasal chunking task can be seen as a sequential prediction problem, where we predict the chunk tag t_i associated with every token w_i in a sequence of tokens. The problem is decomposed into two steps: the estimation of the probability of a given token belonging to a chunk tag class; and the maximization of the token sequence probabilities for the entire sentence. Classifier models are frequently learned using supervised methods (*i.e.*, training examples). A brief overview of the methodology as applied to anatomy phrase chunking is described below, with advancements in phrase chunking occurring naturally along four somewhat dependent lines: 1) context modeling; 2) classifier design; 3) training sample generation; and 4) linear sequence optimization.

Outcome	Definition
B	Token is the beginning of a phrase consisting of more than one token
E	Token is the end of a phrase consisting of more than one token
I	Token is between the beginning and end of a phrase consisting of more than two tokens
S	Token is the lone token of a phrase consisting of only one token
O	Token is outside of the phrase

Table 6.8: Definition of chunk labels (*i.e.*, classifier outcomes).

Context Modeling

Context modeling involves defining the pieces of evidence for which a classification task is executed. The context may involve rich features from linguistic theory (*e.g.*, grammatical function of words, lexical dependency information, semantics, and common-knowledge pragmatics). Modeling the complete context for a linguistic classification task is difficult given the complexity and variability of human language. As such, many NLP researchers have retreated from formal linguistic language models, relying instead on less complicated (more naïve) models based on surrounding word evidence. *N*-grams are rudimentary models that attempt to capture the constraints of a language by simply conditioning the probability of a tag (*i.e.*, BEISO) on a small, fixed number of surrounding words or word features. Table 6.9 shows an example of feature patterns used to capture this context. The window of lexical information in Table 6.9 cannot capture all necessary contexts to disambiguate the chunk tag assignment for all words. This specialized 5-gram model requires a large amount of training data to acquire sufficient statistics. Estimates of 5-gram probabilities are difficult given the relative sparseness of the majority of 5-grams. A basic feature selection strategy assumes that any feature that occurs fewer than five times within a training set is noisy and hence discarded (*i.e.*, less than five samples leads to poor estimates).

N-gram models are attractive to developers, as the developer needs little formal linguistic knowledge and relatively good results can be obtained given sufficient training sample sizes. Despite the utility of this approach, two core issues remain: 1) the determination of a reasonable number of training samples for a given subject domain and target phrase type; and 2) the inability to capture long-range word-word evidence. An example of this latter problem is shown where the phrase right hemithorax is the object of the preposition, overlie, and the phrase right lung is a modifier to the word mass:

Staples overlie the **right hemithorax** and the **right lung** mass is again seen. *Correct*

Staples overlie the **right hemithorax and the right lung** mass is again seen. *Incorrect*

Variable length semantic constraints. To achieve higher precision, one needs to incorporate longer-range word context, which is often done using a set of high-context, hand-crafted rules based on constraint-based formalisms [89]. In this methodology, phrasal chunk constructions are viewed as objects that can have associated complex sets of properties. We need these constraints because naïve *n*-gram models of grammatical phenomena such as agreement and sub-categorization may not provide sufficient context and can lead to over-generalization. Some of the errors made by *n*-gram classifiers can be avoided in a specialized knowledge-based system as rules

Class/word Position	*i-2*	*i-1*	*i*	*i+1*	*i+2*
B	s_{i-2}, p_{i-2}	$s_{i-1}, p_{i-1}, t_{i-1}$	w_i	w_{i+1}	s_{2+1}, p_{2+1}
E	s_{i-2}, p_{i-2}	w_{i-1}, t_{i-1}	w_i	s_{i+1}, p_{i+1}	s_{2+1}, p_{2+1}
I	s_{i-2}, p_{i-2}	w_{i-1}, t_{i-1}	w_i	w_{i+1}	s_{2+1}, p_{2+1}
S	s_{i-2}, p_{i-2}	$s_{i-1}, p_{i-1}, t_{i-1}$	w_i	$s_{i+1}, p_{i+1}, t_{i+1}$	s_{2+1}, p_{2+1}
O	s_{i-2}, p_{i-2}	$s_{i-1}, p_{i-1}, t_{i-1}$	s_i, p_1	s_{i+1}, p_{i+1}	s_{2+1}, p_{2+1}

Table 6.9: Example of five-gram features utilized for anatomy phrase chunking. Here, w_i is the word string appearing in the ith position, s_i is the semantic tag of w_i, p_i is the part-of-speech tag of w_i, and t_i is the chunk tag label for the ith word.

can refer to words and tags across the scope of an entire sentence rather than a limited window. For example:

Posterior-most aspect of the lesion.	*"aspect" is not part of anatomy phrase*
Posterior-most aspect of the liver.	*"aspect" is part of anatomy phrase*
Posterior-most aspect of the liver lesion.	*"aspect" is not part of anatomy phrase*

A rule-based system may identify particular situations when exact word-level surrounding context is important. An example rule using a constraint grammar is shown:

Given input: "small"	*Word under investigation*
if (+1, "word", "bowel/intestine")	*Next word*
then eliminate('O')	*Eliminate 'O' as possible state*

Constraints are used both to eliminate false positive results (*i.e.,* to eliminate tag candidates that are inconsistent with the context) and to avoid false negative classifications. These rules are very specific and often only apply to a small percentage of training case instances. For complex checks such as anatomy phrase identification, several hundreds of rules may be needed to ensure high recall and precision.

[20, 162] describe a transformation-based learning system that automatically identifies contextual rules required to map a sequence of POS tags to phrase chunk labels. *Transformation-based learning* is a non-probabilistic technique for incrementally learning contextual rules to maximize prediction accuracy. Again, the context for word tagging is centered on variable length runs of surrounding parts-of-speech. An iterative scheme that clusters parse islands to grow the boundaries of target phrases is used. Here, the context is estimated by any tags in previous iterations that have been confidently classified.

[83] describes a variable length adaptive phrase chunking system that automatically generates language rules by accumulating a histogram of each unique 1-context, 3-context, 5-context, and 7-context for a given chunk tag as seen over a training corpus. The knowledge is then applied as follows. Start with the 7-context associated with a given target word. If the 7-context is found in the knowledge-base, assign the most common chunk tag for that context as seen in the training data. If the 7-context is not in the knowledge-base, repeat the search using the 5-context and continue to more local contexts until a match is found.

The *barrier word method* used in the NLM's MetaMap program for noun phrase identification is an efficient algorithm that utilizes the semantic information of words that are outside the boundaries of word phrases [74, 118, 152]. This approach exploits the fact that barrier words (low-information words) serve as separators ("barriers") between words in a multiple-word medical term. Thus, the approach focuses on locating as many words within a sentence that belong to an "outside" tag class. Noun phrases are delimited using a long list of stop words (~24,000), including articles, prepositions, and verbs. A potential nominal phrase is computed as a sequence of words occurring between barrier words. Barrier words are associated with phrasal types (*e.g.,* noun phrases, etc.) and are words in which a target phrase cannot "spill" across.

Parser link information. Syntactic parser information can be used as an additional source of knowledge to constrain the classification of word tokens. Partial parsing methods can be used. [77] use a general purpose dependency parser to generate link features for noun phrase identification in radiology reports. A specialized link grammar parse and a semantic interpretation step between linked word pairs was used in [6] to identify common semantic relations between word features in anatomy phrases in clinical text. These constraints are expressed as feature functions within a MaxEnt classifier that integrates all word tag features.

Classifier Design

Classifier design involves constructing a model that utilizes the features deemed relevant to assign a chunk classification for a particular word. The choice of classifier design affects the weighting, and in some cases, the ordering of applied features. Selection criteria for a classifier hence include [81]: the computational complexity (*i.e.,* estimation of model parameters); the ability to understand decision-making mechanisms; the handling of noisy data; the ability to deal with sparse training data (methods for statistical smoothing); the means to deal with overlapping and/or conflicting evidence; the degree of scalability; and the ease in adapting to new training examples. Classifier designs can be roughly divided into four groups: 1) rule-based systems [47, 83, 162]; 2) memory-based systems [160]; 3) statistical methods [143]; and 4) hybrid systems [90, 139, 168].

Rule-based systems. Rule-based systems utilize deterministic symbolic methods implemented as sequence processors. Rules are often expressed in terms of classic first-order logic with sequencing performed using linear cascades and finite state automata. Transition rules are expressed in terms of some formal language, such as a context-free grammar. The context for a given word is extracted, and this context proceeds in sequence through a catalog of rules, applying each rule that matches. The sequence ends when a rule is fired or no further rules are available. Cascading rule systems can be implemented as a hash table keyed on the feature context. In general, symbolic rule-based systems are best applied to language models that provide comprehensive contexts for decision-making. Rule-based systems can be very effective in applications where there is a consistent usage of style and in resolving low frequency context configurations.

Memory-based systems. A disadvantage of rule-based systems is their inability to handle unseen contexts. Memory-based learning techniques are supervised learning methods that compile the contexts associated with a list of training examples into a memory database. Predictions for a word's classification are based on computing a similarity metric between the context of the word in question, to the contexts in the memory database, returning the *k*-nearest neighbors [133]. Of particular importance is the method of weighting features for estimating the similarity distance. The easiest way is to treat each feature with equal relevance. [161] demonstrated a memory-based system that used five tokens to the left, the word itself, and three tokens to the right for context, and a neighborhood of $k = 3$. Feature relevance was determined by an information gain metric, which assigns weights by observing each feature in isolation and estimating how much information it contributes to the correct class label [127]. One advantage of these systems is that it allows learning to proceed incrementally – each new example in the case base is immediately available for decision-making [25].

Statistical models. Statistical models are data-driven frameworks that utilize training examples to estimate the probability of an outcome variable (*i.e.,* chunk tag). These models can be designed to provide good language modeling coverage due to their ability to make decisions based on partial evidence. Bayesian classifiers are the most prevalent in language modeling, but also include naïve Bayes, HMM, MaxEnt, and maximum margin (*e.g.*, SVM) models. The differences in models lies in their assumptions regarding feature independence (*e.g.*, naïve Bayes assumes independence of word level features within a sentence), types of features (*e.g.*, HMMs assume an *n*-gram model), the functional form of the underlying probability density function (*e.g.*, MaxEnt assumes no bias and no inherent conditional independence assumptions), and the underlying constraints placed on the model (*e.g.*, maximize entropy, maximize classification boundary). The differences in model performance can be seen especially in cases where a limited amount of training data is present.

Hybrid (combined) systems. [139] report a phrase chunking system that achieved higher accuracy by applying a weighted voting algorithm to the output of systems that were trained using distinct chunk representations and different machine learning algorithms. [90] achieved similar results by applying a weighted voting algorithm based on the output of eight SVM classifiers, each also trained using distinct chunk representations.

Generation of Training Samples

The performance of a classifier depends on the interrelationship between sample sizes, number of features, and classifier complexity. That is, for a fixed sample size, as the number of features is increased (with a corresponding increase in the number of unknown parameters), the reliability of the parameter estimates decrease. Consequently, the performance of the resulting classifiers, for a fixed sample size, may degrade with an increase in the number of features [81]. Additionally, the relative importance of features can change as the number of training examples changes [160]. The more complex the classifier, the larger the ratio of sample size to feature dimensions should be used to avoid the "dimensionality curse." Thus, an important area of research has been directed towards methods of efficiently collecting training examples. The quality and number of training examples are by far the most important factors in building a classifier.

Sampling procedure. The number of training samples and the means by which these samples are selected can greatly affect the performance of a phrase chunker classifier.

We start by compiling a large a pool of documents (or sentences) that potentially contain instances of the target phrase type. This collection can be generated by creating a very conservative high-recall classifier that looks at the words of each sentence within a document collection and outputs whether it contains a target semantic phrase. For instance, one could create a high recall anatomy classifier by compiling a list of all words contained in UMLS anatomy expressions, filter out stop words, and compare this list of words to the words within a candidate sentence. Once the sentence pool is formed, we likely need to select from this pool those sentences that will be manually tagged to train the classifier; two issues are considered:

1. Representativeness of a sample. Ideally, selected samples should reflect the underlying distribution of the training corpus, which in turn should reflect the underlying distribution of the future application corpus. In essence, one should sample according to the likelihood of seeing a particular instance within a particular context in the targeted document space. For probabilistic models, this policy is especially important in order to correctly learn the distribution. For symbolic- and memory-based classification schemes, it is important to locate samples from the complete spectrum of patterns (feature space).

2. Uncertainty of a training sample relative to an existing model. Normally in NLP, the possibility space for chunking and parsing is substantial and even a large number of representative samples will still result in poor modeling of the distribution tails. As such, one strategy is to annotate those portions of training data for which an existing classifier is not performing well, and then to actively seek further examples to improve the training process. *Active learning methods* have been explored for *selectively sampling* (vs. *randomly sampling*) a test corpus for tagging phrasal boundaries. Selective sampling uses automated methods for choosing which samples to annotate using various entropy-based scoring criteria derived from an existing (probability) model, searching for the most informative samples (*e.g.,* ones close to a classifier boundary) [150]. Thus, instead of blindly annotating the whole training corpus, only those samples that are most uncertain are selected for tagging [43, 70, 154]. Once tagged, these supplementary training data are added to the initial randomly chosen training set and the classifier parameters are re-estimated. [96] uses a heuristic sample selection approach to retrain a generic part-of-speech tagger for medical reports.

Tagging tools. An example of a phrase chunking interface is shown in Fig. 6.9, which provides the user with the next sentence drawn from a sample pool, and a means of tagging the start and end word boundaries for a targeted semantic phrasal chunk (anatomy). Negative examples are automatically generated based on false hypotheses proposed by the system. For example, in the sentence, "*There is a mass in the lung and the liver is clear,*" the phrasal chunk proposition the lung and the liver is false as the lung and liver are not used as a compound phrase.

Ultimately, a user must be involved in the tagging process to identify and correct examples. For instance, there are many instances within medical text of partial descriptions (*ellipsis*) that require some prior knowledge either expressed within a previous portion of the text or simply understood within the domain. For example, the words tip, end, and apex may (or may not) refer to some landmark on an anatomic organ. Inconsistencies or errors in tagging can cause significant performance degradation in the end classifier; and some classifier models are more susceptible to such training noise. Thus, decisions have to be made on how to handle ambiguous tagging

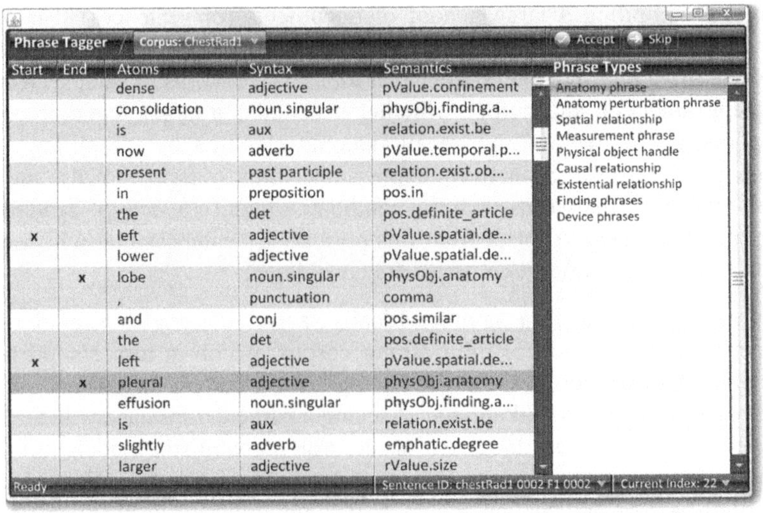

Figure 6.9: Preliminary design of phrase chunker tagging interface.

Example	Example
Loop of *unopacified* bowel	Lower pole of the *transplant* kidney
Loop of *fluid-filled* small bowel	Loops of *presumed* colon
Right true pelvis	Left lower quadrant renal *transplant*
...the descending colon and *filled multiple* loops of *presumed* colon	Loops of *nondistended* bowel
Right renal pelvocaliceal and partial upper ureteric (duplication)	*On the right side*, the femur...

Table 6.10: Examples of phrases for which ambiguous tagging may occur (the word or phrase in question is italicized). A tagging tool should aid the user by suggesting possible choices, ultimately allowing the individual to edit any errors.

assignments. For example, consider the examples in Table 6.10. The interface can include modes where tags are manually specified *de novo*, or the system may suggest the most likely tags, which can then be edited by a user as necessary.

Semi-supervised methods. A popular method, called *co-training*, involves building separate models for the same problem using limited training data [16]. One can then apply these separate models on unlabeled datasets to find examples that each model independently labels with high confidence. New training examples found confidently by one classifier can be added for remodeling to other classifiers. The process is iterated until the unlabeled data is exhausted of any new examples [65]. [124] introduces a moderately supervised variant of co-training in which a human manually corrects the mistakes made during automatic labeling; this approach improves the quality of training examples without unduly increasing the burden of a human annotator.

Linear Sequence Optimization

The above classifiers work by assigning a class probability to a single word. But due to imperfect context modeling, the highest probability chunk tag classification for a word may not maximize the probability of the *sequence* of chunk tags for a given sentence. *Linear sequence optimization*, a probabilistic maximization problem, can be performed efficiently using various forms of dynamic programming techniques [82, 114]. [44] uses a two-pass approach that first locates the top 20 hypotheses for named-entity boundaries within a sentence, and then re-ranks these hypotheses using various global features. The re-ranking methods include boosting and voted perceptron algorithms. Improvements over the baseline of simply selecting the best maximum entropy classifier hypothesis were on the order of 15-17% for relative reduction in error rate. [28] demonstrates the effectiveness of a system that recurrently feeds back results in successive passes to upstream processes, improving disambiguation accuracy across a sequence of tags for a sentence.

Parsing: Relation Extraction and Constituency Parsing

Many medical NLP applications require extracting targeted relations from medical text [75]. Examples of common types of relations include existence, location, interpretation, and appearance of a clinical finding. For instance, the sentence below includes a number of propositions that can be identified and aggregated collectively to form a network of propositions.

> There has been a significant increase in the size of the enhancing 34mm tumor centered in the left thalamus and midbrain.

The goal of sentence level processing is to construct a conceptual representation of the meaning (*i.e.,* information or interpretation) of a sentence within the context of a report and/or patient case. Propositional logic type representations are commonly used and implemented in the form of conceptual graphs, linked frames, or other types of relational models [62, 80, 85]. Generally, an NLP system attempts to create a model of language comprehension by maximizing the likelihood of interpretation given some input text:

$$\underset{I \in \text{All interpretations}}{\operatorname{argmax}} P(\text{interpretation} \mid \text{text}) = \underset{I \in \text{All interpretations}}{\operatorname{argmax}} \frac{P(\text{interpretation}, \text{text})}{P(\text{text})}$$

From this interpretation of the problem, high-level processing issues include:

- Interpretation. First, what is meant by an interpretation? How does a computer map language to meaning? Are we talking only about a "surface" interpretation related to sentence construction, or a "deep" interpretation in which everything that is generally implied is inferred given our understanding of events, causality, and intentionality?
- Large state space. Secondly, how do we deal with the large state space? The number of possible sentences and corresponding interpretations is huge, even for the limited domains seen in medicine. Methods for how to factor and reduce the dimensionality of the NLP problem are required.
- Domain. How domain dependent is an NLP system? A practical system needs to know what the bounds of its capabilities. All of medicine? All of radiology? Only sentences with the word "discharge" in it? Only certain types of propositions?

Compositionality in Language

Mathematically, we can reduce the dimensionality of the NLP problem by factoring the joint probability into smaller compositional elements. Frequently, statistical regularities can be seen by layering of patterns onto each other. Based on this observation, a more advanced and efficient way of encoding language is to organize the information in a hierarchy as a generative model, exploiting such regularities. The ordering of words and the structuring of phrases provide important clues to the meaning and purpose of the text, over and above the meaning one can infer from viewing the sentence as an unstructured bag-of-words. In between our observable input and our desired output, we can conceive of a hierarchical model with many hidden states: we call these *sub-interpretations*. The hidden states can correspond to the layers within a generative theory of language. For example, within formal linguistics, there are various kinds of structuring principles operating over different kinds of primitives relating to phonology, morphology, semantics, and syntax. The exact nature of what these hidden layers are and whether they are an essential part of helping us to make a transformation from text to interpretation is what makes linguistics and cognitive science such an active area of research. But generally, we believe that these intermediate layers of presentation are important in determining the meaning of a text.

In linguistics, we deal with *part-of* hierarchies (combinatorial hierarchies) that have some spatial adjacency order. Grammars describe the constituents of a sentence and rules that describe how these constituents can be constructed. Construction can be seen from various perspectives (*e.g.,* structural, semantics, cognitive). *Structural grammars* deal with generic constituents such as noun phrases, verb phrases, and clauses. These nodes are intended to be general and apply to the primitive constructions of a language (*e.g.,*

English). A sentence is represented as a sort of hierarchically nested set of these syntactic structural parts. Fig. 6.10 shows an example phrase structure syntactic parse for an input sentence that generates a *syntactic parse tree*. The nodes represent phrases that can be composed of other phrases in a recursive manner. Structural parses based on phrase constituency or dependency grammars have been applied to various applications in clinical text processing [23, 76]. Syntactic features provide important evidence for the identification of various relational propositions. In general, there is a strong connection between the structure of a sentence and its meaning: if we know its structure, this gives us strong clues to how we can map this structure to its functional meaning (structure-function mapping; Fig. 6.11).

Note that in syntactic parsing, the main verb of the sentence plays the crucial role in the meaning of the sentence. The verb becomes the head of the sentence. Syntactic parsing performance can be improved by modeling the expected possible arguments for specific verbs (*i.e.,* verbal selectional preferences). We constrain syntactic links based on our knowledge of semantics. For example, [42] have studied the argument structures for verbs that appear in the biomedical literature. Structurally, this approach allows a more "event-centric" modeling approach. For instance, the event could be the

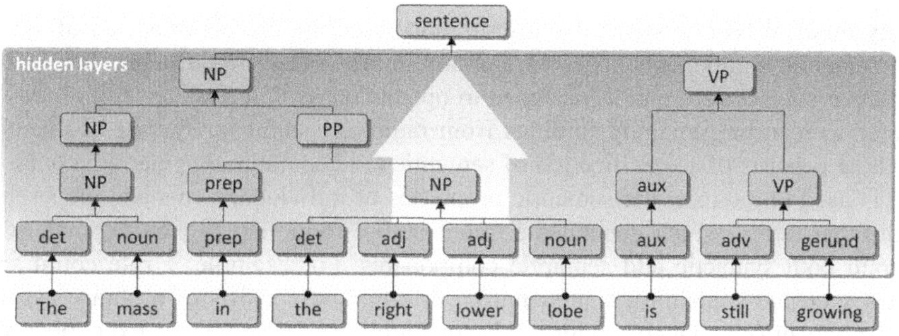

Figure 6.10: A sentence with generic syntactic linguistic structure.

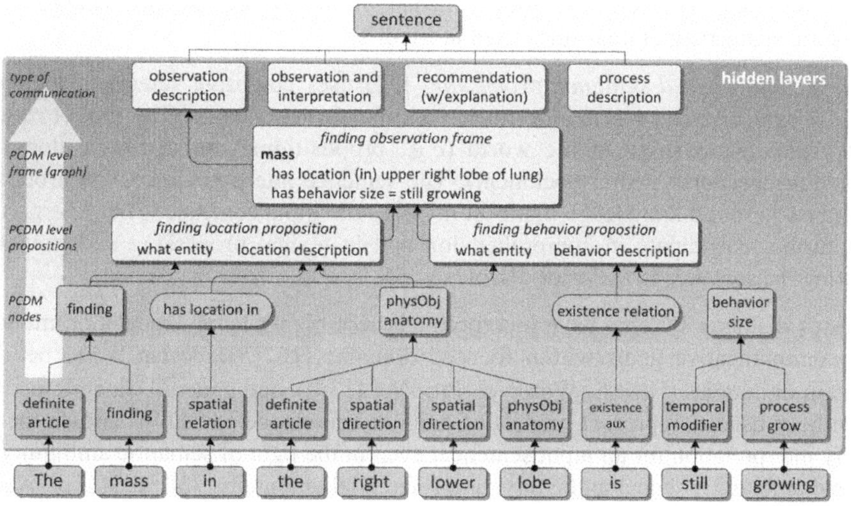

Figure 6.11: A sentence with hidden semantic interpretation, using a phenomenon-centric data model (PCDM) as a foundation (see Chapter 7).

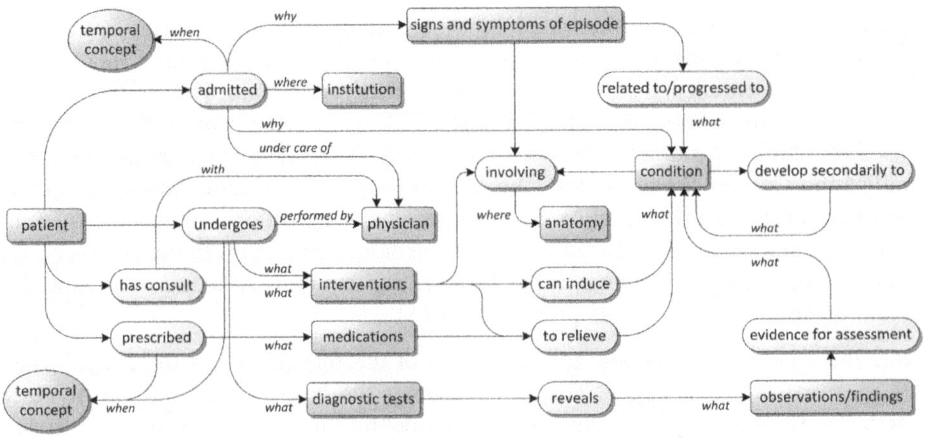

Figure 6.12: An information model for discharge summaries.

act of observing (*e.g.,* "*X* was seen") or the act of discharging a patient from the hospital with arguments describing the specifics of the event (*e.g.,* why, who, where, when, etc.) (Fig. 6.12).

Many medical NLP systems use a grammar based upon a combination of syntactic and semantic components. These grammars improve task performance for which the particular sublanguage type is designed to operate on (*e.g.,* the universe of all discourse related to reporting imaging findings from radiology studies). By way of illustration, MedLee consists of a specification of semantic (and sometimes syntactic) components and is used to interpret the semantic properties of individual terms and their relations with other terms, which generates a target output form. Thus, one grammar rule can contain both syntactic and semantic components. For example, a rule could specify that a sentence containing sign/symptom information consists of a phrase associated with a patient (*i.e.,* patient), with a subsequent evidential verb (*e.g.,* experienced), and followed by a phrase that contains a sign/symptom (*e.g.,* pain in arm). The semantic grammar rules were developed based on co-occurrence patterns observed in clinical text. The reader is referred to [60] a further discussion of semantic categories and semantic syntactic grammar rules used in MedLee.

Interpretation-based sentence processing [21] uses cognitive science principles to build a syntactic and a semantic representation for a sentence. The model relies on background knowledge of the world (*e.g.,* propositional model of the domain or preceding discourse within document). The sentence interpretation is the proposition in the background knowledge that best matches the input sentence. That interpretation can further participate in comprehension and in lexical processing and is vital for relating the sentence to the prior discourse.

Various dialogue systems have incorporated these higher level expectation models of the communicative goals within focused domains [107]. Bayesian belief networks, domain ontologies (UMLS, WordNet, PropNet [136]), and general semantic networks have been used to represent models of domain knowledge and to instantiate the most likely interpretation for an input sentence, even in the light of semantic ambiguity (*i.e.,* lexical ambiguity, scopal ambiguity, and referential ambiguity) [38, 123]. For example, the medical NLP system, MPLUS, encodes semantic and world knowledge of specific phenomena into modular and independent Bayesian belief networks (BBNs) [37]. This BBN is then used for sentence level semantic interpretation. Belief networks have also been used as discourse models to deal with missing information (*e.g.*, ellipsis) and/or spurious concepts (misclassifications) [105].

Notably, the differences in the constituents and parse output representation makes comparison and sharing of training results difficult. The types of constituents within a grammar have ramifications as to the usability of training sets. In some cases there can be possible transformational rules to normalize a given syntactic parse representation (*e.g.*, phrase structure grammars to dependency grammars). For instance, different parsers with heterogeneous representations have been compared for task performance by transforming their results to dependency graphs [39].

Discussion

There remain numerous challenges in developing accurate medical NLP systems that provide deep understanding of clinical text. The requirement for high accuracy means that researchers must develop comprehensive language models that cover the gamut of grammatical (*e.g.*, words, word-word linkages, sentences) and cognitive entities (world objects, medical processes). This intense knowledge requirement necessarily means that practical medical NLP systems will be focused on either specialized domains (*e.g.*, radiology) or types of communications.

Although medical NLP systems borrow heavily from the theories and approaches of general NLP systems, there are a few notable differences, which we summarize here:

- Expectations. In a sublanguage, the target information space is less dense compared to a general language understanding system. What approaches to sublanguage processing should be considered? What are the peculiarities of medicine as a sublanguage that we should take advantage of in order to develop robust systems for the domains in which they are intended to operate? [22] showed that medical text has significant grammatical variation when compared to general text. In general, medical report text includes fewer word types, concepts, and propositions. Expectation models for what information is to be communicated within a given type of report or medical domain are possible [63]. These observations lead to the question of what tasks would be most beneficial for the medical NLP community to tackle beyond the efforts of general NLP researchers that would advance NLP technology to a higher level?
- Framework. Is there a theoretical framework from which we can build as a community? Is there some development or emergent processing architecture that would allow progress to be made steadily and allow for a wide spectrum of applications with greater functionality to be more efficiently developed? Currently, there are no agreed upon "integrated" models for medical NLP, and there is no consensus platform for bringing components together and making them interoperable. It is not clear how to separate general language features from domain specific knowledge such that either is readily integrated and updatable. A framework that allows global optimization of low-level lexical and syntactic evidence as well as high-level patient and phenomenological knowledge would be ideal.
- Sharing of resources. Related to the issue of a shared framework, how should we manage resources such that the collective efforts of research teams can bring NLP closer to fruition in terms of providing a useful service to medical practice, research, and/or education? Interoperability between representational formats, annotations, software, etc. are needed to achieve wide-scale community growth. Eventually, issues

of standardization are important, especially in areas of representation. The development of widespread tools can assist in resource sharing.

- Evaluation. The value of a medical NLP system rests in its ultimate “black box” performance with respect to its end application [61]. The most important aspect of an NLP system is utility and not necessarily the technical accuracy of computing an intermediate text representation.

References

1. Abney S (2002) Bootstrapping. Proc 40th Annual Meeting Assoc Computational Linguistics, pp 360-367.
2. Aho AV, Corasick MJ (1975) Efficient string matching: Aid to bibliographic search. Comm ACM, 18(6):333-340.
3. Aronson AR (2001) Effective mapping of biomedical text to the UMLS Metathesaurus: The MetaMap program. Proc AMIA Symp, pp 17-21.
4. Baneyx A, Charlet J, Jaulent MC (2006) Methodology to build medical ontology from textual resources. Proc AMIA Symp, pp 21-25.
5. Bashyam V (2008) Towards a canonical representation for machine understanding of natural language in radiology reports. Department of Information Studies, PhD dissertation. University of California Los Angeles.
6. Bashyam V, Taira RK (2005) Indexing anatomical phrases in neuro-radiology reports to the UMLS 2005AA. Proc AMIA Symp pp 26-30.
7. Bashyam V, Taira RK (2005) A study of lexical behaviour of sentences in chest radiology reports. Proc AMIA Symp p 891.
8. Bates DW, Evans RS, Murff H, Stetson PD, Pizziferri L, Hripcsak G (2003) Detecting adverse events using information technology. J Am Med Inform Assoc, 10(2):115-128.
9. Baud R (2004) A natural language based search engine for ICD10 diagnosis encoding. Med Arh, 58(1 Suppl 2):79-80.
10. Becker GJ (2005) Restructuring cancer clinical trials. J Am Coll Radiol, 2(10):816-817.
11. Bell GB, Sethi A (2001) Matching records in a national medical patient index. Communications of the ACM, 44(9):83-88.
12. Berger AL, DellaPietra SA, DellaPietra VJ (1996) A maximum entropy approach to natural language processing. Computational Linguistics, 22(1):39-71.
13. Berman JJ (2004) Pathology abbreviated: A long review of short terms. Arch Pathol Lab Med, 128(3):347-352.
14. Berrios DC (2000) Automated indexing for full text information retrieval. Proc AMIA Symp, pp 71-75.
15. Black A, van de Plassche J, Williams B (1991) Analysis of unknown words through morphological decomposition. Proc 5th Conf European Chapter of the Association of Computational Linguistics, pp 101-106.
16. Blum A, Mitchell T (1998) Combining labeled and unlabeled data with co-training. Proc 11th Annual Conf Computational Learning Theory, pp 92-100.
17. Bodenreider O, McCray AT (2003) Exploring semantic groups through visual approaches. J Biomed Inform, 36(6):414-432.
18. Booker DL, Berman JJ (2004) Dangerous abbreviations. Hum Pathol, 35(5):529-531.
19. Bouillon P, Rayner M, Chatzichrisafis N, Hockey BA, Santaholma M, Starlander M, Nakao Y, Kanzaki K, Isahara H (2005) A generic multi-lingual open source platform for limited-domain medical speech translation. Proc 10th Annual Conf European Association for Machine Translation, pp 50-58.

20. Brill E (1995) Transformation-based error-driven learning and natural language processing: A case study in part-of-speech tagging. Computational Linguistics, 21(4):543-565.
21. Budiu R, Anderson JR (2004) Interpretation-based processing: A unified theory of semantic sentence comprehension. Cognitive Science, 28(1):1-44.
22. Campbell DA, Johnson SB (2001) Comparing syntactic complexity in medical and non-medical corpora. Proc AMIA Symp, pp 90-94.
23. Campbell DA, Johnson SB (2002) A transformational-based learner for dependency grammars in discharge summaries. Proc ACL-02 Workshop on Natural language Processing in the Biomedical Domain, vol 3, pp 37-44.
24. Cao H, Markatou M, Melton GB, Chiang MF, Hripcsak G (2005) Mining a clinical data warehouse to discover disease-finding associations using co-occurrence statistics. Proc AMIA Symp, pp 106-110.
25. Cardie C (1994) Domain-specific Knowledge Acquisition for Conceptual Sentence Analysis. Department of Computer Science PhD dissertation. University of Massachusetts, Amherst.
26. Carroll J, Minnen G, Pearce D, Canning Y, Devlin S, Tait J (1999) Simplifying text for language-impaired readers. Proc 9th Conf European Chapter of the Association of Computational Linguistics, pp 269-270.
27. Carter PI (2004) HIPAA Compliance Handbook 2004. Aspen Publishing, Gaithersburg, MD.
28. Chao G (2002) Recurrent probabilistic modeling and its application to part-of-speech tagging. Proc 40th Annual Meeting Assoc Computational Linguistics: Student Research Workshop, pp 6-11.
29. Chapman WW, Bridewell W, Hanbury P, Cooper GF, Buchanan BG (2001) A simple algorithm for identifying negated findings and diseases in discharge summaries. J Biomed Inform, 34(5):301-310.
30. Chapman WW, Chu D, Dowling JN (2007) ConText: An algorithm for identifying contextual features from clinical text. BioNLP 2007: Biological, Translational, and Clinical Language Processing, pp 81-88.
31. Chapman WW, Fiszman M, Dowling JN, Chapman BE, Rindflesch TC (2004) Identifying respiratory findings in emergency department reports for biosurveillance using MetaMap. Stud Health Technol Inform, 107(Pt 1):487-491.
32. Charniak E (2001) Unsupervised learning of name structure from coreference data. Proc North American Chapter Assoc Computational Linguistics, pp 48-54.
33. Chen SF, Goodman J (1999) An empirical study of smoothing techniques for language modeling. Computer Speech and Language, 13(4):359-394.
34. Chinchor N, Marsh E (1998) MUC-7 named entity task definition. Proc 7th Message Understanding Conference (MUC-7).
35. Cho PS, Taira RK, Kangarloo H (2002) Text boundary detection of medical reports. Proc AMIA Symp, pp 155-159.
36. Cho PS, Taira RK, Kangarloo H (2003) Automatic section segmentation of medical reports. Proc AMIA Symp, pp 155-159.
37. Christensen LM, Haug PJ, Fiszman M (2002) MPLUS: A probabilistic medical language understanding system. Proc ACL-02 Workshop on Natural Language Processing in the Biomedical Domain, vol 3, pp 29-36.
38. Ciaramita M, Johnson M (2000) Explaining away ambiguity: Learning verb selectional preference with Bayesian networks. Proc 18th Conf Computational Linguistics, vol 1, pp 187-193.
39. Clegg AB, Shepherd AJ (2007) Benchmarking natural-language parsers for biological applications using dependency graphs. BMC Bioinformatics, 8:24-40.

40. Coates-Stephens S (1992) The analysis and acquisition of proper names for the understanding of free text. Computers and the Humanities, 26(5):441-456.
41. Coden AR, Pakhomov SV, Ando RK, Duffy PH, Chute CG (2005) Domain-specific language models and lexicons for tagging. J Biomed Inform, 38(6):422-430.
42. Cohen KB, Hunter L (2006) A critical review of PASBio's argument structures for biomedical verbs. BMC Bioinformatics, 7 Suppl 3:S5.
43. Cohn A (1996) Calculi for qualitative spatial reasoning. Artificial Intelligence and Symbolic Mathematical Computation, pp 124-143.
44. Collins M (2002) Ranking algorithms for named-entity extraction: Boosting and the voted perceptron. Proc 40th Annual Meeting Assoc Computational Linguistics, pp 489-496.
45. Computational Mdeicine Center (2009) International Challenge: Classifying Clinical Free Text Using Natural Language Processing. http://www.computationalmedicine.org-/challenge. Accessed April 14, 2009.
46. D'Avolio LW, Litwin MS, Rogers SO, Jr., Bui AA (2008) Facilitating clinical outcomes assessment through the automated identification of quality measures for prostate cancer surgery. J Am Med Inform Assoc, 15(3):341-348.
47. Dejean H (2000) ALLiS: A symbolic learning system for natural language learning. Proc CoNLL-2000 and LLL-2000, pp 95-98.
48. DeRose SJ (1988) Grammatical category disambiguation by statistical optimization. Computational Linguistics, 14(1):31-39.
49. Divita G, Browne AC, Rindflesch TC (1998) Evaluating lexical variant generation to improve information retrieval. Proc AMIA Symp, pp 775-779.
50. Dolin RH, Alschuler L, Boyer S, Beebe C, Behlen FM, Biron PV, Shabo Shvo A (2006) HL7 Clinical Document Architecture, Release 2. J Am Med Inform Assoc, 13(1):30-39.
51. Duda RO, Hart PE, Stork DG (2001) Pattern Classification. 2nd edition. Wiley, New York, NY.
52. Eck M, Vogel S, Waibel A (2004) Improving statistical machine translation in the medical domain using the unified medical language system. Proc 20th Intl Conf Computational Linguistics.
53. Eddy SR (2004) What is a hidden Markov model? Nat Biotechnol, 22(10):1315-1316.
54. Eng J, Eisner JM (2004) Radiology report entry with automatic phrase completion driven by language modeling. RadioGraphics, 24(5):1493-1501.
55. Fellbaum C (1998) WordNet: An Electronic Lexical Database. MIT Press, Cambridge, MA.
56. Feng D, Burns G, Zhu J, Hovy EH (2008) Towards automated semantic analysis on biomedical research articles. Proc 3rd Intl Joint Conf Natural Language Processing.
57. Firth JR (1957) Modes of meaning. In: Firth JR (ed) Papers in Linguistics 1934-1951. Oxford University Press, London .
58. Fisk JM, Mutalik P, Levin FW, Erdos J, Taylor C, Nadkarni P (2003) Integrating query of relational and textual data in clinical databases: A case study. J Am Med Inform Assoc, 10(1):21-38.
59. Forney Jr GD (1973) The Viterbi algorithm. Proceedings of the IEEE, 61(3):268-278.
60. Friedman C, Alderson PO, Austin JH, Cimino JJ, Johnson SB (1994) A general natural-language text processor for clinical radiology. J Am Med Inform Assoc, 1(2):161-174.
61. Friedman C, Hripcsak G, Shablinsky I (1998) An evaluation of natural language processing methodologies. Proc AMIA Symp:855-859.
62. Friedman C, Huff SM, Hersh WR, Pattisongordon E, Cimino JJ (1995) The Canon Group's effort: Working toward a merged model. J Am Med Inform Assoc, 2(1):4-18.
63. Friedman C, Kra P, Rzhetsky A (2002) Two biomedical sublanguages: A description based on the theories of Zellig Harris. J Biomed Inform, 35(4):222-235.

64. Friedman C, Shagina L, Lussier Y, Hripcsak G (2004) Automated encoding of clinical documents based on natural language processing. J Am Med Inform Assoc, 11(5):392-402.
65. Goldman S, Zhou Y (2000) Enhancing supervised learning with unlabeled data. Proc 17th Intl Conf Machine Learning (ICML-2000), pp 327-334.
66. Guihenneuc-Jouyaux C, Richardson S, Longini IM, Jr. (2000) Modeling markers of disease progression by a hidden Markov process: Application to characterizing CD4 cell decline. Biometrics, 56(3):733-741.
67. Gundlapalli AV, South BR, Phansalkar S, Kinney AY, Shen S (2008) Application of natural language processing to VA electronic health records to identify phenotypic characteristics for clinical and research purposes. Proc 2008 AMIA Summit on Translational Bioinformatics, pp 36-40.
68. Gupta A, Ludascher B, Grethe JS, Martone ME (2003) Towards a formalization of disease-specific ontologies for neuroinformatics. Neural Networks 16:1277-1292.
69. Gupta D, Saul M, Gilbertson J (2004) Evaluation of a deidentification (De-Id) software engine to share pathology reports and clinical documents for research. Am J Clin Pathol, 121(2):176-186.
70. Hachey B, Alex B, Mecker M (2005) Investigating the effects of selective sampling on the annotation task. Proc 9th Conf Computational Natural Language Processing, pp 144-151.
71. Haug PJ, Christensen L, Gundersen M, Clemons B, Koehler S, Bauer K (1997) A natural language parsing system for encoding admitting diagnoses. Proc AMIA Symp, pp 814-818.
72. Heinze DT, Morsch ML, Sheffer RE, Jimmink MA, Jennings MA, Morris WC, Morch AEW (2001) LifeCode: A deployed application for automated medical coding. AI Magazine, 22(2):76-88.
73. Hersh WR, Campbell EM, Malveau SE (1997) Assessing the feasibility of large-scale natural language processing in a corpus of ordinary medical records: A lexical analysis. Proc AMIA Fall Symp, pp 580-584.
74. Herzig TW, Johns M (1997) Extraction of medical information from textual sources: A statistical variant of the boundary-word method. J Am Med Inform Assoc:859-859.
75. Hripcsak G, Austin JH, Alderson PO, Friedman C (2002) Use of natural language processing to translate clinical information from a database of 889,921 chest radiographic reports. Radiology, 224(1):157-163.
76. Huang Y, Lowe HJ (2007) A novel hybrid approach to automated negation detection in clinical radiology reports. J Am Med Inform Assoc, 14(3):304-311.
77. Huang Y, Lowe HJ, Klein D, Cucina RJ (2005) Improved identification of noun phrases in clinical radiology reports using a high-performance statistical natural language parser augmented with the UMLS specialist lexicon. J Am Med Inform Assoc, 12(3):275-285.
78. Huddleston R (1984) Introduction to the Grammar of English. Cambridge University Press, Cambridge, MA.
79. Humphrey SM, Rogers WJ, Kilicoglu H, Demner-Fushman D, Rindflesch TC (2006) Word sense disambiguation by selecting the best semantic type based on journal descriptor indexing: Preliminary experiment. J Am American Society for Information Science and Technology, 57(1):96-113.
80. Iwanska LM, Shapiro SC (2000) Natural Language Processing and Knowledge Representation: Language for Knowledge and Knowledge for Language. AAAI Press, Menlo Park, CA.
81. Jain AK, Duin RPW, Mao JC (2000) Statistical pattern recognition: A review. IEEE Trans Pattern Analysis and Machine Intelligence, 22(1):4-37.
82. Jelinek F (1999) Statistical Methods for Speech Recognition. 2nd edition. MIT press, Cambridge, MA.

83. Johansson C (2000) A context sensitive maximum likelihood approach to chunking. Proc 2nd Workshop on Learning Language in Logic; 4th Conf Computational Natural Language Learning, vol 7, pp 136-138.
84. Johnson DB, Chu WW, Dionisio JD, Taira RK, Kangarloo H (1999) Creating and indexing teaching files from free-text patient reports. Proc AMIA Symp, pp 814-818.
85. Johnson SB (1998) Conceptual graph grammar: A simple formalism for sublanguage. Methods Inf Med, 37(4-5):345-352.
86. Johnson SB (1999) A semantic lexicon for medical language processing. J Am Med Inform Assoc, 6(3):205-218.
87. Joshi M, Pedersen MJT, Maclin R, Pakhomov S (2006) Kernel methods for word sense disambiguation and acronym expansion. Proc 21st National Conf Artificial Intelligence.
88. Jurafsky D, Martin JH (2000) Speech and Language Processing: An Introduction to Natural Language Processing, Computational Linguistics, and Speech Recognition. Prentice Hall, Upper Saddle River, NJ.
89. Karlsson F (1990) Constraint grammar as a framework for parsing running text. Proc 13th Annual Conf Computational Linguistics, pp 168-173.
90. Kudo T, Matsumoto Y (2001) Chunking with support vector machines. Proc 2nd Meeting North American Chapter Assoc Computational Linguistics on Language Technologies, pp 192-199.
91. Lafferty J, McCallum A, Pereira F (2001) Conditional random fields: Probabilistic models for segmenting and labeling sequence data. Proc 18th Intl Conf Machine Learning, pp 282-289.
92. Le Moigno S, Charlet J, Bourigault D, Degoulet P, Jaulent MC (2002) Terminology extraction from text to build an ontology in surgical intensive care. Proc AMIA Symp, pp 430-434.
93. Lee DL, Chuang H, Seamons K (1997) Document ranking and the vector-space model. IEEE Software, 14(2):67-75.
94. Li L, Chase HS, Patel CO, Friedman C, Weng C (2008) Comparing ICD9-encoded diagnoses and NLP-processed discharge summaries for clinical trials pre-screening: A case study. Proc AMIA Symp, pp 404-408.
95. Lindberg DA, Humphreys BL, McCray AT (1993) The Unified Medical Language System. Methods Inf Med, 32(4):281-291.
96. Liu K, Chapman W, Hwa R, Crowley RS (2007) Heuristic sample selection to minimize reference standard training set for a part-of-speech tagger. J Am Med Inform Assoc, 14(5):641-650.
97. Lovis C, Michel PA, Baud R, Scherrer JR (1995) Word segmentation processing: A way to exponentially extend medical dictionaries. Proc MedInfo, vol 8 Pt 1, pp 28-32.
98. Lyman M, Sager N, Tick L, Nhan N, Borst F, Scherrer JR (1991) The application of natural-language processing to healthcare quality assessment. Med Decis Making, 11 (4 Suppl):S65-68.
99. Manning CD, Schütze H (1999) Foundations of Statistical Natural Language Processing. MIT Press, Cambridge, MA.
100. Marcus MP, Marcinkiewicz MA, Santorini B (1993) Building a large annotated corpus of English: The Penn Treebank. Computational Linguistics, 19(2):313-330.
101. McCallum A, Freitag D, Pereira F (2000) Maximum entropy Markov models for information extraction and segmentation. Proc 7th Intl Conf Machine Learning, pp 591-598.
102. McCray AT, Bodenreider O, Malley JD, Browne AC (2001) Evaluating UMLS strings for natural language processing. Proc AMIA Symp, pp 448-452.
103. McDonald DD (1993) Internal and external evidence in the identification and semantic categorization of proper names. Acquisition of Lexical Knowledge from Text: Proc Workshop Sponsored by the Special Interest Group on the Lexicon of the ACL, pp 32-43.

104. McDonald DD (1996) Internal and external evidence in the identification and semantic categorization of proper names. In: Boguraev B, Pustejovsky J (eds) Corpus Processing for Lexical Acquisition. MIT Press, Cambridge, MA, pp 21-39.
105. McRoy SW, Ali SS, Haller SM (1997) Uniform knowledge representation for language processing in the B2 system. Natural Language Engineering, 3(2):123-145.
106. Melton GB, Hripcsak G (2005) Automated detection of adverse events using natural language processing of discharge summaries. J Am Med Inform Assoc, 12(4):448-457.
107. Meng H, Lam W, Low KF (1999) Learning belief networks for language understanding. Proc ASRU.
108. Meystre S, Haug PJ (2005) Automation of a problem list using natural language processing. BMC Med Inform Decis Mak, 5:30.
109. Meystre S, Haug PJ (2006) Natural language processing to extract medical problems from electronic clinical documents: Performance evaluation. J Biomed Inform, 39(6):589-599.
110. Mikheev A (2000) Tagging sentence boundaries. Proc 1st North American Chapter Assoc Computational Linguistics Conf, pp 264-271.
111. Miller GA, Beckwith R, Fellbaum C, Gross D, Miller KJ (1990) Introduction to WordNet: An on-line lexical database. Intl J Lexicography, 3(4):235-244.
112. Miller JE, Torii M, Vijay-Shanker K (2007) Adaptation of POS tagging for multiple biomedical domains. BioNLP 2007: Biological, Translational, and Clinical Language Processing, pp 179-180.
113. Minsky ML, Papert S (1988) Perceptrons: An Introduction to Computational Geometry. Expanded edition. MIT Press, Cambridge, MA.
114. Molina A, Pla F (2002) Shallow parsing using specialized HMMs. J Machine Learning Research, 2(4):595-613.
115. Nadkarni P, Chen R, Brandt C (2001) UMLS concept indexing for production databases: A feasibility study. J Am Med Inform Assoc, 8(1):80-91.
116. Navigli R (2009) Word sense disambiguation: A survey. ACM Computing Surveys, 41(2):1-69.
117. Neamatullah I, Douglass MM, Lehman LWH, Reisner A, Villarroel M, Long WJ, Szolovits P, Moody GB, Mark RG, Clifford GD (2008) Automated de-identification of free-text medical records. BMC Medical Informatics and Decision Making, 8(32):1-17.
118. Nelson SJ, Olson NE, Fuller L, Tuttle MS, Cole WG, Sherertz DD (1995) Identifying concepts in medical knowledge. Proc MedInfo, vol 8, pp 33-36.
119. Nguyen N, Guo Y (2007) Comparisons of sequence labeling algorithms and extensions. Proc 24th Intl Conf Machine Learning, pp 681-688.
120. Pakhomov S, Pedersen T, Chute CG (2005) Abbreviation and acronym disambiguation in clinical discourse. Proc AMIA Symp, pp 589-593.
121. Pedersen MJT, Banerjee S, Patwardhan S (2005) Maximizing semantic relatedness to perform word sense disambiguation (Technical Report). University of Minnesota Supercomputing Institute.
122. Penz JF, Wilcox AB, Hurdle JF (2007) Automated identification of adverse events related to central venous catheters. J Biomed Inform, 40(2):174-182.
123. Pestian JP, Itert L, Duch W (2004) Development of a pediatric text-corpus for part-of-speech tagging. In: Wierzchon ST, Trojanowski K (eds) Intelligent Information Processing and the Web. Springer, pp 219-226.
124. Pierce D, Cardie C (2001) Limitations of co-training for natural language learning from large datasets. Proc 2001 Conf Empirical Methods in Natural Language Processing, pp 1–9.
125. Polackova G (2008) Understanding and use of phrasal verbs and idioms in medical/nursing texts. Bratisl Lek Listy, 109(11):531-532.

126. Pyper C, Amery J, Watson M, Crook C (2004) Patients' experiences when accessing their on-line electronic patient records in primary care. British Journal of General Practice, 54(498):38-43.
127. Quinlan JR (1993) C4.5: Programs for Machine Learning. Morgan Kaufmann, San Mateo, CA.
128. Rabiner LR (1989) A tutorial on hidden Markov models and selected applications in speech recognition. Proc IEEE, 77(2):257-286.
129. Radiological Society of North America (2009) RadLex: A Lexicon for Uniform Indexing and Retrieval of Radiology Information Resources. http://www.rsna.org/radlex/. Accessed April 14, 2009.
130. Ratnaparkhi A (1996) A maximum entropy model for part-of-speech tagging. Proc Conf Empirical Methods in Natural Language Processing, pp 133-142.
131. Ratnaparkhi A (1998) Maximum Entropy Models for Natural Language Ambiguity Resolution. Department of Computer and Information Science PhD dissertation. University of Pennsylvania.
132. Rind DM, Kohane IS, Szolovits P, Safran C, Chueh HC, Barnett GO (1997) Maintaining the confidentiality of medical records shared over the Internet and the World Wide Web. Ann Intern Med, 127(2):138-141.
133. Roth D (1999) Memory based learning (Technical Report UIUCDCS-R-99-2125). Department of Computer Science, University of Illinois at Urbana-Champaign.
134. Ruch P, Baud R, Geissbuhler A (2003) Using lexical disambiguation and named-entity recognition to improve spelling correction in the electronic patient record. Artificial Intelligence in Medicine, 29(1-2):169-184.
135. Ruch P, Baud RH, Rassinoux AM, Bouillon P, Robert G (2000) Medical document anonymization with a semantic lexicon. Proc AMIA Symp, pp 729-733.
136. Ruppenhofer J, Ellsworth M, Petruck M, Johnson C (2005) FrameNet II: Extended Theory and Practice (Technical Report). ICSI, Berkeley, CA.
137. Sager N, Lyman M, Nhan NT, Tick LJ (1995) Medical language processing: Applications to patient data representation and automatic encoding. Methods Inf Med, 34(1-2):140-146.
138. Salton G (1988) Automatic Text Processing: The Transformation, Analysis, and Retrieval of Information by Computer. Addison-Wesley, Reading, MA.
139. Sang EFTK, Buchholz S (2000) Introduction to the CoNLL-2000 shared task: Chunking. Proc 2nd Workshop on Learning Language in Logic; 4th Conf Computational Natural Language Learning, vol 7, pp 127-132.
140. Savova GK, Coden AR, Sominsky IL, Johnson R, Ogren PV, de Groen PC, Chute CG (2008) Word sense disambiguation across two domains: Biomedical literature and clinical notes. J Biomed Inform, 41(6):1088-1100.
141. Schulz S, Hahn U (2000) Morpheme-based, cross-lingual indexing for medical document retrieval. Int J Med Inform, 58-59:87-99.
142. Schulz S, Honeck M, Hahn U (2002) Biomedical text retrieval in languages with a complex morphology. Proc Workshop on NLP in the Biomedical Domain, pp 61-68.
143. Skut W, Brants T (1998) Chunk tagger: Statistical recognition of noun phrases. Proc ESSLLI-1998 Workshop on Automated Acquisition of Syntax and Parsing.
144. Smith B, Ceusters W, Klagges B, Kohler J, Kumar A, Lomax J, Mungall C, Neuhaus F, Rector AL, Rosse C (2005) Relations in biomedical ontologies. Genome Biol, 6(5):R46.
145. Smith L, Rindflesch T, Wilbur WJ (2004) MedPost: A part-of-speech tagger for biomedical text. Bioinformatics, 20(14):2320-2321.
146. Strzalkowski T (1999) Natural Language Information Retrieval. Kluwer Academic, Boston, MA.

147. Sweeney L (1996) Replacing personally-identifying information in medical records: The Scrub system. Proc AMIA Symp, pp 333-337.
148. Taira R, Bui AA, Hsu W, Bashyam V, Dube S, Watt E, Andrada L, El-Saden S, Cloughesy T, Kangarloo H (2008) A tool for improving the longitudinal imaging characterization for neuro-oncology cases. Proc AMIA Symp, pp 712-716.
149. Taira RK, Bui AA, Kangarloo H (2002) Identification of patient name references within medical documents using semantic selectional restrictions. Proc AMIA Symp, pp 757-761.
150. Tang M, Luo X, Roukos S (2002) Active learning for statistical natural language parsing. Proc 40th Ann Meeting Assoc Computational Linguistics, Philadelphia, PA, pp 120-127.
151. Taskar B, Klein D, Collins M, Koller D, Manning C (2004) Max-margin parsing. Proc Empirical Methods in Natural Language Processing.
152. Tersmette S, Moore M (1988) Boundary word techniques for isolating multiword terminologies. Proc Ann Symp Computer Applications in Medical Care, pp 207-211.
153. Thede SM, Harper MP (1999) A second-order hidden Markov model for part-of-speech tagging. Proc 37th Annual Meeting ACL on Computational Linguistics, pp 175-182.
154. Thompson CA, Califf ME, Mooney RJ (1999) Active learning for natural language parsing and information extraction. Proc 16th Intl Machine Learning Conf, Bled, Slovenia, pp 406-414.
155. Tjong EF, Sang K (2000) Noun phrase recognition by system combination. Proc 1st Meeting of the North American Chapter for the Association for Computational Linguistics, pp 50-55.
156. Tolentino HD, Matters MD, Walop W, Law B, Tong W, Liu F, Fontelo P, Kohl K, Payne DC (2007) A UMLS-based spell checker for natural language processing in vaccine safety. BMC Med Inform Decis Mak, 7:3.
157. Tomanek K, Wermter J, Hahn U (2007) A reappraisal of sentence and token splitting for life sciences documents. Stud Health Technol Inform, 129(Pt 1):524-528.
158. Trieschnigg D, Kraaij W, de Jong F (2007) The influence of basic tokenization on biomedical document retrieval. Proc 30th Ann Intl ACM SIGIR Conf Research and Development in Information Retrieval, pp 803-804.
159. Uzuner O, Luo Y, Szolovits P (2007) Evaluating the state-of-the-art in automatic de-identification. J Am Med Inform Assoc, 14(5):550-563.
160. van den Bosch A, Buchholz S (2001) Shallow parsing on the basis of words only: A case study. Proc 40th Annual Meeting Assoc Computational Linguistics, pp 433-440.
161. Veenstra J, Van den Bosch A (2000) Single-classifier memory-based phrase chunking. Proc CoNLL, Lisbon, Portugal, pp 157-159.
162. Vilain M, Day D (2000) Phrase parsing with rule sequence processors: An application to the shared CoNLL task. Proc CoNLL-2000 and LLL-2000, pp 160-162.
163. Weeber M, Mork JG, Aronson AR (2001) Developing a test collection for biomedical word sense disambiguation. Proc AMIA Symp, pp 746-750.
164. Xiao J, Wang X, Liu B (2007) The study of a nonstationary maximum entropy Markov model and its application on the pos-tagging task. ACM Trans Asian Language Inforamtion Processing, 6(2):1-29.
165. Yarowsky D (1995) Unsupervised word sense disambiguation rivaling supervised methods. Proc 33rd Annual Meeting Assoc Computational Linguistics, pp 189-196.
166. Yu H, Hripcsak G, Friedman C (2002) Mapping abbreviations to full forms in biomedical articles. J Am Med Inform Assoc, 9(3):262-272.
167. Zeng QT, Tse T (2006) Exploring and developing consumer health vocabularies. J Am Med Inform Assoc, 13(1):24-29.
168. Zhou GD, Su J, Tey TG (2000) Hybrid text chunking. Proc 2nd Workshop on Learning Language in Logic; 4th Conf Computational Natural Language Learning, vol 7, pp 163-165.

169. Zitouni I (2007) Backoff hierarchical class n-gram language models: Effectiveness to model unseen events in speech recognition. Computer Speech and Language, 21(1):88-104.
170. Zou Q, Chu WW, Morioka C, Leazer GH, Kangarloo H (2003) IndexFinder: A method of extracting key concepts from clinical texts for indexing. Proc AMIA Symp, pp 763-767.

Chapter 7
Organizing Observations: Data Models

ALEX A. T. BUI AND RICKY K. TAIRA

Thus far, discussion has focused on issues related to collecting and analyzing clinical data. Yet central to the challenge of informatics is the organization of all of this information to enable a continuum of healthcare and research applications: the type of attributes supported in characterizing an entity within a data model and the scope of relationships defined between these objects determine the ease with which we can retrieve information and ultimately drive how we come to perceive and work with the data. This chapter overviews several data models that have been proposed over the years to address representational issues inherent to medical information. Three categories of data models are covered: *spatial models,* which are concerned with representing physical and anatomical relations between objects; *temporal models* that embody a chronology and/or other time-based sequences/patterns; and *clinically-oriented models*, which systematically arrange information around a healthcare abstraction or process. Notably, these models no longer serve the sole purpose of being data structures, but are also foundations upon which rudimentary logical reasoning and inference can occur. Finally, as translational informatics begins to move toward the use of large clinical datasets, the context under which such data are captured is important to consider; this chapter thus concludes by introducing the idea of a *phenomenon-centric data model* (PCDM) that explicitly embeds the principles of scientific investigation and hypotheses with clinical observations.

Data Models for Representing Medical Data

A *data model* provides a formal abstraction for the storage, organization, and access (querying) of data. Though used for a number of purposes, data models are best known for their roles: 1) in logical- and physical-level database schemas; and 2) in specifying conceptual schemas. A diversity of logical data models have become common over the years: *hierarchical*, which organizes records in a tree-like structure to describe the nesting of elements; *network*, where records are grouped together into sets, and the relationship between sets are specified; *relational*, a framework rooted in first order predicate logic to describe variables and constraints; *object-oriented* and *object-relational*, based on the idea of classes, inheritance, and method encapsulation and its mapping to a relational scheme. In contrast, a conceptual data model aims to describe an information space, separate from how the data is arranged in a database: the well-known *entity-relational* (ER) model is a prime example [24]. It is these latter conceptual-level data models with which we concern ourselves here. As seen in prior chapters, the nuances of patient data and medical knowledge precipitate representational challenges: there are often unique semantics that must be handled to accommodate spatial, temporal, and clinically-oriented constructs. Conceptual-level data models in these three areas are covered below, addressing: the representation of individual entities; the description of the relationships between entities; and the methods provided to ask questions about the model and its instances (*i.e.,* querying).

A.A.T. Bui and R.K. Taira (eds.), *Medical Imaging Informatics*,
DOI 10.1007/978-1-4419-0385-3_7, © Springer Science + Business Media, LLC 2010

Spatial Data Models

Observations of the world are intermingled with physical characterizations of the objects around us: the shape and size of an object (geometry), the arrangement (orientation, topology); and the proximity of one object to another (distance, direction) are all spatial properties that color our descriptions. This fact has led to multiple disciplines' investigation of spatial representations and techniques for reasoning with spatial descriptions: a spectrum of applications, including computer vision; geographic information systems (GIS); imaging databases; and natural language processing (for dealing with spatial prepositions) have all dealt with spatial data models. These ideas have been adapted to medicine – and in particular, radiology – to represent structural concepts and coordinate systems, especially when dealing with imaging data: anatomical descriptions and the appearance of findings, for instance, are often given in terms of shape and relative location. GIS are also beginning to be applied to public health questions, for monitoring disease outbreak and surveillance. We start with an overview of general work in spatial modeling, progressing toward more refined anatomical and imaging-based paradigms.

Spatial Representations

Foundational work in spatial and image databases throughout the 1980s and early 1990s provides much of the basis for today's GIS and the implementation of spatial operators in many relational database systems. Central to these efforts was the question of what spatial information needs to be recorded about a given domain (and hence, the data types used to store information). Apart from the issue of dealing with dimensionality (*i.e.*, 2D vs. 3D), [60] defined two key perspectives: the *representation of objects in space*, where each entity is arranged in space and we are interested in the object's own geometric description; and *the representation of space* itself, where each point in space is of interest and thus described. By analogy, consider Fig. 7.1a, where

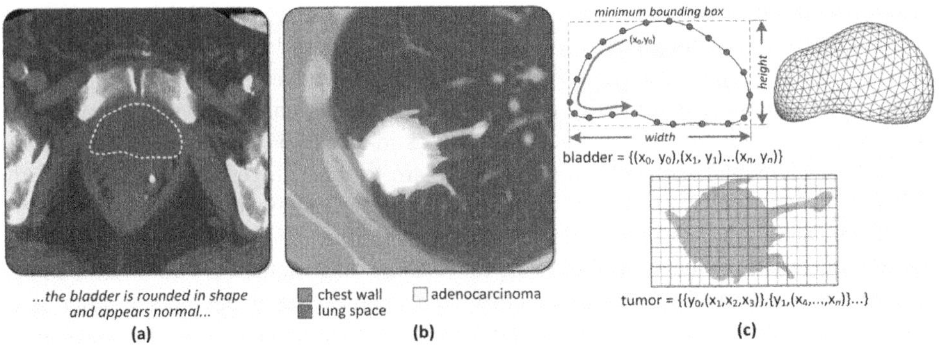

Figure 7.1: **(a)** Computed tomography (CT) image of the abdomen, with the bladder outlined in yellow. Here, the shape over the anatomy is of importance, as opposed to other geometric properties. **(b)** Lung CT showing a tumor. Color overlays separate the chest wall, tumor, and normal lung space. In this case, the entire image is split into three categories, and the relative location and spatial relationships between the entities is of interest. **(c)** Vectors can be used to represent an object's boundaries and its internal region, such as shown with the bladder: a counterclockwise set of points defines a polyline. Geometric properties (*e.g.*, height, width) can also be included in the description. In 3D, a polygonal mesh can be used to represent the shape. Rasters can also be used to represent regions, such as with the tumor.

the intent is to describe the shape of the bladder, while in Fig. 7.1b the aim is to distinguish normal tissue vs. tumor. The former can be captured through the use of points, lines, curves, and areas; the latter typically involves the partitioning of a space into disjoint segments. Detailed reviews of spatial data types and operators are given in [27, 96]; however, we mention two representational issues:

- Spatial data type and location. Related to the question of object vs. space representations is how spatial information is represented, being either raster- or vector-based. *Rasters* impose a discrete grid over the space, thereby setting a fixed spatial resolution to the information. With rasterized data, the grid's cells are the smallest describable element and regions are formed from sets of cells; as such, raster-based systems are well-suited for describing spatial regions. In comparison, *vectors* describe shapes in terms of geometric functions, providing for interpolation across multiple spatial resolutions. The choice of raster vs. vector establishes the underlying spatial data types and a coordinate system. For instance, rasters can be represented using run length encoding schemes, where contiguous spans of a given row are represented by start/end columns. Vector representations are based on ordered sets of connected points, lines (polylines), curves (*e.g.,* b-splines), and basic geometric shapes (*e.g.,* boxes, circles/ovals). Vectors provide compact representations of solid objects, but are less convenient for complex geometries (*e.g.,* regions with holes). Noting that spatial data is sometimes subject to incomplete or vague definitions, fuzzy spatial data types have also been proposed, such that points, lines, and area boundaries are represented with indeterminacy over a range [97]. In 3D, spatial representations also include *polygonal meshes*, which specify the surface boundary of an object through a collection of vertices and edges defining convex polygons (faces) of a polyhedral shape.
- Extra shape and spatial features. Raster and vector-based data types are low-level primitives that are used to construct shapes, and often form the basis for database representations. Spatial modeling applications also use global shape features to present higher-level characterization of geometry, with common descriptors being: area and perimeter (or surface area and volume in 3D), representing the number of pixels enclosed within the shape and its boundary; circularity, or how "compact" the shape appears; minimum bounding box, the extents of the smallest rectangle that can fully enclose the shape; eccentricity, the ratio of the shape's major and minor axes (*i.e.,* the ratio of the height and width of the minimum bounding box); and shape moments. The occurrence of these shape descriptors is exemplified in part by their inclusion within the MPEG-7 (Moving Picture Experts Group) multimedia standard [11]. Also, metrics based on the transformations of an area or boundary to another space (*e.g.,* Fourier, wavelet), have been detailed. Surveys of methods of shape features and analyses are given in [13, 69, 71].

Spatial Relationships and Reasoning

How we represent spatial relationships has been a longstanding inquiry in language, philosophy, and applications involving scene descriptions; [57] provides an early perspective. Spatial relationships can be split into three groups: topological operators; directional; and quantitative relationships. *Topological operators* pertain to spatial relationships between two objects and return a Boolean value. [51] enumerates all possible topological situations for solid shapes in 2D, resulting in eight named operators (Fig. 7.2): disjoint, where two objects are completely separate spatially; meets, in which

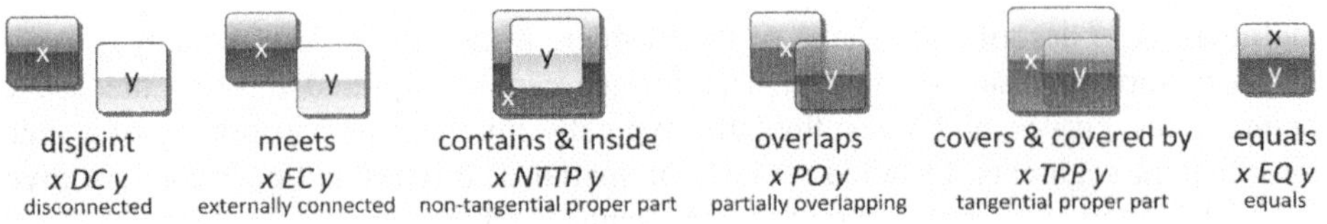

Figure 7.2: The eight basic topological spatial relationships [51]. Similar relationships also form the basis for the region connected calculus (RCC8) [89]: the equivalent relations and names are given in italics.

two objects boundaries touch; contains and inside, wherein one object's boundaries completely encompass a second object's edges (and its converse description); overlaps, where the boundaries of two objects intersect, but there exists regions outside the intersection; covers and covered by, which is similar to contains/inside, but the objects share some edge; and equals, where the two shapes are equal. Analysis of the more general situation of objects with holes resulted in 52 possible binary relationships involving points, lines, and areas. *Directional relationships* describe locations in terms of cardinal directions (*e.g.*, north/south, east/west) or other semantic relations (*e.g.*, above/below; right/left). Lastly, *quantitative (metric) relationships* involve the derivation of some value, such as the distance between two objects.

The above spatial relationships are computable based on the definition of objects' boundaries. A different perspective based on directional relationships is used in *2D string* representations and the use of "symbolic" images [23]. 2D strings encode spatial knowledge about the contents of images through the use of two one-dimensional strings that represent the order of occurrence of codified objects based on projections in the horizontal and vertical axes. Fig. 7.3 demonstrates 2D string encodings for anatomical landmarks. 2D strings have been generalized and extended to 3D, and to different coordinate systems (polar coordinates, variably spaced grids) [22]. Another method for representing spatial relationships can be seen with *scene graphs*, which define a hierarchically organized set of parent/child nodes that decompose an image (*i.e.*, scene) into constituent objects/shapes: each child inherits its parent's coordinate system and specifies its location and boundaries within this scope. Scene graphs are found in 3D computer graphics, VRML (virtual reality modeling language), and are used as the basis for scene descriptions in the MPEG-4 standard [102].

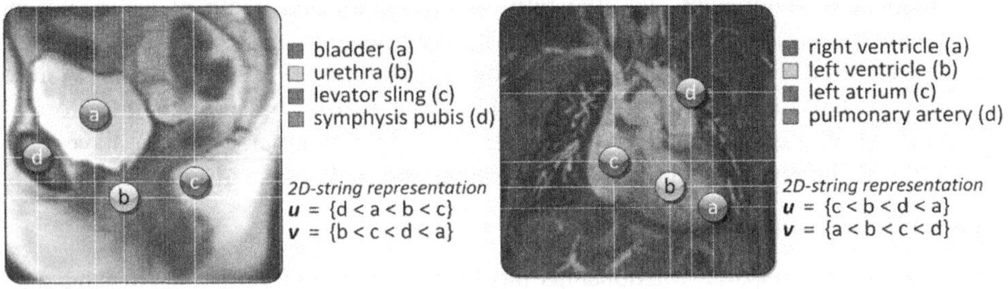

Figure 7.3: Spatial relationships represented using 2D strings. The left shows a mid-sagittal T2 magnetic resonance (MR) pelvic region image obtained in a female patient; the right shows a slice from a coronal T1 cardiac MR series. Two 1D vectors are used to capture information about the relative spatial layout of objects as encountered, in order, along the horizontal and vertical axes.

Spatial queries. In medicine, one might ponder questions such as, "*What brain structures are being impacted because of cerebral edema?*" or, "*From what direction should radiation therapy occur to minimize healthy tissue exposure?*" A significant amount of study has gone into understanding the operators commonly used in spatialqueries. For instance, *geometric operators* involve computational geometry, functioning on two (or more) shapes to obtain a new object. Simple geometric operators include calculating the union, intersection, and difference of the space occupied by two entities. Examples of more complicated geometric operators are calculating the convex hull (the boundary of the smallest convex shape in a plane that encompasses a set of points), a Voronoi diagram, or the shortest distance between two shapes' boundaries. These operators are embedded within query languages, with two approaches seen:

1. Augmenting SQL. Many works have enhanced SQL (structured query language) with spatial data types (points, lines, regions/surfaces) and query syntax [20, 52, 94, 110]. Markedly, many relational databases presently implement the Open Geodata Interchange Standard (OGIS) [62], an extension to SQL3 that adds a base geometry spatial data type with methods for determining topological relationships and spatial analysis. As a case in point, consider the following hypothetical query in SQL3/OGIS syntax to find the anatomical location closest to a tumor based on comparing the distance between segmented image regions:

   ```
   SELECT T.name, A1.name
   FROM Tumor T, Anatomy A1
   WHERE DISTANCE(T.shape, A1.shape) ≤ ALL(SELECT DISTANCE(A2.shape)
     FROM Anatomy A2 WHERE A1.name ≠ A2.name)
   ```

2. Alternative query languages. Another means of spatial query expression includes the adaptation of the form-based query-by-example (QBE) paradigm to manage spatial information in pictures [18, 21]. Moreover, given the nature of spatial relationships, it is often easier to draw a representation of the query rather than expressing it declaratively. Several systems implement visual query languages, which parse spatial relationships from examples drawn by users. Query-by-sketch is one model, where users are able to draw the target objects and the application infers a set of relationships as part of the query constraints [53, 56]. Similarly, iconic representations of objects have been used to enable the sizing and layout of queryable objects [47] (see Chapter 4).

Spatial reasoning. The capacity to automatically reason about objects in space has been delved into via formal spatial algebras and calculi, put forth as logical frameworks for deducing spatial relationships. Methods based in computational geometry are, of course, well-grounded in mathematics and derive their power through quantitative analysis of spatial information. A universal coordinate system is employed to facilitate global comparisons between shapes; and the exact position and extent of objects are known, so distances and other geometric measures can be made. The relativity of spatial references, however, has given rise to the field of *qualitative spatial reasoning* [30, 91] wherein there is qualified and potentially incomplete knowledge about the layout of objects – the full set of spatial relationships must be deduced (*e.g.*, for instance, as with 2D strings). For example, in Fig. 7.4, knowing that b is inside of a, and c is inside of b, one can infer that c is also inside of a. Likewise, conclusions about directions can be drawn in a transitive manner without knowing the precise boundaries of an object. Qualitative spatial reasoning is typified by the region connection calculus (RCC) [89], a topological approach to spatial representation. RCC defines seven

Overlaps $\mathbf{O}(x,y) = \exists z(\mathbf{P}(z,x) \wedge \mathbf{P}(z,y))$

Externally connected $\mathbf{EC}(x,y) = \mathbf{C}(x,y) \wedge \neg\mathbf{O}(x,y)$

Proper part $\mathbf{PP}(x,y) = \mathbf{P}(x,y) \wedge \neg\mathbf{P}(y,x)$

Non-tangential proper part $\mathbf{NTPP}(x,y) = \mathbf{PP}(x,y) \wedge \neg\exists z(\mathbf{EC}(z,x) \wedge \mathbf{EC}(z,y))$

$$\mathrm{NTPP}(a,b) \wedge \mathrm{NTPP}(b,c) \Rightarrow \mathrm{NTPP}(a,c)$$

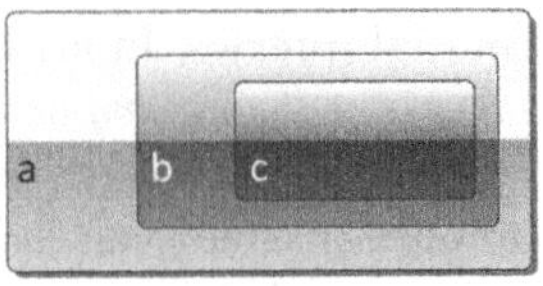

Figure 7.4: Region connection calculus (RCC) theory. Let **P**(x,y) represent the fact that region x is a part of region y; and **C**(x,y) signify that x is connected to y. The topological relations of Fig. 7.2 can be described using logical statements: for instance, overlap and externally connected are defined here. Using RCC, spatial inferences can be made based on the transitivity of relationships and implied meaning. In this basic example, the statement that b is inside a, and c is inside b, implies that c is inside a.

binary relationships on spatial regions (Fig. 7.2), along with the inverses and equality operator. RCC additionally defines Boolean functions for working with regions (sum(x,y), the union of two spaces, x and y; compl(x), the complement of a region; prod(x,y), the intersection of two regions; and diff(x,y), the difference between two regions). A complete listing of inferable relationships (termed the composition table) from a given RCC spatial proposition is defined. Similar work has been proposed with directional relationships, broadened to deal with 3D projections in 2D with in front of/behind semantics: a set of rules dealing with transitivity, symmetry, and other inferences enables deduction of all potential relations [103].

Anatomical and Imaging-based Models

Though the prior discussion illuminates spatial modeling in the context of medicine, the concepts were largely generalized. A considerable body of work exists in terms of providing common frames of spatial reference for human anatomy. The motivation behind these models is to standardize representation such that references to an anatomical entity's geometry and location can be shared. We split our discussion of these works into three parts: 1) *coordinate systems*, which are based on physical landmarks to impose a (conceptual) grid on an anatomical region; 2) *ontological approaches*, which abstract spatial relationships, both in terms of physical and linguistic references; and 3) *shape models*, which draw from geometric and imaging concepts to specify anatomical shapes.

Coordinate systems. The chief impediment in mapping human anatomy lies in the simple fact that the size and shape of every physical body and anatomical region varies from person to person. Furthermore, although we may be able to identify physical landmarks, we are often attempting to assign higher-level physiologic (*i.e.*, functional) concepts to structural areas that may be imprecisely defined. To overcome this issue, a recurrent idea is to define an axis/plane (*e.g.*, given by relatively invariant anatomical points), which serves as a referent origin from which distance can be measured. A well-known example of this approach is the *Talairach* (stereotaxic) *coordinate system* for neuroanatomy. The Talairach coordinate system is based on two points, the anterior and posterior commissures, to define an axis line in the mid-sagittal plane (Fig. 7.5a). The anterior commissure serves as the origin, and 3D coordinates comprise the distance in the x/y/z planes from this point. Comparable landmark-based methods are used for other anatomy (*e.g.*, breast, upper and lower extremities, pelvis), primarily in imaging assessment or biomechanics/kinematic studies [29, 43, 58]. [6] remarks that most spatial references in biology are not in terms of numerical coordinates, but are

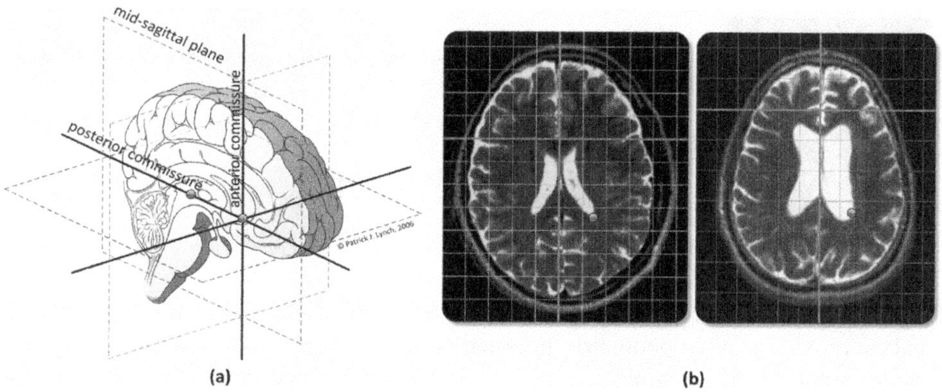

Figure 7.5: **(a)** Talairach coordinate system. The spatial framework is defined by the mid-sagittal plane containing the anterior and posterior commissures. Similar approaches have been used in other anatomical regions. *Brain drawing adapted from Patrick J. Lynch* **(b)** By establishing a shared coordinate system, it is possible to transfer locations from one reference image (*e.g.,* from an atlas) to another image. In this example, the T2 axial MR images are chosen at approximately the same level, and a point on the left lateral ventricle is transferred after applying affine transforms.

directional and suggests instead the idea of domain-specific *natural coordinate systems* based on the orientation and layout of anatomy within an entity.

Introducing a coordinate system opens the door to the development of *morphological imaging atlases* that map physical regions to anatomical labels. In theory, if a coordinate system can be established on a representative anatomical region, then labels from sub-regions may be transferred from one image set to another via the shared coordinate system (Fig. 7.5b). In practice, however, complications occur: the transformation needed to map between two anatomical regions is frequently non-linear, so 3D warping techniques are needed to accurately match volumetric regions; and mapping methods can fail in the presence of gross pathologies (*e.g.,* a tumor that introduces anatomical shift). More sophisticated approaches thus move from the concept of static atlases to *probabilistic atlases* that better represent both the normal variation occurring in a population, and allow flexibility in correlating spatial boundaries (see Chapter 5).

Anatomical shape models. Interrelated with imaging atlases, shape models provide anatomical spatial descriptions based on geometry. Also described in depth in Chapter 5, we only briefly mention these techniques here to show how spatial information is represented. Deformable models cover a broad set of well-known methods that start with a template geometry and iteratively refine (*i.e.,* deform) in response to some force or constraint (*e.g.,* minimizing some energy function) [73] (Fig. 7.6a). Deformable geometries include curves/contours in 2D (e.g., snakes [67]), and surface and solid models in 3D. An important variant to using standard 3D meshes to represent the boundary of deformable surfaces, m-reps align medial atoms along a surface [84], providing a compact data structure for demarcating a (solid) shape (Fig. 7.6b). Active shape and appearance models (ASMs, AAMs) can be seen as an extension of atlases, wherein a composite representation generated from multiple samples is used to statistically represent the degree of variation across a population (*e.g.,* a representative set of images from a group of normal subjects can be co-registered to establish normative spatial distributions) [36] (Fig. 7.6c). Spatial relationships can hence be inferred with ASMs/AAMs by comparing a given subject to the statistical model.

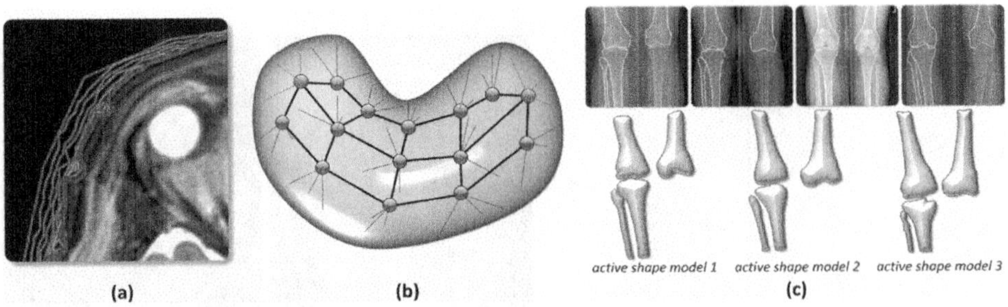

Figure 7.6: Different anatomical shape models. **(a)** Deformable models, such as snakes, provide explicit geometric information about shapes and locations. In this example, a portion of a template (green) is iteratively refined (yellow) to identify the boundary of the head in an MR image (purple). **(b)** An example m-rep, consisting of a 3 x 5 array of medial atoms that specify the surface of the object. **(c)** An active shape model from knee x-rays. 16 x-rays were contoured for the right/left inferior femur and right tibia and fibula (four examples shown on top). The top three eigenshapes are shown below, accounting for the variation seen across the different contours.

Ontological approaches. Coordinate systems and shape models are geared toward imaging and physical models of anatomy. Yet a large portion of our communication of anatomical concepts includes spatial constructs. In addition to standardizing anatomical nomenclature, ontological approaches may be seen as characterizing our conceptual models of anatomy and the language we use to describe spatial relationships. Current ontologies and controlled vocabularies such as UMLS (Unified Medical Language System), SNOMED (Systematized Nomenclature of Medicine), and RadLex codify spatial relations (Table 7.1). Although improving, the coverage proffered by these vocabularies is often incomplete relative to natural language: for example, [7] evaluated spatial references from an anatomical textbook, finding that only a portion of expressed concepts could be mapped to UMLS. Hence, research to improve ontologies falls into two areas of spatial representation, along with methods for reasoning [10]:

1. Topological spatial relationships. Anatomical descriptions can entail a range of spatial relationships, including relative location (*e.g.*, right of the heart), containment (*e.g.*, in the liver), and connectivity/paths (*e.g.*, bronchi branch into bronchioles). For example, based on radiographic findings, an ontology of semantic spatial relationship operators has been created (Fig. 7.7) to provide relaxation/specialization of concepts in queries [25].
2. Mereological relationships. Anatomical models follow an increasingly hierarchical organization (*e.g.*, body → organ systems → organs → constituent tissues and components → cells). Described by is-a and part-of relationships, these associations imply knowledge about the relative spatial properties of anatomical parts [77]. The formal theory of parthood relationships, *mereology*, has been applied to anatomical models. Mereology uses predicate logic to describe parts and their respective wholes; and from this underpinning, spatial reasoning can be performed. The Foundational Model of Anatomy (FMA) [92, 93] best illustrates the mereological approach based on is-a and part-of relations, with the formalization of anatomical spatial entities [74]. UMLS and RadLex provide basic support for mereological relationships as part of their hierarchical structuring. [107] provides a discussion of the semantics of part-of relationships in SNOMED (*e.g.*, the

Ontology/Vocabulary	Spatial descriptors	Examples
Unified Medical Language System (UMLS)	Hierarchical specification of mereological relationships in UMLS semantic network. UMLS Spatial Concept (T082).	physicallyRelatedTo partOf, consistsOf, contains, connectedTo, interconnects
Systematized Nomenclature of Medicine, Clinical Terminology (SNOMED-CT)	Handled by QUALIFIER values for spatial and relational concepts (309825002) and RELATIVE SITES (272424004). Anatomic-specific descriptors are given under anatomical parts/relationships, and general site descriptors.	apical (43674008) left lower segment (264068005) panlobular (263831002) endobronchial (260544000)
RadLex	Mereological relationships for containment, part-of	contains, continuousWith (branch, tributary), memberOf, partOf (segment)
National Cancer Institute Common Data Elements (NCI CDE)	Anatomic site location descriptions (2019174), directional and anatomical descriptors, proximity	

Table 7.1: Examples of how spatial descriptions and relationships are handled in several common ontologies/controlled vocabularies. UMLS and RadLex provide mereological relationships, whereas SNOMED-CT and NCI CDE are principally based on combinations of anatomic-specific descriptors.

mitral valve is part of the heart, which in turn is part of the pulmonary-cardiac system). We note the dual meanings of part-of relationships here: part-of can refer to spatial components (*e.g.*, the left aorta is part of the heart); or it can refer to functionality as part of a system (*e.g.*, the lung is part of the pulmonary system). As a result, the GALEN project defined part-of relationships from several different viewpoints (Fig. 7.8).

Ongoing efforts to unify anatomical ontologies and relationships include the Common Anatomy Reference Ontology (CARO) [61], the Relations Ontology for mereological relationships [104], and the related Spatial Ontology – all hosted as part of the Open

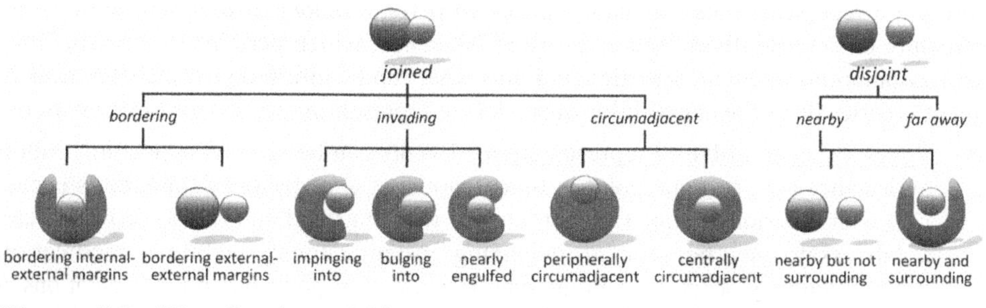

Figure 7.7: Visualization of binary semantic spatial relationships based on the description of radiological findings, organized as an ontology [25].

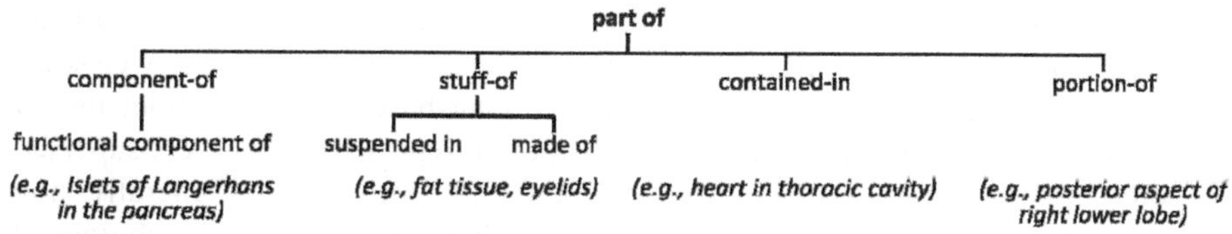

Figure 7.8: The meaning of part-of relationships is expanded in the GALEN project to reflect the complexity of its semantics and usage.

Biological Ontologies (OBO) project. Adopting the top-level classes of the FMA, CARO endeavors to provide a single abstraction for spatial references to anatomy (and temporal changes) across differing levels of spatial granularity (*e.g.,* cellular vs. tissue vs. organism) and species. The Spatial Ontology defines anatomical locations (axes, sides, regions) and relative spatial relationships (*e.g.,* oppositeTo, orthogonalTo).

Temporal Data Models

The world is continuously changing, and as such time is an intrinsic part of the observational process when we record data: knowing *when* events occurred helps to impose order on a sequence of events and to elucidate trends. Furthermore, modeling time allows us to deduce causal relationships: *how* things happen (*i.e.,* behavior) is just as important as the end state. This constant flux is particularly true in considering any biological process, be it a normal condition (*e.g.,* human growth from childhood to adulthood) or a medical problem (*e.g.,* the genesis of disease, its subsequent course, and its resolution in a patient). Indeed, the longitudinal (electronic) medical record reflects this view, aggregating information that documents the evolution of an individual over his lifetime. One of the earliest incarnations of clinical databases, the Time-oriented Database (TOD) System supported basic querying of timestamped data [113]. Today, research across several different fields including temporal databases, multimedia formats, and artificial intelligence has resulted in formalized time-based constructs that have been adapted to clinical data and medical knowledge representation. Temporal modeling is central to a spectrum of topics, including the organization of data within medical information systems; prediction and therapy planning; and automated reasoning with clinical guidelines to name a few [99]. A discussion of these temporal models consists of two distinct issues: the model abstractions that impose a chronology (or other temporal pattern) on the data; and the methods that are used to query and to reason with temporal data.

Representing Time

Temporal representations are made up of two parts: a *theory of time* that defines the structure of the primitives; and a *theory of incidence*, which pertains to the properties and assumptions about an assertion and thus what can be inferred [109]. Differences in these theories drive the capabilities of models in representing time-based information:

- <u>Point- vs. interval-based representations</u>. Events can be seen as happening within an instance (*i.e.,* a single point in time), capturing a snapshot of state; or as occurring over a period of time (*i.e.,* an interval with start and end times). To illustrate, a lab assay may be seen as occurring instantaneously, while a medication prescription may occur over an interval of weeks. For point-based representations, a single value is associated with that instance in time. In comparison, interval-based models can assign a single value for the entire duration, or a range of values (*e.g.,* either explicitly stated or through a function). Both models require quantization of temporal scale, the former using the smallest quanta as the basis for all timestamps, and the latter coalescing one or more quanta into a set. The well-known Arden syntax [64], for instance, employs a point-based representation: intervals are not supported as primitives. Notably, interval representations are a superset of point-based models (by making the start and end times equal). Generally, point-based time models are easier to manage, providing straightforward techniques for sorting and indexing of temporal data – but at the cost of representational power.

- Continuous vs. discrete time models. A dichotomy exists with regard to the quantization of time: a model can represent temporal information either in terms of *discrete* values, in which the associated state for an event is only known for a given point in time with no assumptions about an object's state outside of the explicit time points; or *continuous*, wherein the assertion is that an event and its state occur over a period of time without interruption[1]. For example, consider data from an intensive care unit (ICU) with information collected every 15 minutes: a discrete time model emphasizes that our knowledge between the 15 minute intervals is not truly known, whereas a continuous model provides a method to assume or to compute values between the intervals. Typically, continuous models provide a means of *interpolation* so that values are estimable (although such methods must clearly be domain dependent).
- Supporting temporal abstractions and imprecision. As with spatial representations, time is often referred to in abstract or relative terms that are not directly amenable to point- or interval-based timestamp primitives. For instance, a patient's statement that his symptoms, "*started sometime today,*" or, "*recently*" are imprecise references that can only be framed loosely within a certain period of time or in a qualified manner. In some cases, such temporal abstractions can be defined easily within the scope of a query [88]. More sophisticated scenarios involve a multi-step approach to ascertain the contextual task for a temporal concept and its linkage to lower-level and similar temporal constructs: the RÉSUMÉ project develops this idea in terms of *knowledge-based temporal abstraction* [98, 101]. Other models instead relax absolute time constraints so that relativity between events is allowed (as opposed to fixing events to an absolute time scale, only the ordering of events is represented or a relative event is anchored to another even for which a specific timestamp is given). Clinical protocols and guidelines that are action- or procedure-based represent chronologies that can be perceived as sets of relative events (*e.g.,* do X before Y). Similarly, fuzzy sets and fuzzy logic have been proposed to represent an interval during which a given event definitively occurs, but the exact start and/or end points are unclear [2, 50]. A fuzzy theory approach, for example, has been used to model the evolution of diseases given imprecise temporal information [82]; and has also been used to extend Arden syntax [108].
- Valid and logical time. There are several dimensions of time that characterize what is known about an event. *Valid time* symbolizes the period of time for which a given event exists in the real-world *(i.e.,* a specific date of timestamp). Valid time marks the behavior of the real world object with respect to time *(e.g.,* Medication A was taken by the patient between January and March of 1998), and is often used to model the semantics of time series data sets. *Logical time* denotes when information about an event is actually recorded. For instance, logical time only deals with the time at which data is committed to the database *(e.g.,* the fact that Medication A was taken was recorded in a database on August 12, 1998). Logical time is hence also often referred to as *transaction time*. Temporal databases that model both valid and logical time are said to be *bitemporal*.

The reader is referred to [106] for further discussion of these and additional temporal representational issues, including historic perspective. The basis for most temporal

[1] Ultimately, one can argue that all computerized temporal models are discrete, even with hierarchical time models, as they reach a unit of time that can no longer be subdivided or interpolated given the limitations of precision and representation.

models is a *linear* representation of time – that is, an imposed partial ordering of a set of events. Time series are a well-known linear representation, where a sequence of observations is made over a finite series of time points. In linear time models, each moment takes on only one possible value. In contrast, *branching* time models represent multiple hypothetical timelines where events may (or may not) occur based on transition probabilities: a time point may thus be split across many possible states (*e.g.,* such as in a finite state automata, Petri nets, or Markov model). The result of generalized branching models is a directed graph of possible event sequences [42]. Notably, in both linear and branching time models, temporal ordering entails causes preceding effect. Many refinements to the basic linear temporal model have been developed to enable richer representations; we mention two examples pertinent to medicine: *temporal evolutionary models*; and *cyclic models*.

Temporal evolutionary models. Most temporal models represent an object's initial state and its changes over a given temporal interval: the values may change, but the properties attributed to an object are constant. Yet clear examples exist within medicine wherein the nature of an object evolves, potentially into one or more new entities with fundamentally different properties. For example, normal human skeletal maturation involves the joining of bones (*e.g.,* the epiphysis and phalange merge together): two previously distinct objects come together to create a new entity. Embryologic development illustrates the process of objects splitting with differentiation of tissues: for instance, the cephalad region of the urogential sinus forms the bladder, and the caudal portion the urethra; and growing from the three- to five-vesicle stage, the prosencephalon gives rise to the telecephalon and diencephalon in the neural tube. A *temporal evolutionary data model* (TEDM) is a spatiotemporal framework that explicitly models the *fusion, fission,* and *evolution* of entities in an object-oriented model (Fig. 7.9) [26][2]. Fusion represents the aggregation of two (or more) previously distinct objects fusing together to create a completely new object[3]. Fission is the converse operation, where a single object divides into two or more objects. Evolution accounts for transformation

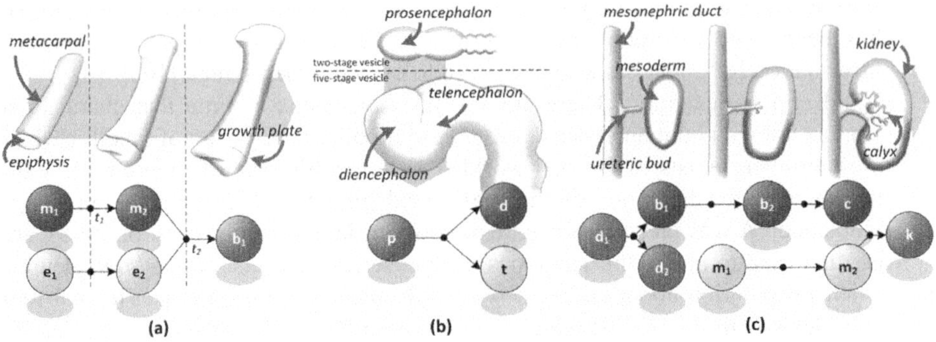

Figure 7.9: Examples of the temporal evolutionary data model. Three temporal constructs are supported: fusion, the joining of two entities into a new object; fission, the

[2] Many ontologies, such as based on OBO, instead use the singular concept of developsFrom to model entity changes/evolution.

[3] Fusion should not be confused with the modeling concept of *aggregation*. This latter concept reflects a larger object composed of a group of smaller objects – however, each constituent element maintains its own identity. In fusion, the participant fused objects cease to exist once joined, with the entity taking their place.

splitting of an object into two new entities; and evolution, as an object's properties change over time. **(a)** Evolution and fusion with bone growth. The pediatric epiphysis (e) and metacarpal (m) slowly change size and shape ($e_1 \rightarrow e_2$, $m_1 \rightarrow m_2$) and fuse together at time t_2 to create the adult bone (b_1). **(b)** In embryology, the prosencephalon in the neural tube divides into the telencephalon and diencephalon ($p \rightarrow d + t$). **(c)** A more complex scenario of fusion, fission, and evolution in the growth of a kidney (k) from the ureteric bud (b) and mesoderm (m).

of an object's properties across conceptual stages. TEDM formalizes the lifespan of an entity, from genesis through to the end of its existence. Although the original TEDM is predicated upon spatial proximity for interaction between entities, it is possible to abstract these concepts further in representing medical phenomena: by way of illustration, the etiology of a cancer, the development of highly differentiated tumor cells, and the progression to necrosis can all be modeled in light of a TEDM. A similar set of evolutionary operators to TEDM are also suggested in [40].

Cyclic models. Biological processes and clinical practice often make reference to repeating patterns of behavior. A cell reproduces according to a cycle defined by stages including mitosis; a dosing schedule for a drug may be specified as twice daily; and a clinical guideline may advise periodic investigations – these examples exemplify repetitious actions. *Cyclic* or *circular* models encapsulate these repeating patterns, providing a succinct representation of a set of events. Various schemes have been proposed to describe cycles, including active rule languages that permit the descriptive composition of complex patterns evolving over time [75]; temporal constraints [4, 19]; and graphical models. Largely, cyclic models focus on characterizing intervals of activity punctuated by periods of inactivity that are qualitatively defined through temporal relationships or quantitatively given by constraints. Different systems have explored the use of cycles and patterns within clinical data [81, 86].

Applying multimedia models to clinical data. Early effort investigating temporal representations form the basis for today's multimedia file formats. Popular standards such as QuickTime, MPEG-4, and Windows Media are based on the chronological ordering of data elements to present a synchronized presentation of auditory and visual information. Likewise, the Synchronized Multimedia Integration Language (SMIL) is an eXtensible Markup Language (XML) syntax for interactive multimedia presentations that also embeds temporal sequences and synchronization [116]. Although different terminology is used, these formats are largely based on the idea of ***streams***. A stream models data based on the sequential nature of constituent elements ordered on the basis of time, providing a structure for expressing the temporal relationships between objects [59]. Stream-based data models abstract the physical organization of raw data while still allowing flexibility in accessing and presenting multimedia information; as such, a stream is itself a chronological entity with its own attributes and relationships. [37] provided the first adaptation of streams to model clinical data; and [45] presents an extended entity-relationship (ER) model and visual notation for streams that are demonstrated in medical applications [1, 46]. [15, 16] extends this work with further constructs for manipulating generalized streams (Fig. 7.10), and [33] provides similar means via an XML syntax for streams:

- Substreams. Often, a stream can be divided into smaller components based on temporal or logical constraints, much as a view is defined in a database. For example, a video stream represents a set of images (the elements) shown at a given frequency (*e.g.*, 30 images/second), and different video segments can be identified (*e.g.*, based on time indices, content, etc.). Similarly, a stream representing a patient's clinical documents can be decomposed based on the report's originating department (*e.g.*, radiology, pathology, surgery, etc.). A *substream* is based on the use of a query condition on the stream's elements to filter a temporal sequence: this constraint is termed a *filter attribute*. A filter attribute falls into one of four categories: *temporal*, which operates on the stream element's temporal attributes; *element attribute*, where the constraint is declared on a defined attribute of the stream element class type; *derived*, which represents a qualification upon a derived attribute of the stream element; and *complex*, in which two or more of the prior three constraint types are combined using Boolean operations. Substreams are considered proper streams and can thus be further subdivided, resulting in a directed acyclic graph of filtered and combined streams.

- Aggregated streams. An *aggregated stream* combines the contents of two or more streams into a single temporal entity. Different types of merging are supported by the definition of *aggregating stream functions*, which provide a mapping between the input streams and the joined representation. For example, a union-aggregating stream function combines all stream elements irrespective of timestamps; whereas an intersection-aggregating stream function would represent the aggregation of streams for which timestamps are shared.
- Derived streams. *Derived streams* associate a function with one or more streams to generate a new data stream. An example derived stream is the application of an image processing algorithm to an image/video series to segment key regions.

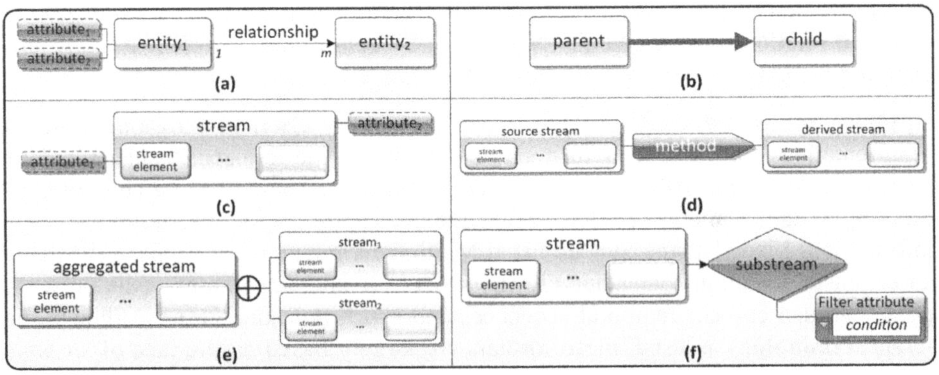

Figure 7.10: The M2 visual notation for multimedia streams. An object-oriented paradigm is used in an extended ER model to represent temporal information. **(a)** Iconic representation for *entities* (a yellow box). Entities can have *attributes* that describe the object, and can participate in named *relationships* with other entities. **(b)** Being object-oriented, the M2 model supports *inheritance* (is-a) relationships, denoted by a thick purple arrow. **(c)** Informally, a *stream* is defined as a temporally ordered, finite sequence of entities; these entities are called *elements* of the stream. Streams are entities, and can thus have their own attributes and relationships. **(d)** *Derived streams* apply a method to the stream's elements to generate new data. **(e)** *Aggregated streams* are the result of amalgamating two or more temporal sequences together. **(f)** *Substreams* provide a logical partitioning of a stream based on filter attribute constraints.

Temporal Relationships and Reasoning

Implicit to the above discussion of temporal representations is the set of relationships between two or more events. Allen's thirteen temporal interval relationships are perhaps the most cited (Fig. 7.11), defining a seminal set of six invertible binary operations (before, during, meets, overlaps, starts, finishes) and a comparator for equality [3]. In establishing relativity between intervals, logical constraints can automatically be deduced (*e.g.*, transitivity of event relationships). [17] extends these relations to enable comparison and reasoning on temporal cycles.

These basic relations are adopted in different medical representations. The Arden syntax, along with the concept of now to model current time, supports several of these temporal relationships (after; before; ago, which subtracts a period of time from now; from, which adds a period of time from a given time point). GLIF3 (Guideline Language Interchange Format) [12] supports further capabilities as part of the GLIF Expression Language (GEL), such as within (equivalent to Allen's during operation) and interval comparators. [70] describes additional ways of characterizing causal-temporal relationships in support of clinical reasoning: immediacy, when Event Y immediately follows Event X (A meets B in Allen's interval relationships) and implying X causes Y; delayed, similar to immediacy but with a determinable period of time between two events (*i.e.*, X causes Y but after a defined passage of time); progressive, such that once a phenomena starts, the trend continues (*e.g.*, a worsening condition); accumulative, where an event only occurs if a given phenomena takes place over an extended period of time (*e.g.*, accrual of urine in the renal pelvis due to an adynamic segment, ultimately causing hydronephrosis); intermittent, when a causal relationship is known, but with irregular frequency; and corrective, where Event X creates a given normal state. Work has also been done to uncover temporal relationships semantically expressed in the text of clinical reports [34, 65]. Based on a review of discharge summaries, [118] describes medical natural language processing to derive time-based relations in terms of a *temporal constraint structure*, comprising: an event description; event points (*e.g.*, start, end) and anchor points (*i.e.*, a fixed point in time); relations between event points (*e.g.*, within, before); interval operators; and fuzziness.

Temporal querying. Research in the 1990s on relational temporal databases led to the consensus creation of the TSQL2 (Temporal Structured Query Language) specification [105], an adaptation of the well-known SQL standard to handle temporal querying. TSQL2 supported queries involving valid and transaction time with methods for casting between timestamps represented at different temporal granularities and period representations. To illustrate, consider a simple table in a patient's medical record that captures the prescription drugs the individual has been taking. The query, "*What drugs has the patient taken for more than three months*," might be specified in TSQL2 as:

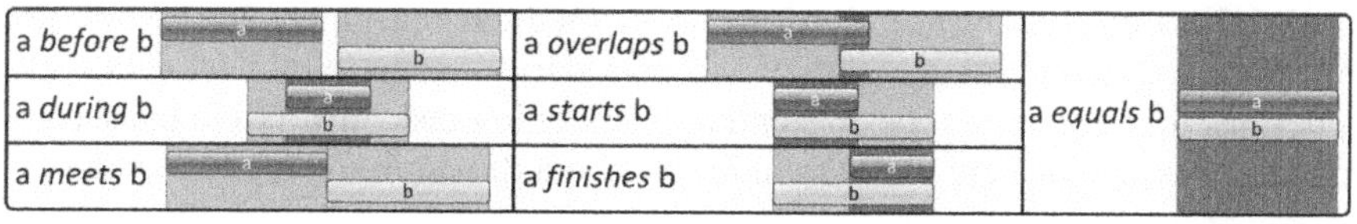

Figure 7.11: Graphical representation of Allen's interval temporal relationships [3]. Gray regions specify the duration of a given temporal interval; darker regions occur when the compared temporal regions overlap. The left and middle columns define binary operations that are invertible.

```
SELECT Drug, Dosage
FROM Prescription(Drug, Dosage) AS P
WHERE CAST(VALID(P) AS INTERVAL MONTH) > INTERVAL '3' MONTH
```

Those familiar with SQL will note the additional syntax added by TSQL2 to handle temporal information. Unfortunately, although TSQL2 was the basis for a formal extension to the core SQL3 effort, it was not fully ratified and to date has not been implemented in a widespread fashion[4]. Despite this issue, the principles set out by TSQL2 are useful and it is considered by many the *de facto* querying model: many of the attributes and temporal relationships described prior can be expressed using TSQL2 queries. Building from the TSQL2/SQL3 undertaking, [32] recently proposed a new temporal query language, T4SQL, to handle additional (user-defined) temporal dimensions and temporal grouping semantics.

For the most part, clinically-oriented databases providing valid-time temporal querying either implement a subset of TSQL2 or have support for Allen's interval operators. Several projects have explored temporal querying of patient data: Chronus is a temporal database mediator that supports the transformation of time-based queries, helping solve issues of temporal granularity and ambiguity [38, 81]; point-based timestamps in medical records are extended to queries with interval operators [78] and temporal comparators [49]; and an entity-attribute-value model for clinical trials is supplemented with interval operators [44]. KNAVE II also provides a rich environment for the expression and exploration of temporal information [72, 100].

Reasoning with time. The formalization of temporal representations and relationships allows one to reason with the information. [117] provides a thorough review of temporal reasoning issues in medicine, suggesting three categories for classifying works in the area. First are the base applications of models stemming from the field of artificial intelligence, including logic-based representations of temporal events (*e.g.*, Allen's interval operators); probabilistic models; and graphical- and constraint-based representations of temporal events. Many of these representations are coupled with formal algebras or mechanisms for automated reasoning: by bridging clinical data with these systems, the intent is to provide a formal environment for analysis. For example, *situation calculus* is a logical formalism for representing and reasoning about dynamic situations described in terms of actions, *fluents* (a condition whose truth state may change), and situations. In reasoning about actions, fluents can be represented in terms of first-order logic by predicates that are dependent on a time argument. *Event calculus* is another logical paradigm based on actions and fluents, defining several axioms to assert the validity of a statement at a point in time. [106] also demonstrates how a Petri net representation can be transformed into predicate logic. The second category of medically-oriented temporal reasoning comprises those frameworks that are driven by the needs of clinical applications, and include clinical temporal databases, temporal abstraction tasks, and data visualization. Lastly, the third group of medical reasoning endeavors aims to resolve issues of temporal uncertainty and granularity innate to clinical data. Notably, [35] defines four common types of temporal reasoning tasks that fall in the second and third categories, and that are useful for thinking about decision-making in medicine:

[4] Most of today's relational database management systems offer some support for temporal variables, having proprietary calendar and timestamp operators. However, the power of TSQL2 is arguably missing in the majority of these implementations.

1. Projection. *Projection* involves computing the likely future consequence of some set of current conditions and an action. In this situation, the prediction is general in nature. For instance, the statement, "Giving medication *X* will decrease the patient's blood pressure over the course of a week," illustrates a projection.
2. Forecasting. More specific than a projection, *forecasting* attempts to provide an exact calculation of some property in the future, given current information and an intended action. Thus, "Giving 50 mg metoprolol b.i.d. will decrease the patient's blood pressure from 130/90 to 110/70 mmHg in two weeks," involves the forecasted assertion that the individual's blood pressure will be an explicit value after receiving the beta-blocker.
3. Planning. Rather than the singular cause-effect prediction seen with projection and forecasting, *planning* involves producing a chain of events that will realize a desired objective. For instance, recognizing that for the goal of Event Z to occur, Event Y must happen, and for Event Y to take place, Event X must also occur, a planning task infers that Event X → Event Y → Event Z. Simple planning may entail a serial sequence of projections, but more complex scenarios can include concurrent events to achieve a goal.
4. Interpretation. Finally, the task of interpretation concerns the abstraction of temporal data into trends and the discovery of patterns comprising higher-level concepts. As an example, a sudden and rapidly increasing creatinine level may indicate acute renal failure. This reasoning task includes finding temporal associations between events that may further signify causality.

Some Open Issues in Temporal Modeling

Temporal representations and reasoning are still active areas of research with many unresolved and open questions [2, 5, 83]. Eluded to in prior sections and chapters, we summarize three outstanding issues here.

Temporal mismatches and granularity. A recurrent theme that arises in dealing with clinical data is semantic heterogeneity. In this case, temporal information from different data sources may be recorded at varying levels of granularity (*e.g.*, a lab with a minute-level accurate timestamp vs. a drug prescription with a date reference). One solution is to create a "universal" timeframe and to map the multitude of encountered temporal models to this one representation, thereby allowing comparisons to take place [39]. TSQL2's casting mechanism also aids in this process. The problem becomes more complicated when considering the gap between semantic-level abstractions and implicit domain knowledge a user may have and the timestamp representations used within databases. Present approaches draw on *temporal ontologies* to establish the appropriate mappings: [80] exploits semantic web constructs (*e.g.*, web ontology language, OWL) in an application with temporal constraints defined by clinical trials, noting the disconnect between high-level trial guidelines that can be ambiguous when attempting to validate low-level temporal constraints; [112] propose a generalized temporal ontology for tasks, also in the context of clinical trial protocols.

Temporal synchronization. Differing temporal granularities present a secondary issue with respect to the synchronization of data sources and the potential loss of precision when combining information. For instance, consider two streams of ICU data: a patient's blood pressure, sampled every ten minutes; and the same patient's blood oxygen saturation levels (SpO_2), captured every minute. If the streams are combined to show information at the lower frequency (*i.e.*, ten minutes), then how should SpO_2 be statistically (and meaningfully) collapsed for the given interval (*e.g.*, average, median, min/max)? [76] outlines a basic algorithm for joining two streams together via *temporal scaling* (a projection of the desired time span), *concatenation* (linking stream data with a set of unique timestamps for the target duration), and *alignment* (selecting the desired temporal granularity) followed by the application of statistical aggregates. Ultimately, the semantics of coalescing temporal sequences together must be context-sensitive.

Temporal similarity. There remains significant challenge in defining metrics to compare the degree of similarity between two temporal sequences. Consider matching one pattern (c-d-e-f) to other sequences (*e.g.*, c-c-d-e-e-f, c-d-d-e, c-c-c-d-f, d-d-e-f-f, c-d-g-e-f, etc.), which may differ because of non-linear scaling along the time axis, or the insertion/deletion of an element: how does one quantify the amount of overlap between patterns? Two classes of methods are presently used [87]: *dynamic time warping* (DTW) algorithms [9] and *transformation-based methods*. DTW finds the optimal alignment between two time series by finding the minimal set of stretches/shrinkages of event subsequences (*i.e.*, warps) to convert one sequence into the other. Transformation-based methods entail the computation of a function on the time series to approximate the temporal pattern (*e.g.*, a Fourier transform); comparisons are then made in this space. Debatably, these approaches work for smaller temporal sequences (*e.g.*, comparing electrocardiograms); it is unclear how they will perform over larger clinical datasets and can be adapted to account for clinical temporal abstractions. In comparison, [66] puts forward a technique that uses a temporal constraint network such that the relationships between nodes represent uncertainty in sequence similarity (similarity is thus a measure over the overall graph): [31] demonstrates this method to compare clinical guideline tasks.

Clinically-oriented Views

The impetus behind clinically-oriented data models is to support physician cognition and workflow, facilitating information retrieval tasks. The organization of patient medical records – both traditional paper chart and electronic medical record (EMR) – is surely a reflection of these models. We often see the record as just being a repository of information on a patient. An alternate view suggests that rather than being a passive reflection of work performed, the medical record is an active tool that aids in memory and communication [8]. The difference is perhaps subtle but significant: in the latter, the medical record and its formulation influence the healthcare process, structuring thinking, interaction, and decision-making. Two methodologies shape current thinking about the medical record:

1. Source-oriented views. Historically, patient paper charts were organized by the origin or type of data (*e.g.*, laboratory, imaging, physician notes, physical exams, etc.), with each section chronologically ordered. Under ideal circumstances, a single physician is responsible for the management of a given individual, and is thus aware of the diverse problems faced by the individual, integrating and distilling the information gathered from each source into appropriate, documented actions. This source-oriented perspective provides a logical indexing scheme for adding and finding recent information. Many of today's EMR implementations follow the source-oriented model, especially to handle the assemblage of data from distributed databases (*e.g.*, hospital and lab information systems, HL7; see Chapter 3).

2. Problem-oriented medical record. Commenting on the lack of scientific rigor in performing clinical diagnosis and treatment, Weed introduced the idea of a *problem-oriented medical record* (POMR) [111]. With increasing sub-specialization of medical care and sophistication in disease management, the "single physician" caring for a patient is arguably less common. The context for clinical observations is often lost (*e.g.*, the reason why a given test was ordered; the interpretation of a test result in light of other evidence); and today's increase in chronic disease often results in individuals with a range of medical issues. Moreover, source-oriented medical records rely upon the clinician to maintain a mental picture of the patient's health status and history; and also to correctly remember what information must be referenced with respect to a given medical condition. Thus, the POMR advocates that all collected patient information be organized around a running list of ongoing and resolved patient issues (*i.e.*, the *medical problem list*). Often associated with the POMR is the use of the structured *SOAP note* (subjective, objective, assessment, and plan) used today for clinical documentation. The *subjective* component describes the patient's presentation (*e.g.*, pertinent positive/negative symptoms, medical, family, and social history). The *objective* section reviews labs, imaging, vital signs, and physical exam findings. *Assessment* summarizes the subjective/objective portions, resulting in a differential diagnosis (or definitive diagnosis): notably, the assessment is meant to be referent to the entries of the medical problem list. Finally, the *plan* covers the physician's actions with respect to the (differential) diagnosis and the patient. Purported advantages of the POMR include better continuity of care (as it becomes easier for a given physician to understand the state of the patient by reconstructing the history of a given problem); and improved context for understanding why a given observation was made.

Early criticisms of the POMR included [54, 55]: 1) the perceived increase in physician work to appropriately classify patient data; 2) the inability for data to be readily included in multiple problems (and hence the redundancy of recorded data); 3) the absence of standardized vocabulary for describing medical problems; 4) the lack of critical thinking and clinical judgment about clinical observations (as fostered by source-oriented views); and 5) the resultant lack of integration of views across medical problems. However, many of these issues have been at the heart of medical informatics research for the past three decades, with strides made in ontologies (*e.g.*, UMLS) and automated methods to help with the maintenance of the medical problem list and organization of data within the EMR across medical problems. Although debate continues as to the merits of the POMR, especially given that clinicians are largely indoctrinated to source-oriented views [85, 95], there remains positive proof that problem-based data organization expedites the information search patterns of the physician and can reduce the time and effort necessary to grasp complicated, detail-intensive abstractions [41, 79].

Somewhat in contrast to the definitive nature of the presentation in both source and problem-oriented medical records, [68] submits that the medical record is best perceived as a clinical narrative constructed by a physician (*i.e.*, a story told by the clinician). The medical record is seen as a book, and they compare the diagnostic action of the physician to the interpretive action of the literary critic who approaches text from a particular perspective and with a particular critical or interpretive purpose. They emphasize that the medical record, as written by a physician, is like a critical view of a literary piece: it is always from a particular perspective, conditioned by the conceptual framework of the clinician and therefore, not truly objective in any absolute sense.

Alternative Views and Application Domains

Both the source-oriented and POMR organization of the medical record are geared toward the physician in clinical practice. Yet the role of patient records now has expanded secondary uses, each with different organizational requirements:

- Personal medical records. There is a progressive empowerment of individuals to access their own medical records (see Chapter 3). While the patient is increasingly savvy and informed with respect to his own healthcare, it is evident that the physician-centric organization of information in the EMR is ill-matched to the needs of the layperson. [115] cites some dissimilarities between patients and physicians, including the former's absence of objectivity in interpreting the record; the need for additional learning sources to contextualize and/or simplify results; and the potential for distinctive access patterns originating from a differing task model (see Chapter 4). The non-physician will fundamentally perceive medical problems differently from a clinician (*e.g.,* as a constellation of symptoms, more holistically). Inclusion of the patient's perspective is thus encouraged in a *patient-centered medical record* (PCMR) [48], which replaces POMR's SOAP with HOAP (history, observations, assessment, and plan) to better capture the individual's presentation of concerns and understanding of the medical problem.
- Clinical trials. EMRs provide an opportunity to capitalize on the organization and structured representation of clinically-derived data, repurposed for research. Leading several aspects of the data standards and models for clinical trials and its integration with the EMR is the Clinical Data Interchange Standards Consortium (CDISC) [28]. The core data model is the *study data tabulation model* (SDTM), which defines a table of timestamped *observations*. Here, an observation is given by discrete pieces of information collected over the course of the trial study. Observations are defined by sets of variables, being one of five types: identifiers (*e.g.,* Subject *X*), topics (*e.g.,* nausea), qualifiers (*e.g.,* mild), timing (*e.g.,* Day 5), and rule-based variables (used for looping conditions, such as the number of times to repeat a step in the study design). Collections of observations can be associated with a given domain, broadly categorized as belonging to interventions, events, or findings. For instance, the observation that, "*Subject X experienced syncope on the tenth day of the study,*" can be connected to the adverse events domain. A consensus-based common set of variables and definitions across different domains is given by the *clinical data acquisition standards harmonization* (CDASH) standard. SDTM is used to submit clinical trials datasets to the United States' Federal Drug Administration (FDA) as part of the approval process. Associated with SDTM is the *operational data model* (ODM) framework that defines a standard XML schema for the interchange and archiving of clinical trials data. ODM includes clinical data from case report forms as well as associated metadata, administrative, reference and audit data. Working with the FDA, HL7, and the National Institutes of Health (NIH) National Cancer Institute (NCI), CDISC is collaborating on the *Biomedical Integrated Research Domain Group* model, which aims to establish a shared representation of the semantics of clinical and pre-clinical protocol-driven research.

Discussion and Applications

Ultimately, a data model represents an information system's conceptual view of the relationships between data. The primary question that drives the design and construction rules of a data model is: *how will it be used*? Clearly, a clinically-oriented data model should assist physicians in medical problem-solving tasks such as:

- What is the underlying medical problem for a given patient?
- What are the best explanations and the evidence for the cause of the problem?
- What are the manifestations (*i.e.*, effects) of this underlying problem?
- What is the extent and seriousness of the problem?
- What are the behaviors of the findings associated with the problem?
- What is the effect of various interventions?

These are core questions any clinician asks on a regular basis. The investigative quality of the questions above is mirrored in the fact that the practice of medicine has long sought to follow scientific methods. As a result, today's diagnostic process can be seen to parallel aspects of a scientific study. Likewise, treatment is akin to a titration experiment, where some intervention (drugs, surgery, etc.) is iteratively applied and assessed to reach some measurable (therapeutic) goal. Indeed, the push for *evidence-based medicine* (EBM) promotes the scientific method in medicine, emphasizing the need for objective measures and the use of validated techniques (*e.g.*, randomized clinical trials) to support decisions. Partly because of this conceit, data models for organizing clinical observations have focused on a declarative framework for collecting assertions: findings, diagnoses, and treatments are provided as sets of facts. What is absent in these models is the context for understanding the interpretation of the findings and the trail of reasoning behind decisions. Current data models are *explanative*, capturing what has been discovered, without provisions for the *investigative* aspects of the healthcare process. Science is exploratory, going from the unknown to the known; to follow its principles, the medical record – and its organization – should act as a scientific notebook documenting the paths of investigation, allowing the reader to fully reconstruct the diagnostic/therapeutic course [111]. To this end, we conclude this chapter by introducing the idea of a *phenomenon-centric data model* (PCDM).

A Phenomenon-centric View: Supporting Investigation

To devise the PCDM we first consider (from albeit a high level) what constitutes a scientific experiment. At the beginning of any scientific inquiry is some *phenomenon* that we wish to explicate. We may have some initial observations about this unknown phenomenon – call it P – and several questions naturally surface. What is P? Why is P happening? Will P change, and if so, when and how? P is thus the center of investigation, and experiments are designed to describe P in richer detail. Based on the observations from the experiment, theories are formulated to explain the occurrence of P, elevating our level of understanding by formally linking together the different observations through a possible (root) cause. Such theories may additionally make use of other experiments/theories in order to provide support for reasoning. Past theories related to the phenomenon may conflict with the new information, in which case they are refuted and replaced with a new theory that better explains the observations. Finally, based on extrapolation of the observations and/or past experience with similar phenomenon, the course of P is predicted.

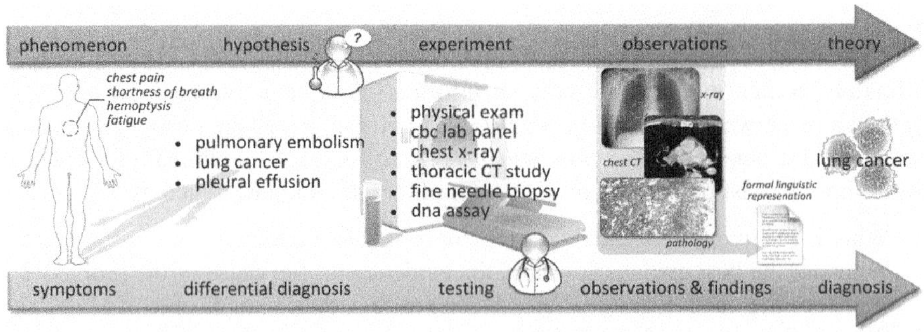

Figure 7.12: The medical exploratory process (bottom arrow) following the scientific method (top arrow). A patient's presentation results in an initial differential diagnosis to explain the symptoms. Testing is performed to rule in/out a given diagnosis until a degree of confidence is reached based on the observations and findings. Note that the transition from left to right moves from the physical world to progressively more abstract concepts.

Applying the systematic process of an experiment to medicine (Fig. 7.12), we construe the phenomenon as the initial presentation of a patient: his set of symptoms and complaints are what we try to explain. In some cases, the phenomenon is immediately recognizable with a high degree of certainty (*e.g.*, a common cold) and so testing is not required. In other situations, the symptoms may derive from a number of equally viable etiologies: a physician treating the patient thus formulates a differential diagnosis and selects tests (labs, imaging, pathology, etc.) to rule out each possible cause. Here, the differential diagnosis is analogous to a hypothesis. Each test gives physical observations that measure properties of the phenomenon; and each type of observation is then interpreted as a finding. Collectively, the findings are then considered as evidence and the differential diagnosis is updated: the certainty of some etiologies decrease, while others increase. Additional tests are performed until there is one high probability cause to explain all the symptoms; this transitions the phenomenon to a (known) medical problem.

What is a Mass? An Exercise in Separating Observations from Inferences

For the knowledge derived from experiments to be vetted, the scientific method stresses that the observations must be detailed enough so that results are reproducible. In this light, the astute reader may discern that in order for the PCDM to meet this criterion of reproducibility, sufficient context and granularity must be modeled. To motivate these aspects of the PCDM, we use as a working example the task of capturing information about a "mass" object in thoracic radiology.

Identifying attributes. A conventional, if not logical strategy to uncover the terminology and concepts used to describe a mass is to examine: 1) how a mass is mentioned within textual clinical reports and textbooks; and 2) what questions experts (radiologists, pathologists, oncologists) ask about a mass (*i.e.*, what do they need to know?). Based on a corpus of thoracic radiology reports containing linguistic references to a "mass" object and discussion with physicians, Tables 7.2 & 7.3 summarize the review's results. The thirteen identified attributes and "mass" description are not surprising. But what can be said about the relationship between the attributes? Can the values be taken as a 13-tuple to describe the state of the mass, or do the values only apply within some

Mass attribute	Description	Typical values
Existence certainty	How certain we are of its existence based on observation	Definite yes, possibly, unlikely, definitely no
Location	Spatial description of mass location	Posterior aspect of RUL
Structural size	The reported/measured size	5.0 x 7.0 cm
Structural shape	The prototypical shape of the object	Round, nodular, triangular, etc.
Lobularity	Does the mass appear similar to an aggregation of bubbles?	Lobulated, not lobulated
Extensions	Does the mass have any tentacle-like extensions?	Linear strands, "rabbit ear," "tail sign"
Homogeneity	How homogenous in appearance does the mass look?	Homogeneous, non-homogeneous
Margination	How distinct is the margin of the mass from its surrounding environment.	Well-circumscribed, poorly demarcated
Margin shape	How smooth is the margin of the mass?	Smooth, spiculated
Radiological density	Density calibrated vs. different tissue types	Soft tissue, fluid/fluid-like, fat, air
Calcification	Description of any calcifications contained within the mass.	Yes, no, scattered small foci of calcification, calcific deposits
Trend	What is its behavior?	Increasing, decreasing, no change
Malignancy	Is the mass malignant?	Malignant, benign, unknown

Table 7.2: Compilation of attributes associated with a mass, as given by analysis of a corpus of textual statements from thoracic radiology reports and reference texts.

Common questions about a "mass"	
What is the histology of the lesion?	What are the consequences of the lesion?
Where is it located?	What is the cause of the lesion?
Which image slice is the lesion best seen?	What is the size/extent of the lesion?
What is the appearance of the lesion (size, border, calcification pattern, density, etc.)?	What are the co-occurring findings associated with the lesion (*e.g.*, lymph node involvement)?
How many lesions are seen?	When was the lesion first observed?
How is it changing (or not changing) with respect to size and distribution?	

Table 7.3: Summary of questions from an array of clinical experts from radiology, pathology, and oncology when asked to pose question about a mass object.

context (*e.g.*, time period)? How do we know the certainty of each value? What do we mean by *trend*? Do we know how each value was obtained? Can we trust these values? And what is meant when we say the mass is of size 5 x 7 cm?

The radiological experiment. The straightforward ontological approach leaves us with several questions; hence, we instead look to the source process of performing a radiological experiment and the tools we use to characterize a mass. Fig. 7.12 shows the physical reality of a typical radiology study, wherein we have a patient with a mass. An imaging study such as CT generates a series of sequential 2D axial images through the mass. It is common in radiology to assume that these images are faithful representations of reality: but the image representation of the mass appears as a round-shaped region of unusual optical density on film or screen – an arguably "warped" representation of the physical truth. A radiologist observes this unusual round density and infers that this density is consistent with a mass. Of course, many abstractions from the image to physical world are routinely precise (*e.g.*, "*this area of the radio-*

graph represents the heart"). But what evidence is there that allows the radiologist to make the jump from opacity to mass? Are all opacities with this shape and density, as seen on CT, masses? Has a surgeon seen the mass directly or a biopsy conducted? Whether his (indirect) interpretation of the density is true or not may need further confirmation from other (direct) experiments. Direct and indirect observations are often not well differentiated. Within the medical record, it is important to differentiate objective facts from observed perceived facts and assumptions [90]. Moreover, [68] points out the fact that everyone sees the world through the glasses of their own conceptual framework. So the definition of an observed fact is hard to describe in medicine as most so-called "facts" can be biased by the perspective of the observer. This realization emphasizes the need for a "certainty" property for each fact and/or inference. Indeed, as we move progressively away from the physical world to a conceptual representation of a phenomenon (physical world/phenomena → imaging → linguistic → conceptual), some degree of confidence is needed when mapping between object representations.

Separate from the interpretative process, the nature of a particular imaging procedure (*e.g.*, the physics associated with x-ray, MR, ultrasound imaging, etc.) and the constancy of the imaged object determine the sensitivity (contrast, spatial resolution, etc.) by which various objects can be detected and detailed. The inherent ability of our tests to provide information must also be documented, as it further affects the degree of confidence in characterizing the "mass" phenomenon.

Defining and representing context. Whatever context is necessary to improve the reproducibility of the radiological measurements should be documented. An experiment by definition means we have control and therefore the support context for a physical measurement or observation should be known. As an example, let us examine the property of "size" for our "mass." How should we represent the phrase, "*the mass is 5.0 x 7.2 cm in size?*" Using an elementary semantic network, the statement might be represented as: mass → hasSize → 5.0 x 7.2 cm. In isolation, this representation of "size" is of limited use, having problems with both context and precision. First, in addressing context, we are trying to describe the size of an inherently 3D object – but are only given two measurements (5.0 cm and 7.2 cm). Thus, what do these dimensions refer to? In radiology, dimensions are understood implicitly from the type of study and/or reconstruction view (*i.e.*, from the experiment context). For example, "*the mass is 5.0 x 7.2 cm*" from an axial CT typically refer to 5 cm in the lateral dimension and 7 cm in the anterio-posterior (AP) dimension. But on projectional x-ray, the same size description might instead refer to the AP and craniocaudal dimensions, respectively. A test's context must be fully given; this tenet is already seen in clinical trials and other biomedical assays (*e.g.*, microarray experiments [14]), and should be brought forth in documenting medical practice. Second, in addressing precision, each linear measure provides only a partial constraint toward the total description of size: just like an image is a gross representation of the object imaged, "*5.0 x 7.2 cm*" is a gross representation of the property of size. These linear measurements are highly dependent upon the mass' shape. In this instance, the fundamental problem is that given a 3D object, there are an infinite number of 1D/2D projections though the object's center of mass. The lesson here is that we should be careful not to operate in an under-specified ontology. In any science, precision and completeness is foremost if phenomena are to be understood.

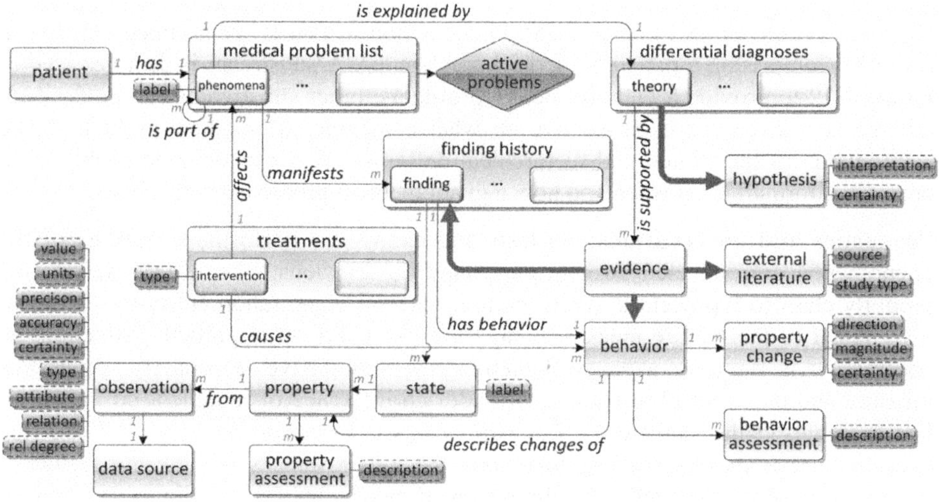

Figure 7.13: Core entities of the phenomenon-centric data model. The PCDM is based on the M2 stream constructs described earlier. For clarity, some attributes and relations have been omitted in order to accentuate the main aspects of the paradigm.

PCDM Core Entities

From the insights gained relative to the modeling of a "mass" we now describe the PCDM. The phenomenon-centric data model revolves around the entities and relationships that are needed to support the diagnostic and therapeutic processes as viewed in terms of scientific conduct. Though a semantic network could be used to represent information, a frame-based approach is used in the PCDM to simplify handling of *n*-ary relationships and to mimic structured data collection (*e.g.*, such as case report forms in clinical trials). Fig. 7.13 shows the mainstay of the PCDM using M2 constructs. Several core abstract classes are defined.

Phenomenon. Looking at the POMR, a central premise is that a medical "problem" can be defined, and that such a definition is agreed upon by the different users of the medical record. However, such agreement is not always clear. For example, episodes of care define problems in terms of symptoms (episode of care), etiology (episode of disease), and seriousness (episode of illness) [63, 114] – all dependent on the context (inpatients *vs.* outpatients; systemic *vs.* localized disease; and chronic *vs.* acute illnesses, respectively). These arbitrary definitions stem mainly from contextual differences regarding inpatients vs. outpatients, systemic vs. localized disease, and chronic vs. acute illnesses, respectively. Moreover, as suggested above, symptoms and abnormalities may not yet be identified as a given medical problem. Therefore, we use the phenomenon class to represent an entity under investigation[5]. The descriptors of a phenomenon change as more knowledge is gained so that it progresses into a medical problem. Importantly, a phenomenon is a hierarchical entity, being recursively associated with lower-level phenomena (via the is part of relationship); through this mechanism, symptoms can be grouped together into singular phenomena. (*e.g.*, sneezing $\wedge$ sinus

[5] The idea behind a phenomenon is not new to medicine, being found in several different medical ontologies. For instance, UMLS defines a phenomenon as a child of the event class; and its subclasses encompass biologic and pathologic functions. However, here we employ a broader, more classical sense of the term.

congestion ∧ cough → common cold). Phenomenon instances are grouped together in a stream, comprising a patient's medical problem list. A substream is demonstrated in Figure 7.12 to provide a view for active problems; other substreams can be defined, tailored to a given user's perspective on what constitutes a "medical problem." The combination of the stream and hierarchical definition of a phenomenon enable us to preserve information on when and how a given medical problem first presents.

Properties and observations. As tests and exams are performed, features of the phenomenon are collected through controlled observations. The PCDM defines the property class to represent a given feature; and the observation entity to document the context of the feature measurement. Here, we make an important distinction: an observation is typically a physical metric that is objective; a property names this measure and based on observations, provides an inference or judgment on the value. For example, in the statement, "*The patient has a moderate fever of 38.7° C*," the measurement is an observation, for which the circumstances of how the feature is quantified must be captured (*e.g.*, the means of measurement); the dimension may be named *body temperature*; and the adjective *moderate* is seen as an assessment of the property. The observation entity defines several attributes to record a value in an exacting manner to fully qualify the feature: the type of observation (qualitative, quantitative); the relation operator (*e.g.*, equals, greater/less than, etc.); the units (*e.g.*, Celsius); the accuracy (*e.g.*, ± 0.2 degrees); the precision (*e.g.*, 0.1 degrees); and the certainty of the measure (*e.g.*, exact, approximate) are all requisite. Being abstract, subclasses are used to implement additional attributes to the core observation entity to add fields needed for contextualizing specific types of measures. The origin of the observation is recorded through a relationship with a data source, which provides the raw data (*e.g.*, an image) or source documentation (*e.g.*, a report) for the observation's derivation.

States. Although observations are ultimately timestamped when linked to a finding (see below), the state entity provides an additional semantic level of categorization for temporal stages commonly referred to in clinical practice. Properties are grouped together by a state to provide a snapshot view of a given phenomenon. As a case in point, during the course of oncological treatment, "baseline" refers to the initial staging of the cancer from which comparisons are then made to gauge treatment response – the size, histologic grade, and appearance of the tumor are evaluated together within a state. Subsequent chemotherapy sessions and re-evaluation may result in new states (*e.g.*, as numbered by the treatment cycles).

Evidence. Evidence is a broadly defined entity that serves the purpose of identifying any source of observation or information in support of explaining the phenomenon. Evidence is a notion used to rationalize our findings, assessments, and theories, and is therefore used within the PCDM to connect objective data to conclusions. In EBM, for instance, external literature is a valid source of knowledge to back a clinical decision. Two subclasses are of particular importance in supporting theories:

- Findings. Findings are the physical manifestations of a phenomenon, documented by tests and the generation of observations. Findings attempt to classify properties into a medical entity. For instance, a single, small region with high attenuation seen on a CT study (an observation) of the mediastinum may be declared a solitary pulmonary nodule (a finding). In the PCDM, findings are tracked in a stream (a finding history), allowing the changes in a finding to be tracked over time.
- Behavior. A behavior is related to the concept of *change*. Change is a relation

between situations and indicates differences in state over time. For example, the response criterion of a mass (tumor) to therapy is often measured as the size of the baseline state with respect to the post-treatment state. Behaviors are often expressed in terms of a single property component of the state (*e.g.,* size behavior), and can be summarized by the magnitude of change and the direction (*i.e.,* increase, decrease). Assessment of a behavior can be either direct (*e.g.,* 10% increase in size) or clinical (*e.g.,* significant increase in size). In addition, the behavior of a property and the associated finding may be given in terms of a function. For instance, a tumor is often described in terms of a doubling time.

Theory. A working conjecture is postulated to explain a patient's symptoms using a range of evidence. In the PCDM, the theory entity represents this idea, with its subclass hypothesis, providing for an interpretation and degree of certainty for belief in the theory (*i.e.,* what is one's confidence in a given diagnosis). Theories explain the facts, but are not the facts themselves, instead representing how facts are related. A final diagnosis (and hence, classification of the related phenomenon as a medical problem) is reached when a theory has a probability nearing 100%. The beliefs we predicate our knowledge upon are continually evolving. As such, the changes in theories are captured over time via another stream, representing the differential diagnoses as it changes in response to new evidence. This stream characterizes how our understanding about the patient and the phenomenon evolves over time.

Interventions. Lastly, intervention is an abstract class that encapsulates information on entities that affect a phenomenon, such as drugs, surgery, or other means of addressing a medical problem. An intervention explicitly causes a behavior. Though not indicated in Fig. 7.13, interventions are also related to evidence, as per EBM, the choice of (optimal) therapy is based on scientific studies verifying efficacy. The temporal sequence of interventions, given by a stream, comprises the patient's treatment history.

Implementing the PCDM

The core entities in the PCDM provide a meta-model for organizing and indexing the information within the medical record. In doing so, it becomes plausible to implement the problem-oriented medical record as originally envisioned by Weed, but with sufficient flexibility to support an array of different organizational viewpoints. However, a gap exists between the PCDM and the ultimate sources of information contained within the EMR. Fig. 7.14 illustrates this mapping process using the finding of an intracranial aneurysm (ICA): as can be seen, the abstract classes defined in the core PCDM are extended into specific entities that enable traversal starting from the finding level, progressively through to the data source of a given observation. In this example, the aneurysm size is a property that is based on the observation, aneurysm measure (representing a linear measurement along some axis). Being a subclass of observation, aneurysm measure inherits all the attributes of its parent (*i.e.,* value, precision, accuracy, etc.). Each aneurysm measure is taken from source; in this case, a CT angiography study in the form of a DICOM image series. A behavior, aneurysm growth, is based on the aneurysm size, tracking the rate of change.

Though a detailed PCDM-based data model can be generated for a given domain, the barrier in implementation presently lies in automatically populating the data model with instances: techniques such as natural language processing (Chapter 6) and content-driven image analysis (Chapter 5) are needed to transform patient data and its current representation into this structure.

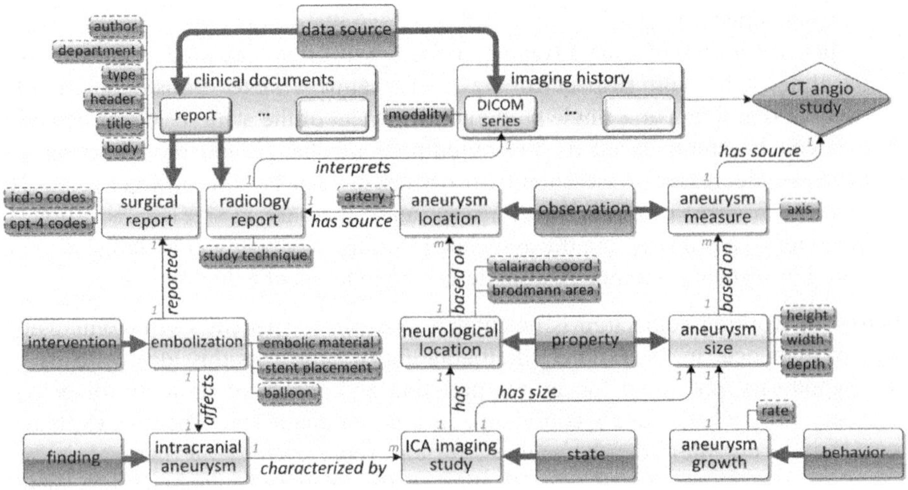

Figure 7.14: A demonstration of linking data sources to the PCDM. A finding of an intracranial aneurysm (ICA) is shown in terms of an imaging study and associated textual reports, showing how clinical information sources are used to derive observations and properties. For convenience, abstract PCDM classes are highlighted in green.

References

1. Aberle DR, Dionisio JD, McNitt-Gray MF, Taira RK, Cardenas AF, Goldin JG, Brown K, Figlin RA, Chu WW (1996) Integrated multimedia timeline of medical images and data for thoracic oncology patients. RadioGraphics, 16(3):669-681.
2. Adlassnig KP, Combi C, Das AK, Keravnou ET, Pozzi G (2006) Temporal representation and reasoning in medicine: Research directions and challenges. Artif Intell Med, 38(2):101-113.
3. Allen JF (1983) Maintaining knowledge about temporal intervals. Communications of the ACM, 26(11):832-843.
4. Anselma L, Terenziani P, Montani S, Bottrighi A (2006) Towards a comprehensive treatment of repetitions, periodicity and temporal constraints in clinical guidelines. Artif Intell Med, 38(2):171-195.
5. Augusto JC (2005) Temporal reasoning for decision support in medicine. Artif Intell Med, 33(1):1-24.
6. Baldock RA, Burger A (2008) Anatomical ontologies: Linking names to places in biology. Anatomy Ontologies for Bioinformatics. Springer, pp 197-211.
7. Bean CA (1997) Formative evaluation of a frame-based model of locative relationships in human anatomy. Proc AMIA Annu Fall Symp, pp 625-629.
8. Berg M (1998) Medical work and the computer-based patient record: A sociological perspective. Methods Inf Med, 37(3):294-301.
9. Berndt D, Clifford J (1994) Using dynamic time warping to find patterns in time series. AAAI-94 Workshop on Knowledge Discovery in Databases, pp 229-248.
10. Bittner T, Donnelly M, Goldberg LJ, Neuhaus F (2008) Modeling principles and methodologies - Spatial representation and reasoning. Anatomy Ontologies for Bioinformatics, pp 307-326.
11. Bober M (2001) MPEG-7 visual shape descriptors. IEEE Trans Circuits Systems Video Technology, 11(6):716-719.

12. Boxwala AA, Peleg M, Tu S, Ogunyemi O, Zeng QT, Wang D, Patel VL, Greenes RA, Shortliffe EH (2004) GLIF3: A representation format for sharable computer-interpretable clinical practice guidelines. J Biomed Inform, 37(3):147-161.
13. Brandt S (1999) Use of shape features in content-based image retrieval. Department of Engineering Physics and Mathematics, MS Thesis. Helsinki University of Technology.
14. Brazma A, Hingamp P, Quackenbush J, Sherlock G, Spellman P, Stoeckert C, Aach J, Ansorge W, et al. (2001) Minimum information about a microarray experiment (MIAME) - Toward standards for microarray data. Nat Genet, 29(4):365-371.
15. Bui AA (2000) A multimedia data model with generalized stream constructs. Computer Science Department, PhD Dissertation. University of California, Los Angeles.
16. Bui AA, Aberle DR, Kangarloo H (2007) TimeLine: Visualizing integrated patient records. IEEE Trans Inf Technol Biomed, 11(4):462-473.
17. Campos J, Hornsby K (2004) Temporal constraints between cyclic geographic events. Proc GeoInfo 2004, Campos do Jordao, Brazil, pp 109-125.
18. Cardenas AF, Ieong IT, Barker R, Taira RK, Breant CM (1993) The knowledge-based object-oriented PICQUERY+ language. IEEE Trans Knowledge and Data Engineering, 5(4):644-657.
19. Chakravarty S, Shahar Y (2000) CAPSUL: A constraint-based specification of repeating patterns in time-oriented data. Annals of Mathematics and Artificial Intelligence, 30(1):3-22.
20. Chan EPF, R. Z (1996) QL/G - A query language for geometric data bases. Proc 1st Intl Conf GIS in Urban Regional and Environment Planning, Samos, Greece, pp 271-286.
21. Chang NS, Fu KS (1979) Query-by-pictorial-example. Proc IEEE 3rd Intl Computer Software and Applications Conference (COMPSAC 79), pp 325-330.
22. Chang SK, Jungert E (eds) (1996) Symbolic Projection for Image Information Retrieval and Spatial Reasoning. Academic Press.
23. Chang SK, Shi QY, Yan CW (1987) Iconic indexing by 2-D strings. IEEE Trans Pattern Analysis and Machine Intelligence, 9(3):413-428.
24. Chen P (1976) The entity-relationship model - Toward a unified view of data. ACM Trans Database Syst, 1(1):9-36.
25. Chu WW, Hsu C-C, Cardenas AF, Taira RK (1998) Knowledge-based image retrieval with spatial and temporal constructs. IEEE Trans Knowledge and Data Engineering, 10:872-888.
26. Chu WW, Ieong IT, Taira RK, Breant CM (1992) A temporal evolutionary object-oriented data model and its query language for medical image management. Proc 18th Intl Conf Very Large Data Bases. Morgan Kauffman, pp 53-64.
27. Clementini E, Di Felice P (2000) Spatial operators. ACM SIGMOD Record, 29(3):31-38.
28. Clinical Data Interchange Standards Consortium (2008) CDISC home page. http://www.cdisc.org. Accessed January 3, 2009.
29. Coburn JC, Upal MA, Crisco JJ (2007) Coordinate systems for the carpal bones of the wrist. J Biomechanics, 40(1):203-209.
30. Cohn A (1996) Calculi for qualitative spatial reasoning. Artificial Intelligence and Symbolic Mathematical Computation, pp 124-143.
31. Combi C, Gozzi M, Oliboni B, Juarez JM, Marin R (2009) Temporal similarity measures for querying clinical workflows. Artif Intell Med, In Press, Corrected Proof.
32. Combi C, Montanari A, Pozzi G (2007) The T4SQL temporal query language. Proc 16th ACM Conf Information and Knowledge Management. ACM, Lisbon, Portugal, pp 193-202.
33. Combi C, Oliboni B, Rossato R (2005) Merging multimedia presentations and semistructured temporal data: A graph-based model and its application to clinical information. Artif Intell Med, 34(2):89-112.

34. Combi C, Pozzi G (2001) HMAP – A temporal data model managing intervals with different granularities and indeterminacy from natural language sentences. The VLDB Journal, 9(4):294-311.
35. Combi C, Shahar Y (1997) Temporal reasoning and temporal data maintenance in medicine: Issues and challenges. Computers in Biology and Medicine, 27(5):353-368.
36. Cootes TF, Edwards GJ, Taylor CJ (2001) Active appearance models. IEEE Trans Pattern Analysis and Machine Intelligence, 23(6):681-685.
37. Cousins SB, Kahn MG (1991) The visual display of temporal information. Artif Intell Med, 3:341-357.
38. Das AK, Musen MA (1994) A temporal query system for protocol-directed decision support. Methods Inf Med, 33(4):358-370.
39. Das AK, Musen MA (2001) A formal method to resolve temporal mismatches in clinical databases. Proc AMIA Symp:130-134.
40. Davidson D (2007) Time in anatomy. Anatomy Ontologies for Bioinformatics: Principles and Practice. Springer, pp 213-247.
41. De Clercq E, Van Casteren V, Jonckheer P, Burggraeve P, Lafontaine MF, Degroote K, France FR (2007) Are problem-oriented medical records (POMR) suitable for use in GPs' daily practice? Stud Health Technol Inform, 129(1):68-72.
42. Del Bimbo A, Rella L, Vicario E (1995) Visual specification of branching time temporal logic. Proc 11th Intl IEEE Symp on Visual Languages. IEEE Computer Society, pp 61-68.
43. Della Croce U, Cappozzo A, Kerrigan D (1999) Pelvis and lower limb anatomical landmark calibration precision and its propagation to bone geometry and joint angles. Medical and Biological Engineering and Computing, 37(1):155-161.
44. Deshpande AM, Brandt C, Nadkarni PM (2003) Temporal query of attribute-value patient data: Utilizing the constraints of clinical studies. International Journal of Medical Informatics, 70(1):59-77.
45. Dionisio JD, Cardenas AF (1998) A unified data model for representing multimedia, timeline, and simulation data. IEEE Trans Knowledge and Data Engineering, 10(5):746-767.
46. Dionisio JD, Cardenas AF, Lufkin RF, DeSalles A, Black KL, Taira RK, Chu WW (1997) A multimedia database system for thermal ablation therapy of brain tumors. J Digital Imaging, 10:21-26.
47. Dionisio JDN, Cardenas AF (1996) MQuery: A visual query language for multimedia, timeline and simulation data. J Visual Languages and Computing, 7(4):377-401.
48. Donnelly WJMD (2005) Viewpoint: Patient-centered medical care requires a patient-centered medical record. Academic Medicine, 80(1):33-38.
49. Dorda W, Gall W, Duftschmid G (2002) Clinical data retrieval: 25 years of temporal query management at the University of Vienna Medical School. Methods Inf Med, 41(2):89-97.
50. Dutta S (1988) Temporal reasoning in medical expert systems. Proc Symp Engineering of Computer-Based Medical Systems, pp 118-122.
51. Egenhofer MJ (1994) On the equivalence of topological relations. Intl J Geographical Information Systems, 8(6):133-152.
52. Egenhofer MJ (1994) Spatial SQL: A query and presentation language. IEEE Trans Knowledge and Data Engineering, 6(1):86-95.
53. Egenhofer MJ (1997) Query processing in spatial-query-by-sketch. J Visual Languages and Computing, 8(4):403-424.
54. Feinstein AR (1973) The problems of the "problem-oriented medical record". Ann Intern Med, 78(5):751-762.

55. Fletcher RH (1974) Auditing problem-oriented records and traditional records: A controlled comparison of speed, accuracy and identification of errors in medical care. N Engl J Med, 290(15):829-833.
56. Flickner M, Sawhney H, Niblack W, Ashley J, Huang Q, Dom B, Gorkani M, Hafner J, Lee D, Petkovic D (1995) Query by image and video content: The QBIC system. Computer, 28(9):23-32.
57. Freeman J (1975) The modelling of spatial relations. Computer Graphics and Image Processing, 4(2):156-171.
58. Georgsson F (2003) Anatomical coordinate system for bilateral registration of mammograms. Image Analysis, pp 83-90.
59. Gibbs S, Breiteneder C, Tsichritzis D (1994) Data modeling of time-based media. Proc 1994 ACM SIGMOD Intl Conf Management of Data. ACM, Minneapolis, Minnesota, United States.
60. Guting RH (1994) An introduction to spatial database systems. The VLDB Journal, 3(4):357-399.
61. Haendel MA, Neuhaus F, Osumi-Sutherland D, Mabee PM, Mejino JL, Mungall CJ, Smith B (2008) CARO - The Common Anatomy Reference Ontology. Anatomy Ontologies for Bioinformatics. Springer, pp 327-349.
62. Herring JR (2006) OpenGIS implementation specification for geographic information - Simple feature access, Part 2: SQL option. Open Geospatial Consortium Inc. http: //www.opengeospatial.org/standards/sfs. Accessed December 15, 2008.
63. Hornbrook MC, Hurtado AV, Johnson RE (1985) Health care episodes: Definition, measurement and use. Med Care Rev, 42(2):163-218.
64. Hripcsak G (1994) Writing Arden syntax medical logic modules. Comput Biol Med, 24(5):331-363.
65. Hripcsak G, Zhou L, Parsons S, Das AK, Johnson SB (2005) Modeling electronic discharge summaries as a simple temporal constraint satisfaction problem. J Am Med Inform Assoc, 12(1):55-63.
66. Juarez JM, Guil F, Palma J, Marin R (2009) Temporal similarity by measuring possibilistic uncertainty in CBR. Fuzzy Sets Syst, 160(2):214-230.
67. Kass M, Witkin A, erzopoulos D (1988) Snakes: Active contour models. Intl J Comp Vision, 1(4):321–331.
68. Kay S, Purves IN (1996) Medical records and other stories: A narratological framework. Methods Inf Med, 35(2):72-87.
69. Loncaric S (1998) A survey of shape analysis techniques. Pattern Recognition, 31(8):983-1001.
70. Long W (1996) Temporal reasoning for diagnosis in a causal probabilistic knowledge base. Artif Intell Med, 8(3):193-215.
71. Marshall S (1989) Review of shape coding techniques. Image and Vision Computing, 7(4):281-294.
72. Martins SB, Shahar Y, Goren-Bar D, Galperin M, Kaizer H, Basso LV, McNaughton D, Goldstein MK (2008) Evaluation of an architecture for intelligent query and exploration of time-oriented clinical data. Artif Intell Med, 43(1):17-34.
73. McInerney T, Terzopoulos D (1996) Deformable models in medical image analysis: A survey. Medical Image Analysis, 1(2):91-108.
74. Mejino JL, Jr., Rosse C (1999) Conceptualization of anatomical spatial entities in the Digital Anatomist Foundational Model. Proc AMIA Symp, pp 112-116.
75. Motakis I, Zaniolo C (1997) Formal semantics for composite temporal events in active database rules. Journal of Systems Integration, 7(3):291-325.

76. Nadkarni PM (1998) CHRONOMERGE: An application for the merging and display of multiple time-stamped data streams. Comput Biomed Res, 31(6):451-464.
77. Neuhaus F, Smith B (2008) Modeling principles and methodologies - Relations in anatomical ontologies. Anatomy Ontologies for Bioinformatics, pp 289-305.
78. Nigrin DJ, Kohane IS (2000) Temporal expressiveness in querying a timestamp-based clinical database. J Am Med Inform Assoc, 7(2):152-163.
79. Nygren E, Henriksson P (1992) Reading the medical record. Part I: Analysis of physicians' ways of reading the medical record. Comput Methods Programs Biomed, vol 39, pp 1-12.
80. O'Connor MJ, Shankar RD, Parrish DB, Das AK (2008) Knowledge-data integration for temporal reasoning in a clinical trial system. International Journal of Medical Informatics, In Press, Corrected Proof.
81. O'Connor MJ, Tu SW, Musen MA (2002) The Chronus II temporal database mediator. Proc AMIA Symp, pp 567-571.
82. Palma J, Juarez JM, Campos M, Marin R (2006) Fuzzy theory approach for temporal model-based diagnosis: An application to medical domains. Artif Intell Med, 38(2):197-218.
83. Pani AK, Bhattacharjee GP (2001) Temporal representation and reasoning in artificial intelligence: A review. Mathematical and Computer Modelling, 34(1-2):55-80.
84. Pizer SM, Fletcher PT, Joshi S, Thall A, Chen JZ, Fridman Y, Fritsch DS, Gash AG, Glotzer JM, Jiroutek MR (2003) Deformable m-reps for 3D medical image segmentation. Intl J Comp Vision, 55(2):85-106.
85. Post A, Harrison J, Jr. (2006) Data acquisition behaviors during inpatient results review: Implications for problem-oriented data displays. Proc AMIA Annu Fall Symp:644-648.
86. Post AR, Harrison JH, Jr. (2007) PROTEMPA: A method for specifying and identifying temporal sequences in retrospective data for patient selection. J Am Med Inform Assoc, 14(5):674-683.
87. Post AR, Harrison Jr JH (2008) Temporal data mining. Clinics in Laboratory Medicine, 28(1):83-100.
88. Post AR, Sovarel AN, Harrison JH, Jr. (2007) Abstraction-based temporal data retrieval for a Clinical Data Repository. Proc AMIA Symp, pp 603-607.
89. Randell DA, Cui Z, Cohn A (1992) A spatial logic based on regions and connection. In: Nebel B, Rich C, Swartout W (eds) Principles of Knowledge Representation and Reasoning: Proc 3rd Intl Conf. Morgan Kaufmann, pp 165-176.
90. Rector AL, Nowlan WA, Kay S, Goble CA, Howkins TJ (1993) A framework for modelling the electronic medical record. Methods Inf Med, 32(2):109-119.
91. Renz J (2002) Introduction. Qualitative Spatial Reasoning with Topological Information. Springer-Verlag, pp 1-11.
92. Rosse C, Mejino JL, Jr. (2003) A reference ontology for biomedical informatics: The Foundational Model of Anatomy. J Biomed Inform, 36(6):478-500.
93. Rosse C, Mejino JL, Modayur BR, Jakobovits R, Hinshaw KP, Brinkley JF (1998) Motivation and organizational principles for anatomical knowledge representation: the digital anatomist symbolic knowledge base. J Am Med Inform Assoc, 5(1):17-40.
94. Roussopoulos N, Faloutsos C, Sellis T (1988) An efficient pictorial database system for PSQL. Software Engineering, IEEE Transactions on, 14(5):639-650.
95. Salmon P, Rappaport A, Bainbridge M, Hayes G, Williams J (1996) Taking the problem oriented medical record forward. Proc AMIA Annu Fall Symp, pp 463-467.
96. Schneider M (1997) Spatial data types - A survey. In: Schneider M (ed) Spatial Data Types for Database Systems. Springer-Verlag, pp 11-83.
97. Schneider M (1999) Uncertainty management for spatial data in databases: Fuzzy spatial data types. Advances in Spatial Databases, pp 330-351.

98. Shahar Y (1999) Timing is everything: Temporal reasoning and temporal data maintenance in medicine. Artificial Intelligence in Medicine, pp 30-46.
99. Shahar Y, Combi C (1998) Timing is everything. Time-oriented clinical information systems. West J Med, 168(2):105-113.
100. Shahar Y, Goren-Bar D, Boaz D, Tahan G (2006) Distributed, intelligent, interactive visualization and exploration of time-oriented clinical data and their abstractions. Artif Intell Med, 38(2):115-135.
101. Shahar Y, Musen MA (1996) Knowledge-based temporal abstraction in clinical domains. Artif Intell Med, 8(3):267-298.
102. Signes J, Fisher Y, Eleftheriadis A (2000) MPEG-4's binary format for scene description. Signal Processing: Image Communication, 15:321-345.
103. Sistla AP, Yu C, Haddad R (1994) Reasoning about spatial relationships in picture retrieval systems. Proc 20th Very Large Data Base (VLDB) Conf, Santiago, Chile, pp 570-581.
104. Smith B, Ceusters W, Klagges B, Kohler J, Kumar A, Lomax J, Mungall C, Neuhaus F, Rector AL, Rosse C (2005) Relations in biomedical ontologies. Genome Biol, 6(5):R46.
105. Snodgrass RT (ed) (1995) The TSQL2 Temporal Query Language. Kluwer Academic Publishers.
106. Sowa JF (2000) Processes. Knowledge Representation: Logical, Philosophical, and Computational Foundations. MIT Press, pp 206-264.
107. Spackman K (2008) SNOMED CT Style Guide: Body Structures - Anatomy. International Health Terminology Standards Development Organization.
108. Tiffe S (2002) Defining medical concepts by linguistic variables with fuzzy Arden Syntax. Proc AMIA Symp:796-800.
109. Vila L (2005) Formal theories of time and temporal incidence. In: Fisher M, Gabbay D, Vila L (eds) Handbook of Temporal Reasoning in Artificial Intelligence. Elsevier.
110. Viqueira JRR, Lorentzos NA (2007) SQL extension for spatio-temporal data. The VLDB Journal, 16(2):179-200.
111. Weed LL (1968) Medical records that guide and teach. N Engl J Med, 278(11):593-600.
112. Weng C, Kahn M, Gennari J (2002) Temporal knowledge representation for scheduling tasks in clinical trial protocols. Proc AMIA Symp, pp 879-883.
113. Wiederhold G (1981) Databases for Health Care. Springer-Verlag New York, Inc..
114. Wingert TD, Kralewski JE, Lindquist TJ, Knutson DJ (1995) Constructing episodes of care from encounter and claims data: Some methodological issues. Inquiry, 32(4):430-443.
115. Winkelman WJ, Leonard KJ (2004) Overcoming structural constraints to patient utilization of electronic medical records: A critical review and proposal for an evaluation framework. J Am Med Inform Assoc, 11(2):151-161.
116. World Wide Web Consortium (W3C) (2008) Synchronized Multimedia Integration Language (SMIL 3.0). http://www.w3.org/TR/SMIL3/smil30.html#smil-introduction. Accessed December 5, 2008.
117. Zhou L, Hripcsak G (2007) Temporal reasoning with medical data - A review with emphasis on medical natural language processing. J Biomedical Informatics, 40(2):183-202.
118. Zhou L, Melton GB, Parsons S, Hripcsak G (2006) A temporal constraint structure for extracting temporal information from clinical narrative. J Biomed Inform, 39(4):424-439.

PART IV
Toward Medical Decision Making

Wherein we consider methods to reach conclusions with our medical data. The escalating amount of electronic medical information provides a unique opportunity to structure and to mine this data for new insights into disease management, drawing upon a large observational dataset. Techniques once seen as computationally intractable are now providing us the tools to handle these datasets and to inform the medical decision making process. We describe one particular method of growing popularity – graphical models – as a means of modeling a disease and for answering prognostic questions. The first chapter in this last section provides an introduction to graphical models, with a particular emphasis on Bayesian belief networks and issues related to causality. Subsequently, the next chapter overviews methods to answer queries posed to belief networks, and the applications that can be realized given the power of these models. Finally, we conclude with an introduction to evaluation, covering core concepts in biostatistics, study design, and decision making; we demonstrate these principles in the context of two common informatics areas: information retrieval and usability studies.

- **Chapter 8 –** Disease Models, Part I: Graphical Models
- **Chapter 9 –** Disease Models, Part II: Querying & Applications
- **Chapter 10 –** Evaluation

Chapter 8
Disease Models, Part I: Graphical Models

ILYA SHPITSER

Scientists building models of the world by necessity abstract away features not directly relevant to their line of inquiry. Furthermore, complete knowledge of relevant features is not generally possible. The mathematical formalism that has proven to be the most successful at simultaneously abstracting the irrelevant, while effectively summarizing incomplete knowledge, is probability theory. First studied in the context of analyzing games of chance, probability theory has flowered into a mature mathematical discipline today whose tools, methods, and concepts permeate statistics, engineering, and social and empirical sciences. A key insight, discovered multiple times independently during the 20th century, but refined, generalized, and popularized by computer scientists, is that there is a close link between probabilities and graphs. This link allows numerical, quantitative relationships such as conditional independence found in the study of probability to be expressed in a visual, qualitative way using the language of graphs. As human intuitions are more readily brought to bear in visual rather than algebraic and computational settings, graphs aid human comprehension in complex probabilistic domains. This connection between probabilities and graphs has other advantages as well – for instance the magnitude of computational resources needed to reason about a particular probabilistic domain can be read from a graph representing this domain. Finally, graphs provide a concise and intuitive language for reasoning about causes and effects. In this chapter, we explore the basic laws of probability, the relationship between probability and causation, the way in which graphs can be used to reason about probabilistic and causal models, and finally how such graphical models can be learned from data. The application of these graphs to formalize observations and knowledge about disease are provided.

Uncertainty and Probability

The creation of disease models in medicine poses certain challenges: 1) the uncertainties inherent to medical knowledge must be captured; 2) the models must be sufficiently intuitive so that domain experts (*e.g.,* physicians) can understand the explanations proposed by the system; and 3) the models must be practically analyzable by algorithms to support queries. One approach, probabilistic modeling, comes from a long tradition in medicine [36, 44, 63, 76, 84]. In this section, we give the case for using probability theory for abstraction, and introduce its basic laws, consider what probabilities mean, and discuss the relationship between probability and change.

Why Probabilities?

Building models of the world and using them to reach useful conclusions involves constructing a knowledge-base of relevant facts. Facts are typically represented by various kinds of logical languages. For instance, in propositional logic facts are built up of simple assertions joined together by connectives such as and, or, if, and so on. We can use propositional logic to build a "diagnostic database" listing in which symptoms imply a given disease. For example, as we know that a cough is a symptom of a cold, pneumonia and tuberculosis, we may have the following three rules in our database: if *Cold* then *Cough*; if *Pneumonia* then *Cough*; if *Tuberculosis* then *Cough*.

A.A.T. Bui and R.K. Taira (eds.), *Medical Imaging Informatics*,
DOI 10.1007/978-1-4419-0385-3_8, © Springer Science + Business Media, LLC 2010

One problem with such a database is that it does not handle exceptions well. A cough is not always a result of the above diseases, or indeed of any disease. Further, unknown symptoms or diseases will simply be absent from the database. Exceptions and gaps in our knowledge will cause misdiagnosis. Early expert systems that were typically rule-based were augmented to deal with these problems [11, 27]. There were three main approaches. One way to deal with exceptions is to augment the logic itself to handle them, resulting in non-monotonic logic where facts are not absolute but could be retracted when additional evidence becomes available [73, 74]. Though a number of useful kinds of reasoning with exceptions is possible in non-monotonic logics, in general this approach runs into difficulties [66]. Another approach to exceptions is to replace the hard truth value associated with each sentence in our diagnostic database by a kind of generalized numeric truth value that captures our confidence in the veracity of the rule (the well-known MYCIN project termed this *certainty factors*). The process of logical reasoning, which ordinarily would combine truth values of sentences to infer new truth values, would in this case combine "confidence scores." By way of illustration, if we had a rule if *Cough* and not *Fever* then *Allergy* in our database, and we attached a confidence value of 0.8 to *Cough* and confidence value of 0.7 to *Fever*, we could combine these numbers using some function associated with the rule into a new confidence for the conclusion Allergy. This approach is attractive as such computations are not any more expensive than dealing with truth values themselves, while allowing more refined distinctions to be made. Nevertheless, this approach, too, suffers from several problems. The first problem is that local update rules for confidence scores are incapable of capturing a common pattern of reasoning known as "explaining away," where increased plausibility of one explanation for an observed symptom should result in the decreased plausibility of another. The second problem is that exceptions are not handled very well in this approach. Colds result in a cough unless a cough suppressant is taken (or in general if some set of exceptions holds) – but the above approach, being based on classical, rather than non-monotonic logic, does not have a good way of representing this even though truth values are no longer absolute. Finally, a single piece of erroneous evidence in our database can be amplified via multiple local updates of our confidence based on this evidence. For instance, if a blood test concluded (falsely) that the patient is HIV (human immunodeficiency virus) positive, this observation would increase our confidence in the presence of a number of respiratory ailments due to the compromised immune system associated with AIDS (acquired immune deficiency syndrome). Further, the likely presence of these respiratory ailments would push our confidence of seeing respiratory symptoms like a cough to a near certainty. The problem here is that our rules for changing our confidence in any given fact do not depend on the history of obtaining said fact. As such, we are unable to notice that we should discount our confidence based on the fact that our multiple sources of evidence ultimately come from the same source. This problem is sometimes called *rumor propagation* because seeing two different people repeat the same rumor should not necessarily cause us to increase our confidence in its truth – both these people may have gotten the rumor from the same (faulty) source.

The only general way of handling the above problems is to abandon the logical properties of locality (*i.e.,* the truth value of a fact can be determined using small parts of our database that involve the fact directly) and monotonicity (*i.e.,* once the truth of a fact is established, it cannot be revoked). Out of reasoning approaches that abandon these properties, *probability theory* is the most popular, partly due to its clear semantics in terms of beliefs or frequencies, and partly because probability-based approaches allow us to take advantage of a problem's structure to reduce the computational burden of reasoning without giving up clarity and correctness.

Laws of Probability: A Brief Review

Probability theory formalizes how to reason about events with random outcomes, like dice throws or coin flips. Such events are called *random variables*. Here, we will consider random variables with a finite set of outcomes, although it is also possible to reason about infinite outcomes using the tools of measure theory [6, 16]. For a finite random variable X[1] with outcomes $\{x_1,\ldots, x_k\}$, associated with each outcome x_i is a real number $P(X = x_i)$ (sometimes shortened as $P(x_i)$) called a *probability*. A valid assignment of probabilities to a random variable with n outcomes has two properties:

$$\forall x_i P(x_i) \in [0,1] \tag{1}$$

$$\sum_{i=1}^{n} P(x_i) = 1 \tag{2}$$

Assignments that obey the above two properties are called *probability distributions*.

Typically, we are interested in a set of random variables. For instance in the game of backgammon, moves are governed by the outcomes of two dice throws (call them D_1 and D_2). In such cases, we define an assignment from a set of outcomes, one outcome for each random variable in our set of interest, to a probability. We comma-separate the set of outcomes in a probability expression, for instance $P(D_1 = 1, D_2 = 1) = {}^1\!/_{36}$ states that the probability of getting two ones in a throw of two dice is ${}^1\!/_{36}$. Joint distributions over a set of random variables $X_1, \ldots, X_k$ where variable X_i has n_i outcomes, obey a generalization of equations (1) and (2):

$$\forall (x_1,\ldots,x_n) P(x_1,\ldots,x_n) \in [0,1]$$

$$\sum_{i_1=1}^{n_1} \cdots \sum_{i_k=1}^{n_k} P(x_{i_1},\ldots,x_{i_k}) = 1$$

A *joint probability distribution* expresses how likely we are to observe a particular combination of outcomes. A *marginal distribution* is obtained from a joint distribution by summing over some random variables we no longer care about. For example, if we are interested in the outcome of only one of two dice, we would consider $P(D_1) = \sum_{D_2=1}^{6} P(D_1, D_2)$. This operation of "summing out" D_2 is called *marginalization*, and it can be justified as collapsing together all the possible worlds that only differ by outcomes we do not care about. In other words, if the first die comes up a 6, then we aren't interested in distinguishing the possible worlds where the second die comes up a 6 or a 5, as we only care about the first die. Thus, we just "lump together" all these possible situations. By analogy, in clinical medicine, marginal distributions can be used to express how frequently we would expect to see a given disease together with a given symptom. Potentially more diagnostically useful is ascertaining the likelihood of a particular disease given that we observe a particular symptom. The concept that captures this inquiry is a *conditional probability distribution*. A conditional probability distribution has a set of outcomes we are interested in, and a set of outcomes that are

---

[1] We follow the standard notation of capitalizing random variables and of using lowercase letters for outcomes. Bold characters symbolize sets or vectors of variables.

given. These sets are separated by a conditioning bar, so our question can be written as $P(\text{disease} \mid \text{symptom})$, and is read as, "the probability of disease given a symptom." This quantity is derived from joint and marginal distributions as follows:

$$P(\text{disease} \mid \text{symptom}) = \frac{P(\text{disease}, \text{symptom})}{P(\text{symptom})}$$

We can justify this definition by rearranging it slightly: $P(\text{disease} \mid \text{symptom})P(\text{symptom}) = P(\text{disease}, \text{symptom})$. This equation states that we should expect to see the disease and symptom occurring together as often as the disease occurs given that we observe the symptom, weighted by the probability of making that observation. In general, a conditional probability is defined as follows:

$$P(x_1,\ldots,x_k \mid x_{k+1},\ldots,x_m) = \frac{P(x_1,\ldots,x_m)}{P(x_{k+1},\ldots,x_m)} \tag{3}$$

From this definition stem two useful properties. The first is known as the *chain rule* of probabilities, the second as *Bayes' rule*:

$$P(x_1,\ldots,x_k) = \prod_{i=1}^{k} P(x_i \mid x_{i+1},\ldots,x_k) \tag{4}$$

$$P(x_1 \mid x_2) = \frac{P(x_2 \mid x_1)P(x_1)}{P(x_2)} \tag{5}$$

Bayes' rule in particular, though it follows straightforwardly from the definition of conditional probability, is quite useful in practical probabilistic reasoning from effects to causes. Consider the standard problem of diagnosis in medicine. We have a list of symptoms and diseases, such that we know how likely these symptoms and diseases are to occur (*e.g.*, we know the marginal probability $P(\text{disease})$ and $P(\text{symptom})$ for each disease and symptom), and we know how likely a given symptom is to occur if a given disease is taking place, in other words $P(\text{symptom} \mid \text{disease})$. What we are interested in is the probability of a disease given an observed symptom. Bayes' rule is what lets us compute the answer.

In general, if we can estimate the joint probability distribution for a set of variables of interest, we can use the above properties to compute any probabilistic query of interest. However, the problem with this approach in practice is that as the number of variables increases, the amount of space and computational effort needed to handle a joint distribution grows exponentially. This growth occurs because a joint distribution is a table that assigns a probability to any possible combination of values of the variables we care about: as the number of variables grows, the number of possible value combinations grows very quickly. Luckily, it is possible to take advantage of redundancy in some joint distributions to reduce the computational overhead of probabilistic reasoning. A crucial notion for capturing this redundancy is *conditional independence*. Two random variables X, Y are conditionally independent given Z if $P(x, y \mid z) = P(x \mid z)P(y \mid z)$. The definition generalizes in a straightforward way to sets, so $\{X_1,\ldots, X_k\}$ is conditionally independent of $\{Y_1,\ldots, Y_m\}$ given $\{Z_1,\ldots, Z_n\}$ if:

$$P(x_1,\ldots,x_k, y_1,\ldots y_m, z_1,\ldots,z_n) = P(x_1,\ldots,x_k, \mid z_1,\ldots,z_n)P(y_1,\ldots y_m \mid z_1,\ldots,z_n)$$

If X is conditionally independent of Y given Z, we will denote this by an abbreviation $(X \perp\!\!\!\perp Y \mid Z)$, invented in [22]. If the set of Z variables is empty, the X and Y variables are marginally independent, written $(X \perp\!\!\!\perp Y)$. Taking systematic advantage of conditional independence is crucial: a joint probability distribution $P(x_1,\ldots, x_n)$ over n binary variables is a table with 2^n entries, while if these variables are independent, the table will only have n distinct entries (as $P(x_1,\ldots, x_n) = \prod_i P(x_i)$).

Interpreting probabilities. Probabilities are real numbers, and for the purposes of developing the theory of their behavior it is sufficient to treat them as such. However, in order to assign probabilities to actual events, and to judge whether the theory developed adequately captures reasoning about such events, we need to interpret what probabilities mean. There are two major interpretations, broadly termed *objectivist* and *subjectivist*. The objectivist interpretation states that probabilities are propensities of occurrence of certain outcomes in the world – for instance, a fair coin's propensity is to land heads about half the time it is thrown. According to this interpretation, probabilities are properties of objects and events themselves. The objectivist interpretation gives rise to frequentism, namely the enterprise of attempting to approximate objective probabilities of events by repeated trials. In contrast, the subjectivist interpretation states that probabilities describe degrees of beliefs of agents with minds, and that probability theory describes precisely (via Bayes' rule) how beliefs of rational agents should be modified as new evidence comes to light. According to subjectivism, probabilities are properties of our minds and are only indirectly related to the external world. The application of subjectivist interpretation in statistics resulted in the birth of Bayesian inference [70]. These differing interpretations do have certain subtle implications for the practical use of probabilities; however, reasonable conclusions can be obtained under either interpretation, as we expect rational beliefs to mirror the frequencies observed in nature. In fact, given enough trials or experiments (such as coin tosses), the conclusions reached under either interpretation will be the same.

Probability and Change

Probability theory is a very flexible way of handling evidence that requires a drastic reevaluation of existing beliefs – a conditional probability $P(y \mid x)$ after we observe x can be very different from the original belief $P(y)$. In some sense, probabilities give us a principled way of changing our minds as uncertainty is removed.

However, aside from having to contend with the world changing our minds, we must also contend with the world itself changing. If a patient is admitted to a hospital with a set of symptoms we can, given a joint distribution expressing the way symptoms and diseases occur typically, diagnose this patient. However, a myriad of things could happen that would make this new patient atypical. What if he fails to respond to treatment? What if he is accidentally given the wrong medicine? Further, it may be that the doctors themselves may want to impose changes on the patient, for instance novel surgical techniques or experimental drugs, which are sufficiently different from "typical" that the standard probabilistic model for diagnosis/prognosis will be insufficient to represent the results of these changes.

Going beyond medicine, modeling responses of systems to experiments is crucial to empirical science, because the goal of empirical science is constructing causal theories of the world. The very notions of cause and effect refer implicitly to a hypothetical change and results of this change. Systematically inducing changes and recording the results is a large part of what many scientific disciplines are about (as per the phenomenon-centric data model, see Chapter 7).

Perhaps surprisingly, despite the importance of the notions of change, cause, and effect; and despite the intuitive consensus people seem to form about these notions, the efforts to formalize these notions have began only relatively recently [47, 58, 77, 78, 92, 100]. In this chapter, we will represent changes using the formalism developed by Pearl [67]. Pearl introduces a new piece of notation, $do(x)$, which stands for a hypothetical experiment or *intervention* where a random variable of interest X is fixed to a specific value x, regardless of what its ordinary behavior might be. The notion of interventions is surprisingly general and can be used to model changes in many probabilistic settings. For instance, an intervention can be used to represent a controlled trial, where half of the patients are given a new drug and half are given a placebo. The patients' response Y to the drug is modeled as an interventional distribution, written $P(y \mid do(x))$. It is important to note that this distribution is not, in general, equal to $P(y \mid x)$. Consider a very simple study where X is a cough and Y is the common cold. We expect the probability of having a common cold given an observed cough, in other words $P(y \mid x)$, to be relatively high. At the same time, if we make sure our patients do not exhibit a cough by giving them a cough suppressant – in other words if we perform $do(x)$ – our belief in the likelihood of the cold should not be affected, as we have not treated the disease, only the symptom. Although it is clear in this example that $P(y \mid do(x))$ is not equal to $P(y \mid x)$, it is less clear that $P(y \mid do(x))$ is equal to $P(y)$. In some sense this distinction is not surprising: conditional distributions model the way our beliefs change as our uncertainty about a static probabilistic system is removed (*e.g.*, as we obtain more evidence about it). On the other hand, interventional distributions model what happens when we "tweak" the system itself. However, the notions of change in the mind versus change in the world are often confused, especially when probabilistic and causal notions are used together. The above example shows how such confusion can result in trivially wrong conclusions; in more complex cases the difference can be subtle and easily obscured. The study of the interplay between these two notions of change, and of interventions and interventional distributions is known as *causal inference*.

Graphical Models

Probabilistic and causal inference may not have achieved their popularity were it not for a surprising but deep and powerful connection between probabilistic notions and graph theory. In this section, we describe how to take advantage of graphs to represent conditional independencies and causal assumptions systematically, provide a qualitative language for probabilistic and causal notions, and in general replace quantitative, algebraic manipulations of probability distributions with more qualitative, visual ones based on graphs. We further introduce a specific type of graphical framework, the Bayesian network, and demonstrate its application as the basis for disease models.

Graph Theory

Graphs are mathematical structures which are used for visually representing relationships between objects. Graphs consist of nodes or vertices, which represent objects we want to model, and edges that represent a relationship between two nodes by connecting them. Graphs are typically classified by the kind of edges they contain (Fig. 8.1). *Directed graphs* contain exclusively directed edges, *undirected graphs* contain undirected edges, and *mixed graphs* contain multiple kinds of edges, possibly with more than one edge connecting distinct nodes.

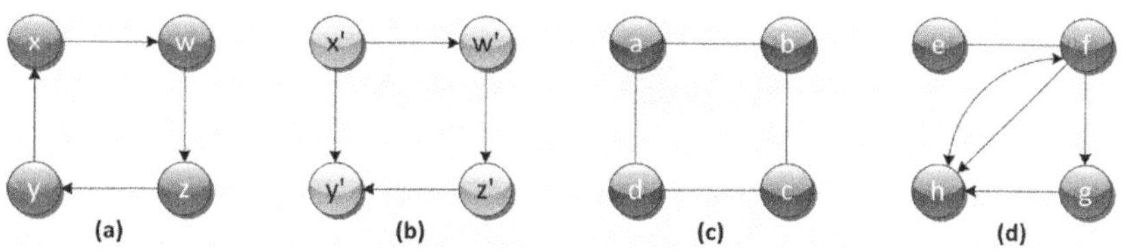

Figure 8.1: Various kinds of graphs. **(a)** A directed cyclic graph. **(b)** A directed acyclic graph (DAG). **(c)** An undirected graph. **(d)** A "mixed" graph.

A path in a graph connecting nodes x and y is a sequence of nodes $x, v_1, \ldots, v_k, y$ such that there is an edge connecting x to v_1, v_k to y , and v_i to v_{i+1} for every $i = 1, \ldots, k$. For instance, $x \rightarrow w \rightarrow z \rightarrow y$ is a path from x to y in the graph shown in Fig. 8.1a, while a–b–c is a path from a to c in the graph shown in Fig. 8.1c. In directed graphs, a node x is called an *ancestor* of y if there is a path from x to y such that all edges on this path point towards y. So the path $x \rightarrow w \rightarrow z \rightarrow y$ in the graph in Fig. 8.1a implies that x is an ancestor of y in that graph. A directed graph is said to contain a *directed cycle* if there is a node x that is an ancestor of y, and there is an arrow from y to x. Directed graphs containing cycles are called *cyclic*, while those that do not are called *acyclic*. It is simple to see that the graph in Fig. 8.1a is cyclic, while Fig. 8.1b is acyclic. In any graph if two nodes are connected by an edge, they are called *adjacent*. Further, if a set of n nodes are all pairwise adjacent in a graph, this set is called a *clique*. In keeping with the "family" theme of interpreting edges in directed graphs, if there is an edge $x \rightarrow y$ in a directed graph, then x is called the *parent* of y, and y a *child* of x. Finally, if two parents x_1, x_2 of a single node y are not themselves adjacent, these three nodes form what is called an *immorality* (*i.e.*, the parents aren't "married"), or a *collider* (as the arrows "collide" at y). For example, in the graph shown in Fig. 8.1b, x', z', and y' form an immorality.

Mathematicians study graphs because human beings tend to be better at comprehending and making use of information arranged in visual form, and graphs provide such a visual representation in a wide variety of settings (see Chapter 4 for a discussion of graph visualization methods). A good example is the one that inspired graph theoretic notions we described above, genealogic trees. Other examples of the use of graphs include phylogenetic trees in evolutionary biology, org charts in business, wiring diagrams in electrical engineering, network connectivity diagrams in computer science, and so on. It turns out there is also a deep connection between graphs and probability theory, allowing us to use graphs to reason about stochastic and causal domains.

Graphs and Probabilities

Consider a joint probability distribution $P(\mathbf{v})$ with an ordering $V_1, V_2, \ldots, V_k$ established over variables in $\mathbf{V}$. This ordering may be temporal, causal, or even arbitrary. By the chain rule of probabilities, $\prod_i P(\mathbf{v}) = P(v_i \mid v_{i-1}, \ldots, v_1)$. A number of conditional independence statements may hold in $P(\mathbf{v})$. In particular, for every V_i, there may be a subset of $\{V_1, \ldots, V_{i-1}\}$, call it $pa(V_i)$ such that $P(v_i \mid v_{i-1}, \ldots, v_1) = P(v_i \mid pa(v_i))$. If we find such a set for every V_i, we can express $P(\mathbf{v})$ as $\prod_i P(v_i \mid pa(v_i))$. So far, we just have a product expression for $P(\mathbf{v})$ derived from the chain rule using independence information. This expression is called the *Markov factorization* of $P(\mathbf{v})$. It is possible to represent the Markov factorization by means of a directed acyclic graph in the following way: we create a node for each variable V_i in $\mathbf{V}$, and add a directed edge from every element of $pa(V_i)$ to V_i, for all i. As an example, if the joint distribution $P(x, w, y, z)$

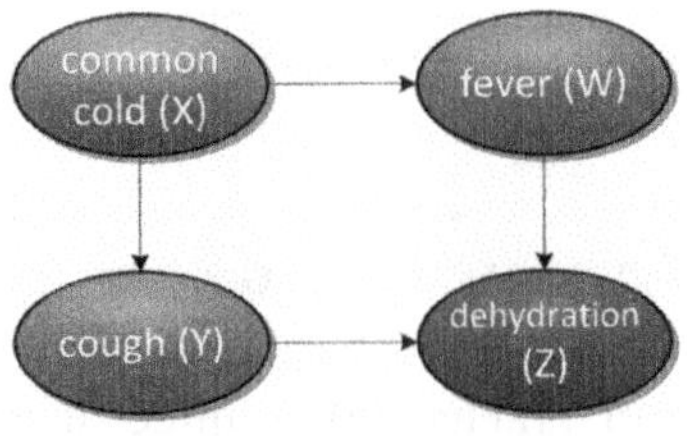

Figure 8.2: A BBN representing a simplified view of the common cold (X) and resultant symptoms (W, Y, and Z).

can be expressed as a product of four factors, $P(z \mid y, w)P(w \mid x)P(y \mid x)P(x)$, then it can be represented by the graph in Fig. 8.2.

An intuitive interpretation of this graph is that nodes are variables of interest, and directed arrows represent direct causal relationships. The relationships between disease and symptoms in the graph in Fig. 8.2 can be interpreted in this way. However, this interpretation only works in some domains, and only if the ordering $V_1, \ldots, V_k$ we picked has effect variables following cause variables in the ordering. Regardless of interpretation, directed graphs of this sort, which are called *Bayesian networks*[2], form a graphical representation of conditional independence of the distribution from which they are constructed. The first set of conditional independence statements represented by Bayesian networks follows by construction: in a Bayesian network, every variable is independent of its non-descendants given its parents (this is known as the *local Markov property*). In fact, Bayesian networks provide a graphical representation for conditional independence in a more general way. The way to think about this representation is to view paths in the graph as "pipes," conditional dependence as "water" flowing along these pipes, and conditional independence as a "blocked" flow. To check if two variables X, Y are independent given a set of variables $\mathbf{Z}$ in a Bayesian network, one checks if all paths from X to Y block the flow of influence. Path blocking is defined in terms of a notion called *d-separation*. Formally, *a path from X to Y is blocked or* d-separated *given Z if it contains one of the node triples in Fig. 8.3.* A path which is not d-separated is called *d-connected*. For instance, in Fig. 8.2, Y is independent of W given X, but not given X and Z.

To determine if X is d-separated from Y given Z, we simply check if all paths from X to Y are blocked by Z using the previous definition. A set X is d-separated from a set Y given Z, if every pair of nodes X, Y in $\mathbf{X}$, $\mathbf{Y}$ is d-separated by $\mathbf{Z}$. The power of Bayesian networks comes from the fact that d-separation in the graph always implies conditional independence in the corresponding distribution.

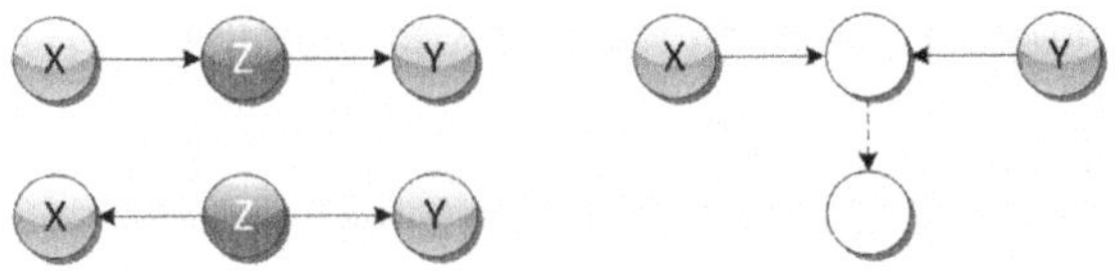

Figure 8.3: Patterns used to determine d-separation between two variables, X and Y. Darkly shaded nodes are elements of $\mathbf{Z}$. A node triple with converging arrows (a collider) cannot have any descendants in $\mathbf{Z}$.

---

[2] Bayesian networks are also referred to as *Bayesian belief networks* (BBNs), or *belief networks*. We use all three terms interchangeably throughout this book.

As an example, consider the Bayesian network shown in Fig. 8.4. This network describes the relationship between osteoporosis, its risk factors, and its consequences. Assuming this network is the correct representation of the domain (*i.e.,* assuming it was constructed using the Markov factorization of the domain joint distribution by the procedure described on the previous page) we can use d-separation on its structure to determine whether a conditional independence statement holds in this distribution. For example, according to d-separation, kidney function and hormone usage are marginally independent as there are only two paths between these nodes: kidney function → osteoporosis ← activity level ← age → hormone usage (*Path 1*), and kidney function → osteoporosis → fracture ← hormone usage (*Path 2*). Both of these paths contain a *collider* such that no descendant node of this collider is observed. In the first path, the arrows of the collider converge on osteoporosis; in the second path on fracture. On the other hand, kidney function and hormone usage are dependent once we observe fracture and osteoporosis. Again, we consider the two paths. The first path, *Path 1*, is d-separated as activity level is observed and lies on the chain age → activity level → osteoporosis. *Path 2* is d-connected as fracture is at the center of a collider triple and observed; thus, this path is d-connected as are kidney function and hormone usage given the observation of fracture.

Independence in probability distributions can also be represented using undirected graphs, where each vertex is a variable, and neighboring nodes are dependent. In this case, the distribution $P(v)$ is also decomposed into small pieces. However, while in the case of directed graphs these pieces represent conditional distributions of a node given its parents, in the case of undirected graphs these pieces are not even proper probability distributions. Rather, they are functions $f_i(\mathbf{c}_i)$ from value assignments of nodes $\mathbf{C}_i$ to real numbers, where $\mathbf{C}_i$ is a maximal clique in the undirected graph. The entire distribution is represented as $P(\mathbf{v}) = {}^1\!/_Z \prod_i f_i(\mathbf{c}_i)$, where Z is a normalizing factor such that $P(\mathbf{v})$ is a true probability distribution. An undirected graph which corresponds to a probability distribution in this way is called a *Markov network* [40]. Markov networks also have a graphical representation of conditional independence, and though their definition is a little less straightforward, this graphical representation is much simpler. In fact, a node in a Markov network is independent of its non-neighbors given its neighbors.

Representing Time

Graphical models can be used to represent domains that extend through time. Typically, the flow of time is modeled as a series of discrete steps, with variable interactions in

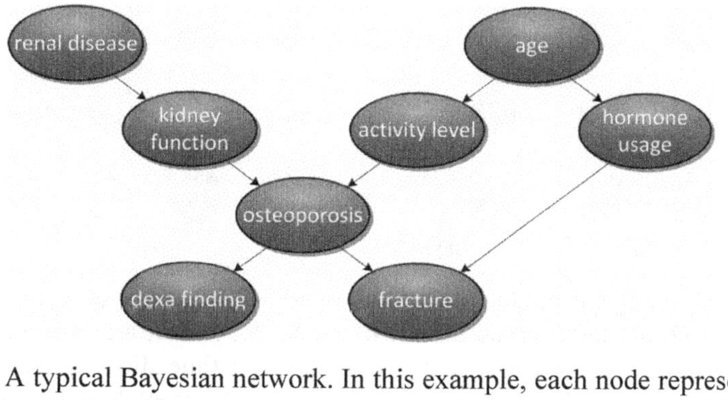

Figure 8.4: A typical Bayesian network. In this example, each node represents a variable related to osteoporosis.

any given step represented by the same graph, while interactions among variables in subsequent time steps encoded by another graph. As an example, consider the problem of a time-varying treatment: a patient with particular diseases (with some observable symptoms) is prescribed a regimen of medication to be taken each day, with each day's dose dependent on the measured symptoms of the previous day. The effect of the regimen is measured by considering the change of the symptoms on the last day compared to the first. A graphical model for this problem is shown in Fig. 8.5.

In this scenario, there are two unobserved diseases, D_1, and D_2, which result in three observable symptoms, S_1, S_2, and S_3 that are observed once daily with a single treatment, M, also administered daily based on the symptom measurements of the prior day. The Bayesian network representing variable interactions within a single day are shown in Fig. 8.5a, while the network representing interactions from the previous day to the next are shown in Fig. 8.5b. Using these two networks, it is possible to represent the course of a time-varying treatment for an arbitrary number of days, by simply "unrolling" the model. That is, we construct a copy of the network in Fig. 8.5a for every day the treatment is given and we connect nodes in a day t to nodes in a day $t + 1$ in a way given by the network in Fig. 8.5b. An example of the kind of "unrolled" model that results for a two day treatment is given in Fig. 8.5c. These temporal models are known as *dynamic Bayesian networks* (DBNs) [55].

A well-known special case of models of this type contains two nodes in each time slice, an unobserved parent, sometimes called the *state*, and an observed child, sometimes called the *sensor* (Fig. 8.6). The state at time step t affects the state at the next time step, $t + 1$, while the sensors affect nothing. A model of this type where sensor and state variables have a finite number of values is called a *hidden Markov model* (HMM). HMMs capture time-varying systems with changes governed by some hidden state. Testing variable independence via d-separation, as well as a variety of probabilistic inference and learning tasks is particularly simple in HMMs due to their special structure [72]. Nevertheless, as the dynamics of the modeled system become more complicated, the size of the hidden state necessary to provide a good model can quickly become intractable. In such cases, it is often possible to decompose the state and sensor variables into a larger set of variables such that additional independencies hold in this set. The resulting model would then become a dynamic Bayesian network. Note that as DBNs have no topology restrictions (other than graph replication across time slices), they can efficiently represent domains that HMMs cannot.

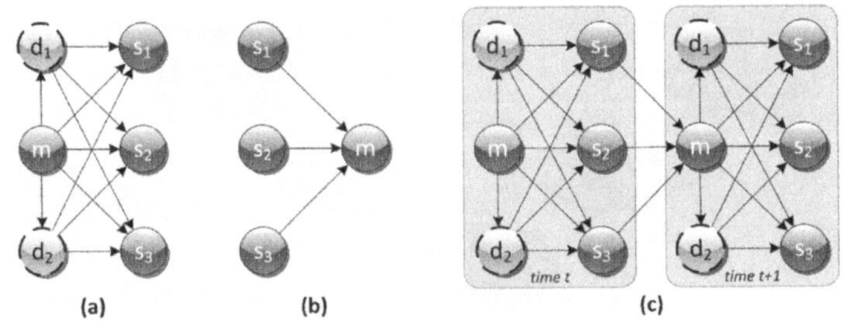

Figure 8.5: A graphical model for the time-varying treatment domain. M is a treatment; D_1 and D_2 are unobserved diseases; and S_1, S_2, and S_3 are observed symptoms. **(a)** The graph showing variable interactions in a single time slice. **(b)** The graph showing variable interactions between two time slices. **(c)** A complete model "unrolled" for two time slices.

Graphs and Causation

In the course of looking over the graphical models shown in the previous section, it may not have escaped your attention that very frequently the parents of a node can be interpreted as its direct causes. This observation is not an accident: a causal interpretation of the local Markov property – that every variable is independent of its non-effects given its direct causes – is true! Moreover, human beings naturally organize their knowledge of the world in terms of cause-effect relationships, and it is very easy for them to provide the set of independences needed by the local Markov property by appealing to their causal knowledge.

Even if our causal knowledge is restricted to just a topological ordering of variables consistent with causal directionality, it is possible to obtain a valid Bayesian network from this ordering alone. The algorithm for doing so is simple, and relies on the definition of Bayesian networks. For every variable V_i in the ordering, we look for a subset of variables $Pa(V_i)$ among those variables preceding V_i, or $V_1, \ldots, V_{i-1}$, such that $P(v_i \mid pa(v_i)) = P(v_i \mid v_{i-1}, \ldots, v_1)$, and no proper subset of $Pa(V_i)$ will have the same property. We make those variables the parents of V_i. In fact, this algorithm will work for any ordering of variables, although in reality causal orderings end up with graphs with fewer edges.

Consider a domain of four variables, X, W, Y, Z where the following independence statements hold: $X \perp\!\!\!\perp Z \mid Y, W$; $Y \perp\!\!\!\perp W \mid X$. We are going to construct Bayesian network for this domain using two variable orderings. First, we will use the ordering X, Y, W, Z. In the first step we add the node for X and no edges as X is the only node in the graph. Then we add Y and make X its parent, as $P(y) \neq P(y \mid x)$. Then we add W and make X its parent, as $P(w) \neq P(w \mid x) = P(w \mid x, y) \neq P(w \mid y)$. Finally, we add Z and make Y, W its parents because $P(z \mid y, w) = P(z \mid y, w, x)$ and for no proper subset of $\{Y, W\}$ is the above equality true. The result is shown in Fig. 8.7a. Now, we use the ordering Z, Y, W, X. In the first step, we add Z and no edges. Then, we add Y and make Z its parent, as $P(y \mid z) \neq P(y)$. Then we add W and make Y, Z its parents, as $P(w \mid y) \neq P(w \mid y, z) \neq P(w \mid z) \neq P(w)$. Finally, we add the node X and make W, Y its parents, given $P(x \mid w, y, z) = P(x \mid w, y)$, and for no proper subset of $\{W, Y\}$ is the above true. The resulting graph is shown in Fig. 8.7b.

In practice, Bayesian networks constructed by interpreting the parents of a variable as its direct causes are not only easier to understand, but tend to have less edges as well. In our example, the ordering X, Y, W, Z was causal for the common cold domain shown in Fig. 8.2. Interpreting edges causally not only results in more concise and understandable probabilistic models, but also allows us to treat the model itself as causal, and use it to answer causal questions such as, "*What is the direct effect of one variable on another*," or, "*Would the patient survive had we not given him the treatment?*"

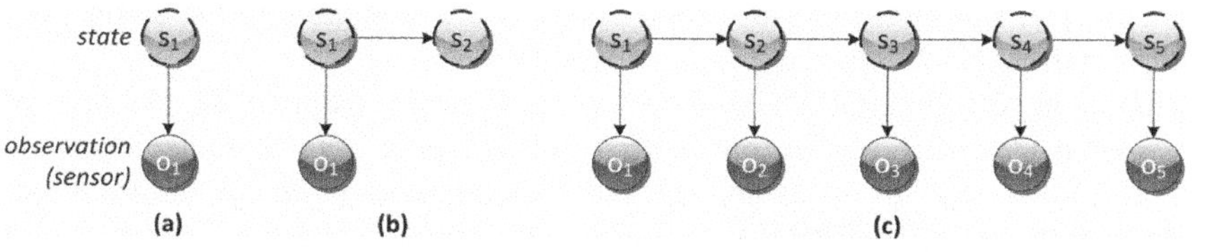

Figure 8.6: A hidden Markov model (HMM). **(a)** Interactions within a single time slice. **(b)** Interactions between consecutive time slices. **(c)** A full model "unrolled" across five time slices.

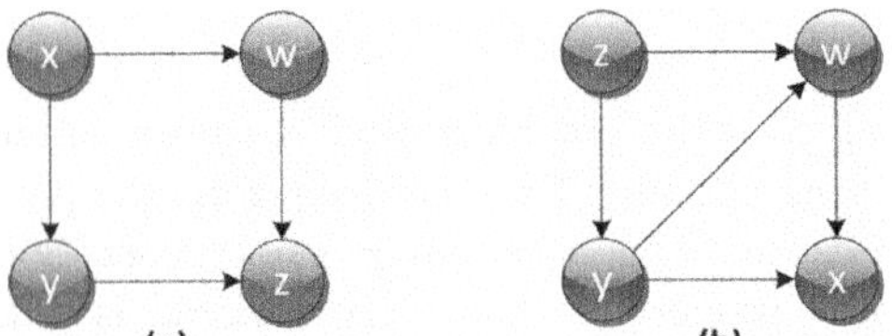

Figure 8.7: Two Bayesian networks for a four variable domain where $X \perp\!\!\!\perp Z \mid Y, W$ and $Y \perp\!\!\!\perp W \mid X$. **(a)** A Bayesian network constructed using the ordering X, W, Y, Z. **(b)** A Bayesian network constructed using the ordering Z, Y, W, X.

To deal with these types of questions, we must formalize what it means for the model itself (rather than our belief about the model) to change. This formalization and the variety of causal inference tasks it allows us to perform are the subjects of a subsequent section.

Bayesian Belief Networks in Medicine

Although a spectrum of approaches exists, Bayesian belief networks are an increasingly popular formalism for representing models of disease. The strength of a BBN-based approach lies in its ability to answer different types of queries, including: the computation of a *posterior probability* (an updated conditional probability given further knowledge about an event) for a specified hypothesis, given only partial evidence; and the support for "most likely" scenario queries, such as maximum *a posteriori* (MAP) hypotheses and *most probable explanations* (MPEs)[3]. Notably, the fact that a BBN can provide answers even in light of missing data is in contrast to conventional clinical decision-making paradigms (*e.g.*, rule-based systems, decision trees, regression analysis) where *all* variables must be known to reach a conclusion. As a rudimentary example, consider an individual for whom we have limited knowledge, such as his demographics, and whether he has high cholesterol and/or diabetes. A BBN constructed of these and other risk factors for stroke (*e.g.*, social and family history, hypertension, level of physical activity, etc.) is capable of estimating how likely a stroke is to occur, even if some of the variables are unknown for this specific person. Moreover, it is possible to update the estimate as new evidence becomes available. If the BBN models treatment information and outcomes, it is further feasible to posit questions regarding which treatment will optimally attain a desired result. [50] overviews applications of BBNs in biomedicine along the following categories:

- Diagnostic and prognostic reasoning. Multiple projects have used BBNs to classify symptoms and clinical findings; work on HEPAR, MUNIN, and Pathfinder exemplify early efforts [3, 4, 36, 49]. Applications have been as broad as internal medicine, to more specific areas such as tumor classification, liver disease, pulmonology, and mammography, among others [5, 12, 39, 41, 51, 59, 61, 62, 71, 84, 102]. A subset of this work supports outcome prediction based on the patient's presentation, providing prognostic capabilities.
- Treatment selection. BBNs have been used to guide treatment selection, allowing a user to view different scenarios to optimize some criteria and select a plan of action [46]. This group of applications encompasses *influence diagrams*, extensions of the traditional BBN to include cost functions, and can be seen as a global optimization problem across the network [52].

---

[3] These and other types of Bayesian queries are covered in further detail in Chapter 9.

- Functional linkage. BBNs are used in bioinformatics to model gene regulatory networks [30], analyze gene expression data [31], and compute genetic linkages [29]. *Genetic linkage analysis* is a statistical method for determining the distance between genes on a chromosome; knowing a gene's location and proximity on a chromosome can help identify individuals with a high probability of disease occurrence. One example of using a BBN to perform genetic linkage analysis is SUPERLINK [28], wherein the BBN is used to model the pedigree and recombination frequencies. [37] demonstrates the ability of Bayesian networks to classify high-dimensional gene expression data.

Belief Network Construction: Building a Disease Model

The process of constructing a Bayesian belief network for modeling a disease can be thought of in three stages: 1) the identification of variables of interest that describe the disease in terms of observable causes and effects; 2) the formalization of the relationships between these variables; and 3) the computation of the conditional probabilities dictated by the relationships. Each step and its challenges are briefly described below. To ground this discussion we first define common terminology with regard to belief networks (Fig. 8.8). The nodes in a BBN are often referred to as *evidence variables.* Although the real-world variable may be a continuous value, the evidence variable is a discretized version (*i.e.,* values are binned), with each discretized value being referred to as a *state.* The edges connecting these nodes are relationships, and the set of nodes and edges together make up the *network topology* or structure of the graph. A *conditional probability table* (CPT) is calculated for each node in the graph, specifying the probability of a given state of the evidence variable given its parents' states.

Variable selection. The first step in the development of a disease model is to identify those variables that can be directly observed as part of the disease process, and those intermediate and output variables that need to be inferred. This collective set of variables is then mapped to nodes within the proposed Bayesian belief network. Implicit to the variable selection task is the question of, *what is the intended use for the disease model?* For research, BBNs and the chosen variables can be used to explore relationships, often in a limited (controlled) setting; whereas for clinical usage, a BBN can be used as a diagnostic or prognostic tool in a broader environment. Here, we take the latter view. Thus, the selection of variables should be comprehensive, allowing

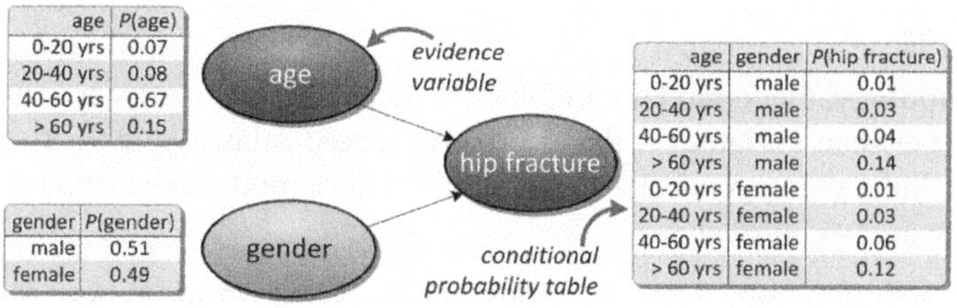

Figure 8.8: A hypothetical Bayesian belief network, showing conditional probability tables (CPTs). Nodes in the graph are referred to as evidence variables. The CPT for a given node is computed based on an enumeration of the possible states that its parents can take on. In this example, the Boolean value for hip fracture is determined by the combination of the age (0-20 years, 20-40 years, 40-60 years, and > 60 years) and gender (male, female). For nodes with no parents, the CPTs are equivalent to the variable's prior probability.

construction of a model, which can be used to make accurate scientific conclusions from the array of evidence available in a real-world patient medical record. Considerations for variable selection include:

- *Can the variable be practically measured as part of the routine management of a patient*? Some types of data may not be widely accessible to the scientific and/or medical community at large, or may simply be difficult to obtain for any number of reasons (*e.g.*, cost). For instance, new advances in imaging and other areas of clinical medicine are making possible new diagnostic insights into disease processes; yet the availability of such methods may be limited to specialized centers. If a variable cannot be collected as part of (routine) testing then its utility within a disease model is debatable: the inclusion of the node will result in a more precise representation, but may confuse the user and make the model more difficult to use. The issue of practicality also entails how consistently the measure can be taken and compared (*i.e.*, is it standardized?). Preferably, variables should be "technology independent" (*i.e.*, not dependent on how it was observed and/or measured). For imaging, for example, the acquisition of image series should be standardized either by the study protocol and/or post-processing (see Chapter 5).
- *What are the possible values that the variable can take on* (e.g., *is it continuous, discrete, or categorical*)? As part of the identification process, the variable's representation must also be chosen. Rigorous definitions for a variable and its measurements can be taken from existing ontologies and controlled vocabularies (*e.g.*, Unified Medical Language System, UMLS; Systematized Nomenclature of Medicine, SNOMED). Continuous variables can be discretized first by determining if an accepted categorical scale exists for mapping the values to a given class; and if no such scale exists, its data distribution should be examined and discretization techniques investigated. While a BBN can mix continuous and discrete variables (*e.g.*, hybrid BBNs), discretizing variables simplifies modeling and inference tasks. A naïve discretization method is to merely divide the quantitative space into n equal bins (*i.e.*, equal-width binning). However, given that the performance of a BBN is sensitive to the method and degree of discretization, an objective of discretization should be to minimize information loss, including the underlying data distribution. Several different discretization methods can be used; [54, 103] review univariate (*e.g.*, quantiles) and class-based methods (*e.g.*, minimum description length processing, attribute grouping). [48] also gives a comparison of discretization techniques.

Network topology. Once model variables are identified, the next step is to specify the association between variables – that is, the presumed relationships between the variables. Such links correspond closely to a scientific hypothesis relating a cause to an effect. There are two approaches:

1. <u>Expert specification</u>. This first approach is based on an expert specifying the relationships between variables (*i.e.*, X causes Y), and echoes the experiential knowledge and insights from clinicians and researchers based on past observations and experiments. Connections from variables can also be extracted from a meta-analysis of hypotheses from published literature. Though an expert-derived topology reflects the best working knowledge of a disease, the models may not account for *hidden variables* (*i.e.*, variables that are not considered in the model) that may be involved in additional d-connected paths not considered by the expert – resulting in a model that makes incorrect statements about conditional independence.

This problem can be addressed by algorithms that attempt to learn the network topology from the data (see below); and also as part of the evaluation of the network topology with pairwise tests for conditional independence. Pairs of d-separated variables can be tested for conditional independence and compared with the real-world dataset for consistency (*e.g.,* a χ^2 test); failure of the independence test will indicate the potential presence of a hidden variable that will require (re)modeling.

2. Automatically learning the structure. In many domains, only general features of the graph are known. With exemplar data, the network topology can also be inferred through algorithms that attempt to deduce the graph structure based on the strength of probabilities [35][4]. The appeal of learning a BBN structure is apparent in experimental efforts, where the relationship between observed variables may not yet be known. Demonstrations of learning BBN structure from clinical data sets can be found in [2, 5, 9, 101]. Apart from an exhaustive search across all possible DAGs for the model (which is generally intractable), several algorithms exist for the purpose of suggesting structures. Two categories of algorithms exist: *scoring-based learning algorithms* [19] that traverse a model search space to optimize some function that measures both how well a proposed BBN fits the data, and how "concise" (in terms of the number of parameters) the BBN is; and *constraint-based* algorithms, which infer structure by ruling out graphs inconsistent with the pattern of constraints observed in the data. Examples of structural learning methods include BENEDICT [1], greedy equivalence search [15], max-min hill climbing [93], fast causal inference (FCI) [85], and structural expectation maximization. The choice of algorithm to learn the BBN is dependent on whether full or partial observability of the data is considered (*i.e.,* are there hidden variables, or variables for which data is unavailable?). A full description of the theory behind these algorithms is beyond the scope of this chapter; and the reader is referred to [56] for a more in-depth discussion. The use of these automated learning techniques is often limited by the lack of availability of sufficient samples for performing either conditional independence tests on which constrained-based algorithms rely, or model selection for search-based approaches. Furthermore, even if sufficient samples are available, the algorithms are often intractable.

Conditional probability calculation. The last stage in creating the BBN is to compute the CPTs as indicated by the network topology; this step is referred to as *parameter estimation* or *parameter learning*. For disease models, the ideal situation is to compute these values using clinical data gleaned from a representative population. The probabilities can then be calculated via the well-known *expectation maximization* (EM) algorithm[5] [24]. In theory, this strategy provides an accurate portrayal of the observations related to a disease (*i.e.,* its presentation within the cohort, its treatment/resolution, etc.). But in some situations, there are not enough samples for reliable estimation of conditional probability parameters. Instead, we can turn to two other

---

[4] For causal models, described later in this chapter, this process is called *inductive causal inference* (also known as *causal discovery*), which is learning the causal graph from data.

[5] Briefly, the EM algorithm consists of two parts: the *E-step*, wherein the missing data are estimated using the *conditional expectation*, based on the observed data and the current estimate of the model parameters; and the *M-step*, where the likelihood function is maximized assuming the missing data are known (the estimated data from the E-step being used in lieu of the actual missing data).

techniques to provide the probabilities. First, reputable resources such as a published randomized controlled trial can potentially be substituted to provide the needed value(s). Notably, the incorporation of experimental results (*e.g.*, from an RCT) alongside the observational data to compute the CPTs must be tempered by issues of selection bias and whether the context of the experiment is compatible with the cohort [26]. Second, if no suitable reported results are found in the literature, a medical expert can also be asked to provide a "best-guess" estimate; while such approximations may be biased and inaccurate, prior studies have demonstrated their general utility [20, 53]. Various methods have been developed to address these issues by: 1) eliciting well-calibrated probabilities from experts [60], such as by using probability assessment scales [99]; 2) validating the accuracy of the elicited probabilities; and 3) reducing the number of probabilities needed to be elicited.

Calculation of the CPTs is usually predicated on the idea that the patient cases are "complete" for each variable. Realistically, however, datasets often contain missing data elements. Missing data within a patient record can arise for a variety of reasons, including incomplete documentation, lost data, or even an outright failure to acquire the information. The missing data problem is also common in controlled experiments, such as RCTs. Assuming that data is missing at random (noting that the issue of observations missing due to specific biases is more problematic), data and the learning of CPTs can be addressed through the use of parameter fitting algorithms. Parametric models can also be used to impose a probability distribution on incomplete data for inferring missing values (*e.g.*, multivariable normal distributions, log-linear models, general location models, and 2-level linear regressions models) [10, 80]. Non-parametric algorithms can also be used to impute missing values [14], with the advantage of being more efficient for moderate sized data sets and less susceptible to fitting errors. Non-trivial joint distributions can be approximated by Monte Carlo sampling methods (see Chapter 9).

Modeling time. Disease processes and the documentation of a patient's state in the medical record are inherently temporal, as are the questions we ask about a patient (see Chapter 7); thus, it makes sense to consider the sequential nature of the information and changes within a disease model and BBN. Indeed, DBNs have been used to model time-variant states such as with fMRI (functional magnetic resonance imaging), gene expression networks, and other domains of interest in bioinformatics [25, 104, 106, 107]. A key issue arises, however, in using such techniques on clinical data: how does one define the temporal granularity and the time slice transition model in the clinical setting? While in other areas the time interval may be naturally given by sampling frequency and/or specific events, the temporal span over which the collection of patient data occurs in a real-world clinical environment is highly variable. We approach this problem from two directions:

1. Semantically-defined temporal clusters. Often, the diagnostic and therapeutic phases of disease management are defined by sentinel events that are recognized clinically. For example, in treating a cancer, there are distinct phases between the diagnosis, subsequent treatments (*e.g.*, surgical resection, radiation therapy, chemotherapy), and cancer remission and survivorship. One strategy therefore is to cluster clinical data around these time points, or to examine the temporal patterns suggested by encounters. This approach further limits the potential problems related to inference across the DBN by constraining the number of time slices that need to be considered.

2. Conditional temporal clusters. The above temporal clustering approach refines the disease model to provide a specific BBN for each "natural" stage of the underlying phenomenon and its progression: while surely better than only considering a static BBN across the entire disease chronology, this strategy may not provide sufficient temporal granularity as sequences of data within a single "cluster" are still considered together. For instance, consider a neuro-oncology patient that initially starts chemotherapy on *drug X*, but switches to *drug Y* due to side effects – but the tumor has not yet progressed. In a DBN that temporally clusters around tumor progression, this fact would be represented as the patient having been on *drug X* and *drug Y* concurrently, or *drug Y* before *drug X*: in effect, the precise temporal relationship of the events is lost. One possible approach to handle this problem is to define "conditional" temporal clusters whereby the DBN transition model is defined on changes of a given variable (or set of variables) deemed significant by an expert (*e.g.,* such as a change in chemotherapy regimen).

Causal Inference

In the previous sections, we have seen how the language of probability theory can help us cope with exceptions, unknowns, and uncertainty in a principled way; how conditional independence can make probability reasoning tractable; and how a combination of probability theory with graphs gives a qualitative, visual representation of conditional independence. Below, we show how directed graphs can be extended to represent models that are not only probabilistic but causal; how complex questions about causal effects and counterfactuals can be effectively formalized and answered in such models; and how such causal inference can be performed in practice.

Causal Models, Interventions, and Counterfactuals

The objects that we will study in this section are called *graphical causal models*. Like Bayesian networks, such models contain a set of variables of interest. However, the values of some variables in such models are determined by means of a function from values of other variables, while variables that are not so determined are random. Such models are represented by directed graphs called *causal diagrams* where variables are nodes, and an arrow leads from node *X* to node *Y* if the value assumed by *X* is used in the function that determines the value of *Y*. See Fig. 8.9a for a typical causal diagram.

In this diagram, there are three observable variables of interest: 1) environmental exposure, represented by smoking; 2) a disease mechanism, represented by tar; and finally 3) the disease itself, represented by cancer. These variables are fully deterministic – their values are determined exactly by the values of their parents in the graph. There are also two unobservable variables, U_1 and U_2, where U_2 represents a common genetic cause of both disease propensity and the likelihood of exposure, whereas U_1 influences

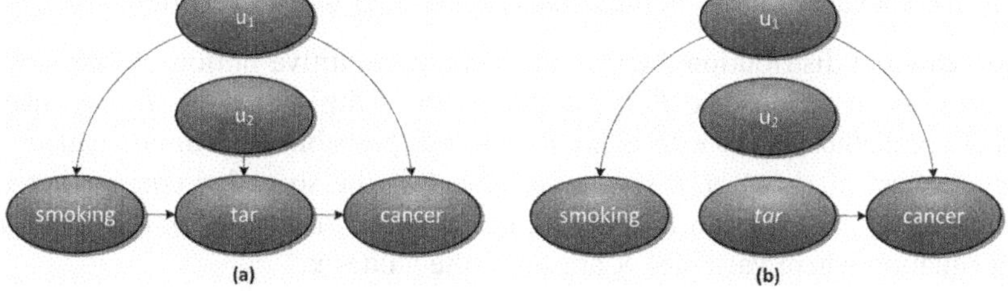

Figure 8.9: (a) A simple causal diagram, illustrating three hypothetical variables (smoking, tar, and cancer) alongside two unknown variables (u_1, u_2). **(b)** The mutilated graph corresponding to the intervention, *do*(tar).

the disease mechanism. U_1 and U_2 are ordinary random variables, representing whatever factors we need to take into account to render their observable children in the graph deterministic given those factors.

It turns out this kind of representation, where some variables are random, and others are determined, is equivalent to the Bayesian network representation discussed earlier where every variable is random, and variable relationships are encoded by means of a CPT of variables given their parents. What this means is that all variables in a causal diagram, even those determined by functions, are random variables and that conditional independence statements between these variables are encoded, just as in Bayesian networks, via d-separation. For instance, the unobserved variables U_1 and U_2 are independent as all paths connecting them are d-separating. However, in addition, the arrows in a causal diagram were defined in terms of a direct functional relationship. As functions in causal models represent causal mechanisms, arrows represent not only probabilistic dependence, but also direct causation.

It is important to keep in mind that the notion of direct cause is relative and dependent on model granularity rather than absolute. In our cancer model, tar serves as a "direct" cause of cancer. Yet this is clearly not true – cancer is a complex progressive disease with many intermediate events between any particular carcinogen such as tar, and eventual growth of tumors. Indeed even a detailed biological description of cancer will still be a model missing certain relevant mechanisms as it abstracts details of genetics, organic chemistry, and ultimately physics. As absolute notions of direct cause are very difficult indeed, we settle for a much simpler model-specific notion of direct cause: an arrow from X to Y in a causal diagram simply means that given the granularity of causal mechanisms we have chosen, X influences the mechanism that determines Y without mediation of any other observable variable at the chosen level of granularity.

The additional meaning carried by directed arrows in causal diagrams allows us to interpret the notion of interventions or experiments graphically. An intervention on a variable X, denoted by $do(x)$ in [67], corresponds to an abstract operation where the value of X in a causal model is set to x, regardless of the normal behavior of X as dictated by the model. In other words, though ordinarily X is a function of its parents in the graph, the intervention replaces this function by one that sets X to a constant x. The distribution of the remaining variables $\mathbf{V} \setminus \{X\}$ is termed an *interventional distribution*, and denoted either as $P(\mathbf{v} \setminus \{x\} \mid do(x))$ or $P_x(\mathbf{v} \setminus \{x\})$. There is a very simple way to graphically represent both conditional independences and direct causal claims encoded by $P_x(\mathbf{v} \setminus \{x\})$. The key observation is that $do(x)$ ignores parents of X, but leaves all other causal mechanisms and random variables as is. Intuitively, the way to represent $do(x)$ graphically would be to remove all arrows pointing to X, while leaving the rest of the causal diagram intact. In fact, it can be shown that the resulting graph, called the *mutilated graph*, is the correct representation of both conditional independences and direct cause claims in the causal model after the intervention.

An interventional distribution $P_{\mathbf{X}}(\mathbf{y})$ represents the intuitive notion of "causal effect," where $do(\mathbf{x})$ is an action, and $P_{\mathbf{X}}(\mathbf{y})$ is the effect of this action on $\mathbf{Y}$. Because causal models are probabilistic, the effects are themselves probabilistic. An alternative notation for causal effect of $do(\mathbf{x})$ on a single variable Y is the so called *counterfactual variable*, denoted by $\mathrm{Y}_{\mathbf{X}}$. $\mathrm{Y}_{\mathbf{X}}$ can be interpreted to mean, "the value Y attains in a hypothetical situation where variables $\mathbf{X}$ are set to the values $\mathbf{x}$."

What is interesting about counterfactual variables is that it is possible to define probability distributions over a set of them, even if two individual (counterfactual) variables in the set disagree on the values of some variables (in the original model). For instance, we may ask what the effect is of smoking on cancer among the sub-population of non-smokers. Intuitively, this effect may be different from that in the general population, as non-smokers may differ in important ways from smokers that would influence their susceptibility to cancer. If X is the variable representing smoking (with values x meaning smoker and x' meaning non-smoker), and Y is the variable representing the presence of cancer (value y), then the above effect can be represented by the *counterfactual probability* $P(Y_{x'} = y \mid X = x)$, in other words the probability of the counterfactual variable "the effect of smoking on cancer" assuming value cancer conditioned on observing non-smoker. Note that the two variables involved in this distribution, X and $Y_{x'}$ disagree on the value of X. Yet despite this conflict, this probability is well-defined from the causal model, and generally is not equal to 0. In general, if you have a set of counterfactual variables $Y^k_{\mathbf{X}^k},\ldots, Y^k_{\mathbf{X}^k}$, the joint probability $P(Y^1_{\mathbf{X}^1} = y^1,\ldots, Y^k_{\mathbf{X}^k} = y^k)$ is defined as:

$$\sum_{\{\mathbf{u} \mid Y^1_{x^1}(\mathbf{u})=y^1 \wedge \ldots \wedge Y^k_{x^k}(\mathbf{u})=y^k\}} P(\mathbf{u})$$

In a causal model there are two kinds of variables: the observed variables, $\mathbf{V}$, whose values are determined by means of a function of other variables; and the unobserved variables, $\mathbf{U}$, whose values are determined by a probability distribution $P(\mathbf{u})$. What this means is that if we fix the values of $\boldsymbol{U}$, the model becomes deterministic. Thus, to figure out the probability of $P(Y^k_{\mathbf{X}^k} = y^1,\ldots, Y^k_{\mathbf{X}^k} = y^k)$, which is just a set of value assignments to counterfactual variables, all we have to do is add up the probabilities of all value assignments of $\mathbf{U}$ that give rise deterministically to the value assignments we want. This summation is precisely the above definition.

Counterfactual probabilities have an intuitive graphical representation in terms of possible "worlds." In our smoking/cancer example from before, there are two possible worlds: one in which a person does not smoke, and another in which a hypothetical intervention making that same person smoke is made. Each such world can be represented by a causal diagram. The first world is represented by the original causal diagram, as shown in Fig. 8.9, while the second is represented by a mutilated graph where arrows incoming to smoker are cut, signifying an intervention at that variable. These two worlds are not completely disjoint: the intuition is they share their histories up until the point where the intervention is performed in one but not the other world.

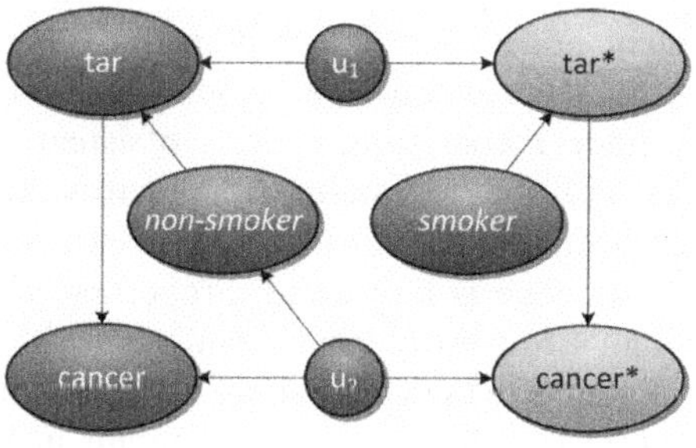

Figure 8.10: A twin network graph for the query, "*The effect of smoking on cancer among non-smokers.*" The (green) italicized variable, non-smoker, is conditioned upon while smoker is fixed.

We represent these shared histories by making sure these two causal diagrams share their unobservable **U** variables. The result in our example, which consists of only two worlds, is known as the *twin network graph* [8], and is shown in Fig. 8.10: the world on the left is the original causal diagram, while the world on the right is the mutilated graph, where the arrow from U_2 to smoker is absent. The distribution we are interested in corresponds to the distribution over the variable cancer* given that we fix smoker on the right, and observe non-smoker on the left. The twin network graph can be generalized to other queries, including those involving more than two worlds, although there are additional subtleties in such cases [82, 83].

Causal inference in graphical models deals with three major problems: representation, deduction, and induction. The representation problem consists of translating informal causal questions of interest to doctors and scientists, such as, "*What is the effect of smoking on cancer,*" or, "*Does vitamin C help fight the common cold?*" into a formal statement in the language of interventions and probability. Typically this formal statement will be an interventional or counterfactual probability distribution. The deduction problem, also known as the *identification problem*, consists of finding estimates for these distributions given certain causal assumptions represented by a causal diagram, and partial knowledge of the causal model itself, typically represented by a probability distribution over observable variables. Lastly, the induction problem consists of learning aspects of the causal diagram from observations, and certain assumptions about the causal model that allow conclusions about the graph to be drawn from the pattern of conditional independences observed in the data.

Latent Projections and their Causal Interpretation

The meaning of arrows in Bayesian networks is closely tied to conditional independence, as a variable in a Bayesian network is independent of its non-descendants given its parents. This probabilistic interpretation of arrows also holds in graphical causal models, because the d-separation criterion is valid in causal diagrams just as it is in Bayesian networks. However causal models entail an additional, causal interpretation of arrows. Informally, this interpretation asserts that an arrow from X to Y means that X is a direct cause of Y. It is possible to translate this statement into the language of interventions. Before doing so, it would be beneficial to introduce a certain canonical form for causal diagrams, known as a *latent projection*.

Causal diagrams, as described thus far, are merely DAGs with certain nodes marked as being latent (*i.e.,* unobserved). This representation, though it closely mirrors Bayesian networks, can be inconvenient as sections of d-connected paths containing only latent variables really act like a single edge between observable nodes, but are cluttered with multiple intermediate latent nodes. The latent projection representation is meant to remedy this problem. In this representation, a d-connected section from an observable node X to an observable node Y consisting entirely of latent variables is replaced by one of two kinds of edges: 1) if the path starts with an arrow away from X and ends in an arrow into Y, the edge is directed; or 2) if the path starts with an arrow into X and ends in an arrow into Y, the edge is bidirected. Properties of d-separation guarantee that if the path starts with an arrow away from X and ends with an arrow away from Y, then the path cannot be d-connected (as we cannot condition on latent variables). Fig. 8.11a shows the latent projection for the causal diagram in Fig. 8.9a. The intuitive interpretation of edges in a latent projection is that a directed edge represents "direct

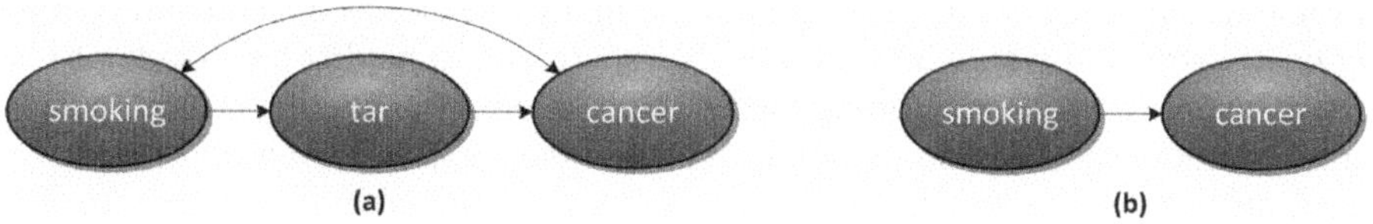

Figure 8.11: **(a)** A latent projection of the smoking/cancer causal diagram. **(b)** A simple causal diagram.

causal influence" given the granularity of the model and a bidirected edge represents "confounding" or "unobserved causes."

Recall that edges in a causal diagram are drawn based on the functional relationships between variables in a causal model. Given that these functional relationships are modified by interventions, we would expect the assumptions encoded by arrows in causal diagrams (and latent projections) to have something to do with interventions. It turns out that the absence of a directed arrow from X to Y in a latent projection implies that $P_{x,\mathbf{Z}}(y) = P_{\mathbf{Z}}(y)$, where $\mathbf{Z}$ is the set of parents of Y, while the absence of a bidirected arrow between X and a set $\mathbf{Y} = \{Y^1,\ldots, Y^k\}$ implies that $X_{\mathbf{Z}} \perp\!\!\!\perp Y^1_{\mathbf{Z}_1},\ldots, Y^k_{\mathbf{Z}_k}$, where $\mathbf{Z}$ is the set of parents of X, and $\mathbf{Z}_i$ is the set of parents of Y^i. Given the interpretation of edges in a latent projection, these implications make sense. If X is not a direct cause of Y, we expect that fixing X will not affect Y once all direct causes of Y are fixed. Similarly, the absence of hidden causes between X and Y implies that fixing their respective observable direct causes should render them independent. Independence in latent projections can be checked by a generalization of d-separation to graphs containing bidirected arcs [75]. Next, we will show that knowledge of the latent projection along with a probability distribution over observable variables can be sufficient for answering a wide range of causal queries of interest.

Identification

Interventional distributions $P(y \mid do(x))$ capture the notion of causal effect of an action. Many clinical questions can be phrased in terms of causal effects (*e.g., Did exposure to substance X cause this patient's bout of asthma?*). These questions can be formalized with results of a particular intervention in a particular (although possibly very complicated) model. Certain other questions, like ascertaining the effect of smoking on non-smokers, can be phrased in terms of distributions of counterfactual variables (which can be more complicated than interventional distributions). It is thus very important to obtain estimates of these distributions. How can this be done?

If we have access to the entire causal model – that is, we know exactly the distribution of every parentless variable and the functions that determine variables from their parents – we can calculate any quantity derived from the model, including both causal effects and counterfactuals. However, this degree of knowledge is often not realistic. Even if we knew all the correct parent-child relationships in the causal model for cancer, for example, we would not know the functional relationships to the level of determinism, and we certainly would not know the exact distributions of all background factors that may contribute to cancer.

An alternative method is to actually intervene in the domain and observe the results, such as in performing a randomized (clinical) control trial (*e.g.,* where a test group is made to smoke and the control group is made not to). Naturally, there are problems with this approach as well – some interventions are very expensive and possibly irreversible, and some are illegal or immoral. Finally, the effects of some interventions may be disastrous. It would be very valuable, therefore, to predict the effects of interventions before actually performing them. The problem of predicting causal quantities of interest from available information is known as the *identification problem*.

What information is typically available? We are certainly free to collect data on observable variables in the system – with enough of such data we will be able to have an estimate of the joint distribution over these variables, which we have denoted as $P(\mathbf{v})$. It is also typically assumed that we know the correct causal diagram. In many domains such causal knowledge may be elicited from experts, or learned by automated causal discovery algorithms [43, 85, 88, 97]. Even if only aspects of the graph can be learned, it is possible to identify many causal effects of interest, although for simplicity we assume that the entire causal diagram is known.

It might seem counterintuitive that the effects of an action can be predicted from passive observations. Fig. 8.11b illustrates a simple example where this kind of prediction is plausible. In this diagram, we assume smoking causes cancer, and there are no hidden common causes. The claim is that in this diagram, $P(\text{cancer} \mid do(\text{smoking})) = P(\text{cancer} \mid \text{smoking})$. To see why this statement might be true, consider the visual interpretation of conditional dependence and causal effect. If we observe the value of a particular variable, say X, in a causal model, this observation influences other variables in the model that are dependent on X. As d-separation captures independence, we can view the changes introduced by the observation as flowing along d-connected paths away from X. This analogy with flow is, in fact, the basis for *message passing algorithms* that efficiently implement inference in Bayesian networks [66]. Some d-connected paths that lead away from X start with arrows pointing towards X, and some start with arrows pointing away from X. The former paths are called *backdoor paths*, while the latter are called *frontdoor paths*.

We can view an effect an intervention $do(x)$ has on other variables in a similar way. The only difference between observing X and fixing X is that the latter ignores the values of parents when setting X, which means the resulting graph does not contain arrows pointing towards X. Thus, the only d-connected paths along which the effect of the intervention flows are frontdoor paths. But in the graph in Fig. 8.11b, all d-connected paths from smoking are frontdoor paths. Thus the effect of observing smoking and fixing smoking is the same.

Consider the more complicated graph in Fig. 8.11a, and assume we are interested in the effect of physical changes in the body due to environmental exposure on cancer. The physical changes here are represented by tar and environmental exposure by smoking. In this graph, $P(\text{cancer} \mid do(\text{tar}))$ is not equal to $P(\text{cancer} \mid \text{tar})$, as there is a backdoor d-connecting path (through smoking) from tar to cancer, so the previous reasoning does not apply. However, we know from d-separation that paths can be blocked by conditioning. In fact the above backdoor path can be blocked by conditioning on smoking. Thus, if we were interested in the conditional effect $P(\text{cancer} \mid \text{smoking}, do(\text{tar}))$, that is the effect of tar on cancer among either *smokers* or *non-smokers*, then we could equate this effect with $P(\text{cancer} \mid \text{smoking}, \text{tar})$, as the only backdoor path is d-separated because we conditioned on smoking. However, we are not interested in the effect in a particular subpopulation with certain smoking habits – we are interested in the effect in the overall population. What we can do is average above conditional effects, weighted by the prior probability of smoking. In other words, we can estimate

$P(\text{cancer} \mid do(\text{tar}))$ to be equal to $\prod_s P(\text{cancer} \mid \text{smoking} = s, \text{tar})P(\text{smoking} = s)$, where s ranges over all possible smoking levels from non-smoker to multiple packs a day.

It turns out that this method generalizes, and whenever we can find a set $\mathbf{Z}$ such that $\mathbf{Z}$ does not contain descendants of X and after $\mathbf{Z}$ is conditioned on, there are no d-connected backdoor paths from X to Y, then $P(y \mid do(x))$ is equal to $\sum_{\mathbf{Z}} P(y \mid \mathbf{z}, x)P(\mathbf{z})$. This is known as the *backdoor criterion* [67]. The set $\mathbf{Z}$ must obey the above restrictions so it can block precisely all backdoor paths from X but no frontdoor paths.

In some cases we may be interested in a causal effect such that no such set $\mathbf{Z}$ can be found. For instance, in the same graph, we may be interested in the effect of smoking on cancer. Smoking and cancer share a bidirected arc which forms a backdoor d-connecting path which is impossible to block by d-separation, since bidirected arcs represented hidden common causes, and hidden variables by definition may not be conditioned on. Surprisingly, even in this case this effect can be identified. The intuition here is that the effect $P(\text{cancer} \mid do(\text{smoking}))$ can be decomposed in this graph into two effects: the effect of smoking on tar, that is $P(\text{tar} \mid do(\text{smoking}))$; and the effect of tar on cancer, that is $P(\text{cancer} \mid do(\text{tar}))$. Specifically, $P(\text{cancer} \mid do(\text{smoking})) = \sum_t P(\text{cancer} \mid do(\text{tar} = t))P(\text{tar} = t \mid do(\text{smoking}))$, where t is possible levels of tar we can find built up in the lungs. We will justify this decomposition based on the assumptions encoded by the given causal graph a little later, but given that this decomposition is valid, it becomes a simple matter to estimate each of the two new effects. $P(\text{tar} \mid do(\text{smoking}))$ is equal to $P(\text{tar} \mid \text{smoking})$ as the only backdoor path from smoking to tar is d-separated (because we do not observe cancer). On the other hand, we can apply the backdoor criterion to identify $P(\text{cancer} \mid do(\text{tar}))$. Putting everything together, we obtain:

$$P(\text{cancer} \mid do(\text{smoking})) = \sum_t P(t \mid \text{smoking}) \sum_s P(\text{cancer} \mid t, s)P(s)$$

In fact, whenever we can find a set $\mathbf{Z}$ such that $\mathbf{Z}$ intercepts all directed paths from X to Y, there is no backdoor path from X to $\mathbf{Z}$, and all backdoor paths from $\mathbf{Z}$ to Y are blocked by X, we can estimate $P(y \mid do(x))$ as $\sum_{\mathbf{Z}} P(\mathbf{z} \mid x) \sum_{x'} P(y \mid x', \mathbf{z})P(x')$. This result is known as the *frontdoor criterion* [67]. Naturally, there are some cases where an effect is identifiable, but neither the frontdoor nor backdoor criteria apply. Consider Fig. 8.12, where we have added a new node representing socioeconomic background, which may have a causal influence on smoking, tar buildup by other means, and cancer by means other than smoking. We are interested in the joint effect of a particular background and tar on cancer in this graph. In fact, there is a general method of computing causal effects from assumptions embedded in the graph, known as *do-calculus* [67]. The do-calculus consists of three rules:

Rule 1. $P_X(y \mid z, w) = P_X(y \mid w)$ if $(\mathbf{Y} \perp\!\!\!\perp \mathbf{Z} \mid \mathbf{X}, \mathbf{W})_{G_{\overline{X}}}$

Rule 2. $P_{X,Z}(y \mid w) = P_X(y \mid z, w)$ if $(\mathbf{Y} \perp\!\!\!\perp \mathbf{Z} \mid \mathbf{X}, \mathbf{W})_{G_{\overline{X},\underline{Z}}}$

Rule 3. $P_{X,Z}(y \mid w) = P_X(y \mid w)$ if $(\mathbf{Y} \perp\!\!\!\perp \mathbf{Z} \mid \mathbf{X}, \mathbf{W})_{G_{\overline{X},\overline{Z(W)}}}$

where $Z(\mathbf{W}) = Z \setminus An(\mathbf{W})_{G_{\overline{X}}}$. Here $G_{\overline{X}}$ denotes the graph obtained from G by removing all arrows pointing towards $\mathbf{X}$, and $G_{\overline{X},\underline{Z}}$ is obtained from G by removing all arrows pointing towards $\mathbf{X}$ and all arrows pointing away from $\mathbf{Z}$. The statement $(\mathbf{Y} \perp\!\!\!\perp \mathbf{Z} \mid \mathbf{X})$ denotes d-separation of $\mathbf{Y}$ and $\mathbf{Z}$ given $\mathbf{X}$ in the appropriate graph.

Though the notation may be daunting, what the rules assert is fairly straightforward. Recall that the graph $G_{\overline{X}}$ is precisely the causal diagram that represents conditional independence statements in the model after an intervention *do*(x) is performed. If $\mathbf{Y}$ is d-separated from $\mathbf{Z}$ given $\mathbf{X}$ and $\mathbf{W}$ in this graph, this implies that $\mathbf{Y}$ is independent of $\mathbf{Z}$ given $\mathbf{X}$ and $\mathbf{W}$ in the post-interventional distribution $P_{\mathbf{X}}(\mathbf{v} \setminus \mathbf{x})$. But by definition of conditional independence this can be rewritten as $P_{\mathbf{X}}(y \mid \mathbf{z}, \mathbf{w}) = P_{X}(\mathbf{y} \mid \mathbf{w})$. Rule 1 thus asserts that d-separation in the post-intervention graph captures post-intervention independence. Rule 2 states that if all backdoor paths from $\mathbf{Z}$ to $\mathbf{Y}$ are blocked by conditioning on $\mathbf{W}$ and fixing $\mathbf{X}$, then it makes no difference for the purpose of $P_{\mathbf{X}}(\mathbf{y} \mid \mathbf{z}, \mathbf{w})$ if we fix or condition on $\mathbf{Z}$. We used this observation in discussing the backdoor criterion; Rule 2 merely codifies it in a more general way. Finally, Rule 3 governs when interventions are irrelevant, though unfortunately the precise conditions are somewhat complex. We illustrate the usage of these rules by identifying *P*(cancer | *do*(tar, background)) in Fig. 8.12. To make the derivation easier to read we shorten variable names in this query as *P*(c | *do*(t, b)), and show the rule used in each derivation above the equality symbol (*P* above equality means the identity follows by rules of probability):

$$P(c \mid do(t,b)) \overset{P}{=} \sum_{s} P(c \mid s, do(t,b)) P(s \mid do(t,b)) \overset{2}{=} \sum_{s} P(c \mid s, b, do(t)) P(s \mid do(t,b)) \overset{2}{=}$$

$$\sum_{s} P(c \mid s,b,t) P(s \mid do(t,b)) \overset{3}{=} \sum_{s} P(c \mid s,b,t) P(s \mid do(b)) \overset{2}{=} \sum_{s} P(c \mid s,b,t) P(s \mid b)$$

As another application of do-calculus, we can justify the use of the frontdoor criterion in the graph shown in Fig. 8.11a by showing that *P*(cancer | *do*(smoking)) = $\sum_t$ (cancer | *do*(tar = t))*P*(tar = t | *do*(smoking)):

$$P(cancer \mid do(smoking)) \overset{P}{=} \sum_{t} P(cancer \mid tar = t, do(smoking) P(tar = t \mid do(smoking) \overset{2}{=}$$

$$\sum_{t} P(cancer \mid do(tar = t, smoking)) P(tar = t \mid do(smoking) \overset{3}{=}$$

$$\sum_{t} P(cancer \mid do(tar = t)) P(tar = t \mid do(smoking))$$

Rules of do-calculus are more general than specific graphical criteria such as the backdoor and frontdoor criteria. One may reasonably wonder whether do-calculus can be used to identify every identifiable causal effect (*i.e.,* is it complete) – and in fact, the answer is yes [38, 81, 82]. Moreover, there exist graphical criteria that precisely characterize identifiable effects, and polynomial algorithms that directly construct expressions for such effects in terms of *P*(**v**), without having to search for a valid do-calculus derivation. Similar algorithms exist for identifying counterfactual queries like, "the

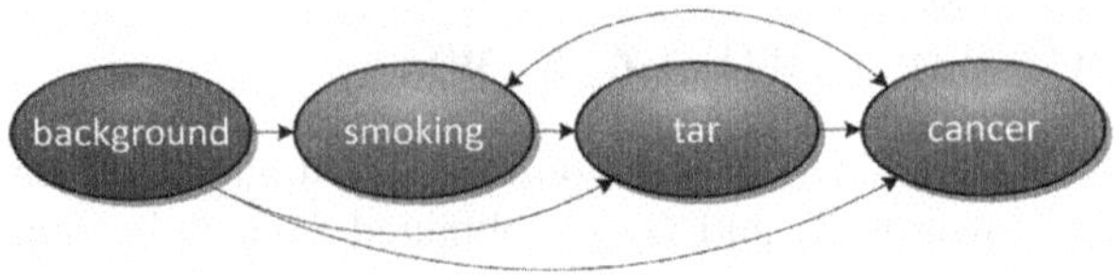

Figure 8.12: A causal graph where *P*(cancer | *do*(tar, background)) is not identifiable by either the backdoor or frontdoor criteria.

effect of smoking on cancer among non-smokers." Further details can be found in the literature [81, 82, 83].

Discussion and Applications

The notion of causality and clinical medicine are inherently intertwined. Physicians are trained to think in terms of causal relationships when examining a patient: the standard diagnostic methodology taught in medical school (*i.e.,* differential diagnosis) asks the physician to mentally develop a list of possible causes (*i.e.,* diseases) and then to narrow the list given observed effects (*i.e.,* a patient's symptoms). But a physician's understanding of causality is primarily based on intuition or so-called "implicit" case-based knowledge from past experiences, rather than on a formal and complete understanding of the underlying disease process, or "explicit" knowledge. Physicians equipped with the knowledge and the tools to communicate using a formal representation for expressing causal assumptions can potentially better organize and explain their thought processes, resulting in improved models of disease and enhancing methods for diagnosis and treating patients.

[45] first proposed the systematization of diagnostic reasoning processes using a combination of set theory and Bayesian reasoning to clearly define factors involved in a disease; and starting in the late 1970s, several attempts were made to computerize this process [64]. One such system, CASNET, used a network of pathophysiological states to describe a disease as a pattern of causally-related states [98]. Diagnosis was performed by first determining which states were valid given certain observations, then matching the pattern of (in)valid states against a disease database. While able to hypothesize the presence of many states concurrently, CASNET could not determine whether a causal relation existed among these states. Later efforts used scoring systems to compute a likelihood value based on clinical observations and matching observed vs. expected findings [65, 68]. And finally, CADUCEUS encoded causal knowledge using four specific types of relationships between diseases and clinical states (cause-of, caused-by, develops-into, complications-of) [69]. During the late-1980s, systems started utilizing a more complex representation that combined symbolic reasoning approaches with the Bayesian approaches explored in earlier works. Although Bayesian belief networks continue to be a commonly used framework for describing dependencies and causal relationships in medical data, care must be taken in using them for drawing causal conclusions: as we saw, their construction does not necessarily lead to directed arrows representing direct causation, as is the case in graphical causal models. A non-causal Bayesian network is only capable of correctly answering questions that can be derived from the joint distribution, such as diagnostic questions phrased as a conditional distribution, association questions between variables of interest, and so on. Causal graphs, on the other hand, permit a wide variety of truly causal questions to be placed on a firm mathematical footing, and estimated given available information.

Presently, one key challenge within the informatics community is in the construction of comprehensive disease models from clinical datasets. The expanding contents of electronic medical records (EMRs) provide a rich opportunity to create robust population-based models that can elucidate disease processes and inform evidence-based medical practice. However, apart from the problems of standardizing patient data access and its contents, using this information to create disease models faces several technical barriers. First, model development must address the issue of completeness. Many models thus far have only addressed a well-circumscribed set of variables within a single domain; but physicians rarely make decisions based on a single source of information. Disease models must come to employ the full range of clinical observations that are made daily by physicians, blending multiple perspectives (radiology, pathology, genetics, etc.) into a single, comprehensive decision-making tool. Likewise, a temporal perspective can enrich the models: constructs such as DBNs can better denote change as it pertains to the course of disease, and assist in the selection of treatments to obtain desired outcomes. In doing so, an explanative framework for observations over time that spans multiple spatial scales – from molecular to organism – can perhaps be realized, wherein phenomena at the micro-level can be used to explain effects at the macro-level. Second, disease models must be made "portable." A common complaint about Bayesian networks, for example, is that the probabilities derived for one site may not be representative of another (*e.g.,* the validity of the BBNs joint probability distribution may be questionable). Thus, the representations used to characterize a domain must either be sufficiently widely used/accepted, or a ready means to recompute CPTs from site-specific data must be supported. In a related vein, methods are needed to validate a given disease model against a given site's population.

We conclude by touching upon several practical issues in creating BBNs and in performing causal inference when dealing with the particulars of clinical patient data.

Building Belief and Causal Networks: Practical Considerations

The procurement of patient case data to be used for creating a disease model must proceed with caution such that measurements are as accurate as possible and specification of findings and their content is formalized and precise. To this end, work has looked at linking a formal data model with a BBN; specifically, the phenomenon-centric data model (PCDM, see Chapter 7) to a graphical model. The relationships delineated in a BBN can be seen as a subset of the overall PCDM. Fig. 8.13 illustrates a bi-level stratification used to connect the two models: the PCDM subsumes the probabilistic

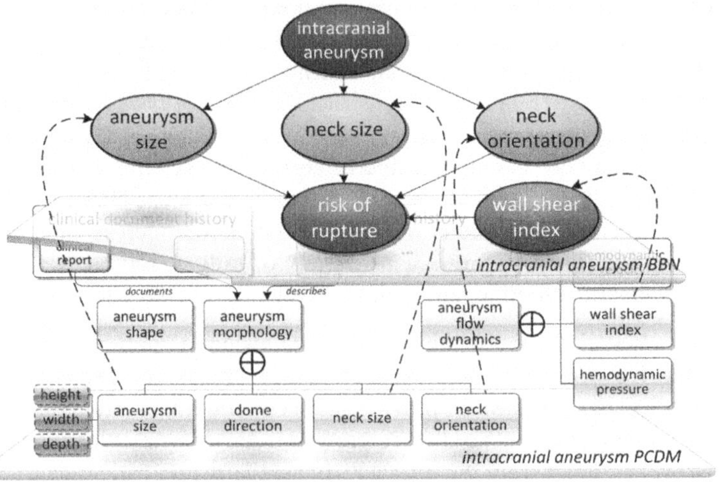

Figure 8.13: Stratification of a Bayesian belief network with the phenomenon-centric data model. Data from the PCDM is used to compute the CPTs defined in the BBN.

graphical model, with specific mappings between findings, theories, and phenomena to evidence variables within the disease model BBN. The concept of evidence and hypotheses within the PCDM are then bound to a directed path within the BBN, establishing the rationale for the observations and clinical findings. In creating the PCDM-BBN connection, the source of information for a given evidence variable is linked to an underlying clinical data source through the PCDM database.

While this strategy helps facilitate computation of probabilities from structured patient data, certain issues remain, including: ensuring that sufficient data is available; taking into account intrinsic uncertainty or error with clinically-derived observations; and handling any potential bias that may arise in the data collection process. Additional discussion regarding the construction of BBNs can be found in [23, 89, 94].

Accruing Sufficient Patient Data

As we have shown, there exist sophisticated methods for solving deductive causal inference problems in the presence of uncertainty, given the knowledge of the joint distribution over observable variables, and the causal assumptions represented by the graph. Unfortunately, in practical settings both of these givens are problematic. In many domains where estimating causal effects is important, such as epidemiology or biomedical informatics, the amount of samples available for estimating the observable joint distribution may be quite small. When few outcomes are available relative to the number of covariates, reliable estimation of many parameters is not possible using maximum likelihood estimates [34], thereby complicating the construction of the CPTs. What this means in practice is that the methods we discussed for estimating causal effects, such as the backdoor criterion, must be replaced by small sample versions using a number of statistical techniques.

Propensity scores. One common way of dealing with a small amount of data is to use *propensity scores* [79]. Assume we are interested in estimating the effect $P(y \mid do(x))$, and the backdoor criterion happens to hold with set $\mathbf{Z}$ (*i.e.*, $P(y \mid do(x)) = \sum_{\mathbf{Z}} P(y \mid x, \mathbf{z})P(\mathbf{z})$). If the number of samples for estimating $P(\mathbf{v})$, and thus the two probabilities $P(\mathbf{z})$, and $P(y \mid x, \mathbf{z})$ is small, the resulting effect estimate will not be very good. In particular, if the cardinality of the Cartesian product of the value sets of variables in $\mathbf{Z}$ is much larger than the number of samples, it does not make sense to use $P(\mathbf{z})$ or $P(y \mid x, \mathbf{z})$ directly. A statistical trick that has been developed to handle this situation is to make use of the propensity score $L(\mathbf{z})$, which is defined as the conditional distribution, $P(x \mid \mathbf{z})$. Note that if X and $\mathbf{Z}$ are discrete, $L(\mathbf{z})$ is a table mapping values of $\mathbf{Z}$ into a real number between 0 and 1, namely the probability of X assuming value x given that we observe values $\mathbf{z}$. As $L(\mathbf{z})$ is a function of $\mathbf{Z}$, a set of random variables, it is itself a random variable. Knowing the value of this random variable makes X and $\mathbf{Z}$ independent, that is $X \perp\!\!\!\perp \mathbf{Z} \mid L(\mathbf{z})$. If plotted, $L(\mathbf{z})$ will look like a set of points between 0 and 1, with the number of points equal to $\prod_{i=1}^{k} |Z_i|$, where $\{Z_1,\ldots, Z_k\} = \mathbf{Z}$. If the function is "nice enough," these points will cluster into a few large groups, which can be well-approximated by a small set of real numbers, $l_1,\ldots, l_m$, which can be estimated from limited samples. It can be shown using the above independence that $\sum_{\mathbf{Z}} P(y \mid x, \mathbf{z})P(\mathbf{z}) = \sum_{i=1}^{m} P(y \mid x, l_i)P(l_i)$.

Structural equations models. Another strategy that can have implications even in the large sample case is to make parametric assumptions about the graphical causal model – that is, rather than assume the functions relating variables and distributions over unobserved variables are arbitrary, we can restrict functions and distributions to

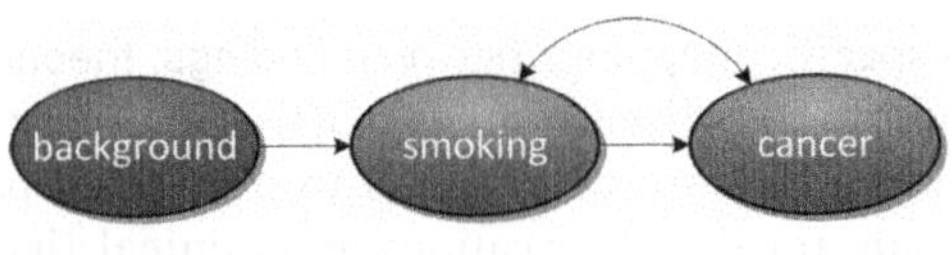

Figure 8.14: A causal graph where P(cancer | do(smoking)) is not identifiable in general, but is identifiable using linear Gaussian models.

certain parametric families. Typically, linear functions and normal distributions are assumed, which results in a class of graphical causal models known as *structural equation models* (SEMs) [33, 42, 100]. The observable distribution $P(\mathbf{v})$ in SEMs is always a multivariate normal distribution and can always be renormalized to have a 0 mean. Thus, this distribution can be fully characterized by the covariance matrix, Σ, which is determined by the number of parameters which grows quadratically with the number of observable variables, rather than exponentially as is the case in general. To be more precise, each function in an SEM is of the form $Y_j = \sum_i c_{ji} Y_i + \varepsilon_j$, where ε_j is a noise term, and $Cov(\varepsilon_i, \varepsilon_j)$ is the i, j entry in a matrix that, together with the matrix of coefficients c_{ji}, result in the covariance matrix Σ over observable variables. The coefficient c_{ji} itself can be thought of as representing the direct effect of Y_i on Y_j. As all functions are linear, this direct effect is a single number, not dependent on the assignments of values to other parents of Y_j. An SEM is said to be identified if every direct effect coefficient c_{ji} can be computed in terms of the covariance matrix Σ. SEMs are convenient not only because it takes less samples to obtain reliable estimates of causal effects, but it is also possible to identify certain causal effects that cannot be identified in general. Consider for instance Fig. 8.14, where we are interested in computing the direct effect of smoking on cancer. In an arbitrary non-parametric causal model, this corresponds to identifying P(cancer | do(smoking)). It is not difficult to show that this effect is not identifiable in this graph. All we have to do is find two causal models that agree on the observable distribution $P(\mathbf{v})$, but disagree on the effect in question. There are many counterexamples with these properties; here we give an uncomplicated one where every variable is binary. The way in which background and smoking are related is not important for this example, so we omit discussion of the background variable. Let the hidden parent of smoking and cancer be a fair coin, and assume the function relating that parent and smoking is identity (*i.e.,* smoking is just equal to its parent), and let the value of cancer be determined to be the exclusive or (bit parity) of its parents in one model, and equal 0 in the other. It is easy to see that the observational behavior of the two models is identical, as the bit parity of equal values is 0. However, intervening on smoking cuts the link from the hidden parent, and reveals the different functional relationship between cancer and its parents in the two models.

In SEMs, by contract, it can be shown that the direct effect coefficient relating smoking to cancer is identifiable from Σ. The intuition is that the correlation between two variables in an SEM can be decomposed, due to the linearity of the model, into a sum of terms where each term corresponds to the portion of this correlation that "flows" along a particular d-connected path between these variables. Each term consists of a product of coefficients corresponding to edges on this path. This method is known as *Wright's rule of path analysis* [100]. In our example, if we let the direct effect coefficient of background on smoking equal α, and the direct effect coefficient of smoking on cancer equal β, then the correlation of background and cancer, that is $\sigma_{B,C}$ is equal to $\alpha * \beta$, by Wright's rule, while the correlation of background and smoking, that is $\sigma_{B,S}$ is equal to α. Hence, $\beta = \sigma_{B,C}/\sigma_{B,S}$. In general, we can always identify the direct effect coefficient of X on Y if we can find a variable Z that is dependent on X, but independent of all error terms that have an influence on Y not mediated by X. Such a Z is called an *instrumental variable* (it is an instrument by which the causal effect may be identified) and the method is called the *method of instrumental variables*. Instrumental variables can be used in non-linear models as well, though appropriate use then becomes a rather technical matter. Another strategy for dealing with non-identifiable effects, if making parametric assumptions is not faithful to the domain, is to try to find bounds for the effect, and hope that these bounds either restrict the effect sufficiently to permit causal conclusions, or collapse to a point estimate entirely [7, 91].

Dimension reduction techniques. There are many more methods for estimating causal effects from limited samples. Statistical learning theory indicates that learning probabilities in a lower-dimensional (feature) space can improve results while using less data [96]. To this end, dimensional reduction techniques can be used, in effect joining several variables together to provide a more compact representation for which the CPTs can be computed. Notably, the use of dimensionality reduction comes at the potential expense of some obfuscation as the combination of variables can remove the intuitiveness of key variables within a graphical model. Examples of linear dimensional reduction methods are the well-known principal component analysis and linear discriminant analysis (PCA, LDA). Non-linear dimensional reduction methods include kernel PCA, Isomap [90], and multidimensional scaling (MDS). Comprehensive reviews of these and other techniques can be found in [13, 95].

Handling Uncertainty in Data

Clinical data serves as a proxy for underlying phenomena and thus is subject to both uncertainty and error. *Qualitative uncertainty* results from a lack of definitive knowledge (*e.g.*, a patient is unsure if he had a fever). In such cases, which are frequently documented in medical reports, the degree of certainty expressed in the statement should be maintained as part of the feature extraction process (see Chapters 6 & 7 with regards to natural language processing and the use of the PCDM to record uncertainty). The use of "uncertain" evidence and the issues in interpreting evidential statements can then be taken into account within a disease model. For instance, BBN variables with qualitative uncertainty can model an unknown state that represents this uncertainty or use a parent binary variable representing a certain/uncertain state to condition the variable. *Quantitative measurement error* arises from inherent limitations of the precision of an instrumentation. [86] provides an example of this problem using microarray data in gene expression networks, wherein conditional independence may not hold because of instrumental error bounds; however, it is noted that if the noise/error is sufficiently small relative to the observations, then conditional independence may hold approximately. In cases where measurement error is known or can be modeled, it is also possible to adopt the use of a "measurement idiom" that combines an observed value with estimation accuracy [57]; this paradigm is commonly encountered in sensor networks using a noisy sensor model [21]. For example, if a discrete measured variable, X, is modeled by a noisy observation, X', which in turn is given by a continuous distribution (*e.g.*, Gaussian), we can interpret X' as "soft evidence" for X with a Bayes factor.

Noisy-OR. One method for handling error within belief networks is to model it explicitly as noise. The noisy-OR construct is frequently used for this purpose, representing a single observed effect being the result of one or more possible causes with an adjunctive noise factor (called a *leak variable*) added to represent non-modeled (perhaps unknown) causes (Fig. 8.15). A logical operation table is used to dictate the underlying behavior of the node, similar to the classical logical-OR operator used in digital circuits. Three assumptions are made: 1) that conditional independence exists between each cause; 2) that all possible causes are modeled, with a leak variable used to represent the non-explicitly modeled component (such as noise/error); and 3) that any element that is inhibitory to one cause does not inhibit any other cause (*e.g.*, if X causes Y, and Z causes Y, but A inhibits X, then A does not also inhibit Z) and that these "inhibitors" are not modeled as nodes but rather as "noise parameters" within the probabilities. Hence, in a noisy-OR component, if none of the parents is true, then the effect is not true with 100% certainty unless the leak variable is true. If one of the parents is true, then the probability that the event is true is equal to the probability associated with that parent's noise parameter. Consider Fig. 8.15: to implement this noisy-OR model, four probabilities are needed (the number of causes, three; plus one for the leak variable). Let $P(c_1) = 0.75$, $P(c_2) = 0.67$, $P(c_3) = 0.9$, and $P(l) = 0.05$. The CPT for the noisy-OR effect node can be compactly represented by $P(e \mid \alpha) = (1 - P(l))\prod \theta_{q_i}$, where e is the effect, α are causes (c_1, c_2, c_3), and θ_{q_i} is the probability that the suppressor parameter for a given cause is active (*i.e.*, $\theta_{q_i} = 1 - P(c_i)$) (Fig. 8.15c).

Handling Selection Bias

Non-experimental data such as from clinical observations can be subject to selection bias, resulting in a dependency between variables due to some selection criteria [32]. From a statistical viewpoint, tests can be performed to assess the population dataset: for example, a Mantel-Haenszel test can be used to compare covariate distributions; or a Kolmorgorov-Smirnov test to compare marginal distributions. From the perspective of constructing a BBN or causal model, several approaches have been proposed to identify selection bias in causal frameworks. [17] describes conditions under which structure and parameters can be learned from conditional independence tests given selection bias; [18] extends the use of a selection variable along with the number of (un)sampled cases, combined with prior beliefs, to compute a posterior probability. [87] outlines the fast causal inference (FCI) algorithm to detect selection bias in the presence of latent variables. We note, however, that selection bias is still an ongoing challenge – for instance, the majority of results on identification of causal effects assume no selection bias. Though recent work provided some identification results in graph structures inferred under possible selection bias [105], the problem in general remains open.

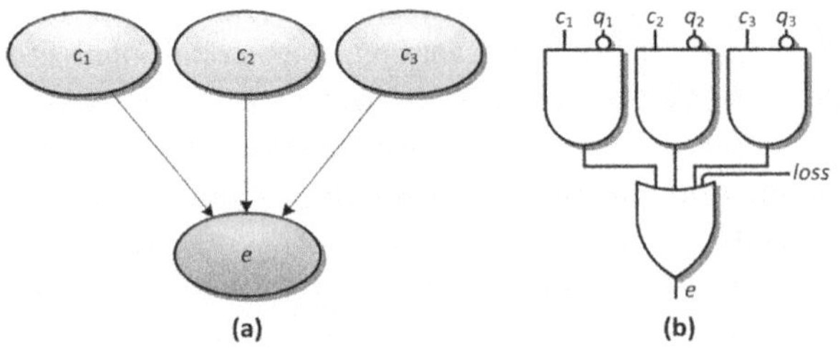

c_1	c_2	c_3	$P(e)$
true	true	true	(1-0.05)x(0.25)(0.33)(0.1)
true	true	false	(1-0.05)x(0.25)(0.33)
true	false	true	(1-0.05)x(0.25)(0.1)
true	false	false	(1-0.05)x(0.25)
false	true	true	(1-0.05)x(0.33)(0.1)
false	true	false	(1-0.05)x(0.33)
false	false	true	(1-0.05)x(0.1)
false	false	false	(1-0.05)

(a) (b) (c)

Figure 8.15: (a) The noisy-OR can be used to model parent nodes all contributing to a given child, such as this BBN, with some associated noise or error. **(b)** The idea is to model the child node's CPT like a logical-OR gate with added noise, represented as the leak variable, *l*. **(c)** The calculation of the resultant CPT for the child node is shown based on the prior probabilities of the parent and the leak variable's probability.

References

1. Acid S, de Campos LM (2001) A hybrid methodology for learning belief networks: BENEDICT. Intl J Approximate Reasoning, 27(3):235-262.
2. Acid S, de Campos LM, Fernandez-Luna JM, Rodriguez S, Maria Rodriguez J, Luis Salcedo J (2004) A comparison of learning algorithms for Bayesian networks: A case study based on data from an emergency medical service. Artif Intell Med, 30(3):215-232.
3. Andreassen S, Suojanen M, Falck B, Olesen K (2001) Improving the diagnostic performance of MUNIN by remodelling of the diseases. Artificial Intelligence in Medicine, pp 167-176.
4. Andreassen S, Woldbye M, Falck B, Andersen SK (1987) MUNIN: A causal probabilistic network for interpretation of electromyographic findings. Proc 10th Intl Joint Conf on Artificial Intelligence, pp 366-372.
5. Antal P, Fannes G, Timmerman D, Moreau Y, De Moor B (2004) Using literature and data to learn Bayesian networks as clinical models of ovarian tumors. Artif Intell Med, 30(3):257-281.
6. Ash RB, Doleans-Dade CA (2000) Probability & Measure Theory. 2nd edition. Academic Press, San Diego, CA.
7. Balke A, Pearl J (1994) Counterfactual probabilities: Computational methods, bounds, and applications. Proc 10th Conf Uncertainty in Artificial Intelligence (UAI), pp 46-54.
8. Balke A, Pearl J (1994) Probabilistic evaluation of counterfactual queries. Proc 12th American Assoc Artificial Intelligence (AAAI), pp 230-237.
9. Brown LE, Tsamardinos I, Aliferis CF (2004) A novel algorithm for scalable and accurate Bayesian network learning. Stud Health Technol Inform, 107(Pt 1):711-715.
10. Bryk AS, Raudenbush SW (1992) Hierarchical linear models: Applications and data analysis methods. Sage Publications, Newbury Park.
11. Buchanan BG, Shortliffe EH (1984) Rule-based expert systems: The MYCIN experiments of the Stanford Heuristic Programming Project. Addison-Wesley, Reading, Mass..
12. Burnside ES, Rubin DL, Fine JP, Shachter RD, Sisney GA, Leung WK (2006) Bayesian network to predict breast cancer risk of mammographic microcalcifications and reduce number of benign biopsy results: Initial experience. Radiology, 240(3):666-673.
13. Carrerira-Perpinan MA (1997) A review of dimension reduction techniques (Technical Report). Dept Computer Science, University of Sheffield. www.dcs.shef.ac.uk/intranet/re-search/resmes/CS9609.pdf. Accessed February 5, 2009.
14. Caruana R (2001) A non-parametric EM-style algorithm for imputing missing values. Proc 8th Intl Workshop Artificial Intelligence and Statistics, Key West, FL.
15. Chickering DM (2002) Optimal structure identification with greedy search. J Mach Learn Res, 3:507-554.
16. Chung KL (2001) A Course in Probability Theory Revised. 2nd edition. Academic Press, San Diego, CA.
17. Cooper GF (1995) A Bayesian method for learning belief networks that contain hidden variables. J Intell Inf Sys, 4(1):71-88.
18. Cooper GF (2000) A Bayesian method for causal modeling and discovery under selection. Proc 16th Conf Uncertainty in Artificial Intelligence (UAI), pp 98-106.
19. Cooper GF, Herskovits E (1992) A Bayesian method for the induction of probabilistic networks from data. Machine Learning, 9:309-347.
20. Coupé VM, Peek N, Ottenkamp J, Habbema JD (1999) Using sensitivity analysis for efficient quantification of a belief network. Artif Intell Med, 17(3):223-247.
21. Darwiche A (2009) Modeling and reasoning with Bayesian networks. Cambridge University Press, New York.

22. Dawid AP (1979) Conditional independence in statistical theory. J Royal Statistical Society, 41(1):1-31.
23. Dekhtyar A, Goldsmith J, Goldstein B, Mathias KK, Isenhour C (2009) Planning for success: The interdisciplinary approach to building Bayesian models. International Journal of Approximate Reasoning, 50(3):416-428.
24. Dempster AP, Laird M, Rubin D (1977) Maximum likelihood from incomplete data using the EM algorithm. J Royal Statistical Society, 39(1):1-38.
25. Dojer N, Gambin A, Mizera A, Wilczynski B, Tiuryn J (2006) Applying dynamic Bayesian networks to perturbed gene expression data. BMC Bioinformatics, 7:249.
26. Druzdel MJ, van der Gaag LC (2000) Building probabilistic networks: "Where do the numbers come from?" (Guest editorial). IEEE Trans Knowledge and Data Engineering, 12(4):481-486.
27. Duda RO, Hart PE, Nilsson NJ (1976) Subjective Bayesian methods for rule-based inference systems. Proc Natl Computer Conf (AFIPS), pp 1075-1082.
28. Fishelson M, Geiger D (2002) Exact genetic linkage computations for general pedigrees. Bioinformatics, 18(S1):189-198.
29. Fishelson M, Geiger D (2004) Optimizing exact genetic linkage computations. J Comput Biol, 11(2-3):263-275.
30. Friedman N (2004) Inferring cellular networks using probabilistic graphical models. Science, 303(5659):799-805.
31. Friedman N, Linial M, Nachman I, Pe'er D (2000) Using Bayesian networks to analyze expression data. J Comput Biol, 7(3-4):601-620.
32. Greenland S (2003) Quantifying biases in causal models: Classical confounding vs collider-stratification bias. Epidemiology, 14(3):300-306.
33. Haavelmo T (1943) The statistical implications of a system of simultaneous equations. Econometrica, 11:1-12.
34. Harrell FE, Jr., Lee KL, Mark DB (1996) Multivariable prognostic models: Issues in developing models, evaluating assumptions and adequacy, and measuring and reducing errors. Stat Med, 15(4):361-387.
35. Heckerman D (1999) A tutorial on learning with Baysesian networks. In: Jordan M (ed) Learning in Graphical Models. MIT Press, Cambridge, MA.
36. Heckerman DE, Horvitz EJ, Nathwani BN (1992) Toward normative expert systems: Part I. The Pathfinder project. Methods Inf Med, 31(2):90-105.
37. Helman P, Veroff R, Atlas SR, Willman C (2004) A Bayesian network classification methodology for gene expression data. J Computational Biology, 11(4):581-615.
38. Huang Y, Valtorta M (2006) Pearl's Calculus of intervention is complete. Proc 22nd Conf Uncertainty in Artificial Intelligence (UAI), pp 217-224.
39. Kahn CE, Jr., Roberts LM, Shaffer KA, Haddawy P (1997) Construction of a Bayesian network for mammographic diagnosis of breast cancer. Comput Biol Med, 27(1):19-29.
40. Kindermann R, Snell JL (1980) Markov Random Fields and their Applications. American Mathematical Society.
41. Kline JA, Novobilski AJ, Kabrhel C, Richman PB, Courtney DM (2005) Derivation and validation of a Bayesian network to predict pretest probability of venous thromboembolism. Ann Emerg Med, 45(3):282-290.
42. Kline RB (2005) Principles and Practice of Structural Equation Modeling. The Guilford Press, New York, NY.
43. Lam W, Bacchus F (1994) Learning Bayesian belief networks: An approach based on the MDL principle. Computational Intelligence, 10(4):269-293.

44. Lavrac N, Keravnou E, Zupan B (2000) Intelligent data analysis in medicine. In: Kent A, et al. (eds) Encyclopedia of Computer Science and Technology, vol 42, pp 113-157.
45. Ledley RS, Lusted LB (1959) Reasoning foundations of medical diagnosis. Science, 130(3366):9-21.
46. Leibovici L, Fishman M, Schonheyder HC, Riekehr C, Kristensen B, Shraga I, Andreassen S (2000) A causal probabilistic network for optimal treatment of bacterial infections. IEEE Trans Knowledge and Data Engineering, 12(4):517-528.
47. Lewis D (1973) Counterfactuals. Harvard University Press, Cambridge, MA.
48. Liu H, Hussain F, Tan CL, Dash M (2002) Discretization: An enabling technique. Data Mining and Knowledge Discovery, 6(4):393-423.
49. Lucas PJ, Segaar RW, Janssens AR (1989) HEPAR: An expert system for the diagnosis of disorders of the liver and biliary tract. Liver, 9(5):266-275.
50. Lucas PJ, van der Gaag LC, Abu-Hanna A (2004) Bayesian networks in biomedicine and healthcare. Artif Intell Med, 30(3):201-214.
51. Luciani D, Marchesi M, Bertolini G (2003) The role of Bayesian networks in the diagnosis of pulmonary embolism. J Thromb Haemost, 1(4):698-707.
52. Meyer J, Phillips MH, Cho PS, Kalet I, Doctor JN (2004) Application of influence diagrams to prostate intensity-modulated radiation therapy plan selection. Phys Med Biol, 49(9):1637-1653.
53. Monti S, Carenini G (2000) Dealing with the expert inconsistency in probability elicitation. IEEE Trans Knowledge and Data Engineering, 12(4):499-508.
54. Monti S, Cooper GF (1998) A multivariate discretization method for learning Bayesian networks from mixed data. Proc 14th Conf Uncertainty in Artificial Intelligence (UAI), pp 404–413.
55. Murphy K (2002) Dynamic Bayesian networks: Representation, inference, and learning. Department of Computer Science, PhD dissertation. University of California, Berkeley.
56. Neapolitan RE (2003) Chapter 8, Bayesian structure learning. Learning Bayesian Networks. Prentice Hall, London.
57. Neil M, Fenton N, Nielson L (2000) Building large-scale Bayesian networks. The Knowledge Engineering Review, 15(3):257-284.
58. Neyman J (1923) Sur les applications de la thar des probabilities aux expereince agaricales: Essay des principles. (Excerpts reprinted and translated to English, 1990). Statistical Science, 5:463-472.
59. Nikiforidis GC, Sakellaropoulos GC (1998) Expert system support using Bayesian belief networks in the prognosis of head-injured patients of the ICU. Med Inform, 23(1):1-18.
60. O'Hagan A, al. E (2006) Uncertain Judgements: Eliciting Experts' Probabilities. John Wiley & Sons, London.
61. Ogunyemi OI, Clarke JR, Ash N, Webber BL (2002) Combining geometric and probabilistic reasoning for computer-based penetrating-trauma assessment. J Am Med Inform Assoc, 9(3):273-282.
62. Onisko A (2003) Probabilistic causal models in medicine: Application to diagnosis in liver disorders. Institute of Biocybernetics and Biomedical Engineering, PhD dissertation. Polish Academy of Science.
63. Parker RC, Miller RA (1987) Using causal knowledge to create simulated patient cases: The CPCS Project as an extension of INTERNIST-1. Proc Ann Symp Computer Applications in Medical Care, pp 473-480.
64. Patil RS (1987) Causal reasoning in computer programs for medical diagnosis. Comp Methods and Programs in Biomedicine, 25(2):117-124.

65. Pauker SG, Gorry GA, Kassirer JP, Schwartz WB (1976) Towards the simulation of clinical cognition: Taking a present illness by computer. Am J Med, 60(7):981-996.
66. Pearl J (1988) Probabilistic Reasoning in Intelligent Systems: Networks of Plausible Inference. Morgan Kaufmann Publishers, San Mateo, CA.
67. Pearl J (2000) Causality: Models, Reasoning, and Inference. Cambridge University Press, New York.
68. Pople H (1977) The formation of composite hypotheses in diagnostic problem solving: An exercise in synthetic reasoning. Proc 5th Intl Joint Conf Artificial Intelligence, Cambridge, MA, pp 1030-1037.
69. Pople H (1982) Heuristic methods for imposing structure on ill-structured problems: The structuring of medical diagnostics. In: Szolovits P (ed) Artificial Intelligence in Medicine. Westview Press, Boulder, CO, pp 119-190.
70. Press SJ (2003) Subjective and Objective Bayesian Statistics: Principles, Models, and Applications. John Wiley & Sons, Hoboken, NJ.
71. Price GJ, McCluggage WG, Morrison MM, McClean G, Venkatraman L, Diamond J, Bharucha H, Montironi R, Bartels PH, Thompson D, Hamilton PW (2003) Computerized diagnostic decision support system for the classification of preinvasive cervical squamous lesions. Hum Pathol, 34(11):1193-1203.
72. Rabiner LR (1989) A tutorial on hidden Markov models and selected applications in speech recognition. Proc IEEE, 77(2):257-286.
73. Reiter R (1980) A logic for default reasoning. Artificial Intelligence, 13:81-132.
74. Reiter R (1981) On interacting defaults. Proc 4th Intl Joint Conf Artificial Intelligence (IJCAI), pp 270-276.
75. Richardson T, Spirtes P (2002) Ancetral graph Markov models. Annals of Statistics, 30:962-1030.
76. Riva A, Bellazzi R (1996) Learning temporal probabilistic causal models from longitudinal data. Artif Intell Med, 8(3):217-234.
77. Robins JM (1987) A graphical approach to the identification and estimation of causal prameters in mortality studies with sustained exposure periods. J Chronic Disease, 2:139-161.
78. Rubin D (1974) Estimating causal effects of treatments in randomized and non-randomized studies. J Educational Psychology, 66:688-701.
79. Rubin DB (1997) Estimating causal effects from large data sets using propensity scores. Ann Intern Med, 127(8 Pt 2):757-763.
80. Schafer JL, Olsen MK (1998) Multiple imputation for multivariate missing-data problems: A data analyst's perspective. Multivariate Behavioral Research, 33:545-571.
81. Shpitser I, Pearl J (2006) Identification of conditional interventional distributions. Proc 22nd Conf Uncertainty in Artificial Intelligence (UAI).
82. Shpitser I, Pearl J (2006) Identification of joint interventional distributions in recursive semi-Markovian causal models. Proc 21st National Conf Artificial Intelligence, p 1219.
83. Shpitser I, Pearl J (2007) What counterfactuals can be tested. Proc 23rd Conf Uncertainty in Artificial Intelligence (UAI).
84. Shwe MA, Middleton B, Heckerman DE, Henrion M, Horvitz EJ, Lehmann HP, Cooper GF (1991) Probabilistic diagnosis using a reformulation of the INTERNIST-1/QMR knowledge base. Part I: The probabilistic model and inference algorithms. Methods Inf Med, 30(4):241-255.
85. Spirtes P, Glymour C, Scheines R (1993) Causation, Prediction, and Search. Springer, New York, NY.

86. Spirtes P, Glymour C, Scheines R, al. E (2001) Constructing Bayesian network models of gene expression networks from microarray data. Proc Atlantic Symp Computational Biology, Duke University.
87. Spirtes P, Meek C, Richardson T (1995) Causal inference in the presence of latent variables and selection bias. Proc 11th Conf Uncertainty in Artificial Intelligence (UAI), pp 499-506.
88. Suzuki J (1993) A construction of Bayesian networks from databases based on an MDL scheme. Proc Conf Uncertainty in Artificial Intelligence (UAI), pp 266-273.
89. Tabachneck-Schijf HJM, Geenen PL (2009) Preventing knowledge transfer errors: Probabilistic decision support systems through the users' eyes. International Journal of Approximate Reasoning, 50(3):461-471.
90. Tenenbaum JB, da Silva V, Landford JC (2000) A global framework for nonlinear dimensionality reduction. Science, 29:2319-2321.
91. Tian J, Pearl J (2000) Probabilities of causation: Bounds and identification. Annals of Mathematics and Artificial Intelligence, 28(1):287-313.
92. Tinbergen J (1937) An Econometric Approach to Business Cycle Problems. Hermann Publishers, Paris, France.
93. Tsamardinos I, Brown L, Aliferis C (2006) The max-min hill-climbing Bayesian network structure learning algorithm. Machine Learning, 65(1):31-78.
94. van der Gaag LC, Tabachneck-Schijf HJM, Geenen PL (2009) Verifying monotonicity of Bayesian networks with domain experts. Intl J Approximate Reasoning, 50(3):429-436.
95. van der Maaten LJP, Postma EO, van den Jerik HJ (2007) Dimensionality reduction: A comparative review. Maastricht University. http://tsam-fich.wdfiles.com/local-files/apunt-es/TPAMI_Paper.pdf. Accessed February 5, 2009.
96. Vapnik VN (1998) Statistical Learning Theory. Wiley, New York.
97. Verma TS, Pearl J (1990) Equivalence and synthesis of causal models (Technical Report). Computer Science Department, UCLA.
98. Weiss S, Kulikowski C, Amarel S, Safir A (1978) A model-based method for computer-aided medical decision making. Artificial Intelligence, 11(2):145-172.
99. Witteman CL, Renooij S, Koele P (2007) Medicine in words and numbers: A cross-sectional survey comparing probability assessment scales. BMC Med Inform Decis Mak, 7:13-21.
100. Wright S (1921) Correlation and causation. J Agricultural Research, 20(7):557-585.
101. Wu X, Lucas P, Kerr S, Dijkhuizen R (2001) Learning Bayesian network topologies in realistic medical domains. Proc 2nd Intl ACM Symp Medical Data Analysis, pp 302-308.
102. Xiang Y, Pant B, Eisen A, Beddoes MP, Poole D (1993) Multiply sectioned Bayesian networks for neuromuscular diagnosis. Artif Intell Med, 5(4):293-314.
103. Yang Y, Webb GI (2002) A comparative study of discretization methods for naive-Bayes classifiers. Proc Pacific Rim Knowledge Acquisition Workshop (PKAW), pp 159-173.
104. Yu J, Smith VA, Wang PP, Hartemink AJ, Jarvis ED (2004) Advances to Bayesian network inference for generating causal networks from observational biological data. Bioinformatics, 20(18):3594-3603.
105. Zhang J (2006) Causal inference and reasoning in causally insufficient systems. Department of Philosophy, PhD dissertation. Carnegie Mellon University.
106. Zhao W, Serpedin E, Dougherty ER (2006) Inferring gene regulatory networks from time series data using the minimum description length principle. Bioinformatics, 22(17):2129-2135.
107. Zou M, Conzen SD (2005) A new dynamic Bayesian network (DBN) approach for identifying gene regulatory networks from time course microarray data. Bioinformatics, 21(1):71-79.

Chapter 9
Disease Models, Part II: Querying & Applications

WILLIAM HSU AND ALEX A.T. BUI

In the previous chapter, the mathematical formalisms that allow us to encode medical knowledge into graphical models were described. Here, we focus on how users can interact with these models (specifically, belief networks) to pose a wide range of questions and understand inferred results – an essential part of the healthcare process as patients and healthcare providers make decisions. Two general classes of queries are explored: *belief updating*, which computes the posterior probability of the network variables in the presence of evidence; and *abductive reasoning*, which identifies the most probable instantiation of network variables given some evidence. Many diagnostic, prognostic, and therapeutic questions can be represented in terms of these query types. For models that are complex, exact inference techniques are computationally intractable; instead, approximate inference methods can be leveraged. We also briefly cover special classes of belief networks that are relevant in medicine: probabilistic relational models, which provide a compact representation of large numbers of propositional variables through the use of first-order logic; influence diagrams, which provide a means of selecting optimal plans given cost/preference constraints; and naïve Bayes classifiers. Importantly, the question of how to validate the accuracy of belief networks is explored through cross validation and sensitivity analysis. Finally, we explore how the intrinsic properties of a graphical model (*e.g.,* variable selection, structure, parameters) can assist users with interacting with and understanding the results of a model through feedback. Applications of Bayesian belief networks in image processing, querying, and case-based retrieval from large imaging repositories are demonstrated.

Exploring the Network: Queries and Evaluation

Inference: Answering Queries

The usefulness of a belief network (and other graphical models) lies in the ability to ask questions of the model. The output of such queries is a probability that assesses some likelihood of the states across the variables and modeled joint probability distribution, and can provide diagnostic/prognostic guidance and/or classification. *Inference* is the process of computing the probabilities of each variable based on evidence that has been specified. The inference process begins when the user *instantiates* the model by assigning one or more variables to a specific state. Dependent on the provided evidence and the nature of the query, a model can invoke methods for belief updating or abductive inference to compute the probabilities needed to provide an answer. This section describes algorithms involved in both types of queries, and several of the issues surrounding the efficient computation of query probabilities.

A.A.T. Bui and R.K. Taira (eds.), *Medical Imaging Informatics*,
DOI 10.1007/978-1-4419-0385-3_9, © Springer Science + Business Media, LLC 2010

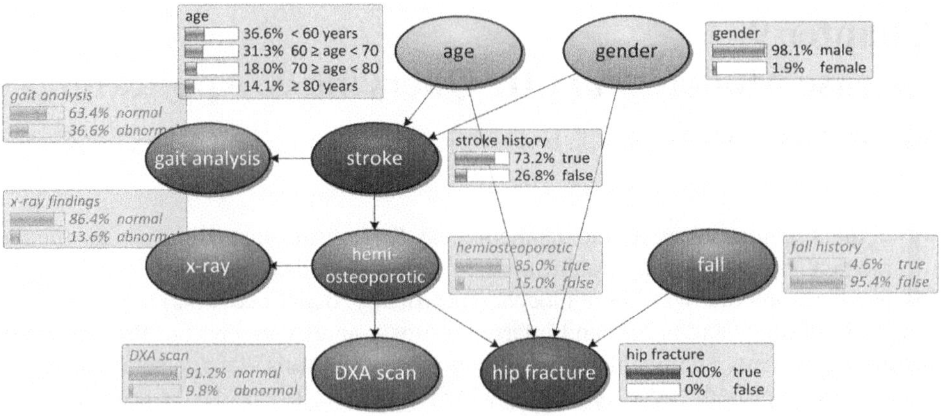

Figure 9.1: Hypothetical Bayesian belief network relating causes of stroke and hip fracture. The boxes shown per variable are called node monitors, and graphically indicate the potential values taken on by the variable, along with the current probability. In this case, the BBN shows the calculation for a posterior marginal for age, gender, and stroke given the evidence that the patient has a hip fracture; grayed-out node monitors are inactive.

Belief Updating

Belief updating involves the computation of a posterior probability for one or more variables in the network, given the instantiation of other nodes in the model (*i.e.*, evidence). Several types of queries are associated with belief updating, described below.

Probability of evidence. The simplest query that can be posed to a BBN is to ask for the probability of some variable, X[1], being instantiated to a specific value x, as represented mathematically by the statement, $P(X = x)$. By way of illustration, using the model in Fig. 9.1, we may be interested in knowing the probability of an individual having a hip fracture, H ($P(H = true)$), given without having a stroke, S ($P(S = false)$). Here, the set of variables $E = \{H, S\}$ are considered evidence variables, and the query, $P(e)$, is known as a *probability of evidence query*. Though computing the probability of a single variable instantiated in the model is useful, most queries involve instantiating multiple variables: often, we want to examine a logical combination of variables (*e.g.*, the probability of a propositional sentence). For example, if we are interested in finding the probability of stroke or hip fracture occurring, the statement may be written as $P(S = true \vee H = true)$. The answer can be computed indirectly using one of two techniques. First, the *case analysis method* can be used to rewrite the original statement as a combination of instantiations of the evidence variables, $P(S = true \vee H = true) = P(S = true, H = true) + P(S = true, H = false) + P(S = false, H = true)$. By summing these terms, the original probability can be calculated accordingly. Alternatively, the *auxiliary-node method* adds an additional node, E, to the network with S and H as its parents and a conditional probability table (CPT) as follows:

[1] As in Chapter 8, we follow standard notation with uppercase letters representing a random variable; lowercase letters indicating instantiations/specific values of the random variable; and bold characters symbolizing sets or vectors of variables.

S	H	E	P(E | s, h)
true	true	true	1
true	false	true	1
false	true	true	1
false	false	true	0

With this CPT, the event, $E = true$, is equivalent to the statement $S = true$ or $H = true$.

Posterior marginals. To see how the addition of evidence by instantiating certain variables in the model affects all of the other variables, the *posterior marginal* may be calculated. Given a joint probability distribution, $P(X_1,\ldots,X_n)$, the *marginal distribution* is the probability over a subset of the variables, $P(X_1,\ldots,X_m)$ where $m < n$. The marginal distribution can thus be viewed as a projection of the joint distributions onto a potentially smaller set of variables. Marginal distributions are also called *prior distributions*, as no evidence is given to affect their values. From the marginal distribution, the posterior marginal is computed by summing the entire joint probability distribution over the instantiated variables given the evidence, *e*:

$$P(x_1,\ldots,x_m \mid e) = \sum_{x_{m+1},\ldots,x_n} P(x_1,\ldots x_n \mid e)$$

Continuing with the previous BBN, an example of such a computation would be to answer a query such as, *what are the probable states of* age, gender, *and* stroke *given that the patient experienced a hip fracture*? This query is depicted in Fig. 9.1; the boxes that visualize the probabilities for each state are called *node monitors* and are updated to reflect updated probabilities as the user inputs a new piece of evidence. For this query, the hip fracture variable is set to true (100%) and the remaining variables are accordingly computed. In general, the computation of posterior marginals in a belief network is considered to be NP-hard [11].

Relative likelihood queries. In some cases, we only wish to know the comparative difference between two variables given some evidence. To illustrate, consider the basic network shown in Fig. 9.2, consisting of Boolean variables: if we observe that an individual is coughing and wish to know whether the cough (*C*) is more likely due to emphysema (*E*) or asthma (*A*), Bayes' rule can be applied to compute the conditional probability of each explanation from the conditional probability tables:

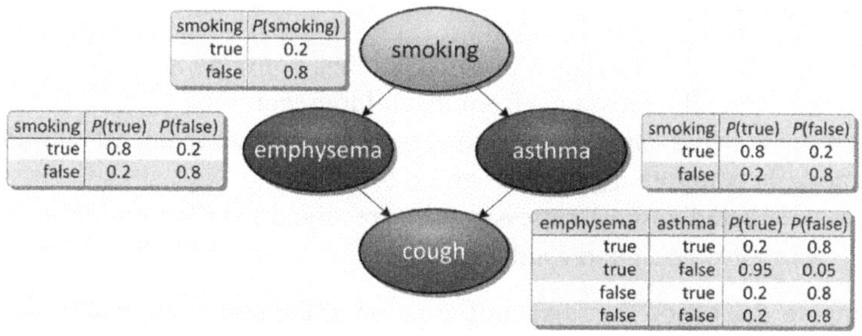

Figure 9.2: Example belief network with conditional probability tables shown. In some queries, the need for certain probabilities can be ignored if two variables are being compared, such as in computing the relative likelihood of two causes.

$$P(E|C=T)=\frac{P(E=T\wedge C=T)}{P(C=T)}=\frac{\sum_{S,A}P(S=s\wedge E=T)\wedge P(A=a\wedge C=T)}{P(C=T)}=0.575$$

$$P(A|C=T)=\frac{P(A=T\wedge C=T)}{P(C=T)}=\frac{\sum_{S,E}P(S=s\wedge A=T)\wedge P(E=e\wedge C=T)}{P(C=T)}=0.200$$

$$P(C=T)=\sum_{S,E,A}P(S=s\wedge E=e\wedge A=a\wedge C=T)=0.32$$

Computing the likelihood ratio of the two conditional probabilities (*i.e.*, $^{0.575}/_{0.200}$), the cough is much more likely due to emphysema rather than asthma by a factor of 2.8. Note that the calculation of $P(C = true)$ is not required if only the ratio is desired.

Computing the probabilities. The most direct way to perform inference is to calculate the marginalization over non-instantiated variables. However, the number of terms involved in the marginalization exponentially grows with the number of variables. A range of efficient algorithms thus exist for answering queries involving marginals, including summing out, cutset conditioning, and variable/bucket elimination [18]. Still, in larger, more complex networks with limited resources, exact computations to answer queries may be taxing, if not computationally intractable; therefore a variety of techniques may be used to instead approximate the desired probability. This difference gives rise to *exact inference* vs. *approximate inference* algorithms. We briefly describe some key techniques in both areas; for a more detailed discussion, the reader is referred to [4, 15].

Belief propagation (BP)[2] is an iterative algorithm that was originally intended for the exact computation of marginals on graphical models and polytrees [57]. The core idea is as follows: each node, X, computes a belief, $BEL(x) = P(x \mid E) = P(x \mid e^+, e^-)$, where E is the observed evidence contributed by evidence from the node's parents (e^+) and children (e^-). Expanding the last term, $BEL(x)$ can be determined in terms of a combination of messages from its children, $\lambda(x) = P(e^- \mid x)$, and messages from its parents, $\pi(x) = P(x \mid e^+)$, so that $BEL(x) = \alpha\lambda(x)\pi(x)$ where α is a normalization constant equal to $(\sum_X\lambda(X)\pi(X))^{-1}$. To start, the graph is first initialized such that: $\forall x_i \in E$, $\lambda(x_i) = \pi(x_i) = 1$ if $x_i = e_i$ and 0 otherwise; for nodes without parents, $\pi(x_i) = P(x_i)$; and for nodes without children, $\lambda(x_i) = 1$. Next, the algorithm iterates until convergence such that for each node, X:

- If X has received all π messages from its parents, compute $\pi(x)$.
- If X has received all λ messages from its children, compute $\lambda(x)$.
- If $\pi(x)$ is calculated and all λ messages are received from all children except parent node Y, compute $\pi_{XY}(x)$ and send it to Y.
- If $\lambda(x)$ is calculated and all π messages are received from all children except child node U, compute $\lambda_{XU}(x)$ and send it to U.

Finally, compute $BEL(x)$ on the final configuration of the nodes. BP can be implemented using dynamic programming methods. For the specific case of polytrees, BP provides exact inference in at most linear time relative to the diameter of the tree. The amount of computation performed per node is proportional to the size of the node's CPT. [57]

[2] Belief propagation is sometimes also referred to as the *sum-product algorithm*.

modifies this approach to provide approximate inference for general networks that may contain cycles; in this situation, the algorithm is often referred to as *loopy belief propagation.* It remains unclear as to under what situations loopy BP will converge (though empirical evidence supports its utility). Several variants of BP have been developed, including generalized BP and Gaussian belief propagation [76]. These newer approaches focus on restricting the set of messages being passed (*e.g.,* only passing messages that are likely to convey useful information), and can be seen in terms of approximating the graph structure via a simpler graph on which computation is more feasible.

Although a BBN permits one to compactly represent a distribution, its direct formulation is not suited for obtaining answers to arbitrary probabilistic queries. Instead, many (exact) inference algorithms compile an intermediate representation that can be used to more efficiently answer queries. A widespread construct for this purpose is the *junction tree* or *join tree* [33, 43], which also handles the problems associated with using BP on general graphs. The construction of a junction tree from a belief network can be abstracted in four steps:

1. An undirected graph is constructed from the BBN, termed the *moral graph,* wherein edges become undirected and nodes with a common child are connected.
2. Edges are added to the moral graph to *triangulate* the graph such that any two non-adjacent nodes on a cycle have an edge connecting them. Note that a graph can be triangulated in several ways (*i.e.,* the solution is not necessarily unique). The choice of triangulation greatly affects the end result such that inferences on the junction tree may go from being polynomial to exponential time in some cases; and the challenge of determining the optimal triangulation for a BBN is known to be NP-hard [62].
3. The cliques are identified in the triangulated graph, along with a potential function obtained by multiplying $P(X \mid Pa(X))$ for each node X in the clique and where $Pa(X)$ represents the parents of X.
4. From the graph constructed by the clique identification step, a minimum spanning tree can be constructed, resulting in the final junction tree.

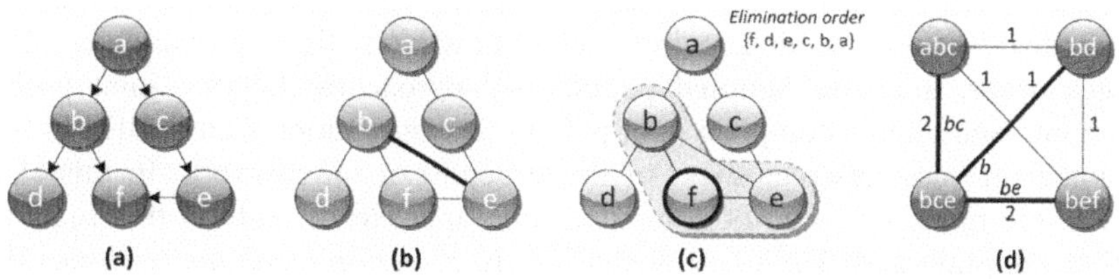

Figure 9.3: Transformation of a directed graph into a junction tree. **(a)** The original belief network. **(b)** Edge directions are removed and edges between nodes sharing children are created, establishing the moral graph (bold line); the graph is then triangulated as needed. In this case, the moral graph is already triangulated. **(c)** An elimination ordering of the variables is determined, and each node is considered sequentially to create cliques. In the first step, node f is examined, resulting in a node bef **(d)** The cliques are arranged in a graph, and a minimum spanning tree is determined using edge weights based on common variables. The final junction tree and labeled edges are shown with bold lines.

Central to Steps 3 & 4 is an elimination ordering that considers each node in sequence and determines a set of immediate nodes not yet seen in order to form cliques; the choice of variable order affects the final tree. Fig. 9.3 shows an example of this process. Given this tree, BP can then be applied to compute a probability using the calculated potential functions. The standard junction tree process is structure-based, and the size of the final structure is dependent only on the network topology. In practice, if the network topology is loosely connected, then junction tree algorithms work well; but when a network is densely connected, then this framework is less optimal. This observation has triggered research for alternative methods that can exploit local structure as well as network topology; for instance, exact inference using arithmetic circuits has been developed, taking advantage of local regularities within a BBN [14].

In addition to loopy BP, two other methods exist that perform approximate inference: *sampling methods* and *variational methods*; the former set of approaches is described here. In general, sampling methods operate on the premise that samples can be taken of a probability of a variable being assigned a specific state. The basic operation involves sampling each variable in topological order according to the conditional probability over its parents. If we represent $P(X_1,\ldots,X_n)$ as a BBN, the model can be sampled according to its structure by writing the distribution using the chain rule and sampling each variable given its parents. This process is called *forward sampling* (also known as direct Monte Carlo sampling). For each root node X, with probabilities $P(X = x_i)$, a random number r is drawn uniformly from the interval [0, 1]. To illustrate how forward sampling works, we refer to the example BBN in Fig. 9.2. We first sample the value of the variable smoking where P(smoking) = <0.2, 0.8> and assume that we obtained the result smoking = *true*. We then sample the value of the variable emphysema. As smoking = *true*, we are limited to using the corresponding conditional probability: P(emphysema | smoking = *true*) = <0.8, 0.2>. Let us next assume that the sample returns emphysema = *false*. We then proceed to sample the variable asthma using P(asthma | smoking = *true*) = <0.8, 0.2>. Again, let us assume that the sample returns asthma = *true*. We finally sample the value of the variable cough using the conditional probability P(cough | emphysema = *false*, asthma = *true*) and obtain cough = *true*. Through this first iteration, we thus obtain the event <smoking, emphysema, asthma, cough> = <*true*, *false*, *true*, *true*>. If we perform this process over multiple iterations while keeping track of how many times a specific combination of states occur, then the sampled population approaches the true joint probability distribution. Sampling a complete joint probability distribution from a BBN is linear in the number of variables regardless of the structure of the network. In this example, however, the marginals are not computed; two approaches address this requirement: rejection sampling and likelihood weighting. In *rejection sampling*, samples that are randomly drawn but do not agree with the specified evidence (*i.e.*, $e_i \neq x_i$) are thrown out. The problem with this approach is that many samples are potentially rejected, resulting in a largely inefficient process. *Likelihood weighting* addresses this pitfall by fixing the evidence variables and sampling only the remaining variables. To avoid biasing the sampling process, each sample is associated with a weight that expresses the probability that the sample could have been produced if the evidence variables are not fixed. Weights initially have the value of 1, but with each iteration in which the evidence variable is assigned a state, this probability of assignment is multiplied with the existing weight of the sample.

Rather than independently creating each sample as done in forward sampling, suppose we generate new samples by altering the previous one. To achieve this, we use a general class of methods called *Markov Chain Monte Carlo* (MCMC) sampling [50]. MCMC sampling is based on the premise that if all neighbors in the Bayesian network of X_i have assignments, their values must be accounted for before sampling X_i. The idea is based on the property of Markov chains, which are sequences of discrete random variables such that knowing the present state makes past and future states independent of one another: subsequent states are generated by sampling a value from one of the non-evidence variables after instantiating the variables in its Markov blanket using their current states. In order for a Markov chain to be useful for sampling from $P(x)$, we require for any starting state X_0, that $\lim_{t\to\infty}P_t(x) = P(x)$, and the stationary distribution of the Markov chain must be $P(x)$. Given these constraints, we can start at an arbitrary state and use the Markov chain to do a random walk over a specified number of iterations, and the resulting state will be sampled from $P(x)$. One popular sampler implementing this process is *Gibbs sampling* [22]. The process of Gibbs sampling can be understood as a random walk in the space of all instantiations, **e**, and can be used when the joint distribution is not known explicitly, but the conditional distribution of each variable is known – a situation well-suited for BBNs. To illustrate using Fig. 9.2, Gibbs sampling may be used to estimate the posterior probability, *P*(asthma | emphysema = *true*, cough = *true*). Given that emphysema and cough are set to *true*, the Gibbs sampler draws samples from *P*(asthma, smoking | emphysema = *true*, cough = *true*) and proceeds as follows. In the initialization stage, say we arbitrarily instantiate asthma = *true*, smoking = *true* as our X_0. Then, for each iteration (t = 1, 2,…) we pick a variable to update from {asthma, smoking} uniformly at random. If asthma is picked, sample asthma from *P*(asthma | smoking = s_{t-1}, emphysema = true, cough = true) and set X_t = (asthma = a_t, smoking = s_{t-1}), where s_{t-1} represents the value for smoking from the previous iteration, and a_t is the value of asthma for the current iteration. If smoking is picked, then perform a similar computation as in the case of asthma, but instead, sample smoking from *P*(smoking | asthma = a_{t-1}, emphysema = *true*, cough = *true*), where a_{t-1} is the value for asthma from the previous iteration. The sequence of samples being drawn by relying on the immediate prior is a Markov chain. This process can be further simplified by computing the distribution on X_i that is only part of the Markov blanket of X_i. Gibbs sampling is a one instance of a broader class of methods known as *Metropolis-Hastings* algorithms. The reader is referred to [50] for additional discussion.

Lastly, we briefly mention here the inference issues with respect to dynamic Bayesian networks (DBNs). For DBNs with a minimal number of time slices, the DBN can be recast as a hidden Markov model and exact inference methods applied via unrolling. For larger DBNs, where such techniques are computationally intractable, approximate inference is applied as described above. Key work in this area includes the Boyen-Koller algorithm and its variants [5, 6, 49]; particle filtering, which use sampling methods [38]; and more recently, a hybrid approach called factored sampling [52].

Abductive Reasoning

Unlike the previous class of queries that computes the probabilities of variables in the presence of given evidence, *abductive inference* identifies the most probable instantiation of network variables given some evidence. Abductive inference, also sometimes referred to as *inference to the best explanation*, is a common type of query asked by physicians in clinical practice: for instance, given the symptoms presented, what is the most likely diagnosis; or given the diagnosis, what is the most likely state of the patient? There are two types of abductive inference: most probable explanation and maximum *a posteriori*.

Most probable explanation queries. The objective of a *most probable explanation* (MPE) query is to identify the most likely instantiation of the entire network (*i.e.,* the state of all evidence variables) given some evidence [57]. If $\{X_1,\ldots,X_n\}$ are network variables, and **e** represents the set of available evidence, the goal of MPE is to find the specific network instantiation, $\mathbf{x} = \{x_1,\ldots,x_n\}$, for which the probability of $P(x_1,\ldots,x_n \mid \mathbf{e})$ is maximized. More concisely, MPE queries solve: $\text{argmax}_{\mathbf{x}} P(\mathbf{x} \mid \mathbf{e}) = \text{argmax}_{\mathbf{x}} P(\mathbf{x}, \mathbf{e})$. Consider the following query, again based on Fig. 9.1: *given that the patient is a 65-year old male and has had a stroke, but has a normal x-ray, what is the most likely state of the other variables in the network* (gait analysis, DXA scan, hemiosteoporotic, hip fracture, *and* fall)? There lies a certain subtlety to an MPE calculation, as it cannot be obtained directly from individual conditional probabilities: if $\{x_1,\ldots,x_n\}$ are chosen to maximize each $P(x_i \mid \mathbf{e})$ rather than the global problem, then the choice of x_i is not necessarily the most probable explanation. Also, given the nature of an MPE query, the result may not be unique: there may in fact by several configurations of the network's variables that result in the same maximal probability. For the special case of a hidden Markov model (HMM; see Chapter 8), the MPE problem is solved by Viterbi decoding, where the most likely sequence of states is determined. However, in general, one can see that the search space for MPE is potentially enormous. As such, while an exhaustive set of permutations can be examined for smaller networks, most MPE algorithms employ approximate inference methods and can be divided between stochastic sampling methods and search techniques. In particular, the latter category include best-first search, AND/OR search [19], and genetic algorithms [46]. The efficiency of MPE algorithms is considered in terms of a *treewidth* metric that measures the number of graph nodes mapped onto a tree node in the decomposition.

Maximum *a posteriori* queries. Unlike MPE queries, a more general type of query that attempts to find the most probable instantiation for a subset of network variables is called a *maximum a posteriori* (MAP) query. MPE is hence a specific instance of a MAP query where the subset is defined as the entire set of evidence variables in the network. Let **M** represent some subset of variables in the belief network, and **e** some given evidence; the objective of a MAP query is to find an instantiation of **m** such that $P(\mathbf{m} \mid \mathbf{e})$ is maximized. MAP queries are an optimization problem, with the resulting probability as the objective function that one tries to maximize. As such, the MAP problem can be stated as: $\text{argmax}_{\mathbf{m}} P(\mathbf{m} \mid \mathbf{e}) = \text{argmax}_{\mathbf{m}} \sum_{\mathbf{Z}} P(\mathbf{m}, \mathbf{z} \mid \mathbf{e})$ where **Z** is equal to the set of variables remaining once evidence and the query variables in M are removed from X (*i.e.,* $\mathbf{Z} = \mathbf{X} - \mathbf{E} - \mathbf{M}$). From Fig. 9.1, one may ask the following: *what is the most likely state for* hip fracture *and* stroke *given that the patient is female and that she fell*? Note that this query does not attempt to provide information on gait analysis, x-ray, DXA scan, age, or the hemiosteoporotic states.

Variable elimination can be used to solve a MAP query by marginalizing non-MAP variables, thereby simplifying the problem to a MPE query. The key is to decide on an elimination order of the variables such that the MAP variables, **M**, are marginalized last. The process is summarized by the following equation: $\sum_{X_1}\sum_{X_2}\cdots\sum_{X_m}\prod_j \theta_{X_j|\text{Pa}(X_j)}$. Intuitively, this equation states that the probability of the query variables **M** is computed by implicitly constructing the joint probability distribution induced by the Bayesian network and summing over each non-query variable. Variable elimination utilizes the notion of *factors*, which enable variables to be summed out while keeping the original distribution. The use of factors helps to overcome the exponential complexity seen with the brute-force method of simply summing out variables. Factors are tables that contain two components: an instantiation and a number. The instantiation is an assignment of values to variables; the number represents the probability of the corresponding instantiation. Two operations can be performed on factors: multiplication and summing out. Multiplication can be likened to a natural join (Cartesian product) on two database tables: the set of variables in the product of two factors is the union of the sets of variables in the operands. Summing out is the same as the process of marginalization (see Chapter 8). Variable elimination commences with each factor represented as a CPT in the model. To compute the marginal over **M**, the algorithm iterates over each variable X_i in the given elimination order. Next, every factor f_i that mentions variable X_i is multiplied together to generate a new factor, f. We then proceed to sum out variable X_i from f and replace factors f_i by factor $\sum_{Xi} f$. After going through each variable in the elimination order, only the set of factors over variables **M** will remain. Multiplying these factors produces the answer to the desired query, $P(\mathbf{M})$.

To ground this discussion, we refer back to the example of the osteoporosis BBN illustrated in Fig. 8.4. Assume that we are interested in finding the probability that a patient is at risk of getting a fracture and are given a predefined elimination order of {renal disease (R), DXA finding (D), age (A), kidney function (K), activity level (L), hormone usage (H), osteoporosis (O)}. While determining the optimal elimination order is outside the scope of this chapter, the reader may refer to [15] for additional discussion on the topic. The MAP query is written mathematically as:

$$P(F) = \sum_{O,H,L,K,A,D,R} \theta_{F|O,H}\theta_{F|K,L}\theta_{H|A}\theta_{L|A}\theta_{A}\theta_{D|O}\theta_{K|R}\theta_{R}$$

The process starts by eliminating the first variable in our elimination order, R. Writing out the operation, $\sum_R \theta_{K\,|\,R}\theta_R$, we see that two terms mention R and involve variable K. We then compute the product of each value of K and summarize the result as factor, f_1, which can in turn be substituted into the summation to remove R from the network:

$$P(F) = \sum_{O,H,L,K,A,D} \theta_{F|O,H}\theta_{F|K,L}\theta_{H|A}\theta_{L|A}\theta_{A}\theta_{D|O} f_1(K)$$

The next variable for elimination is D. From the equation, we find that only one term involves D, and it also involves O. So for each value of O, we compute the sum over D of $P(D \mid O)$. However, if we fix O and sum over D, the probabilities need to add up to 1, and therefore D can be removed from the network without adding a new factor to the expression. The process of identifying the next elimination variable, multiplying factors, and summing over variables continues for all variables in the elimination order. We then multiply the remaining factors together, resulting in the exact answer for $P(F)$. The prior marginal is a special case of the posterior marginal query where the evidence set is empty. To compute the posterior marginal, a similar process is followed but prior to eliminating variables, rows in the joint probability distribution that are inconsistent with the evidence are zeroed out.

In examining the complexity of variable elimination, the algorithm runs in time exponential in the number of variables involved in the largest factor. The elimination order is critical because a bad order potentially can generate large factors; finding the best elimination order is itself an NP-hard problem. Computationally, MPE queries are easier to compute than MAP: the decision problem for MPE is NP-complete, whereas the corresponding MAP problem is considered NP^{PP}-complete [54]. Because of this intractability, most software implementations answering MAP queries provide only an approximate answer. A variety of approaches have been explored for approximate MAP inference, including genetic algorithms [16], MCMC with simulated annealing [77], and hill climbing [55] to name a few. More recently, exact methods employing search have been developed [30, 56].

Inference on Relational Models

Standard probabilistic models are said to be *propositional*, not permitting quantification over an object. In some domains, this limitation results in an unwieldy number of statements that must be explicitly made to represent an instantiated dataset, especially when dealing with similar entities that may arise in slightly different configurations. For example, consider the problem of trying to correlate radiographic imaging features, gene expression, and end outcomes in brain tumor patients. Fig. 9.4 presents a portion of a hypothetical relational schema that links these elements of data together. Though each type/grade of tumor (*e.g.,* astrocytoma, glioblastoma multiforme, etc.) presents different gene expressions, appearances on imaging, and responds to different chemotherapies, there is some commonality. Capturing such variation is relatively straightforward in a relational model and can be expressed as tables within a database. Imagine, however, that a BBN is to be created from the same entities and attributes: the number of variables needed to express all of the variations will increase dramatically, thereby creating an overly complex network.

Thus, efforts to augment probabilistic models with quantifiable operators have led to the development of frameworks that take advantage of relational and/or first-order probabilistic constructs to extend graphical models. [17] provides a recent survey of the efforts to link BBNs with first-order logic, including a discussion of relational Bayesian networks [32] and *probabilistic relational models* (PRMs) [23]; we use the latter here as an example. A PRM consists of two parts: a relational component that describes the domain schema; and a probabilistic component modeling the distribution of the attributes of the entities in the domain. From the PRM specification, a Bayesian

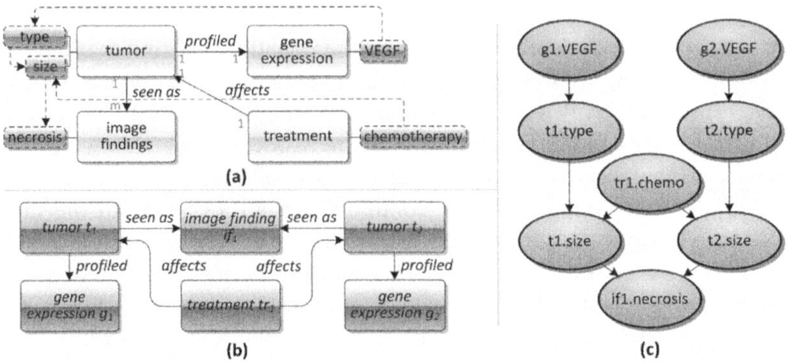

Figure 9.4: Translation of a relational schema into a BBN via a probabilistic relational model. **(a)** An entity-relational (ER) model showing a part of a relational schema in M2 notation (see Chapter 7). Standard ER relationships are shown with solid arrows; connectors shown with dashed arrows between the attributes represent BBN linkages. **(b)** A database instantiation of the schema. **(c)** The resultant generated BBN.

belief network can be compiled. One of the simplest advantages of PRMs over propositional models is that of compactness: a large number of propositional variables and models can result from instantiating the same piece of "generic knowledge" over a large domain of objects. The succinctness of relational models can provide a more intuitive representation. Furthermore, statistical techniques have been developed to learn PRMs directly from a database. Building from plate models and PRMs, a probabilistic entity-relation (PER) model has been described [26].

Inference in PRMs and other relational models can be categorized twofold: 1) approaches that transform the relational model into a propositional graphical model, permitting the inference algorithms discussed previously to be used; and 2) approaches developing a new set of algorithms that operate directly on the relational representation. The former in effect constructs the BBN associated with the PRM. [37] remarks that in some cases, the use of PRMs can actually aid in the inference process, given: that unlike standard BBNs, the relational model encapsulates influence by grouping influencing attributes together within the same object; and that the relational model lends itself to reuse in terms of multiple objects being of the same entity (and thus the same inference methods can be used). Both factors can be exploited within an inference algorithm to speed computations. Still, efficient reasoning and inference is a major challenge for PRMs. Systems have been demonstrated for exact inference across relational models, PRIMULA being a prime example [9]. Lifted inference methods also provide another approach to computations on PRMs [58].

Diagnostic, Prognostic, and Therapeutic Questions

As demonstrated by the various queries described thus far, many types of questions familiar to clinical care can be answered via a disease model represented by a BBN. Four categories of BBN querying have been suggested and are useful to bear in mind in the context of medicine and decision-making:

1. Diagnostic/evidential. The first category employs bottom-up reasoning to deduce a cause when an effect is observed. For instance, a patient presents with a symptom (*e.g.,* cough) and the physician is trying to find the most likely cause (*e.g.,* bronchitis, asthma). Abductive inference can be seen to fall into this category.
2. Causal. In contrast to diagnostic queries, this second category involves top-down reasoning to determine, given a known cause, the probabilities of different effects. In essence, a prognostic query is posited. For example, given the flu, what is the chance of experiencing a headache? Or given an intervention or drug, what is the likely outcome for a given patient? (*e.g.,* if we give the patient a bronchodilator, will the coughing go away?). Belief updating and causal inference with counter-factuals (see Chapter 8) comprise this group of queries.
3. Explaining away. Sometimes referred to as intercausal queries, explaining away is a common reasoning pattern that looks to contrast causes with a common effect, often deducing one cause as being the reason for an event (as opposed to another cause) given some evidence. For example, two diseases may be suspected; however, given evidence of some symptom, the probability of one cause increases while the other is lowered.
4. Mixed. Lastly, one can consider queries that combine any of the above three techniques into a single inquiry.

Influence diagrams. A subset of clinical decisions often involves the selection of a treatment plan for an individual or course of action to optimize some criteria. Unto themselves, BBNs do not provide these answers directly, providing only tools for reasoning under uncertainty; instead, an important class of models known as *influence diagrams* aids in decision making in uncertain situations. Rudimentarily, a decision is aimed at selecting a strategy that maximizes the chance of a desired outcome occurring given our knowledge of the domain (as represented by a model). Originally framed as a compact alternative to decision trees, influence diagrams permit different configurations of this model and potential choices to be considered in terms of quantifiable values supplied via a *utility function,* $U(a)$, where a represents an action. The aim, therefore, is to identify the configuration and actions that maximize the utility functions that solve $\mathrm{argmax}_{\mathbf{A}}\sum U(\mathbf{x},\mathbf{a})P(\mathbf{x} \mid \mathbf{e})$. Influence diagrams consist of nodes and edges like their graphical model counterparts, but reclassify the nodes into three types:

1. Chance nodes. *Chance nodes* are random variables, similar to the evidence variables in a BBN. Like evidence variable nodes, CPTs are associated with chance nodes. Chance nodes can have both decision and other chance nodes as parents.
2. Decision nodes. *Decision nodes* represent those points in the state/process where a choice of actions can be made; the result of a decision is to influence the state of some other variable (*e.g.,* a chance node). An influence diagram will have one or more decision nodes.
3. Utility nodes. Utility nodes are a measure of the overall "outcome" state, with the goal of optimizing the utility (*i.e.,* maximizing) based on the contributing chance, decision, and causal factors. Utility nodes may not have children in the graph.

Additionally, some types of influence diagrams include *deterministic nodes*, defined as nodes with constant values or algebraically calculated from parent nodes' states – once the parent nodes are known, the child node's state is definitively assigned. Fig. 9.5 shows an example of a simple influence diagram, where the decision points involve the use of COX-2 inhibitors to relieve knee pain due to inflammation, or the use of MR imaging to further diagnose a problem before doing endoscopic surgery. It is important to note the implications of influence diagrams with respect to evidence-based medicine (EBM). An underlying principle of EBM is that decisions take into consideration an individual's preferences (*e.g.,* with respect to diagnostic and treatment options): by fixing the selection within a decision node, an influence diagram can view a patient's preferences as an explicit constraint within the optimization problem. The utility node can be seen as being related to a patient's quality of life (*e.g.,* for decisions involving substantial risk, quality-adjusted life years, QALY) in addition to considering cost and other factors. [51] gives additional examples on the use of influence diagrams in medicine.

A basic algorithm for querying the influence diagram instantiates the entire network based on the given constraints/evidence; each possible decision is analyzed, examining the output of the utility nodes. The decision that maximizes the utility node is deemed the best decision and returned. For influence diagrams with only a single decision node, selection of the decision that maximizes the utility node is straightforward; however, the challenge is more profound when multiple decision nodes exist and/or require explicit sequential modeling of actions (*i.e.,* action X before action Y) – resulting in large search spaces. [10] thus shows how influence diagrams can be transformed

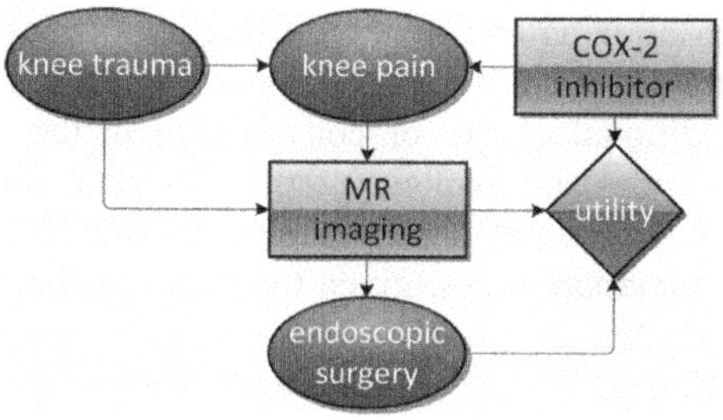

Figure 9.5: Example of an influence diagram. Chance nodes are drawn as ovals, decision nodes are rectangles, and utility nodes are illustrated as diamonds.

into a belief network, incorporating decision theory. The method effectively translates all of the nodes in an influence diagram into chance nodes, making the structure equivalent to a BBN: CPTs are assigned to decision nodes with an equal chance distribution; and utility nodes are changed into binary variables with probabilities proportional to the node's parents' utility functions. From this transformed BBN, the inference algorithms described prior can be applied to select decision nodes' states based on the desired utility (*e.g.*, as MPE/MAP queries). For further discussion of decision making theory and influence diagrams, the reader is referred to [63].

Evaluating BBNs

Inference results are only useful if the underlying BBN is capable of representing the real world. The question then naturally occurs as to how to assess the ability of a belief network to provide true answers; this issue is perhaps particularly significant given the use of approximate inference techniques. BBN verification can be performed with respect to different criteria. We touch upon two strategies: examining predictive power, where the BBN's diagnostic/prognostic capabilities are compared against known outcomes; and sensitivity analysis, which aims to determine what aspects of the model have the greatest impact on the probabilities of query variables (and therefore must be accurate).

Predictive Power

In healthcare applications, classic BBN evaluation compares the predictions of the model *vs.* known outcomes (or expert judgment). A test set of cases is compiled and used as a benchmark for ground truth; precision and accuracy metrics are often reported. For instance, as an aid to classification or as a diagnostic tool, a BBN can be given partial evidence per test case and asked to infer the remaining values (or a subset of values, as per a MAP query); the BBN result is then compared to the true value stated in the case. A confusion matrix can then be composed to identify the rate of (in)correct classifications (see Chapter 10). [68] also details a method for estimating the variance associated with each query result, in effect determining "error bars" for each point estimate. Though a BBN is capable of answering a gamut of queries, given the size of some belief networks it is untenable to test all variables against all combinations of partial evidence in a test set. Rather, there is usually a specific purpose in its construction, and the queries that the BBN is designed primarily to answer are evaluated.

An automated Monte Carlo-based technique is described in [59] to discover those portions of the model responsible for the majority of predictive errors; and to identify changes that will improve model performance. Briefly, the algorithm consist of three steps: 1) labeling each node as one of three categories (observations, such as labs or image findings; phenomenon, such as the underlying etiology of a tumor; and ancillary, providing additional clarity or convenience in the model); 2) selecting a subset of phenomenon nodes and explicitly setting the state, and using a Monte Carlo simulation to determine the state of observation nodes; and 3) computing the posterior probability for all phenomenon nodes, given the states of the observation nodes. The second step in this algorithm uses normal Bayesian inference techniques to calculate the posterior distribution of a node, with Monte Carlo sampling of the posterior distribution to assign the node's state.

Depending on how a BBN is constructed, the test set must be carefully specified to avoid bias and overfitting. For example, if the network topology and the CPTs are both derived from experts (*i.e.,* the structure and its parameters are not learned), then any reasonably derived test data can be used (assuming it is representative of the population against which the BBN is targeted). If either (or both) the structure and probabilities are learned, then the training data and test set must be separated. For instance, an *n*-fold cross-validation study is reasonable if sufficient data is available: the dataset is randomly partitioned into *n* groups; training is performed using *n*-1 groups, with the resultant model tested on the remaining group; and this train-test process is then repeated a total of *n* times until each group has been used once for evaluation. Unfortunately, as with any framework using training data, *overfitting* of the model can still be a concern when the amount of training data set is small or when the number of parameters in the model is large. As such, a 10-90 test can be used to examine model stability: reversing a 10-fold cross validation pattern, 10% of the test set is instead used to train and 90% of the data is used to test in each iteration. In theory, a well-formulated model will provide consistent results per iteration of a 10-90 test; the results can also be compared to the tenfold cross-validation to ascertain if overfitting has occurred in the latter. Ultimately, the most convincing evaluation is one that uses a *holdout set*, wherein a portion of the dataset is withheld from training (and testing, in a cross-validation study) so that unbiased performance metrics are computed on a "clean" set of data. Markedly, a common complaint about BBNs is that the trained probabilities (and to a lesser extent, structure) are developed and assessed relative to a single environment, and thus subject to local operational bias: simply exporting a BBN from one locale to another often fails to achieve the same degree of performance. Hence, if the holdout set is instead obtained from an outside source (*e.g.,* a published national database or public data repository), then evaluation bias can potentially be overcome.

Comparison to other models. Belief networks are only one means of providing classification and/or diagnostic/prognostic insights – a range of statistical and probabilistic methods also exist. As such, it is worthwhile to evaluate a BBN's performance relative to these other techniques. For example, if the primary application of the BBN is to answer causal or predictive queries, then a comparison against a decision tree or a logistic regression model may be appropriate (see Chapter 10); a discussion of current methods for predictive modeling is given in [3].

Sensitivity Analysis

In broad terms, *sensitivity analysis* is concerned with how variations in the model or inputs to a BBN impacts the quality of decision making [2, 13, 42]. [35] discerns between *evidence sensitivity analysis*, in which the sensitivity of results is examined in light of variations in evidence; and *parametric sensitivity analysis*, which looks at how changes in model parameters (*i.e.,* the CPTs) affect query results. Both capabilities are instrumental in giving users a handle on the models they build, and can be critical in model validation and debugging. For example, through sensitivity analysis, we can evaluate false positive/negative rates of a diagnostic test on the quality of decision making; as a corollary, it is possible to search for the appropriate false positive and negative rates that would be necessary to confirm a hypothesis at a certain level of confidence. We can assess the utility of information through sensitivity analysis, allowing a user to decide what additional evidence is needed in order to gain useful insights. Given the scope of this text, we limit our discussion to a description of parametric sensitivity analysis. Formally, parametric sensitivity analysis is concerned with three types of questions:

1. What guarantees can be made on the sensitivity/robustness of a query, q, to changes in parameter values, $\theta_1, \ldots, \theta_n$?
2. What are the necessary and sufficient changes to parameters $\theta_1, \ldots, \theta_n$ that would enforce some integrity constraints, $q_1, \ldots, q_m$, assuming that these integrity constraints are violated by the current model?
3. What guarantees can be offered on the sensitivity/robustness of some decision, d, to changes in parameter values $\theta_1, \ldots, \theta_n$, where the decision is computed as a function of some probabilities?

As shown by these questions, sensitivity analysis can elucidate the stability of a BBN relative to specific inquiries. In *single parameter sensitivity analysis*, the influence of one parameter within a query is examined by fixing all other parameters and "perturbing" the selected parameter in the network. Sensitivity analysis tools permit inspection of how secondary evidence variables change in response to alterations in the parameter. Single parameter sensitivity analysis can be used as a type of query to BBNs: by specifying a single constraint or condition on a conditional probability, it is possible to derive what other network variables must be changed to satisfy this constraint (*e.g.,* given that we want to diagnose with 99.5% certainty, what other tests would need to be performed, and/or what probabilities must be changed?). Additionally, this type of sensitivity analysis can be used in conjunction with model building tasks to identify those variables that have a high degree of influence over results (and therefore, the CPTs must be as accurate as possible) [12]. *Multi-parameter sensitivity analysis* perturbs *n*-pair combinations of variables simultaneously, either within the same CPT, or across different CPTs. This technique is much more computationally expensive – but it is able to estimate the training error for the associated statistics and calculate a generalization error for the entire network [8]. Software packages such as HUGIN, NETICA, and SAMIAM provide graphical tools for conducting sensitivity analysis.

Using the model depicted in Fig. 9.1, let us consider an example of how sensitivity analysis helps us determine what improvements are needed to existing tests. In the model, we are interested in determining whether the combination of x-ray and DXA scan is capable of accurately diagnosing whether the patient is hemiosteoporotic. When performing sensitivity analysis, the question to be posed to the model is *which network parameters can we change, and by how much, to ensure that the probability of the patient being hemiosteoporotic is above 95% given that the patient has positive x-ray and DXA scan tests*? Currently, the model states that the specificity of the DXA scan is 66% and the specificity of the x-ray is 64%. If we run a sensitivity analysis on the model with the variables x-ray and DXA scan instantiated to abnormal, we would see that the results return three possible changes that each satisfy the constraint $P(\text{hemiosteoporotic} = \text{true}) \geq 0.95$:

1. If the true negative rate for the DXA scan was 92% instead of 66%.
2. If the true negative rate for the x-ray scan was 91% rather than 64%.
3. If the probability of being hemiosteoporotic given that the patient did not have a stroke was greater than or equal to 0.768 rather than 0.27.

As making the third change would not be feasible, we could act on one of the first two suggestions by investing in a better DXA or x-ray scan. If we are willing to compromise on the constraint and be satisfied with $P(\text{hemiosteoporotic} = \text{true}) \geq 0.90$, then we can find tests that achieve true negative rates of 83% (DXA scan) or 82% (x-ray) instead. The same approach can also be used to determine what changes are necessary to make the model fit the beliefs of a domain expert. For instance, if an expert believes that the probability of a hip fracture given that the patient has fallen after having a stroke is greater than the result that the model returns, we can identify which variables (*e.g.*, age, gender, hip fracture) need to be modified such that his beliefs holds true.

Interacting with Medical BBNs/Disease Models

The focus of the prior sections has been on the underlying concepts and algorithms that permit inference and other computational analyses on BBNs. We now turn to the secondary issue of interacting with the belief network, enabling a user to specify the queries and explore the model. Today's BBN graphical user interfaces (GUIs) typically employ the directed acyclic graph (DAG) as a pictorial representation upon which queries are made and results presented. Visual cues and animation (*e.g.*, highlighting nodes, changing colors, motion) are used to denote key structures and altered values in response to queries [7, 25, 29, 78]. While providing sophisticated querying, two problems arise: 1) as the complexity of the BBN grows, understanding the nuances of variable interaction is difficult at best; and 2) the variables are visually abstracted and thus lose contextual meaning – a concern for clinically-oriented BBNs when interpreting a patient's data.

In general, the challenges arising in interacting with larger BBNs handle two areas: 1) methods for building and exploring the graphical structure, along with the model's parameters (*i.e.*, the CPTs); and 2) methods for posing queries and seeing the resultant response.

Defining and Exploring Structure

The most obvious difficulty with BBN visualization lies in the organization of a large number of variables in a constrained amount of space. An array of methods has been developed for general graph visualization, including: various geometric layouts (*e.g.*, radial, 3D navigation); hyperbolic trees; and distortion techniques (*e.g.*, fisheye views). An overview of these approaches is given in Chapter 4. Here, we highlight some challenges specific to BBNs. One issue in BBN visualization is the depiction of the linkages between nodes, emphasizing those variables that are clustered together through a high degree of connectivity; and variables that in particular are dependent on a large number of parents, or conversely, serve as a parent to a large number of other dependent variables (*i.e.*, the number of incoming edges, the *in-degree*; and the number of outgoing edges, the *out-degree*). A common graphical method of emphasizing

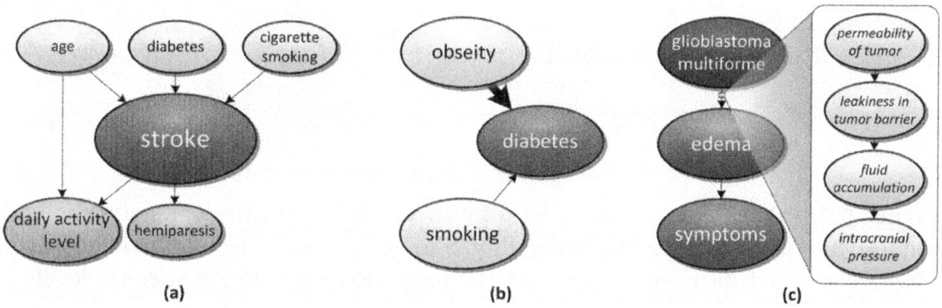

Figure 9.6: Different methods for belief network visualization. **(a)** To emphasize the in- and/or out-degree of a given node (and hence its connectivity), the node size can be varied. In this example, stroke is deemed important, and so is rendered as the largest node. **(b)** The strength of a relationship (*e.g.,* based on sensitivity analysis or conditional probabilities, for instance) is often depicted via line thickness. **(c)** Grouping of node clusters or the collapsing of variable chains into edges can help to compact space.

the importance of such nodes is through size: larger nodes represent a higher number of connections (Fig. 9.6a). However, not all relationships are of equal importance: therefore, some systems render graph edges using line thickness in proportion to the strength of the relationship between the two variables (*i.e.,* a thicker line indicates a stronger link; Fig. 9.6b). Object-oriented paradigms can also be applied to present related entities together, subsuming related variables into a single visual representation (*e.g.,* a super-node); or to collapse chains of variable into one edge (Fig. 9.6c). The causal semantics between variables have also been visualized using animation [34]. [70] also considers the problem of navigating the conditional probability tables: as the number of possible states and dependencies grows, the depiction of the CPT itself can outgrow the available visual space, thus requiring scrolling or other means to change focal points. Methods including the use of a treetable widget for hierarchical presentation of CPTs, and the dynamic hiding/reveal of parent/child relationships within a CPT are discussed.

Expressing Queries and Viewing Results

Query formulation broadly consists of two steps: 1) stating the type of query that is desired (*e.g.,* MAP), if not the variables of interest; and 2) specifying the evidence available as part of the query (*i.e.,* the query constraints). Most GUIs provide a means of choosing variables by directly selecting and highlighting nodes from the DAG, with options to invoke the corresponding type of inference. As mentioned earlier, node monitors provide a direct view of a variable's state, graphically depicting associated probabilities. Node monitors can be made interactive, permitting a user to directly manipulate the values to set evidence: numeric scroll wheels, sliding bars, and *probability wheels* are used to elicit probabilities across a variable's different states (Fig. 9.7). Rather than use the graph, form-based and checklist querying approaches have been explored as front-ends to BBNs [71], but arguably become untenable given a large number of variables.

Once an inference result is computed, the probabilities across the network are updated and displayed to the user, who looks to see to what extent a variable may change and/or the end state of some node. Although results can be tabulated into a separate table, node monitors are commonly employed for displaying the information directly.

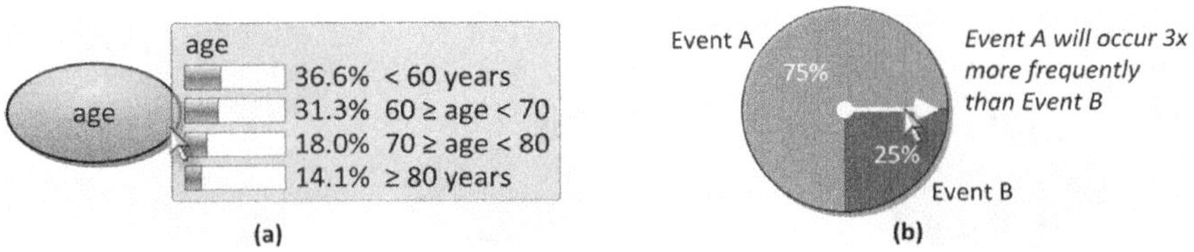

Figure 9.7: (a) A node monitor. For a given variable, the different states the node can take are shown alongside a probability. Sliding bars are often used to help provide a sense of visual distribution, and can be made interactive to set specific levels of evidence. **(b)** A probability wheel can be manipulated by users to specify a value for the likelihood of an event. The idea is that based on some question of an event's occurrence, the wheel's wedges are proportioned accordingly.

However, the use of node monitors can prove problematic: the user must actively search for changes in values, so subtle differences between states or variables can be lost in reviewing results. Furthermore, given limited space to show information, the overuse of node monitors leads to a cluttered interface. To ameliorate the identification of changes, visual changes are often made to the underlying DAG rendering: nodes are colored to indicate the degree of change (*e.g.*, shades of red and blue are used to indicate positive/negative changes in values, with the intensity of the color proportional to the magnitude of the change); transparency/opacity is changed; or highlights are added to nodes to indicate updated statuses. Similarly, edges can be colored or adjusted based on changes in the conditional probability tables.

In the healthcare domain, an alternative strategy is to create problem- or disease-specific applications that tailor the visualization and querying capabilities to a target domain. The literature contains many examples of such problem-specific interfaces (*e.g.*, diabetes, oncology) and there is evidence that such problem-oriented data visualization can enhance the cognitive processes of physicians (see Chapter 4). Unfortunately, many of these interfaces sacrifice flexibility: the displays restrict the types of queries that can be posed by the user, limiting discovery and "what if" questions. A new class of visualizations has been developed to make interacting with probabilistic disease models more intuitive by providing tools to pose queries visually. One such system is TraumaSCAN [53], in which the user interacts with a 3D model of the body to place entry and exit wounds for injuries from gunshots; then, in combination with inputted patient findings, the system performs reasoning on a Bayesian network to predict the most probable symptoms and conditions arising from the specified injuries. However, many of these querying interfaces have been developed for specific diseases; they do not address the long standing problem of creating GUIs that can effectively support the broad spectrum of physician activities (*e.g.*, reviewing a patient for the first time, diagnosis and treatment planning, a follow-up visit, etc.). Part of the difficulty lies in working with the diversity and amount of information that is pertinent to a given clinical task: not all data is of equal importance at any given time, and must be selected and presented appropriately based on the task at hand. The Vista [27] and Lumiere [28] projects are examples of using BBNs to automatically select the most valuable information or program functionality: an influence diagram is used to model the user's background, goals, and competency in working with the software.

Explaining results. One application for which interaction with BBNs has been explored is to explain a recommendation to a user as part of a medical decision-making tool [67]. Explanations are useful: for determining what configuration of unobserved variables provides the most probable outcome for a target variable; for eliciting what information is contained in a model; and for understanding the reasons for a model's inference results. A review of explanation methods can be found in [41].

The majority of approaches to conveying explanations have centered on the use of verbal or multimedia methods. For instance, [21] translates the qualitative and quantitative information of a BBN into linguistic expressions. Probability values are mapped to semantic expressions of uncertainty; for example, the range 0.25-0.4 is mapped to the adjective "fairly unlikely" and the adverb "fairly rarely." These adjectives are then used in combination with the structure of the network to generate meaningful statements. To illustrate, given the model depicted in Fig. 9.2, one statement would be, "*Smoking commonly causes emphysema.*" Visual cues have been used: [45] utilizes color coding and line thickness to support explanations in terms of weight of evidence and evidence flows. One system that combines both graphical and verbal approaches to explaining inference results is BANTER [24]. This system allows the user to enter a scenario by specifying known values for history and physical findings for a disease of interest using standard node monitors in a GUI. Based on this data, the system uses the BBN to assess which tests best determine whether the patient has the identified disease. Explanations are provided in natural language using two methods: identifying the evidence that have the greatest impact on a target variable using mutual information; and describing the path that maximizes the overall impact of evidence variables to the target variable. Alternatively, [69] describes the use of a three-tier system to address the inability of traditional BBNs to provide updated prognostic expectations given new data during the healthcare process: 1) the first tier is a BBN composed of a collection of local supervised learning models, which are recursively learned from the data; 2) the second tier is a task layer that translates the user's clinical information needs to a query for the network; and 3) the third tier is a presentation layer that aggregates the results of the inferences and presents them to the user using a bar graph representation. The novelty of this method allows users to pose new queries at each stage of patient care (*e.g.,* pre-treatment, treatment, post-treatment), having the model explain the changes in the target variable based on updated information at each point.

Research has also been done to utilize the network topology to aid in the generation of explanations. [75] exploits Markov blankets to identify a subset of variables that result in a concise explanation of a target variable's behavior. This approach first restructures the BBN so that the target variable has its Markov nodes as its parents. Next, the target node's conditional probability tables are converted into decision trees. Explanations are finally derived by traversing the decision trees.

Discussion and Applications

As seen here and in Chapter 8, disease models provide a method of extracting the scientific knowledge encoded within routine clinical observations and applying this knowledge to inform decisions related to diagnosis, prognosis, and treatment. Such models, represented as Bayesian belief networks, enable a probabilistic framework upon which a range of queries can be made. Advances in BBN inference techniques are providing the computational means to answer increasingly complex questions over complex models throughout healthcare (*e.g.,* see Chapter 8, and [20] provides a review of BBN applications specific to bioinformatics). But unless these tools are made readily accessible to a broader audience, the translation of the models to routine practice will be limited. To this end, we examine several applications of belief networks:

1) a simplified version used for classification purposes, the naïve Bayes classifier; 2) the use of BBNs in imaging, particularly focusing on its applications for medical image processing and related retrieval tasks; and 3) the use of belief networks to guide the visualization process, and in turn, to serve as a front-end to BBN model interaction.

Naïve Bayes

There is a special case of Bayesian belief networks, called *naïve* Bayesian belief networks (also sometimes called *simple* or *naïve Bayes*), which are often used as classifiers; as such, they have been used extensively in medical image processing, text classification, diagnostic/prognostic queries, and other tasks. Naïve Bayes classifiers are also well-suited to visualization in terms of nomograms [48]. Structurally, a naïve Bayesian classifier consists of one parent node (the class) and multiple children (the attributes) (Fig. 9.8b), and is predicated on a very strong assumption about the independence of the modeled attributes. Often this assumption is unrealistic; but accepting this restriction, naïve Bayesian classifiers can be trained efficiently over both large datasets and a number of attributes. And despite their simple design, naïve Bayes classifiers tend to perform well in real-world scenarios. Studies comparing classification algorithms have found naïve Bayesian classifier to be comparable in performance with classification trees and with neural network classifiers. Analyses have demonstrated possible theoretical reasons for naïve Bayes' efficacy [40, 61].

To appreciate how naïve Bayesian classifiers work, consider discriminating between histological types of lung cancer based on size, lobulation, and margin appearance on computed tomography (CT) imaging. Given a (labeled) dataset, a normal Bayesian classifier operates by stating that, if a tumor is encountered that is between 0.5-6 cm in diameter, *and* with poorly defined margins, *and* is lobulated, what is the most likely classification based on the observed data sample? Unfortunately, to properly estimate these probabilities, a sufficiently large number of observations is needed to capture all possible combination of features (thereby representing the joint distribution). In assuming that the features are independent of one another, naïve Bayes classification instead circumvents this problem: in our example, the probability that the tumor is 0.5-6 cm, has poorly defined margins, and appears lobulated (and is probably an adenosquamous carcinoma) can be computed from the independent probabilities that a tumor is of a given size, that it has a specific type of margin, and that it is lobulated. These independent probabilities are much easier to obtain, requiring less sample data.

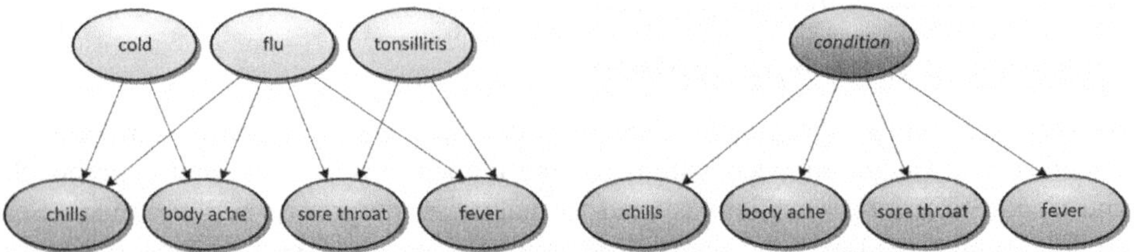

Figure 9.8: **(a)** An example belief network consisting of three diseases and four symptoms. Examination of the topology (via d-separation) shows that the symptoms are not independent. **(b)** A naïve Bayes classifier. This classifier is not equivalent to Fig. 9.8a given that various symptoms (attributes) of the class (*i.e.,* condition) are dependent; the naïve Bayesian classifier requires independence.

More formally, let $\mathbf{A} = \{A_1,\ldots,A_n\}$ be the n attributes used in a classifier, C. For a given instance $\{a_1,\ldots a_n\}$, the optimal prediction is class $C = c$ such that $P(c \mid A_1 = a_1 \wedge \ldots \wedge A_n = a_n)$ is maximized. Using Bayes' rule, this probability can be rewritten as:

$$P(c \mid A_1,\ldots,A_n) = \frac{P(c)P(A_1 = a_1 \wedge \ldots \wedge A_n = a_n \mid c)}{P(A_1 = a_1 \wedge \ldots \wedge A_n = a_n)}$$

where $P(c)$ is readily computed from a given training set. For classification purposes, as the denominator is identical across all values of c, we need only concern ourselves with the numerator term. Using Bayes' rule again, the numerator can be stated as $P(A_1 = a_1 \mid A_2 = a_2 \wedge \ldots \wedge A_n = a_n, c)P(A_2 = a_2 \wedge \ldots \wedge A_n = a_n \mid c)$, recursively rewritten for each corresponding attribute. Given the independence assumption, then: $P(A_1 = a_1 \mid A_2 = a_2 \wedge \ldots \wedge A_n = a_n, c) = P(A_1 = a_1 \mid c)$. The original numerator is thus equal to the product of each independent conditional probability. Accumulating these results, each probability can then be estimated from a training data set such that:

$$P(A_j = a_j \mid c) = \frac{\text{count}(A_j = a_j \wedge c)}{\text{count}(C = c)}$$

where the above equation provides maximum likelihood probability estimates. It can be further shown that naïve Bayes is a non-parametric, nonlinear generalization of the more well-recognized logistic regression. Caution must be used when naïve Bayes techniques are applied. Consider Fig. 9.8, which collapses three different causes into one condition that has four possible effects. In general, these two graphs are not equivalent unless the *single-fault assumption* is made. The single-fault assumption states that only one condition can exist at any given time, as the multiple values of the condition variables are exclusive to each other. Applying the single-fault assumption to Fig. 9.8b, inconsistencies quickly arise; for instance, if a patient is known to have a cold, then whether the patient has a fever does not influence the belief that the patient also has a sore throat (intuitively, this does not make sense, as if the patient has a fever, this should increase our evidence for tonsillitis, which would in turn increase our belief in a sore throat). The single-fault assumption requires that the sore throat and fever be d-separated (*i.e.*, that they are independent variables), but based on the original network, sore throat and fever are not d-separated.

Imaging Applications

Graphical models have become increasingly popular in computer vision, being applied to bridge the gap between low-level features (*e.g.*, pixels) and high-level understanding (*e.g.*, object identification). Given the wide variety of images that exist (*e.g.*, natural images, medical images) and the large number of pixels that compose an image, problems in computer vision benefit from a BBN's ability to integrate domain-specific knowledge and to simplify computations by exploiting conditional independence relationships; a few applications are summarized here:

- Enhancing image processing. [64] creates a geometric knowledge-base using BBNs to provide an efficient framework for integrating multiple sources of information (*e.g.*, various edge detectors). The results of such detectors are typically partial, disjoint shapes; but when used as inputs into the BBN, the model infers the most probable complete geometrical shape based on these available features. [47] uses a BBN to perform real-time semi-automatic object segmentation. First, the image is segmented using watershed segmentation; then, a graph is imposed onto the resulting gradient lines, placing a node where three or more converge and drawing an edge along the section of the watershed line between two nodes. The model's prior probabilities encode the confidence that an edge belongs to an object boundary while the CPTs enforce global contour properties. The image segmentation is determined by computing the MPE. [1] improves on a split-and-grow segmentation approach by using a BBN to integrate high-level information cues. The BBN is used to infer the probability of a region having an object of interest based on image attributes such as location of the region and color. Regions with the highest probabilities are taken as seed points for subsequent region-growing stages.

- Image object identification/classification. [44] demonstrates the ability of using BBNs to perform scene categorization: an image is initially characterized into two sets of descriptors: low-level features (*e.g.,* color and texture) and semantic features (*e.g.,* sky and grass). These features are used as evidence to instantiate a BBN-based inference engine that produces semantic labels for each image region. [72] examines how visual and physical interactions between objects in an image can provide contextual information to improve object detection. A BBN is automatically generated from features detected in the image; this model is then used to generate multiple hypotheses about how the objects interact with each other (*e.g.,* occlusions). The system then generates a synthetic image that represents the most probable configuration of individual image features.

While many of these algorithms have been developed for natural images, they may also be applied to medical images. Many of the low-level features (*e.g.,* texture) are used to characterize medical images and in combination with BBNs, may provide for more accurate indexing and retrieval of images from large biomedical databases to support content-based image retrieval.

Querying and Problem-centric BBN Visualization

An emergent question in applications of BBNs to the medical domain has been how to merge their use into clinical care, abstracting away the underlying complexity while still exposing their utility as tools for decision-making. Rather than use the DAG representation of a BBN to interact with a disease model, one approach is to create an intermediate layer that replaces BBN variables with common graphical representations (icons or visual metaphors) that can be drawn using the patient's own data to compose queries. A user's visual query is then interpreted by the application and translated into a question to the BBN. Additionally, this "visual querying" approach is well-suited to imaging data, where geometric and spatial features (*e.g.,* size, shape, location) are more readily graphically depicted. The BBN itself can also be used as a source of knowledge to guide the display of patient information and results, enabling "problem-centric" BBN visualization: the network topology and conditional probabilities give clues as to which variables (and thus, which data) are closely related and should be presented as part of the query's context. We conclude by presenting two systems that illustrate these techniques.

Visual Query Interface

The first system, called the *visual query interface* (VQI), facilitates inference on a disease model through a graphical paradigm; moreover, the user's querying process itself is guided by the topology and the parameters of the underlying model. This system is designed for radiologists and other physicians who are interested in using image features (*e.g.,* color, texture, shape) to find other similar studies in a large repository (*e.g.,* picture archive and communication systems, PACS), such as in applications for medical content-based image retrieval. For example, consider the request, *retrieve all related patient cases that have nodules with a speckled appearance in the right lower lobe of the lung.* Given the nature of medical images, a visual query-by-example interface is well-suited to the task of query composition: spatial (*e.g.,* location) and morphological attributes (*e.g.,* irregular tumor border) are naturally described by a graphical representation. In point of fact, one usability study showed that when asked to specify complex queries, users' found visual queries to be more intuitive and expressive than traditional text query languages [66]. Yet as the number of queryable features within a domain grows, the tools to facilitate visual expression of the query must be well-organized and guarantees on the logic of the query must be made.

To this end, in VQI the user manipulates a pictographic representation of BBN variables, referred to as *graphical metaphors.* Two types of graphical metaphors exist: 1) a freehand metaphor that allows the user to sketch a query object (*i.e.,* a tumor) and its environment (*e.g.,* surrounding anatomical structures); and 2) a component metaphor that prompts the user to input numerical or categorical values based on fields in the patient record. By combining graphical metaphors in different ways, a variety of diagnostic, prognostic, and treatment-related questions may be posed. For imaging-based variables, graphical metaphors take on the properties of their image feature counterparts, allowing users to alter their sizes, locations, relative geometrical positions, and shapes to obtain the desired query. The metaphors bridge a user's knowledge of a familiar domain (*e.g.,* a radiologist's expertise in image interpretation) to an exploratory framework that may include additional variables. Additionally, the selection of graphical metaphors in VQI is context-specific such that as the query is built, different metaphors are made available (or removed) to enable the user to draw a permissible query. A feedback loop exists between the user and the underlying graphical model, as illustrated in Fig. 9.9: given a disease model, contextual information provided by the variables, structure, and user interaction with the model influence what graphical metaphors or functionality is the displayed to the user. As the user selects metaphors to formulate a query, the inputs provide some context about the types of variables that are of interest to the user and in turn can be used to identify the subsets of variables in the model that are directly related and relevant for the query. This feedback loop provides a form of relevance feedback: as the user chooses a set of variables to be a part of the query, the system uses this information to refine what metaphors are presented next to the user.

VQI supports the use of labeled imaging atlases to provide spatial information about anatomical structures. For example, when the user overlays a tumor metaphor atop a

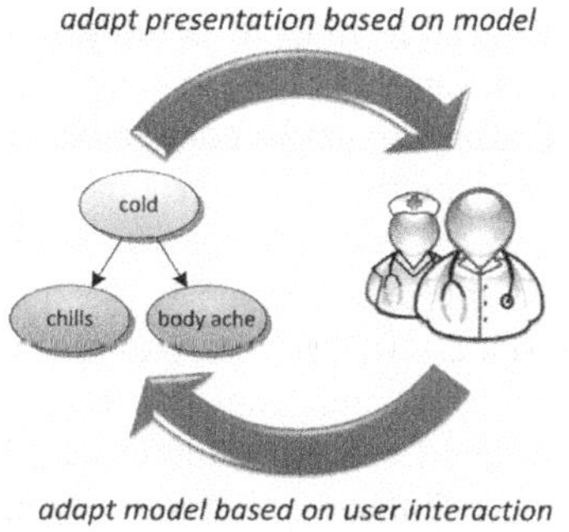

Figure 9.9: VQI's relationship between user interaction and the underlying BBN.

representative slice from a brain atlas, the anatomical information encoded in the atlas is used to determine the location of the metaphor and whether the metaphor affects any surrounding structures (*e.g.*, mass effect on the right ventricles).

Adaptive interfaces using BBNs. To dynamically adapt the interface, the BBN is utilized to perform two tasks: 1) to capture knowledge about a disease in a probabilistic manner so that inference may be performed with instantiations of the information; and 2) to map variables to graphical metaphors and to determine when a metaphor is pertinent to the user's query. The variables, structure, and user interaction with the model are hence used by VQI to determine when a given variable is "relevant" as described subsequently:

1. Variables. In constructing a disease model, a select number of variables are chosen and modeled to characterize a disease process. Each variable is mapped to a unique graphical metaphor. By way of illustration, an age variable would map to a component graphical metaphor that prompts the user to specify a numerical value. In addition, each variable has a number of states that the variable can take on; these states dictate what properties a graphical metaphor can take on. For a variable that models the percentage of tumor removed from a patient, the states may be specified by a range of percentage values (*e.g.*, 90-100% resection); the graphical metaphor is responsible for transforming a user's numerical input and placing it into one of the variable's states. Variable names can also be mapped to a broader knowledge source, such as an ontology, that allows the variable to be defined and placed into the context of other related variables. For example, if a disease model representing brain-related symptoms includes a variable word blindness that represents a loss of the patient's ability to read written text, the variable is mapped to the term *alexia* in the Unified Medical Language System (UMLS) lexicon and assigned to the semantic type *T047 - Disease or Symptom*. After mapping all of variables to UMLS, variables with identical or similar semantic types are grouped and presented together in the query interface.
2. Model structure. The network topology encodes information about the conditional independencies that exist in the model. Based on the Markov assumption, conditional independencies allow the model to be decomposed into small subgroups given evidence about certain variables. For instance, a variable, given information about the parents, children, and children's parents, can be fully explained by these variables and therefore isolated from the rest of the network. This specific property is called a *Markov blanket*. VQI leverages this property to identify those subsets of variables in the model related to a given variable of interest. When a variable of interest is selected, VQI examines the variable's Markov blanket to identify additional graphical metaphors to be presented in the interface. Also, the in- and out-degree of a variable help to determine the relative importance of a variable: highly connected variables can be considered more crucial to a disease process than variables that are sparsely connected. In VQI, the connectedness of a variable is used to determine the initial group of metaphors that is presented to the user.
3. Query. Information about the user's goals is gleamed from the query itself. The variables that the user selects to be a part of the query elucidate the types of information that the user is seeking from the model. As an example, if the user selects several imaging-related variables, the probability that the user is interested in determining how imaging features affect the outcome of the patient is increased.

Therefore, the model increases the weight of other imaging-related variables in the model so that they are visually highlighted or presented prior to other metaphors in the interface.

This adaptive presentation of relevant graphical metaphors not only simplifies the process of creating a query by reducing visual (selection) clutter, but also enforces logical rules regarding the order that metaphors are selected to formulate a query. For instance, in neuroradiology, contrast enhancement, if present, appears around certain image features of a tumor, such as a cyst or necrosis. Therefore, the option to add a rim contrast metaphor is only applicable when a cyst or necrosis metaphor is already present in the query.

Formulating a query. The process of posing a visual query is as follows: from a normal or patient imaging study, the user selects a representative slice or location to pose the query; the user iteratively constructs a query by drawing upon the available set of presented metaphors to represent visual features of the disease; and the final

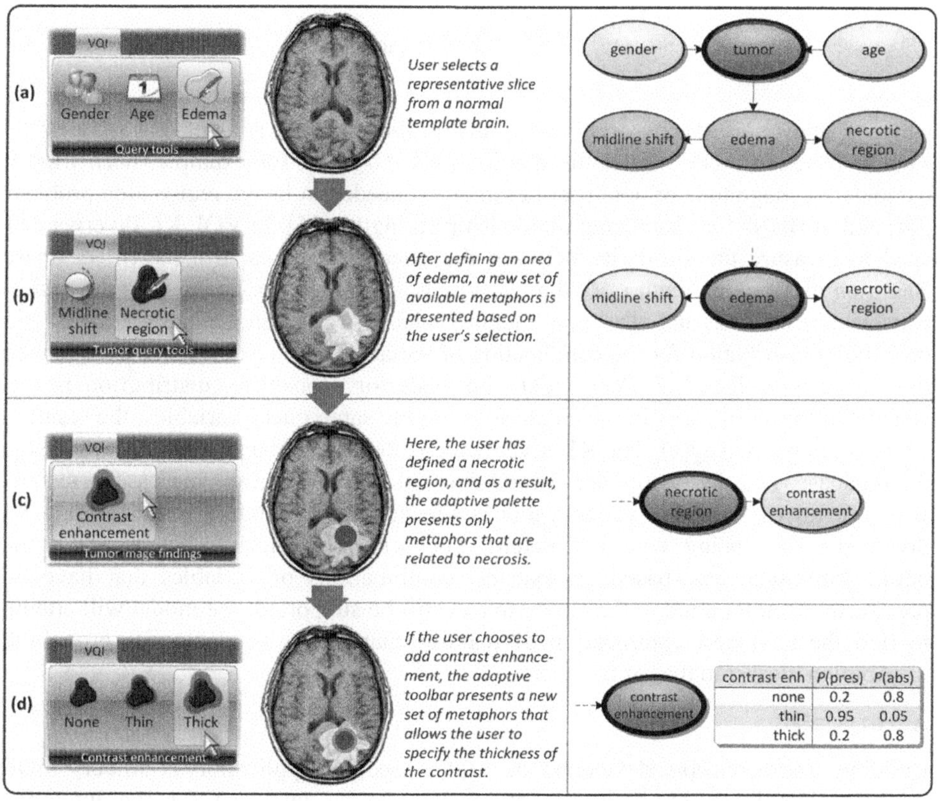

Figure 9.10: Demonstrating query formulation using VQI and how the adaptive interface uses the model to determine the presentation of graphical metaphors. **(a)** The user initially selects a representative slice from an atlas to place a tumor object. **(b)** After drawing an edema metaphor in the query; the model then identifies which metaphors to present next based on the structure of the model. **(c)** After adding a necrotic metaphor, the next relevant metaphor is contrast enhancement. **(d)** The user specifies properties of the contrast enhancement based on the states defined in the variable.

query is translated into an object representation that is used to set the states of variables in the BBN as the basis of a MAP query. Fig. 9.10 demonstrates how VQI's adaptive interface works in the context of posing a query in the domain of neuro-oncology: users are presented with a normal brain atlas (ICBM452 [60]), from which axial, coronal, or sagittal slices can be selected (Fig. 9.10a). An adaptive toolbar presents available metaphors based on context: as the user selects structures (*e.g.,* white matter) or metaphors (*e.g.,* edema metaphor) in the editor, related metaphors are presented in the toolbar (and unrelated metaphors are removed). For instance, when the contrast enhancement metaphor is selected, the user is prompted to define whether the border is thick or thin. A user progressively composes a visual query, which is then translated to values posed against the BBN and inference can then take place.

Case-based retrieval. VQI supports case-based retrieval by using the *Kullback-Leibler (KL) divergence* (D_{KL}). Originally posed as an information theoretic measure [39], D_{KL} assesses the difference between two probability distributions (over the same event space) and is formally defined for discrete random variables as:

$$D_{KL}(P,Q) = \sum_{x \in \chi} P(x) \log \frac{P(x)}{Q(x)}$$

where P and Q are the two probability distributions, and χ is a set of variables: the smaller D_{KL}, the more similar the distributions. D_{KL} has, for example, been used to compute the magnitude of nonlinear deformation needed in image registration problems [74] and in BBNs for visualizing relationship strengths [36]. In VQI, KL divergence is used to measure the similarity between the query and cases in a patient database. Based on the imaging features of the query (*e.g.,* size, location, geometric relationships between objects, etc.) and other non-imaging values specified in the query, the posterior probability distribution for the combination of variables given as evidence is computed; this value is assigned as $P(x)$. Next, the posterior probability distribution is then calculated for all of cases in the database using the same query variables; the resulting value is assigned as $Q(x)$. The KL divergence is iteratively calculated for each case in the database, and the results are ranked from lowest to highest. The case associated with the lowest KL divergence value is the "closest" matching case (with a KL divergence of 0 being a perfect match). The benefit of using this approach is that unlike traditional case-based approaches, combinations of variables that have not previously been inputted in the database can still be supported; the model will attempt to find the next best combination of features that result in a posterior probability distribution closest to that of the query.

AneurysmDB

Building from concepts developed in VQI, a second application is AneurysmDB. AneurysmDB is an ongoing project to develop an interface for the integrated visualization and querying of a clinical research database for intracranial aneurysms (ICAs) (Fig. 9.11). ICAs are a relatively common autopsy finding, occurring in approximately 1-6% of the general population; this statistic suggests that up to 15 million Americans have or will develop this potentially debilitating, if not deadly problem [65, 73]. Yet little is known about the true etiology of intracranial aneurysms and optimal treatment is still largely debated.

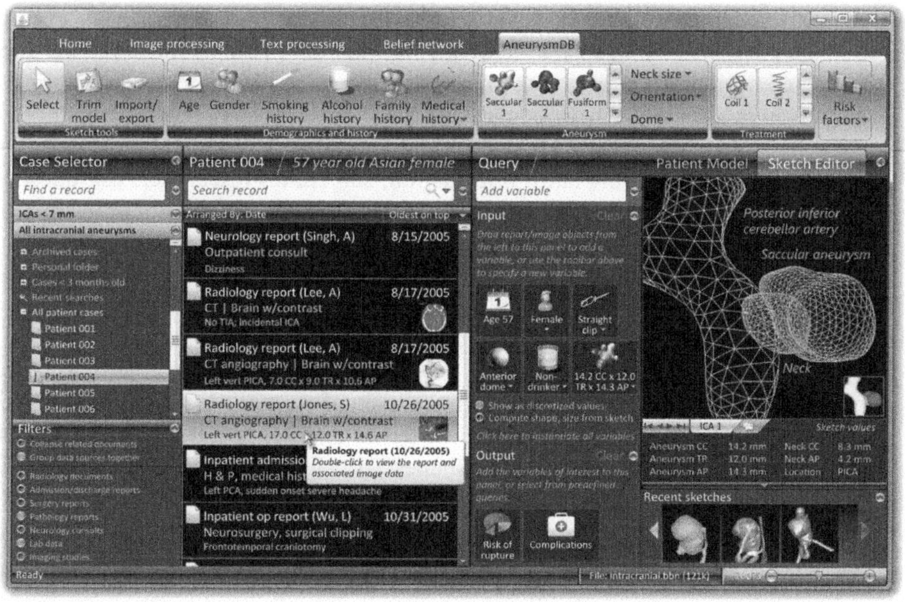

Figure 9.11: Example AneurysmDB interface, shown with a fictitious research patient. A BBN can be automatically populated with information extracted from the patient's electronic medical record. The user can then select which variables to query on and formulate a belief update or MPE/MAP query. Additionally, a 3D sketch interface is integrated to enable the user to draw an aneurysm shape/location as part of the query process. Elements in this display (*e.g.*, the extracted text components, task ribbon, overall layout) are also driven by the topology of the BBN.

Like VQI's application domain of neuro-oncology, this application's motivation is to support prognostic "what if" queries to an underlying disease model and the retrieval of similar cases (and hence, potential outcomes for a given individual). Additionally, the ribbon toolbar presented at the top of the interface is also guided by examination of the BBN and current query to identify likely variables to include. Unlike VQI, where the predominant focus is on guiding image-oriented queries, AneurysmDB aims to expand the querying process to all clinical variables extracted from the electronic medical record. Some key differences are highlighted:

- <u>Linkage to a phenomenon-centric data model</u>. As outlined toward the end of Chapter 8, a BBN can be connected to a phenomenon-centric data model (PCDM) to drive computation of the CPTs. Moreover, the classification of the variables and relations expressed in the data model provide additional semantic information that can be used to control what aspects of data are presented in the user interface. In this case, AneurysmDB is linked to a PCDM for ICAs, facilitating access to patient-level records for perusal; moreover, the elements that are shown (*e.g.*, the summaries for each document or imaging study; the grouping and ordering of elements in the task ribbon) are decided through a weighting of the relationships in the PCDM and the BBN. For instance, in addition to using Markov blankets, path length is used to determine the opacity and size of a graphical element as it is rendered in the interface. Path length is measured by the minimum number of arcs required to go between output variables and the given query input variables in both the BBN and PCDM: as the path length increases, the graphical representation for that variable would be rendered with reduced opacity and/or size. An example of adjusting opacity/size is in showing the list of findings for a report: those extracted elements that are most influential (*e.g.,* ICA size) should be highlighted and given in a summary before more ancillary variables (*e.g.,* smoking history).

- Temporal modeling. While clinical care is governed by making decisions about patient treatment with the latest information, sometimes researchers ask questions that are driven by retrospective analysis (*e.g.,* given the information up till a certain point in time, such as one year ago, how would the probabilities change compared to now?). To answer such a question, we must carefully separate out the data elements and inferences made over time. Connection to the PCDM allows us to pose queries based on different time points in the patient's history.
- Integration of a 3D sketch interface. Aneurysms are 3D entities, the morphology of which is critical to understanding the risk of rupture. In contrast to the 2D interface in VQI, a 3D sketch interface based on [31] is introduced, along with standard templates for common ICA shapes.

References

1. Alvarado P, Berner A, Akyol S (2002) Combination of high-level cues in unsupervised single image segmentation using Bayesian belief networks. Proc Intl Conf Imaging Science, Systems, and Technology, Las Vegas, NV, pp 235-240.
2. Bednarski M, Cholewa W, Frid W (2004) Identification of sensitivities in Bayesian networks. Engineering Applications of Artificial Intelligence, 17(4):327-335.
3. Bellazzi R, Zupan B (2008) Predictive data mining in clinical medicine: Current issues and guidelines. Intl J Medical Informatics, 77(2):81-97.
4. Bishop CM (2006) Graphical models. Pattern Recognition and Machine Learning. Springer, New York, pp 359-418.
5. Boyen X (2002) Inference and learning in complex stochastic processes. Department of Computer Science, PhD dissertation. Stanford University.
6. Boyen X, Koller D (1998) Tractable inference for complex stochastic processes. Proc 16th Conf Uncertainty in Artificial Intelligence (UAI), pp 313-320.
7. Breitkreutz BJ, Stark C, Tyers M (2003) Osprey: A network visualization system. Genome Biol, 4(3):R22.
8. Chan H, Darwiche A (2004) Sensitivity analysis in Bayesian networks: From single to multiple parameters. Proc 20th Conf Uncertainty in Artificial Intelligence (UAI), pp 67-75.
9. Chavira M, Darwiche A, Jaeger M (2006) Compiling relational Bayesian networks for exact inference. Intl J Approximate Reasoning, 42(1-2):4-20.
10. Cooper GF (1988) A method for using belief networks as influence diagrams. Proc 12th Conf Uncertainty in Artificial Intelligence, pp 55-63.
11. Cooper GF (1990) The computational complexity of probabilistic inference using Bayesian belief networks. Artificial Intelligence, 42:393-405.
12. Coupé VM, Peek N, Ottenkamp J, Habbema JD (1999) Using sensitivity analysis for efficient quantification of a belief network. Artif Intell Med, 17(3):223-247.
13. Coupé VMH, van der Gaag LC (2002) Properties of sensitivity analysis of Bayesian belief networks. Annals of Mathematics and Artificial Intelligence, 36(4):323-356.
14. Darwiche A (2003) A differential approach to inference in Bayesian networks. Journal of the ACM, 50(3):280-305.
15. Darwiche A (2009) Modeling and reasoning with Bayesian networks. Cambridge University Press, New York.

16. de Campos LM, Gámez JA, Moral S (1999) Partial abductive inference in Bayesian belief networks using a genetic algorithm. Pattern Recognition Letters, 20(11-13):1211-1217.
17. de Salvo Braz R, Amir E, Roth D (2008) A survey of first-order probabilistic models. In: Holmes DE, Jain LC (eds) Innovations in Bayesian Networks: Theory and Applications. Springer, pp 289-317.
18. Dechter R (1999) Bucket elimination: A unifying framework for probabilistic inference. Learning in Graphical Models, pp 75-104.
19. Dechter R, Mateescu R (2007) AND/OR search spaces for graphical models. Artificial Intelligence, 171(2-3):73-106.
20. Donkers J, Tuyls K (2008) Belief networks for bioinformatics. Computational Intelligence in Bioinformatics, pp 75-111.
21. Druzdzel MJ (1996) Qualitiative verbal explanations in Bayesian belief networks. AISB Quarterly:43-54.
22. Geman S, Geman D (1987) Stochastic relaxation, Gibbs distributions and the Bayesian restoration of images. Readings in Computer Vision: Issues, Problems, Principles, and Paradigms:564-584.
23. Getoor L, Friedman N, Koller D, Pfeffer A, Taskar B (2007) Probabilistic relational models. In: Getoor L, Taskar B (eds) Introduction to Statistical Relational Learning. MIT Press, Cambridge, MA, pp 129-174.
24. Haddaway P, Jacobson J, Kahn CE, Jr. (1997) BANTER: A Bayesian network tutoring shell. Artif Intell Med, 10(2):177-200.
25. Heckerman D, Chickering DM, Meek C, Rounthwaite R, Kadie C (2001) Dependency networks for inference, collaborative filtering, and data visualization. J Machine Learning Research, 1:49-75.
26. Heckerman D, Meek C, Koller D (2004) Probabilistic Models for Relational Data (MSR-TR-2004-30). Microsoft Research. http://research.microsoft.com/pubs/70050/tr-2004-30.pdf. Accessed March 3, 2009.
27. Horvitz E, Barry M (1995) Display of information for time-critical decision making. Proc 11th Conf Uncertainty in Artificial Intelligence (UAI), pp 296-305.
28. Horvitz E, Breese J, Heckerman D, Hovel D, Rommelse K (1998) The Lumiere Project: Bayesian user modeling for inferring the goals and needs of software users. Proc 14th Conf Uncertainty in Artificial Intelligence (UAI), pp 256-265.
29. Hu Z, Mellor J, Wu J, Yamada T, Holloway D, DeLisi C (2005) VisANT: Data-integrating visual framework for biological networks and modules. Nucleic Acids Res:W352-357.
30. Huang J, Chavira M, Darwiche A (2006) Solving MAP exactly by searching on compiled arithmetic circuits. Proc 21st Natl Conf Artificial Intelligence (AAAI-06), Boston, MA, pp 143-148.
31. Igarashi T, Hughes JF (2003) Smooth meshes for sketch-based freeform modeling. Proc ACM Symp Interactive 3D Graphics (ACM I3D 2003), pp 139-142.
32. Jaeger M (1997) Relational Bayesian nets. Proc 13th Conf Uncertainty in Artificial Intelligence (UAI), pp 266-273.
33. Jensen FV, Lauritzen SL, Olesen KG (1990) Bayesian updating in recursive graphical models by local computation. Computational Statistics Quarterly, 4:269-282.
34. Kadaba NR, Irani PP, Leboe J (2007) Visualizing causal semantics using animations. IEEE Trans Vis Comput Graph, 13(6):1254-1261.
35. Kjærulff UB, Madsen AL (2008) Sensitivity analysis. Bayesian Networks and Influence Diagrams, pp 273-290.
36. Koiter JR (2006) Visualizing inference in Bayesian networks. Department of Computer Science, PhD dissertation. Delft University of Technology (Netherlands).

37. Koller D (1999) Probabilistic relational models. Inductive Logic Programming, vol 1634. Springer, pp 3-13.
38. Koller D, Lerner U (2001) Sampling in factored dynamic systems. In: Doucet A, de Freitas JFG, Gordon N (eds) Sequential Monte Carlo Methods in Practice. Springer-Verlag, pp 445-464.
39. Kullback S, Leibler RA (1951) On information and sufficiency. Annals Mathematical Statistics, 22:79-86.
40. Kuncheva LI (2006) On the optimality of naïve Bayes with dependent binary features. Pattern Recognition Letters, 27(7):830-837.
41. Lacave C, Dez FJ (2002) A review of explanation methods for Bayesian networks. Knowl Eng Rev, 17(2):107-127.
42. Laskey KB (1995) Sensitivity analysis for probability assessments in Bayesian networks. IEEE Trans Syst Man Cybern, 25:901-909.
43. Lauritzen SL, Spiegelhalter DJ (1988) Local computations with probabilities on graphical structures and their application to expert systems. J Royal Statistical Society, 50(2):157-224.
44. Luo J, Savakis AE, Singhal A (2005) A Bayesian network-based framework for semantic image understanding. Pattern Recognition, 38(6):919-934.
45. Madigan D, Mosurski K, Almond RG (1996) Graphical explanation in belief networks. J Comput Graphical Statistics, 6:160-181.
46. Mengshoel O, Wilkins D (1998) Genetic algorithms for belief network inference: The role of scaling and niching. Evolutionary Programming VII, vol 1447. Springer, pp 547-556.
47. Mortensen EN, Jin J (2006) Real-time semi-automatic segmentation using a Bayesian network. IEEE Proc Conf Computer Vision and Pattern Recognition, vol 1, pp 1007-1014.
48. Mozina M, Demsar J, Kattan MW, Zupan B (2004) Nomograms for visualization of naive bayesian classifier. Proc Principles Practice of Knowledge Discovery in Databases (PKDD-04), Pisa, Italy, pp 337-348.
49. Murphy K, Weiss Y (2001) The factored frontier algorithm for approximate inference in DBNs. Proc 18th Conf Uncertainty in Artificial Intelligence (UAI), pp 378-385.
50. Neal R (1993) Probabilistic inference using Markov chain Monte Carlo methods (CRG-TR-93-1). Department of Computer Science, University of Toronto.
51. Nease RF, Owens DK (1997) Use of influence diagrams to structure medical decisions. Med Decis Making, 17(3):263-275.
52. Ng BM (2006) Factored inference for efficient reasoning of complex dynamic systems. Computer Science Department, PhD dissertation. Harvard University.
53. Ogunyemi OI, Clarke JR, Ash N, Webber BL (2002) Combining geometric and probabilistic reasoning for computer-based penetrating-trauma assessment. J Am Med Inform Assoc, 9(3):273-282.
54. Park J (2002) MAP complexity results and approximation methods. Proc 18th Conf Uncertainty in Artificial Intelligence (UAI), pp 388-396.
55. Park J, Darwiche A (2001) Approximating MAP using local search. Proc 17th Conf Uncertainty in Artificial Intelligence (UAI), pp 403-410.
56. Park JD, Darwiche A (2003) Solving MAP exactly using systematic search. Proc 19th Conf Uncertainty in Artificial Intelligence (UAI), pp 459-468.
57. Pearl J (1988) Probabilistic Reasoning in Intelligent Systems: Networks of Plausible Inference. Morgan Kaufmann Publishers, San Mateo, CA.
58. Poole D (2003) First-order probabilistic inference. Proc 18th Intl Joint Conf Artificial Intelligence, pp 985-991.
59. Przytula KW, Dash D, Thompson D (2003) Evaluation of Bayesian networks used for diagnostics. Proc IEEE Aerospace Conf, pp 1-12.

60. Rex D, Ma J, Toga A (2003) The LONI pipeline processing environment. Neuroimage, 19(3):1033-1048.
61. Rish I (2001) An empirical study of the naive Bayes classifier. Workshop on Empirical Methods in Artificial Intelligence; Proc Intl Joint Conf Artificial Intelligence, vol 335.
62. Romero T, Larrañaga P (2009) Triangulation of Bayesian networks with recursive estimation of distribution algorithms. Intl J Approximate Reasoning, 50(3):472-484.
63. Russell SJ, Norvig P (2003) Artificial Intelligence: A modern approach. 2nd edition. Prentice Hall/Pearson Education, Upper Saddle River, NJ.
64. Sarkar S, Boyer KL (1993) Integration, inference, and management of spatial information using Bayesian networks: Perceptual organization. IEEE Trans Pattern Analysis and Machine Intelligence, 15(3):256-274.
65. Schievink WI (1997) Intracranial aneurysms. N Engl J Med, 336(1):28-40.
66. Siau K, Chan H, Wei K (2004) Effects of query complexity and learning on novice user query performance with conceptual and logical database interfaces. IEEE Trans Syst Man Cybern, 34(2):276-281.
67. Suermondt HJ, Cooper GF (1993) An evaluation of explanations of probabilistic inference. Comput Biomed Res, 26(3):242-254.
68. Van Allen T, Singh A, Greiner R, Hooper P (2008) Quantifying the uncertainty of a belief net response: Bayesian error-bars for belief net inference. Artificial Intelligence, 172(4-5):483-513.
69. Verduijn M, Peek N, Rosseel PMJ, de Jonge E, de Mol BAJM (2007) Prognostic Bayesian networks: I: Rationale, learning procedure, and clinical use. J Biomedical Informatics, 40(6):609-618.
70. Wang H, Druzdel MJ (2000) User interface tools for navigation in conditional probability tables and elicitation of probabilities in Bayesian networks. Proc 16th Conf Uncertainty in Artificial Intelligence (UAI), pp 617-625.
71. Wemmenhove B, Mooij J, Wiegerinck W, Leisink M, Kappen H, Neijt J (2007) Inference in the Promedas medical expert system. Artificial Intelligence In Medicine, pp 456-460.
72. Westling M, Davis L (1997) Interpretation of complex scenes using Bayesian networks. Computer Vision - ACCV'98. Springer, pp 201-208.
73. Wiebers DO, Whisnant JP, Huston J, 3rd, Meissner I, Brown RD, Jr., Piepgras DG, Forbes GS, Thielen K, Nichols D, O'Fallon WM, Peacock J, Jaeger L, Kassell NF, Kongable-Beckman GL, Torner JC (2003) Unruptured intracranial aneurysms: Natural history, clinical outcome, and risks of surgical and endovascular treatment. Lancet, 362(9378):103-110.
74. Yanovsky I, Thompson PM, Osher S, Leow AD (2006) Large deformation unbiased differomorphic nonlinear image registration: Theory and implementation. UCLA Center for Applied Mathematics (Report #06-71).
75. Yap GE, Tan AH, Pang HH (2008) Explaining inferences in Bayesian networks. Applied Intelligence, 29(3):263-278.
76. Yedidia JS, Freeman WT, Weiss Y (2003) Understanding belief propagation and its generalizations. In: Lakemeyer G, Bernhard N (eds) Exploring Artificial Intelligence in the New Millennium. Elsevier Science, pp 239–236.
77. Yuan C, Lu T-C, Druzdzel MJ (2004) Annealed MAP. Proc 20th Conf Uncertainty in Artificial Intelligence (UAI), Banff, Canada, pp 628-635.
78. Zapata-Rivera JD, Neufeld E, Greer JE (1999) Visualization of Bayesian belief networks. Proc IEEE Visualization '99 (Late Breaking Topics), pp 85-88.

Chapter 10
Evaluation

EMILY WATT, COREY ARNOLD, AND JAMES SAYRE

Evaluation is a cornerstone of informatics, allowing us to objectively assess the strengths and weaknesses of a given tool. These insights ultimately provide insight and feedback for the improvement of a system and its approach in the future. Thus, this final chapter aims to provide an overview of the fundamental techniques that are used in informatics evaluations. The basis upon which any quantitative evaluation starts is with statistics and formal study design. A review of inferential statistical concepts is provided from the perspective of biostatistics (confidence intervals; hypothesis testing; error assessment including sensitivity/ specificity and receiver operating characteristics). Under study design, differences between observational investigations and controlled experiments are covered. Issues pertaining to population selection and study errors are briefly introduced. With these general tools, we then look to more specific informatics evaluations, using information retrieval (IR) systems and usability studies as examples to motivate further discussion. Methods for designing both types of evaluations and endpoint metrics are described in detail.

Biostatistics and Study Design: A Primer

Central to any evaluation is an understanding of statistics and the systematic methods used to design experiments that are unbiased and that will correctly answer questions of efficacy and impact. The focus of statistical analysis is the interpretation of a collection of data describing some phenomena. *Descriptive statistics* (*e.g.*, mean, median, mode) provide a summary of the collection, whereas *inferential statistics* aim to draw inferences about a population from a (random) sample. We start this chapter with a brief review of biostatistical concepts common to evaluation in biomedical informatics, leading into a discussion of study design and decision-making methods. Note that this section is not intended to be an instructional resource for statistics, but rather assumes some basic statistical knowledge on the part of the reader. For more detailed coverage of foundational concepts, the reader is referred to [15].

Statistical Concepts

Inferential statistics is concerned with the estimation of parameters that describe a population. Common tasks include: point estimates from a distribution (*e.g.*, calculating the mean from a random sample); interval estimates (*e.g.*, confidence intervals); hypothesis testing; and prediction (or, in the context of biostatistics, medical decision making). Interval estimates and hypothesis testing are covered in the sections immediately below; and medical decision making is covered in a separate section.

Confidence Intervals

When inferring values about a population, there is an inherent question of how "good" the estimate might be. *Confidence intervals* indicate the reliability of an estimate, providing an upper and lower bound around an estimated parameter. For instance, assume that a drug test shows that 40% of subjects experience improvement; a 95% confidence interval on this statistic would mean that in the general population (assuming

A.A.T. Bui and R.K. Taira (eds.), *Medical Imaging Informatics*,

DOI 10.1007/978-1-4419-0385-3_10,

a normal distribution), between 36-44% of the public would likely see benefits. The width of the confidence interval is driven by the degree of confidence: a higher confidence results in a smaller interval around the estimate.

The computation of a confidence interval is dependent on the parameter and whether a standard error can be calculated (*e.g.*, based on the standard deviation). In general, a statistic's confidence interval is given by *statistic* $\pm (z)\sigma_{stat}$, where σ_{stat} represents the *standard error*, and z is the critical value determined from the confidence level for a normal distribution. The standard error is defined as the standard deviation of the sampling distribution. For example, if the mean (μ) of a population is being estimated with a known standard deviation (σ) and normal distribution, then the interval is computed as follows:

$$\bar{x} - z\frac{\sigma}{\sqrt{n}} < \mu < \bar{x} + z\frac{\sigma}{\sqrt{n}}$$

where $\bar{x}$ is the sample mean and n is the sample size. However, if the standard deviation is unknown, then a t-distribution is substituted for z; and σ_{stat} is replaced by s_{stat}, the standard error/deviation as computed from the sample (*i.e.*, *statistic* $\pm (t)s_{stat}$).

Significance and Hypothesis Testing

In evaluating a tool or system, one must measure the strength of the evidence supplied by the data: is the data sufficiently strong to draw a conclusion? A *significance test* (or *hypothesis test*) is meant to help answer this question; by performing such tests, it is possible to ascertain whether the difference between an observed value and expected (*i.e.*, hypothesized) result is attributable to the claim or due to chance. To demonstrate, consider a diagnostic tool to test for the presence of X, and a pool of subjects, half of whom have X and we know definitively the state of each individual. Next, we randomly select from the pool 16 times (say, using the flip of a fair coin) and apply the diagnostic test to the chosen subject. In comparing the results, if the test correctly identifies 13 of the 16 subjects, is it probable that the test is able to truly discern X or it is simply chance that it "guessed" these classifications[1] correctly? Formally, a significance test evaluates in terms of a probability whether an apparent effect is due to chance: this hypothesis is called the *null hypothesis* (H_0) and denotes that the test or intervention had no effect. In contrast, the *alternative hypothesis* (H_a) positively states what the tool is meant to do. Continuing the example, the null hypothesis can be stated as, "*The diagnostic tool cannot detect X*," whereas the alternative hypothesis would be, "*The diagnostic tool can detect X*." In testing a group, the hypotheses are often stated in terms of a parameter of the population, such as the mean, or using some other hypothesis test (see below). H_a is said to be *one-sided* if the hypothesis states that the parameter is greater or smaller than some value defined in H_0 (*e.g.*, if H_0 states $\mu = k$, H_a states $\mu > k$ or $\mu < k$). H_a is said to be *two-sided* if it states that the parameter is simply not equal to the value chosen in the null hypothesis (*i.e.*, H_a states $\mu \neq k$). If the probability returned by a significance test, referred to as the *p-value*, is sufficiently low, then we reject the null hypothesis (therefore accepting H_a as true) and the result is deemed *statistically significant*. Although the choice of what is "sufficiently low" is arbitrary, *significance levels* (denoted by α) of 0.05 and 0.01 are traditionally used.

---

[1] In this particular example, if we assume a binominal distribution, then the probability of the test guessing correctly 13 of 16 times is ~0.01. Given this low probability, it is unlikely that the tool's results are due to chance.

Note that the p-value is the probability of observing data like the actual outcome when the null hypothesis is true: a small p-value hence indicates that the observed data are unlikely under the null hypothesis. One should not see the p-value as the probability that the null hypothesis is true; rather, the null hypothesis is rejected because an event has occurred that is unlikely if H_0 is true.

Test name	Usage and assumptions	Calculation	Non-parametric
One-sample z-test	Normally distributed population, σ is known. μ_0 represents the hypothesized population mean or specified value to be tested.	$z = \frac{\sqrt{n}(\bar{x} - \mu_0)}{\sigma}$	
One-sample t-test	Normally distributed population, σ is unknown. μ_0 represents the population mean or specified value to be tested. s is the standard deviation of the sample.	$t = \frac{\sqrt{n}(\bar{x} - \mu_0)}{s}$	Wilcoxon test
Paired t-test	A set of paired observations from a normal population (*e.g.*, before-after study); σ is unknown. s is the standard deviation of the sample. d represents the sample mean of the differences, d_0 is the population mean difference.	$t = \frac{\sqrt{n}(d - d_0)}{s}$	Wilcoxon signed-rank test
Pearson's chi-square test (distribution comparison)	Examines frequency distribution a univariate variable for the observed data vs. the expected data values via a goodness-of-fit test. The chi-square test looks at the difference between each observed value, O_i, and a computed expected value, E_i, based on a cumulative distribution function.	$\chi^2 = \sum_{i=1}^{n} \frac{(O_i - E_i)^2}{E_i}$	
Chi-square test (independence)	Tests for independence between two (nominal) variables, and can be likened to comparing the two variables in a row/column table, where each cell counts the number of occurrences from the sample distribution and compares it to the expected frequency for that variable combination.	$E_{i,j} = \sum_{k=1}^{c} \frac{O_{i,k}}{n} \sum_{j=1}^{r} \frac{O_{i,j}}{n}$ $\chi^2 = \sum_{i=1}^{r} \sum_{j}^{c} \frac{(O_{i,j} - E_{i,j})^2}{E_{i,j}}$	
Two-sample t-test	Two normally distributed populations that are compared. x_1 and n_1 are the sample mean and sample size of the first group; x_2 and n_2 of the second group. s_p is the pooled standard deviation. Δ is the difference between the two population's means. If the populations have the same variance, the denominator simplifies to $((s_1/n_1)+(s_2/n_2))^{½}$.	$t = \frac{(\bar{x}_1 - \bar{x}_2 - \Delta)}{s_p \sqrt{\frac{1}{n_1} + \frac{1}{n_2}}}$	Mann-Whitney test

Table 10.1: Examples of hypothesis tests, describing their usage and assumptions. Different tests are used based on the parameter being examined in a null hypothesis. The names of equivalent non-parametric versions of the tests are also shown.

A general procedure for conducting a significance test can be outlined as follows:

1. Specify H_0 and H_a. Choose α and an analysis plan, which determines how the test parameter in the hypothesis will be calculated from the sample data. This test method thus involves a test statistic and a sampling distribution.
2. Perform the study experiment/test and collect the data. Compute the selected test parameter used in the null hypothesis from the data.
3. Use the sample distribution for the chosen test method to find the probability (p-value) of the observed test parameter occurring.
4. Conclude whether the observed data are consistent with the null hypothesis (*i.e.*, is the p-value less than the selected α?).

It is important to understand that significance tests only speak to the statistical perspective. In a medical study, a rejected null hypothesis may be an accurate representation of the test population statistically, but may not be of any real clinical significance. Moreover, even though a not statistically significant finding may occur (*i.e.*, the p-value exceeds α), this does not mean that H_a is untrue – it only implies that the data as such does not support the alternative hypothesis.

Hypothesis tests. Although μ was suggested as a parameter, the null hypothesis may involve different estimates: the appropriate hypothesis test is needed to evaluate H_0 based on the quantity being estimated and the assumptions about the sample distribution. Examples of univariate hypothesis tests are summarized in Table 10.1. A *z-test*, for instance, is used to assess whether the difference between a sample mean and population mean is statistically significant given a sufficiently large sample. When a "smaller" number of samples are used to infer parameters on a normal distribution or a population's σ is unknown, the *t-test* is often substituted, referencing a *t*-distribution for n-1 degrees of freedom (where n is the number of samples). For hypotheses involving comparisons of means between two populations, two-sample tests can be used to compare the difference between means. A *paired t-test* is used when there is one measurement variable and two nominal variables; the classic example is a before-after study of some intervention on a group of subjects. Finally, one important hypothesis testing method is based on *chi-square statistics* (χ^2), which can be used to evaluate whether observed data matches the expected distribution[2] or χ^2 to test for independence between two variables. χ^2 tests uses nominal (categorical) data; continuous data can be transformed for χ^2 tests via binning methods if a cumulative distribution function is available.

Analysis of variance (ANOVA). A t-test is useful when there are two groups being assessed; however, situations often arise in which three or more groups are involved (*e.g.*, considering comparisons between multiple sites in a study). The standard statistical technique in this case is *analysis of variance* (ANOVA), which compares the difference in means among several study groups. A *one-way ANOVA* is performed when only one variable defines the groups (*e.g.*, age). More complicated tests involving multiple variables use a multivariate ANOVA (MANOVA). ANOVA involves three assumptions: 1) that the study groups' distributions are normal; 2) that each case is independent; and 3) that the standard deviation is equal across all study groups (homogeneity of variance). We outline the one-way ANOVA here, which is based on the decomposition of the sums of squares:

---

[2] A more rigorous, but complicated goodness-of-fit test is the *Kolmogorov-Smirnov test*, which can also be used to assess whether a sample comes from a population with a given distribution.

$$\sum_{i=1}^{k}\sum_{j=1}^{n_i}(x_{ij}-\bar{x})^2=\sum_{i=1}^{k}(\bar{x}_i-\bar{x})^2+\sum_{i=1}^{k}\sum_{j=1}^{n_i}(x_{ij}-\bar{x}_i)^2$$

where $\bar{x}$ is the mean across all samples, k is the number of groups, and n_i is the number of observations per the ith given group. The leftmost term represents the total sum of squares (total SS), the middle term is the sum of squares of treatments (SST), and the final term is the sum of squares of error (SSE). Based on the SST and SSE, the mean squares for treatment (MST) and error (MSE) are computed as:

$$MST=\frac{SST}{k-1}=\frac{\sum_{i=1}^{k}(\bar{x}_i-\bar{x})^2}{k-1}\qquad MSE=\frac{SSE}{N-k}=\frac{\sum_{i=1}^{k}\sum_{j=1}^{n_i}(x_{ij}-\bar{x}_i)^2}{N-k}\qquad F=\frac{MST}{MSE}$$

where N is the total number of samples across all groups. The F-score is then computed as the ratio of MST to MSE, and looked up in an f-distribution table based on the chosen α and the degrees of freedom associated with the MST and MSE (k - 1, and N - k, respectively). The confidence interval for a one-way ANOVA comparison between the difference of two means is:

$$(\mu_i-\mu_j)\pm(t_{\frac{\alpha}{2},N-k})\sqrt{\sigma^2(\frac{1}{n_i}+\frac{1}{n_j})}$$

where i, j represent the two different groups, t is given by the t-distribution, and σ^2 is equal to MSE. Note that if there are only two groups, a one-way ANOVA test is equivalent to a t-test, with the relationship being that $F = t^2$.

Correlation. *Correlation* reflects the strength of (linear) relationship between two variables – that is to say, how related are the variables? The most common method of quantifying correlation is through the use of Pearson's correlation coefficient, denoted by ρ (although when computed in a given sample, this symbol is denoted as r, hence this statistic is often referred to as Pearson's r):

$$\rho_{X,Y}=\frac{\text{cov}(X,Y)}{\sigma_X\sigma_Y}=\frac{\sum_{i=1}^{n}x_iy_i-n\overline{xy}}{(n-1)\sigma_x\sigma_Y}$$

where X, Y are the two variables of interest; cov(X, Y) represents the covariance between X and Y; and σ_x, σ_y are the standard deviations of the two variables. A correlation of $\rho = 1$ indicates that the two variables are correlated in an increasing manner (*i.e.,* as X increases, Y increases); whereas $\rho = -1$ indicates a decreasing correlation (*i.e.,* as X increases, Y decreases). Two independent variables should have $\rho = 0$ (however, the converse is not true). As Pearson's correlation is dependent on the distribution, a non-parametric assessment of correlation can be computed using Spearman's rank correlation coefficient (Spearman's rho), which converts the values in X, Y to ranks prior to using Pearson's correlate.

Assessing Errors and Performance

An imperfect model will, of course, result in misclassification. A *Type I error* (*α error*) is a false positive (FP), where the null hypothesis is rejected when null is actually true. In contrast, a *Type II error* (*β error*) is a false negative (FN), with the decision to keep the null hypothesis when null is false. A *confusion matrix* (also called a *2 x 2 contingency table*) is one method of visualizing the performance of a given classifier or test, tallying the true positive (TP), true negative (TN), false positive, and false negative rates (Fig. 10.1a). Based on the true/false positive/negative rates, the *accuracy* and *precision* of a test can be computed as:

$$accuracy = \frac{TP + TN}{P + N} \qquad precision = \frac{TP}{TP + FP}$$

where $(P + N)$ is the total number of samples tested. Accuracy is a measure of how well a test correctly classifies both positive and negative cases (*i.e.,* 100% accuracy means that a test classifies all positive and negative cases correctly). Precision refers to the reproducibility of a given test result.

Sensitivity and specificity. Two prevalent performance measures used in validating diagnostic tests are *specificity* and *sensitivity.* Sensitivity (SN) can be defined as the proportion of true positives that are correctly identified as such, whereas specificity (SP) is the proportion of true negatives that are correctly classified. SN and SP are calculated as follows:

$$sensitivity = \frac{TP}{TP + FN} \qquad specificity = \frac{TN}{TN + FP}$$

Sensitivity is also called the *true positive rate*, and specificity the *true negative rate.* In the context of a diagnostic test, sensitivity is the ability of the test to correctly detect disease; a sensitivity of 100% therefore implies the test finds all individuals with the disease, and it can be used to rule out its presence. Similarly, specificity thus assesses how well a diagnostic test determines that an individual is healthy. There is often a trade-off between sensitivity and specificity: choosing a low threshold value will result in identification of more true positive cases and thus increase sensitivity for an evaluation, but conversely identify more false positive cases (resulting in low specificity).

Receiver operating characteristic analysis. Receiver operating characteristic (ROC) analysis is often used to assess system performance, and the efficacy of computational

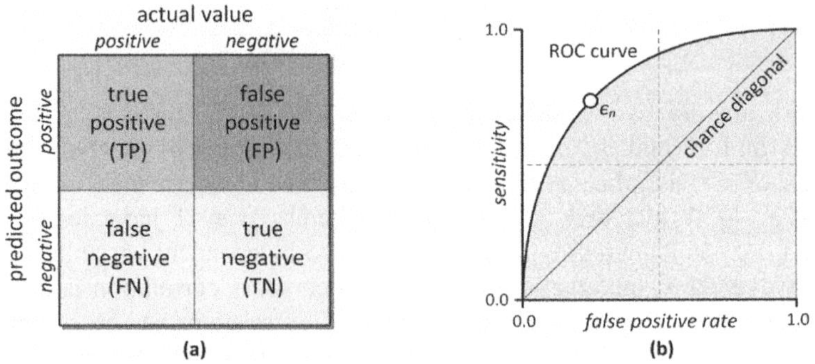

Figure 10.1: (a) A confusion matrix (2 x 2 contingency table) is used to tabulate the true positive, false positive, true negative, and false negative rates for a classifier. **(b)** An example of a receiver operating characteristic (ROC) curve, diagramming the sensitivity vs. false positive rate. ϵ_n are points that define the curve based on varying the threshold value, generating a sensitivity/specificity pair that is plotted.

models and diagnostic tests for decision support [38, 48]. In point of fact, in comparing two or more diagnostic tests, ROC analysis is often the only valid method of comparison. The typical classification metric used by ROC analysis is the *discrete classifier*, or binary test, which yields two discrete and opposing results (*e.g.*, positive and negative) to estimate an unknown. The simplest example of a binary test is the presence or absence of a disease in a population. The accuracy of these tests is quantified using sensitivity and specificity. A *continuous classifier* produces a computed metric based on a scale. To discretize the numerical value into a binary value, a threshold value (sometime referred to as a *cut point*) is chosen such that the test result is positive if the value exceeds the threshold (and negative otherwise).

An *ROC curve* plots a test's sensitivity against its false positive rate (*i.e.*, 1 - specificity) as the threshold is varied over its full range (Fig. 10.1b). That is to say, each data point on the plot is generated by using a different threshold value; as each threshold gives a particular set of TP, FP, TN, and FN counts, a pair of sensitivity and specificity values can be computed for each threshold value. The diagonal line connecting (0,0) and (1,1) is known as the *chance diagonal*, and represents the ROC curve of a diagnostic test with no ability to discern between classes. A *fitted* or *smooth ROC curve* is the result of assumptions made about the distribution of the test results. Using an ROC curve, visual comparison of two or more tests can be done on a common scale across all possible thresholds. A classic measure of ROC analysis is the *area under the curve* (AUC): an AUC value of 1 reflects a perfectly accurate diagnostic test, as sensitivity is 1.0 while the false positive rate is 0; whereas an AUC value of 0 is perfectly inaccurate; and an AUC of 0.5 represents a test that is no better than random guessing (*i.e.*, equivalent to the chance diagonal). The AUC can be interpreted as stating the probability of a diagnostic test as giving a correct positive result for a patient with the disease (relative to an individual without the disease). Lastly, the ROC curve permits one to optimally determine the threshold that balances sensitivity and specificity. For instance, [58] considers the use of an ROC curve to assess a diagnostic test such that the optimal threshold can be found by finding the sensitivity/specificity pair that maximizes the following function:

$$f(m, SN, SP) = SN - m(1 - SP), m = \frac{P(\text{normal})(C_{FP} - C_{TN})}{P(\text{disease})(C_{FN} - C_{TP})}$$

where m is the slope of the ROC curve; $P(\text{normal})$, $P(\text{disease})$ represent the probability of a patient being normal or having the disease (prior to knowledge of the test), respectively; and C represents the cost of a FP, TN, FN, or TP result. For further details on ROC analysis, the reader is referred to [59].

Study Design

Given the above statistical tools, study design is concerned with the specification of an objective experiment to test a hypothesis, and entails at least three steps: 1) deciding upon the study type that will serve as the vehicle for how data is collected; 2) determining the variables of interest that will be studied; and 3) defining the cohort or population, which provides the source for the information. At each point, the researcher's choice will establish the ultimate strength of the study's conclusions and whether it is subject to bias and/or other confounding factors that complicate result interpretation. We briefly consider each of these design steps next.

Types of Study Designs

There are several dimensions along which studies can be classified, the prevailing framework being the dichotomy between *observational* and *controlled experiments*. Both types of studies compare a treatment group (for which an intervention of some sort occurs to affect outcomes) vs. a control group (those subjects in which the intervention does not occur, providing a baseline). Observational studies can be seen as passive studies where the researcher does not intervene in any way; in particular, the researcher has no control over the assignment of subjects into the various test groups. In contrast, in a controlled experiment the researcher explicitly selects how subjects will be split among two (or more) arms of a study.

Traditionally speaking, observational studies are considered "weaker" in terms of the conclusions that can be drawn relative to a controlled experiment[3]. The former is subject to confounding variables and it can be difficult to separate association from true causation (*i.e.,* two study variables, X and Y, may be statistically correlated, but there may be no true physical meaning to the co-occurrence of X and Y in that X does not cause Y). The latter, in theory, permits for the examination of a well-defined set of controlled variables so that variable correlation is more likely to represent a causal relationship. However, the choice of implementing an observational vs. controlled study is driven by some considerations: 1) there may be legal/ethical issues in a study such that it is infeasible to interfere with the population, especially when invoking a harmful effect (*e.g.,* for a study on malnutrition, it would be unethical to starve subjects); 2) there may be practical issues in implementation with respect to cohort size (*e.g.,* a researcher studying a rare disease may not have enough volunteers for a controlled experiment); and 3) there may be cost issues, as controlled experiments are typically more expensive to conduct.

Observational studies. There are several classes of observational studies, described here in terms of increasing quality of evidence (and complexity):

- Cross-sectional study. The simplest of the observational studies, a *cross-sectional study* is a collection of data across multiple subjects (*i.e.,* a subset of a population) taken at the same time point. This type of investigation provides a "snapshot" of the status of a cohort, and can be contrasted with a *longitudinal study* wherein repeated observations are made on a group over a period of time.
- Case series/case report. A *case series* is a study that tracks a set of subjects with known characteristics (*e.g.,* exposure to a substance, a disease). Largely, the number of subjects in a case series is small relative to other types of study design (*e.g.,* case-control) and is thus subject to selection bias (see below). Data may be accrued either retrospectively from medical records, or prospectively. A *case report* is simply a detailed study of a single patient, often of a rare condition. As such, although weak in terms of study design, case series/reports may be the only information available to support a therapeutic strategy or decision.
- Case-control study. A *case-control study* involves both a test group of subjects, selected based on a characterization of a selected outcome factor, and a control

[3] From a statistical viewpoint, this belief may be true – but from a probabilistic perspective, if a sufficiently large cohort is used, an observational study may in fact have equivalent power to a controlled experiment. Knowledge discovery through inductive observational studies can be as conclusive as those obtained from experimental methods [6, 13].

group lacking this same outcome factor. Hence in studying a disease, case-control studies use patients that already have this disease, and retrospectively examine the group's characteristics to see how they differ from those who do not have the disease. Case-control designs facilitate the adoption of strategies to avoid methodological bias by selecting sample cases using factors independent of the variables influencing the effects under study. Subjects between the test and control groups are usually matched, and the selection of each case can be controlled without biasing the type of data that is acquired.

- Cohort study. A *cohort study* is a longitudinal study involving the selection of a group of individuals that share some assigned classification or feature (*e.g.*, a disease/condition; a treatment; an inherent characteristic such as the same birth date). The group is then followed across time to assess subjects' outcomes. A comparison group is often taken from the general population, so as to be separate from the study group classification, but in all other respects is similar to the study group (*i.e.*, matched). For instance, to study if prolonged treatment with *X* correlates with onset of condition *Y*, a cohort study would use a group of individuals treated with *X* over time, and a group with no exposure to *X*; both groups would then be tracked to evaluate the occurrence of condition *Y*. This example represents a *prospective cohort*, where the groups are defined before the data collection. Cohort studies looking to uncover causes of a disease define the group before the onset of disease in the subjects, and follow individuals to see who develops the condition. This case represents a *retrospective cohort*, wherein grouping of subjects occurs after data has been collected. By recording and comparing all characteristics of the group, a cohort study aims to ensure that observed differences between groups are attributable to the study phenomena (*i.e.*, are changes in the study's dependent variable due to the observed independent variable or some outside influence?). The disadvantage of this approach is that careful judgment must be applied to filter out irrelevant causal factors. Potentially the most well-known cohort study is the Framingham Heart Study, which originally followed over 5,000 adults in one town to assess the development of cardiovascular disease.

Controlled experiments. The archetypal controlled experiment in medicine is the *clinical trial*, a prospective study that assigns subjects to one of two or more study categories, referred to as *trial arms*, and then follows the individual longitudinally until some defined endpoint at which time the person is assessed and an end outcome is determined. Over the course of the clinical trial, each arm follows a well-defined study protocol. With the exception of the control group, a trial arm comprises some intervention. Comparisons are then made between each trial arm. Though classically associated with drug and device testing, the clinical trial construct is equally viable for the assessment of informatics tools.

Core to the clinical trial process is how subjects are assigned to the different study arms. A clinical trial can be either *randomized* or *non-randomized*. In the randomized design, the objective is to create an unpredictable allocation of participants; several methods have been described to achieve this effect [33, 64]. In a non-randomized design, the allocation is instead decided by some known parameter (*e.g.*, birth date). The *randomized controlled trial* (RCT) is considered one of the most reliable and scientifically objective types of study, and is the preferred source of knowledge in evidence-based medicine.

Given randomization, the investigators and its participants can be *blinded* or *non-blinded* to their assignment within a clinical trial in an effort to remove potential bias and/or placebo effect. In a *single-blind trial*, the investigator is aware of the assignment, but the subject is not. In a single-blind setup, the investigator's interaction may still bias study results and thus in a *double-blind trial*, both the interacting investigator and the subjects are unaware of the assignment, thereby ensuring impartiality. Because of this, the randomized, double-blinded controlled clinical trial is the gold standard of study design. Unfortunately, the execution of a double-blinded study – and RCTs in general – is largely difficult, given the time, expense, and degree of control needed to properly execute the study protocol.

Paired and crossover designs. In many of the above study types, the studies are divided between intervention and control groups, with the latter drawn from the general population. However, in clinical investigations, patients can also act as their own matched control by comparing an individual's state before and after two sequential tests are performed. Such an experiment defines a *paired study*. In a similar vein, a *crossover study* design [46] entails a subject alternating between a control and intervention state(s) (*e.g.*, a patient may be instructed to take a drug on alternate weeks, with testing each week to assess changes). Paired and crossover studies can be likened to *before-after study* designs, which are often used to evaluate information system deployments and the potential for healthcare impact.

Study Variable Selection and Population Definition

Upon selecting a specific study design, investigators must choose the appropriate features to measure for the sampled population. Clearly, variable selection is driven by the underlying study hypothesis and the need to characterize the population, the intervention, and its observed effects. In a *descriptive study*, which focuses on the general exploration or observation of a population's characteristics, descriptive statistics are used to study one variable at a time (*e.g.*, measuring the presence or absence of a hormone treatment.) An *analytic study* scrutinizes the relationships between two or more variables (*i.e.*, predictors and outcomes) to discover cause-and-effect relationships.

Given the study type and selected variables, the next step involves the definition of the actual populations that will be used in the investigation. As the study population is a microcosm of the real-world, the sample is chosen so that: the subset will control for systematic error (described below); and the sample is large enough to handle random error in generalizing the study findings to the population. *Inclusion criteria* define the main characteristics of the target population, with elements that describe an individual's eligibility. *Exclusion criteria* identify subjects that cannot qualify because of characteristics that may interfere with the intervention and finding interpretation. Different categories exist for describing the sampling process, divided between random and non-random techniques:

- Random sampling. *Random sampling* is frequently used in study designs and is judged the gold standard for ensuring maximal representation of a target population: a rigorous technique is used to estimate the fidelity with which phenomena observed in the sample represent those in the population. Common types of random sampling are *simple random sampling*, *systematic sampling*, *stratified sampling,* and *cluster sampling*. A simple random sample comprises drawing from the general population so that any member of the population is equally likely to be drawn; often this is effective in organizing larger collections of data such as seen in assembling data from a large-scale database (*e.g.*, a national clinical trial). Systematic sampling instead chooses every n^{th} individual from the general population. Stratified sampling involves the use of predefined groups of importance to the overall study (*e.g.*, age, gender); participants are grouped according to these criteria and then selected as sub-populations. Stratified approaches work well when intra-strata variability is minimal, inter-strata variability is maximized, and the strata itself is strongly correlated with the study's dependent variable. Both systematic and stratified sampling attempt to overcome potential bias issues seen with simple random sampling by using *a priori* knowledge of the population to guide the selection process so it more closely models the target.

- Non-random sampling. Any subjects that fit a study's selection criteria form a *convenience study sample*. These subjects typically are readily accessible to the investigator. A variation on convenience sampling is *consecutive selection*, which as its name implies uses all potential subjects within a given time period, and helps to minimize any type of selection bias. *Snowball sampling* involves the recruitment of individuals into a study based on word-of-mouth from subject to subjects in some type of (social) network (*e.g.*, between friends); inherently, however, this sampling method may result in selection bias.

The terms *probability* and *non-probability sampling* are sometimes mentioned. Probability sampling refers to the fact that within a population, every subject has a non-zero chance of selection (and that this probability of inclusion is calculable). The examples of random sampling given above are all instances of probability sampling. In contrast, non-probability sampling means that some individuals within the population have no chance of being included (*i.e.*, a zero-probability), and as such, is non-random.

Population Size: Sample Size and Power Calculations

The number of samples to include in a research study, or the study's *sample size*, is an important consideration in the design of many technical and clinical investigations. Before collecting data, it is essential to determine the sample size requirements for the study design, the measurement scales to be used, and the outcome statistics (*e.g.*, will the study be quantified as a means, mean differences, proportions, odds ratios, or an area under the ROC curve?). Based on these decisions, knowing the estimate of the outcome measures' precision is key: if the study precision is insufficient or the study lacks power, the study squanders resources and effort. Formally, the *statistical power* of a test is the probability of avoiding a Type II error, correctly rejecting a false null hypothesis: if β is the false negative rate, then power is defined as $(1 - \beta)$. In general, statistical power depends on the type of statistical test, the size of the difference being examined (*i.e.*, effect size), and the sensitivity of the data. *A priori* statistical power analysis therefore permits an estimate of the number of samples needed to reach adequate power[4].

Sample size and power for estimating means/mean differences. The sample size requirements for computing a mean or mean difference entail defining an acceptable *margin of error*, ε. This margin of error is equal to half the confidence interval width. When estimating μ with 95% confidence (without hypothesis testing), the sample size, n, can be calculated as $n = 4s^2/\varepsilon^2$ where s is the standard deviation for the variable

[4] In contrast, *post-hoc* power analysis is done subsequent to data collection to compute the study's actual power based on the observed data.

being sampled. Given the importance of s in this calculation, effort should be made to obtain a reasonable estimate of the standard deviation, drawing from prior or pilot studies as possible. This equation can be adapted to estimate a mean difference based on paired samples by substituting the standard deviation of delta (s_d); and for independent samples, the pooled estimate of standard deviation (s_p) can be used as the standard deviation estimate. In general, to test a mean or mean difference taking into account hypothesis testing, the following equation can be used:

$$n = \frac{2\sigma^2 (z_{crit} + z_{power})^2}{\Delta^2}$$

where n is the required sample size; σ represents the standard deviation of the variable as estimated by s, s_d, or s_p depending on whether the data are from a single sample, paired samples, or independent samples; Δ represents the expected mean difference; z_{crit} is the desired significance criterion; and z_{power} the desired statistical power. For a one-sample t-test, $\Delta = \mu - \mu_0$; for a paired-sample t-test, $\Delta = \mu_d$; and for an independent t-test $\Delta = \mu_1 - \mu_2$. The values for Δ are not calculated, but instead arise from some speculated difference. To evaluate the power of a t-test, let $\phi(z)$ represent the area under the curve to the left of z on a standard normal distribution curve (*e.g.*, $\phi(0) = 50\%$; $\phi(1.28) = 90\%$) The power of a t-test comparing means is given by $\phi(-\alpha + (|\Delta|n^{1/2})/\sigma)$, where Δ again denotes the expected mean difference, n denotes the sample size, α is the critical value for the chosen confidence interval (*e.g.*, for a 95% confidence interval, $\alpha = 1.96$), and σ is the standard deviation of the variable (*i.e.*, s, s_d, s_p).

Sample size for estimating a single proportion. To calculate a sample size for proportion p at a given confidence interval with margin of error ε, the equation $n = \alpha^2 p(1 - p)/\varepsilon^2$ is used, where α is the critical value for the chosen confidence level. As p is the variable being assessed, if no estimate is available, it is possible to assume $p = 0.5$ to obtain a sample that is big enough to ensure high precision.

Sample size for testing two proportions. For a study where two proportions are compared with a χ^2-test or a z-test, which is based on the normal approximation to the binomial distribution, the sample size can be computed as:

$$n = \frac{2(z_{crit}\sqrt{2p(1-p)} + z_{power}\sqrt{p_1(1-p_1) + p_2(1-p_2)})}{\Delta^2}, \; p = \frac{p_1 + p_2}{2}$$

where p_1 and p_2 are pre-study estimates of the two proportions to be compared; and $\Delta = |p_1 - p_2|^2$ (*i.e.*, the minimum expected difference). The two groups comprising n are assumed to be equal in number, and it is assumed that two-tailed statistical analysis will be used. Note that n in this case depends not only on the difference between the two proportions, but also on the magnitude of the proportions themselves. Therefore, this equation requires an estimate of p_1 and p_2, as well as their difference.

Sample size for two-rater kappa statistic. Let κ be the estimate of kappa, $\kappa = (p_0 - p_e)/(1 - p_e)$, where p_0 and p_e are respectively the estimates of the actual probability of agreement between the two raters, and the expected agreement when rating independently. The large-sample standard error (SE) is given by [20]:

$$SE(\kappa) = \frac{\tau(\kappa)}{\sqrt{n}} = \frac{1}{\sqrt{n}(1-p_e)^2}\left(p_0(1-p_e)^2 + (1-p_0)^2 \sum_{i=1}^{k}\sum_{j=1}^{k} p_{ij}(p_0+p_e)^2 - 2\lambda\right)^{1/2}$$

$$\lambda = (1-p_0)(1-p_e)\sum_{i=1}^{k} p_{ii}(p_i + p_i) - (p_0 p_e - 2p_e + p_0)^2$$

and where p_{ij} is the estimated proportion of samples that the first rater places into category i but that the second rater places into category j. Based on large sample theory, κ will have an approximately normal distribution with mean equal to the true inter-rater agreement measure and SE as approximated above. Power and confidence interval computations can therefore be based on the upper α percentiles of the standard normal distribution, $z_{1-\alpha}$. To obtain the $(1-\alpha)$ percent confidence interval for κ that has length $\leq d$: $2z_{1-\alpha}SE(\kappa) \leq d$. Replacing $SE(\kappa)$ by its maximum value guarantees that the inequality will be met and results in the sample size determining inequality:

$$n \geq 4(z_{1-\alpha/2})^2 \max\left(\frac{\tau(\kappa)^2}{d^2}\right)$$

For hypothesis testing, the sample size is determined so that a level α test of $H_0 = \kappa \leq \kappa_0$ against the alternative $H_1 = \kappa > \kappa_0$ will have power of at least $(1-\beta)$ when $\kappa = \kappa_1$. Thus, the sample size formula is:

$$n \geq \left(\frac{z_{1-\alpha}\max\tau(\kappa=\kappa_0) + z_{1-\beta}\max\tau(\kappa=\kappa_1)}{\kappa_0 - \kappa_1}\right)^2$$

where κ_0, κ_1 are the comparison kappa values from the null hypothesis being tested.

Sample size for ROC analysis. To determine the sample size for a study using rating data to measure the area under an ROC curve, the standard error estimator is needed and should be based on the binormal distribution. The binormal approximation to the SE for rating data is given as:

$$SE = \left(\frac{1}{2\pi}e^{\frac{-A^2}{W}}\frac{V_1}{W+(AB)^2}\frac{V_2}{W^3}\right)^{1/2} \quad A = \Phi^{-1}(\theta)(1+B^2)^{1/2}, B = \frac{\sigma_N}{\sigma_A}, W = 1+B^2$$

$$V_1 = \frac{1}{n_A} + \frac{B^2}{n_N} + \frac{A^2}{2n_A}, V_2 = \frac{B^2}{2}\left(\frac{1}{n_A} + \frac{1}{n_N}\right)$$

where Φ^{-1} is the inverse of the cumulative normal distribution function; and n_N and n_A represent the number of normals and abnormal, respectively. From the standard error, the sample size estimate can be modeled by:

$$n_A = \frac{(z_\alpha\sqrt{2\sigma_0} - z_\beta\sqrt{\sigma_0 + \sigma_1})^2}{(\theta_0 - \theta_1)^2}$$

where z_α and z_β are the chosen upper α and β percentiles of the standard normal distribution; σ_0 is the variance under the null hypothesis and σ_1 is the variance under the alternative hypothesis assuming that $n_A = n_N$; and θ is the ROC AUC per the corresponding distribution.

Internal pilot studies. There sometimes exists a need to design a study in the absence of any good estimates of variance or other parameters that may inform sample calculations. To overcome this problem, a preliminary pilot study can be used prior to conducting the primary evaluation; such a pilot proffers the chance to check study feasibility and to refine the many facets of the study design (*e.g.*, randomization, data collection, etc.). However, if the researchers are sufficiently confident of a study, they may instead wish to immediately proceed with the primary evaluation. [81] hence advocates the use of an *internal pilot study*, which uses the first portion of the primary study to recalculate the needed sample size using estimates of relevant study parameters. Given the importance of estimating these parameters from the actual study population, a degree of efficiency is gained by using data obtained from an internal pilot study as it need not be discarded (*i.e.*, it can be used in the final analysis).

Consider the following common problem: a lack of knowledge about the variance for a normal distribution at the outset of a study. Suppose that we have samples from two populations and that we want to evaluate whether the population means are different in normal populations $N(\mu_A, \sigma^2)$ and $N(\mu_B, \sigma^2)$ [17, 72]. The null hypothesis is that $\mu_A = \mu_B$, tested against a two-sided alternative with type I error α and power $(1 - \beta)$ at $\mu_A - \mu_B = \delta$. The total sample size per arm, n, is $\max(n, (t_{2n-2,\ 1-\alpha/2} + t_{2n-2,\ 1-\beta})^2(2s^2/\delta^2)$. If n responses on each treatment, the t-test with two-sided type I error rate α rejects H_0 if:

$$\frac{|\bar{x}_A - \bar{x}_B|}{\sqrt{2s^2 / n}} > t_{2n-2,\alpha/2}$$

where s^2 is obtained on $(2n - 2)$ degrees of freedom from the pilot data. Using this information, the primary evaluation's estimates can then be subsequently updated.

Study Bias and Error

In spite of the detailed planning and care that may be used, a study is susceptible to both data errors and bias, impacting the results, their interpretation, and ultimately the conclusions drawn from the study. Formally, error is a departure of the observed value from the expected (true) value. Two types of error afflict research studies: *random error* and *systematic error*. Random error is attributable to unknown sources of variation that are equally likely to distort the sample in either direction. Increasing a study's sample size is the usual solution to reducing the influence of randomness. Bias, however, reflects systematic errors that are introduced into the evaluation as a result of sampling, measurements, or other problems. In this case, increasing sample size does not alleviate the effects of systematic error; instead, improving the accuracy of the estimate will reduce the impact of any bias. [21] categorizes biases affecting study design: 1) *selection bias*, wherein the sampled subjects or phenomenon studied are not representative of the population; 2) *measurement* or *information bias*, where the measurement of the sampled phenomenon is systematically different from that used in the population; and 3) *confounding bias*, which makes the sample or measurements unrepresentative of the population. Specific types of selection bias include: *group membership bias* such that test participants may be naturally classified more heavily in one group (*e.g.*, race, geographic location); *Berkson's bias* (also referred to as state-of-health bias), where subjects are selected from a captive patient pool; and *Neyman's bias* (prevalence/incidence bias), which encompasses the selection of patients who exhibit irregular patterns of health or a specific condition of a study disease. An example of measurement bias comes from the problem of missing data: results may be skewed to different groups because subjects in a different group may not have measured data (*e.g.,* failing to answer questions, dropping out of a survey, etc.). Information bias includes: *recall bias*, in which subjects may provide additional, more complete, or even exaggerated responses when prompted as part of a study; *Hawthorne bias*, where subjects act differently knowing that they are being observed; and *observer bias*, in which the individual collecting data has a preconceived expectation of the results.

One particular type of source of error/bias bears some further discussion. Scientific studies are based on the idea that measurements are repeatable. However, there are many sources of variation that arise, undermining the repeatability of measurements. *Random subject variation*, for instance, corresponds to the fact that a measurement taken in a given individual (*e.g.,* blood pressure) will likely vary, even if taken in rapid succession. In studies involving expert or user assessment, *intra-* and *interrater variability* must be assessed. Intrarater variability (intra-observer agreement) measures the degree to which a given user is consistent in their responses. For example, a radiologist may be asked to read an imaging study and to measure a tumor lesion; after some time (to remove any memory effect), if the same radiologist is asked to repeat the task with the same study, the measurement would ideally be identical – yet often, the measurement, albeit similar, is not exactly the same. The statistical method of *test-retest* (*i.e.,* performing the same test twice with the same subjects to validate test result consistency per subject) is similar to gauging intrarater variability. Interrater variability (inter-observer agreement) represents the variation between experts; in theory, different experts should provide the same response to a task, but again variation can occur. A kappa statistic is typically computed to assess interrater variability (see below). Both intra- and interrater variability are metrics of the reliability of a measurement.

Meta-analysis

Study design is centered about the creation of an experiment to examine data, be it either retrospective or prospective, to answer a hypothesis. Another means of answering a hypothesis is to perform *meta-analysis*, which synthesizes the (quantitative) results of several existing common studies together. Meta-analyses are often used to examine the strength of relationship – that is, the *effect size* – between two variables (*e.g., what is the effect of X on Y*?). Each study independently provides an effect size estimate, which can be modeled alongside the study's characteristics; when combined via meta-analysis, a more powerful estimate of the effect size is derived. Hence, the appeal of meta-analysis lies in its ability to combine several different types of studies together (*e.g.,* RCTs, observational studies) across a large population. Moreover, proponents of meta-analysis argue that this type of assessment allows one to measure the magnitude of an effect across studies, rather than just its (statistical) significance, and to explain the differences between studies.

A meta-analysis consists of four steps: 1) a review of the literature to collect studies related to the hypothesis; 2) a sub-selection of the studies based on some inclusion/ exclusion criteria (*e.g.,* the quality of the study, the particular cohort, degree of perceived publication bias); 3) a selection of the study variables and/or summary measures that will be extracted for analysis (*e.g.,* means, differences, etc.); and 4) analysis via a meta-regression model to compute an effect size across all studies and subsets of the studies. The analysis is performed by converting all of the accepted studies' statistics into a common effect size metric, being one of two categories: a standardized mean difference (*e.g.,* Cohen's d or Hedges' g statistics) or a correlation (*e.g.,* Pearson's r). For instance, a one-way ANOVA F-score can be converted into an r value. Correcting for sample variation and other study characteristics, an aggregate effect size statistic is then computed. A full discussion of the meta-regression models is beyond the scope of this primer; we refer the reader to [50] for further details.

Decision Making

The process of care is predicated upon the fact that given information about a patient, we can make a decision to improve the individual's health and/or quality of life. Medical decision making is thus a major area of informatics investigation, and has longstanding origins in biostatistical approaches that aim to predict an outcome based on some quantitative model derived from observed data. Chapters 8 & 9 provided insight into the Bayesian approach to modeling and decision making; here, we cover two additional techniques rooted in statistical methods that are commonly seen in biomedicine to provide decision-making models: regression analysis and decision trees.

Regression Analysis

Perhaps the most common set of techniques, *regression analysis* encompasses a range of methods that model a dependent variable (the *response variable* or measurement) with one or more independent variables (the *explanatory* or *predictor variables*). The regression equation or model is created based on an analysis of sample data that is deemed representative of the target population; as such, the assumptions require that errors seen in the data are due to randomness and that the predictor variables are linearly independent of each other.

Multiple linear regression. Under linear regression, the relationship between the dependent and independent variables is given by a function that is a linear combination of one or more model parameters (β_i), and is of the form: $Y_i = \beta_0 + \beta_1 X_{i1} + \beta_2 X_{i2} \ldots + \beta_p X_{ip} + \epsilon_i$ where Y is the response variable, X_{ip} are the predictor variables, and ϵ_i is a random variable representing error. The equation, as stated, results in a straight line through the data points; but in general, the equation may be used to fit non-linear curves through the data points (*e.g.*, a parabola by using a quadratic term). The aim of linear regression is to estimate the model parameters, β_0 through β_p, which best fit the data; this estimate is typically given through a least squares fit. Rewriting this problem in matrix notation, then the estimate can be given as $\boldsymbol{\beta}'$:

$$\mathbf{Y} = \mathbf{X}\boldsymbol{\beta} + \varepsilon \qquad \mathbf{X} = \begin{pmatrix} 1 & x_{11} & \cdots & x_{1p} \\ 1 & x_{21} & \cdots & x_{2p} \\ \vdots & \vdots & \ddots & \vdots \\ 1 & x_{n1} & \cdots & x_{np} \end{pmatrix} \qquad \boldsymbol{\beta}' = (\mathbf{X}^{\mathrm{T}}\mathbf{X})^{-1}\mathbf{X}^{\mathrm{T}}\mathbf{y}$$

where x_{ij} represents the data value from the ith observation for the jth predictor variable. In the case of only one dependent (Y) and one independent variable (X), *simple linear regression* reduces to computing a line of the following form, $y = mx + b$, with the slope computed as follows for n samples:

$$m = \frac{\sum_{i=1}^{n} x_i y_i - n\bar{x}\bar{y}}{\sum_{j=1}^{n} x_j^2 - n\bar{x}^2} = r\frac{\sigma_x}{\sigma_y}$$

and equivalently, where r is the correlation coefficient between X and Y; and σ_x, σ_y are the standard deviations of X and Y, respectively. Intuitively, simple linear regression computes the line through the data that minimizes the distance between the line and each data point.

Logistic regression. *Logistic regression* is used to predict the probability of an event by fitting a *logistic curve* (a sigmoid instead of a line) to observed data with predictor variables that may be categorical or numerical in nature [60]. As such, logistic regression is a type of binomial regression, where the event outcome is binary. By way of illustration, a logistic regression model could be used to compute the risk of stroke occurring given an individual's age, gender, and other quantitative or nominal features (*i.e.,* stroke occurs or does not occur with some probability). The logistic function takes the following form:

$$f(z) = \frac{1}{1+e^{-z}}, z = \beta_0 + \beta_1 x_1 + \beta_2 x_2 + ... + \beta_n x_n$$

for n predictor variables. The variable, z, represents a summation of the predictor variables, weighted by the model parameters, while $f(z)$ indicates the probability of the event occurring given z. As in linear regression, the goal is to choose model parameters, β_0 through β_p, which best fit the observed data.

Decision Trees

Decision trees are hierarchical, graph-based classifiers that present a series of questions and choices; by traversing a pathway along the tree's branches and reaching a leaf, a classification is determined. The nodes in the graph present questions regarding an object's features; edges leading away from the node are the possible values for the feature. Terminal nodes in the graph represent a class decision. In some variations, each leaf contains a probability distribution over the classes, estimating the conditional probability that an item reaching the leaf belongs to a given class. As with regression analysis, decision trees are constructed by analyzing a set of training examples for which the class labels are known. Several well-known algorithms are available to automatically construct decision trees from labeled data sets:

- Classification and regression trees (CART). CART is a non-parametric technique that builds a decision tree using binary recursive partitioning [8]: there are no assumptions about the distribution of the predictor or dependent variables. At each level in the tree, the algorithm selects amongst a set of variables to best split a dataset into two different classes; each node thus provides a set of features with values and a binary yes/no decision pathway. Ultimately, the leaves in the tree provide for classification. Trees can still be generated even when predictor variables are not known for all data points by creating surrogate variables interpolated from values of a previous split point.
- QUEST. Like CART, QUEST is a binary-split decision tree algorithm [44]. Differences include methods for unbiased variable selection, imputation of missing data (rather than the use of values from a parent split point), and improved performance in selection on categorical variables with a large number of values.
- ID3 and C4.5. The ID3 algorithm creates decision trees by selecting at each level the predictor variable that provides the largest amount of discriminatory power by minimizing *information entropy* [61]. Once selected, the attribute is removed from future consideration, and the algorithm repeats until all data is classified

(therefore, not all given predictor variables may be used). Note that unlike CART and QUEST, each node only considers a single variable at a time. C4.5 extends the core ID3 algorithm in several ways [62]: allowing for both continuous and categorical variables; managing missing data (by not using the data points as part of the entropy calculations); and pruning of the tree on algorithm completion, replacing branches that can be terminated earlier with leaves.

Classification rules can be created from a decision tree, providing a predictive model.

Informatics Evaluation

To facilitate the adoption of informatics tools into the clinical or research workflow, such technologies must be developed around the needs and capacities of medical professionals. Largely, engineers and informatics researchers approach clinical problems with the goal of developing the tools, systems, and/or algorithms to improve the accuracy and efficiency with which tasks are completed. On the other hand, clinicians are concerned with usability, applicability, and the simplicity of technology to facilitate their daily work. As such there can be a disconnect between the engineer's vision of a robust, functionally-rich system and the basic tool physicians are willing to use in the clinical setting. Evaluation and usability testing thus serve the role of fulfilling the needs for both researchers and healthcare providers in developing these tools: through evaluation we can mediate the differences between the developer and the user, optimizing the design of such systems' abilities. Moreover, formal evaluation and usability provide a formal basis upon which comparisons between systems can be made.

[5] enumerates three perspectives in how evaluation has handled the modeling of change brought about by healthcare information systems: 1) the computer as an external agent, which brings about changes in participant behavior, workflow, and the healthcare organization; 2) the system as a tool explicitly designed to meet user information needs; and 3) the deployed system as a factor in a more complex network of social interactions that decide the usability and impact of the system. The first of these viewpoints is predominated by technical evaluations of a system, assessing elementary performance of the algorithm/system (*e.g.*, accuracy, speed). The second view optimistically looks to informatics tools as problem solvers, inducing change in a logical manner. The last point of view argues that technology (perhaps regardless of its performance) is ultimately subject to the attitudes and interactions within an organization, which influence its acceptance and usage. [31] elaborates on the origins of these models of change within evaluation research, and advocates examination of change within the 4Cs (*i.e.*, changes in *communication*, changes in *care*, issues of *control*, and *context*) as a unifying evaluatory framework that draws upon common threads across each of the perspectives.

Within informatics, [22] also contrasts two fundamental approaches to evaluation studies: the *objectivist approach*, which performs numerical measurement and statistical analysis of performance and outcomes that are considered precise and replicable (*i.e.*, objective); and the *subjectivist approach* that involves a more subjective assessment based on (expert) judgment and interview-based assessment. The former is aimed at determining if a system meets its outcomes-based needs, whereas the latter provides a more qualitative framework for testing with less emphasis on quantitative metrics. Related to the subjectivist approach, *qualitative research* investigates the perspectives and the behavior of people within a given context, and the rationale behind their actions. Central to this type of inquiry is that it is conducted in a natural setting, and tends to use observations, interviews, and source documents to draw inductive conclusions. Notably, qualitative research methods are increasingly used to study the complexities surrounding healthcare information systems [32], teasing out the "how" and "why" of specific outcomes. Both objectivist and subjectivist assessments can be performed as comparative evaluations (*e.g.,* as seen with clinical trials), where each system component is assessed relative to a control (*e.g.,* a normal group, the current information system, etc.).

The following sections divide the questions of how to evaluate informatics tools into two demonstrative parts: 1) methods for testing information retrieval systems, such as content-based medical image retrieval applications; and 2) methods for assessing the usability of a system, focusing on the evaluation of graphical user interfaces (GUIs) and other system aspects. By no means are the discussions of these areas meant to be complete – indeed, in and of themselves, each topic can encompass an entire book. Hence, we only aim to introduce core concepts and terminology, relating the concepts of study design and statistical analysis from the previous section; references to seminal works are given for the reader to pursue in further detail.

Evaluating Information Retrieval Systems

Evaluation is a critical and challenging step in information retrieval (IR) system development. Apart from the mathematical algorithms that drive an IR system, evaluation requires considering the IR problem from a social perspective, addressing subjective issues such as relevance, user experience, and information needs. Early IR evaluations were exemplified by the Cranfield experiments and the SMART system assessment. The Cranfield experiments used a small test collection (1,400 documents), queries, and exhaustive relevance judgments to evaluate the efficiency of different indexing languages and methods [12]. The SMART system proposed the vector-space model for IR and its experimental results were seminal [65]. Early evaluations of SMART involved metrics that considered the ranking of results based on relevance concepts. These influential works paved the way for modern IR systems and evaluation; and today, annual meetings such as TREC (Text Retrieval Conference) and CLEF (Cross Language Evaluation Forum) serve as the premier forums for textual IR and evaluation across multiple information domains.

In medicine, the problem of IR is evident in the extreme growth of information and knowledge being created by clinicians, scientists, and other publishers. Accessing this body of information is a primary concern for all participants in the healthcare domain. Clinicians must evaluate patients and therefore need access to current information on diseases and procedures. Researchers developing new technologies depend on previous literature to learn and refine their methods (*e.g.,* the US National Library of Medicine (NLM) MEDLINE currently services over 60 million queries a month). Patients, who are not experts in the field, require specialized information on their medical problems to help guide decision-making. These different scenarios underscore the importance of IR systems and evaluation in medicine to ensure that the variety of users are able to find the information they need.

Information Needs

Compared to queries, information needs are ambiguous and unstructured. They represent a user's desire to learn more about what they do not know and therefore may poorly describe the corpus of satisfactory information. For example, a clinician may express the following information need: *I need information on current treatments for a patient with chronic lower back pain due to disc herniation.* Such a need may be translated into the following Boolean query: current AND treatment AND chronic AND lower AND back AND pain AND disc AND herniation. Although there may be documents in a given collection that contain these words, they do not necessarily satisfy the clinician's information need. Whether or not a document helps satisfy a need is a matter of *relevance* (see below).

To better understand information needs, consider Taylor's description of the four levels of information need that a user experiences [77]: 1) *visceral need,* or the actual, but unexpressed need for information; 2) *conscious need,* the cognizant, mental description of the need; 3) *formalized need,* the statement of a question to answer the information need; and finally 4) *compromised need,* the query presented to the IR system. The transition of a need from a visceral to a compromised query follows a path of formalization and potential mistranslation. An IR system may perform accurately on answering queries (*e.g.,* a Boolean retrieval system), but it is possible that the underlying information need that led to the query formalism remains unmet. This observation illustrates the importance of defining not only queries, but also information needs when performing IR evaluation.

In the clinical environment, information needs most often relate to therapies and overviews of diseases [25]. Popular methods for translating these needs to queries require a clinician to structure the need using a patient-oriented framework. For example, the PICO structure (problem/population, intervention, comparison, outcome) has been proposed as a method for encoding clinical information needs into machine readable queries to support evidence-based medicine [16]. The PICO structure for posing clinical questions is thus a way to transition between Taylor's conscious and compromised need levels. To demonstrate, consider a clinician with a conscious information need involving back pain, which can be formalized to: *Given a 40-year-old male with a disc herniation, what are the tradeoffs between surgery, non-steroidal anti-inflammatory drugs (NSAIDs), and physical therapy to reduce chronic lower back pain as measured by NIH pain scales?* Such a need can be translated to the PICO format and structured using a controlled terminology (Table 10.2). Evaluation of PICO has found that clinical questions relating to therapies are most likely to fit the framework as they tend to include definite interventions and outcomes. In contrast, questions relating to prognosis and etiology are difficult to structure with PICO as they are more vague in nature. Other notable challenges with PICO include the inability to encode fine-grained relationships between elements, a lack of an explicit temporal/state model, and the inability to capture anatomical relations [29].

PICO field	Description
Problem	[DISEASE OR SYNDROME] "disc herniation", [SIGN OR SYMPTOM] "chronic back pain"
Population	[AGE GROUP] "40-year-old", [POPULATION GROUP] "male"
Intervention and comparison	[THERAPEUTIC OR PREVENTATIVE PROCEDURE] "surgery", [THERAPEUTIC OR PREVENTATIVE PROCEDURE] "physical therapy", [CLINICAL DRUG] "NSAID"
Outcome	[DIAGNOSTIC PROCEDURE] "NIH pain scales"

Table 10.2: Example of structuring and standardizing a query using the PICO format and Unified Medical Language System (UMLS).

Relevance

Given a set of information needs, the evaluation of a system requires *relevance* judgments on the underlying collection of documents. The most common notion of relevance is that of *topical relevance* in which a document is considered relevant to a need if the two share common topics. In contrast, it can be argued that there is no fixed relevance between a need and a document – that all assessments of relevance are instead *situational* with respect to the user [67]. Pursuing topical or situational relevance in an IR system is a design choice where the limitations of each may be limitations of each may be mitigated by clearly defining the user group and their information needs, *i.e.,* when using the notion of topical relevance, control should be placed on situation. It may also be possible to incorporate aspects of both. For clinical applications, topical relevance is often used with judgments made by a panel of experts (*e.g.,* physicians).

Traditionally, evaluations of small collections involved exhaustively testing the relevance of each document to a query, which although time-consuming, is possible to accomplish. For modern collections, however, such a process is infeasible and other methods must be used. *Pooling* provides a way to analyze a subset of a collection's documents for relevance to create an estimate of the total relevant documents in a collection for a query. The pooling process requires searching a collection using several different IR systems and judging the relevance of the first n documents returned. From each used system, relevant documents are placed in a pool, which then forms the gold standard for relevant documents given a query. The hope is that given a variety of systems, the pool will closely represent the actual set of relevant documents. Pooling thus presents a trade-off between exhaustively evaluating each document with the potential of missing relevant documents in the collection.

Judging relevance is not a straightforward task as its subjective nature can cause disagreement between judges – even experts may have different opinions. Therefore, in an IR evaluation it is important to measure the degree of agreement among judges (raters). The most common way to do this is the *kappa statistic* (K), which measures the rate of agreement between raters, correcting for agreement by chance:

$$K = \frac{P(agreement) - P(expected\ agreement\ by\ chance)}{1 - P(expected\ agreement\ by\ chance)}$$

A value of 1 reflects perfect agreement, whereas 0 indicates no agreement. Typically, kappa scores greater than 0.8 are viewed as strong agreement between judges. *Cohen's kappa coefficient* is a special case of this measure for categorical data and when there are only two raters. When there are more than two judges, an average pairwise kappa value is usually calculated.

The discussion thus far that a document is simply relevant or not to an information need is overly binary and fails to convey the true complexity of a searcher reviewing results. Consider the set of documents relevant to a query. It is probable that there is a high degree of redundancy of content, and therefore each retrieved document will be only marginally more relevant as a user progresses through the results [9]. It could also be possible that there are documents in the collection that are only relevant when returned with other documents. These two points help to illustrate that document relevancy is not dependent only on the query, but also the other documents in the collection.

Evaluation Metrics

Selecting an evaluation metric depends on the type of results returned from the IR system. *Recall* and *precision*, as well as *F-measure*, are the standard methods of evaluating unranked sets of documents. In ranked-retrieval these metrics are extended to precision-recall curves as the size of the returned set changes. *Mean average precision* (MAP) is another method for evaluating ranked results.

Unranked retrieval. In unranked retrieval, a document is binary-classified as being relevant or not. For a given query, the number of relevant documents retrieved by the IR system divided by the total number of relevant documents in the corpus is known as *recall* (*R*); and *precision* (*P*) is the number of relevant documents retrieved divided by the number of retrieved documents:

$$recall\ (R) = \frac{\#\ relevant\ documents\ retrieved}{\#\ relevant\ documents} = P(retrieved \mid relevant)$$

$$precision\ (P) = \frac{\#\ relevant\ documents\ retrieved}{\#\ documents\ retrieved} = P(relevant \mid retrieved)$$

From a statistical classification viewpoint, precision and recall may also be calculated in terms of true/false positives/negatives. Figure 10.2 illustrates this relationship, which is given by the following equations:

$$P = \frac{true\ positives}{true\ positives + false\ positives} \qquad R = \frac{true\ positives}{true\ positives + false\ negatives}$$

Note that *P* is identical to the precision equation given earlier, and *R* is equal to sensitivity. These equations illustrate a fundamental tradeoff in IR systems: the balance between the number of results returned and the relevance of those results. A system may achieve 100% recall by returning every document in the collection, but the precision for such a system would be low. Conversely, the system that returns only a few relevant documents will have high precision, but low recall. An IR system's place in this spectrum is a function of the information being retrieved, the domain, and the user's preferences. By way of illustration, a researcher doing a literature search may desire systems with high recall to ensure that every important citation is found; whereas a patient researching a medical condition on the Internet may wish to have high precision, with only a few relevant sources returned outlining the disease rather than an exhaustive search that returns information that is not applicable. In the healthcare setting, a doctor using a clinical system that returns relevant patient cases may have high demands on both recall and precision as he does not want to miss any relevant cases, but also does not have the time to sift through large numbers of irrelevant cases.

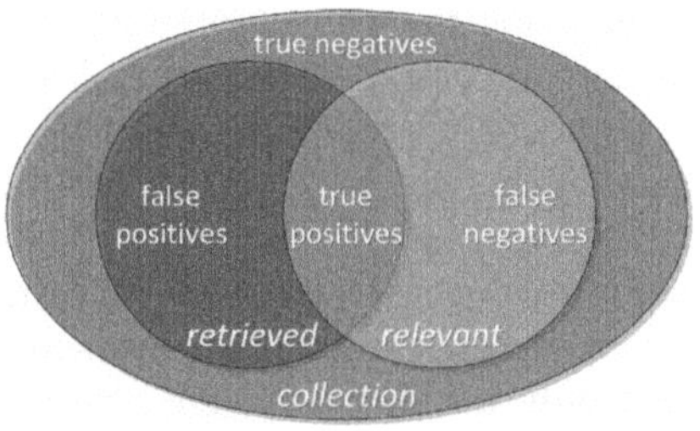

Figure 10.2: Venn diagram showing a document collection with retrieved and relevant sets. Definitions of true/false positive/negatives are illustrated.

7. Analyze the data. For experiments where observations are recorded (*e.g.*, as with videotaping), information must be extracted and logged. As an example, *time motion studies*, which are primarily directed at determining the amount of time needed to perform various tasks, often review videotapes to determine task durations. Quantifiable usability metrics are then computed (see below). For qualitative information, such as user comments, analysis may include categoryzation or codification of statements.
8. Interpret the data. Based on data analysis, usability problems are identified. For example, if there are tasks that seemingly take longer periods of time to complete, or the behavior of the system confused the user, these problems are documented.
9. Iterate input to design. Lastly, the results of the interpreted data are used to inform and refine the design process, solving key problems with the system.

Defining tasks. Usability assessment is dependent on the definition of tasks/goals that the user is meant to achieve using the application. [63] discusses, for instance, the use of task analysis in a qualitative study on an EMR. In general, there are several methods to bring out and characterize the tasks within a domain, thus defining points for usability testing [42] (see also Chapter 4 for a discussion of task models). *Cognitive task analysis* (CTA) provides a means to identify parts of a system that involve decision making, reasoning, and other information processing needs. Hierarchical task analysis can be used to examine a process in terms of increasing granularity, resulting in a decomposition of a task into smaller steps. Applied CTA involves the use of task diagrams to describe a process in steps that highlight cognitive skills and the knowledge needed to complete each step [49]. *Cognitive work analysis* (CWA)[5] [79] has also been applied to the analysis of medical systems, providing progressive levels of abstraction to elucidate design issues in terms of user tasks and responsibilities [19, 24].

Usability metrics. Usability testing involves the observation of users in order to provide (numerical) measures that assess design impact. [27] provides a comprehensive review of different usability metrics that have appeared across recent studies, grouped along the lines of measures for effectiveness, efficiency, and satisfaction; Table 10.3 summarizes the key types of measures. Measures under effectiveness and efficiency tend to be quantitative in nature, directly observing user actions in employing the user interface/system; whereas satisfaction tends to be more qualitative in nature. A review of clinical information system evaluations echoes aspects of this categorization [18] finding that satisfaction, acceptance, and success are frequently measured variables.

User testing. Testing with users often entails statistical analysis of a group of users. Study measures including sensitivity/specificity and ROC analysis are computed across the group, potentially stratified based on their characteristics. For instance, [34] describes a longitudinal study evaluating the usability of a health information system, contrasting novice and expert users and the significant difference seen in the results. As it is impractical to evaluate a system across all potential users, testing is predicated on a sub-sampling of individuals chosen to represent the general user set. Here, the problem of bias can arise. [22] elaborates on general biases applicable to all systems evaluation involving users:

[5] Researchers distinguish CWA from CTA in that CWA is a broader construct for understanding the environment, work domain, and interaction/behavior whereas CTA is directed more to accomplishing a goal in terms of sequential steps.

	Measure	Description
Effectiveness	*Task completion*	Number of tasks successfully completed (potentially within an allotted amount of time), number of failed/incomplete tasks
	Accuracy	Error measures, including the number of tasks completed with error, number of errors per task
	Recall	Time to learn, ability for users to remember the usage of the interface
	Completeness	Completeness of user's solutions to given tasks
	Quality of outcome	Quantifiable changes pre- and post-usage of using the interface (*e.g.*, such as from learning new knowledge)
	Expert/user assessment	Expert/user grading of the end product in using the system
Efficiency	*Task/error completion time*	Timed duration of tasks or sub-tasks, including breakdown of time spent in certain modes of interaction (*e.g.*, using help); time motion study
	Input rate	User throughput in using an input device (*e.g.*, typing, annotation)
	Mental effort	Cognitive load in using the user interface
	Usage patterns	Frequency of usage of system functions, amount of data accessed
	Learning effects	Changes in task completion time over different sessions
Satisfaction	*Standard questionnaires*	Questionnaire for User Interface Satisfaction (QUIS), surveys
	Preferences	Rank comparisons between different GUIs and components, preferred system behavior
	Satisfaction	Rating scales for ease of use, context-specific usage scenarios
	User attitude	User perception of the interaction, end outcome

Table 10.3: Usability metrics, as compiled in [27]. Three categories of measures are defined. Effectiveness and efficiency tend to be quantitative measures of users' performance in doing tasks with the system, whereas satisfaction is more qualitative.

- Assessment bias. Subjects in the experiment allow their own feelings or beliefs about the system to affect how they interact with the intervention.
- Allocation bias. Researchers may informally bypass the randomization process to ensure simpler cases are assigned to the intervention that they prefer and/or geared towards the specific users.
- Hawthorne effect. Mentioned earlier in the chapter, humans tend to improve (or alter) their performance if they know they are under observation/study.
- Checklist effect. When humans are given checklists for a cognitive task, their decision making tends to be more complete and better structure.

Each of these types of bias can affect the sub-sampled user group (*e.g.*, inappropriate selection of individuals; bias in reporting; etc.) and end usability results. Evaluation, therefore, must take these issues into account within usability study design. [71] suggests some approaches to minimize the effects of evaluation bias, including: 1) utilizing qualitative analysis; 2) determining the direction of each type of bias (*i.e.*, does the bias strengthen or weaken the system?); and 3) estimating the magnitude of each effect relative to each other, not on an absolute scale.

Lastly, a longstanding debate also surrounds the number of users that should be employed to conduct usability testing. Although anecdotal evidence and some models advise that five users is sufficient to uncover 80-85% of all usability issues [80] (with subsequent drop off in terms of discovered issues vs. the number of testers), most researchers agree that a larger number of representative users is better, especially if the intended user group is highly heterogeneous. Moreover, it is suggested that rather than focus of the total number of testers, that task coverage is a more important factor in the evaluation process [43].

Questionnaires and surveys. Questionnaires and surveys are a popular means of evaluating user acceptance and system efficacy. Since the early 1980s, this technique has been relied upon as an easy to deploy technique for evaluating system usability.

The intent of a survey is to measure users' rating and (self-reported) usage patterns of the system. For instance, [37, 39, 73] leverage questionnaires to gauge users' satisfaction on the implementation and use of CPOE and EMR systems. *Likert scales* are typically used to capture user responses on a visual analog scale: each question is posed such that responses can be given along a spectrum. The HCI and usability literature both lend strong support to the use of questionnaires to (subjectively) assess user satisfaction. This dependence has led to the development of the popular Questionnaire for User Interface Satisfaction (QUIS). Researchers developed the QUIS as a standardized user evaluation instrument for HCI components [68]. The generalizability of the QUIS was established by having different user populations (*e.g.*, students, computer experts) evaluate different systems (*e.g.*, websites, information retrieval systems). QUIS' reliability was determined by increasing the number of questions and reducing coarseness of the rating scale [11]. QUIS was thus created based on psychological test construction methods to ensure empirical validity and test-retest reliability. Until the development of QUIS, few questionnaires had focused exclusively on evaluating the user's subjective satisfaction of the system and related issues. Issues challenging past studies, such as lack of validation and low reliability, further led to the development of a questionnaire to directly address subjective satisfaction of the system. QUIS has been widely used in the past: for instance, [70] evaluated via QUIS physician satisfaction and interaction with an EMR system, covering areas of clinical results review, the ambulatory medical record and list management; and [55] surveyed physician satisfaction to compare a commercially available order entry system with one deployed at the US Department of Veterans' Affairs (VA). Aside from QUIS, other usability instruments include those designed by IBM [41]. Although questionnaires can be designed readily, they must be sufficiently validated to ensure that results can be interpreted properly.

Discussion

Evaluations of expert systems, teleconsultation frameworks, and general clinical information systems account for almost 60% of all informatics-related assessments in the literature, with an excess of 80% of all evaluations being quantitative [3]. But despite the visible progress made in the evaluation of informatics developments, the area continues to be an active focus of research given the lack of standardization and the complexities that arise in testing such systems. One early effort to promote standardization, the European Union (EU) VATAM project, established guidelines for assessment for telemedicine-related applications [75]; and [2] provides additional review of general EU-based initiatives. However, these efforts have yet to be fully embraced by the informatics community.

Finally, we conclude by noting that questions of evaluation eventually lead to questions of impact and examining healthcare outcomes. Undoubtedly, once a system is deployed, an assessment of the system's use should take place, both to further improvement (*i.e.*, improvements to the underlying algorithms, usability) and to gauge what effect, if any, the tool or system has on users, organizations, and in the end, patients and healthcare processes. [5, 26] put forward several key questions to such an evaluation, summarized

Common system evaluation and impact questions	
Was the system used, and if so, for what?	How much training was needed for system use?
Did the system work as designed?	How well are users employing the system?
Is system used as anticipated?	Were the users satisfied?
What factors were associated with success/failure of system use?	Does the system work better than the procedures it replaced?
Does the system produce the desired results?	Is the system cost effective?
What changes occurred to patient care, the organization, or otherwise because of implementation?	Did the system have an impact in the short-term and/or long-term?

Table 10.4: Compilation of common questions for post-system deployment evaluation of informatics tools and systems, based on [5, 26].

in Table 10.4. Apart from basic inquiries into whether the system worked as envisioned, the questions interrogate usability, user satisfaction, and short- and long-term effects of the implementation. Admittedly, measuring the impact of a system is a complicated issue, especially given that the usage of a system changes over time in response to new technologies, user needs, and evolving perceptions. One conceivable metric is to estimate the number of times the implemented system led to some change in a patient's care and/or the end outcome relative to baseline (*i.e.*, the existing information system or equivalent healthcare process already in place). Ideally, such an evaluation could occur in a real-world randomized controlled trial framework. However, both ethical and practical considerations make such study designs difficult to execute: one must ensure that patient care is not compromised (therefore "test" systems must not be sub-optimal relative to baseline or the standard of care); and given the number of factors that must be accounted for in conducting patient care, outcomes may be ambiguous, making it impossible to separate out confounding variables (and thus conclude to what extent a system is responsible for affecting the quality of care). Recent analyses have thus remarked upon the lack of true RCT evaluations of informatics applications [4]. And arguably, the measure of end outcome variables affected by a given information system or tool is, *per se*, unrealistic: a patient can have complex, multi-organ system disease, and the specific problem addressed by a test system may only be one component of his overall health status. Hence, determination of *intermediate outcomes* may be a better approach, wherein the impact of a system is judged relative to measurable changes in the underlying healthcare process.

References

1. Aisen A, Broderick L, Winer-Muram H, Brodley C, Kak A, Pavlopoulou C, Dy J, Shyu, CR, Marchiori A (2003) Automated storage and retrieval of thin-section CT images to assist diagnosis: System description and preliminary assessment. Radiology, 228(1):265-270.
2. Ammenwerth E, Brender J, Nykanen P, Prokosch HU, Rigby M, Talmon J (2004) Visions and strategies to improve evaluation of health information systems: Reflections and lessons based on the HIS-EVAL workshop in Innsbruck. Int J Med Inform, 73(6):479-491.
3. Ammenwerth E, de Keizer N (2005) An inventory of evaluation studies of information technology in health care trends in evaluation research 1982-2002. Methods Inf Med, 44(1):44-56.
4. Ammenwerth E, de Keizer N (2007) A viewpoint on evidence-based health informatics, based on a pilot survey on evaluation studies in health care informatics. J Am Med Inform Assoc, 14(3):368-371.

5. Anderson JG, Aydin CE (2005) Overview: Theoretical perspectives and methodologies for the evaluation of healthcare information systems. In: Anderson JG, Aydin CE (eds) Evaluating the Organizational Impact of Healthcare Information Systems. Springer, New York, NY, pp 5-29.
6. Benson K, Hartz AJ (2000) A comparison of observational studies and randomized, controlled trials. N Engl J Med, 342(25):1878-1886.
7. Beuscart-Zephir MC, Anceaux F, Crinquette V, Renard JM (2001) Integrating users' activity modeling in the design and assessment of hospital electronic patient records: The example of anesthesia. Intl J Medical Informatics, 64(2):157-171.
8. Breiman L, Friedman J, Stone CJ, Olshen RA (1984) Classification and Regression Trees. Wadsworth International Group, Belmont, CA.
9. Carbonell J, Goldstein J (1998) The use of MMR, diversity-based reranking for reordering documents and producing summaries. Proc 21st Intl ACM SIGIR Conf Research and Development in Information Retrieval, Melbourne, Australia, pp 335-336.
10. Card SK, Moran TP, Newell A (1983) The Psychology of Human-computer Interaction. L Erlbaum Associates, Hillsdale, NJ.
11. Chin JP, Diehl VA, Norman KL (1988) Development of an instrument measuring user satisfaction of the human-computer interface. Proc SIGCHI Conf Human Factors in Computing Systems, Washington DC, USA, pp 213-218.
12. Cleverdon C, Mills J, Keen M (1966) Factors determining the performace of indexing systems. Aslib Cranfield Research Project, College of Aeronautics.
13. Concato J, Shah N, Horwitz RI (2000) Randomized, controlled trials, observational studies, and the hierarchy of research designs. N Engl J Med, 342(25):1887-1892.
14. Daniels J, Fels S, Kushniruk A, Lim J, Ansermino JM (2007) A framework for evaluating usability of clinical monitoring technology. J Clin Monit Comput, 21(5):323-330.
15. Dawson B, Trapp RG (2004) Basic & Clinical Biostatistics. 4th edition. Lange Medical Books/McGraw-Hill, Medical Pub. Division, New York, NY.
16. Demner-Fushman D, Lin J (2007) Answering clinical questions with knowledge-based and statistical techniques. Computational Linguistics, 33(1):63-103.
17. Denne JS, Jennison C (1999) Estimating the sample size for a t-test using an internal pilot. Stat Med, 18:1575-1585.
18. Despont-Gros C, Mueller H, Lovis C (2005) Evaluating user interactions with clinical information systems: A model based on human-computer interaction models. J Biomedical Informatics, 38(3):244-255.
19. Effken JA (2002) Different lenses, improved outcomes: A new approach to the analysis and design of healthcare information systems. Int J Med Inform, 65(1):59-74.
20. Flack V, Afifi A, Lachenbruch P, Schouten H (1988) Sample size determinations for the two rater kappa statistic. Psychometrika, 53(3):321-325.
21. Fletcher RH, Fletcher SW (2005) Clinical epidemiology: The essentials. 4th edition. Lippincott Williams & Wilkins, Philadelphia, PA.
22. Friedman CP, Wyatt JC, Owens DK (2006) Evaluation and technology asessment. In: Shortliffe EH, Cimino JJ (eds) Biomedical Informatics: Computer Applications in Health Care and Biomedicine. Springer.
23. Graham MJ, Kubose TK, Jordan D, Zhang J, Johnson TR, Patel VL (2004) Heuristic evaluation of infusion pumps: Implications for patient safety in intensive care units. Int J Med Inform, 73(11-12):771-779.
24. Hajdukiewicz JR, Doyle DJ, Milgram P, Vicente KJ, Burns CM (1998) A work domain analysis of patient monitoring in the operating room. Proc 42nd Annual Meeting Human Factors and Ergonomics Society, pp 1038-1042.

25. Hersh W (2003) Information Retrieval: A Health and Biomedical Perspective. Springer-Verlag, New York.
26. Hersh W, Hickam D (1998) How well do physicians use electronic information retrieval systems. JAMA, 280(15):1347-1352.
27. Hornbæk K (2006) Current practice in measuring usability: Challenges to usability studies and research. Intl J Human-Computer Studies, 64(2):79-102.
28. Horsthemke WH, Raicu DS, Furst JD (2008) Evaluation challenges for bridging the semantic gap: Shape disagreements on pulmonary nodules in the Lung Image Database Consortium. Intl J Healthcare Information Systems and Informatics, 4(1):17-33.
29. Huang X, Lin J, Demner-Fushman D (2006) Evaluation of PICO as a knowledge representation for clinical questions. Proc AMIA Annu Symp:359-363.
30. Järvelin K, Kekäläinen J (2002) Cumulated gain-based evaluation of IR techniques. ACM Trans Information Systems, 20(4):422-446.
31. Kaplan B (1997) Addressing organizational issues into the evaluation of medical systems. J Am Med Inform Assoc, 4(2):94-101.
32. Kaplan B, Maxwell J (2005) Qualitative research methods for evaluating computer information systems. In: Anderson JG, Aydin CE (eds) Evaluating the Organizational Impact of Healthcare Information Systems. Springer, New York, NY, pp 30-55.
33. Kernan WN, Viscoli CM, Makuch RW, Brass LM, Horwitz RI (1999) Stratified randomization for clinical trials. J Clin Epidemiol, 52(1):19-26.
34. Kjeldskov J, Skov MB, Stage J (2008) A longitudinal study of usability in health care: Does time heal? Int J Med Inform.
35. Kurosu M, Kashimura K (1995) Apparent usability vs. inherent usability. Proc SIGCHI Conf Human Factors in Computing Systems, pp 292-293.
36. Kushniruk AW, Patel VL (2004) Cognitive and usability engineering methods for the evaluation of clinical information systems. J Biomed Inform, 37(1):56-76.
37. Laerum H, Ellingsen G, Faxvaag A (2001) Doctors' use of electronic medical records systems in hospitals: Cross sectional survey. BMJ, 323(7325):1344-1348.
38. Lasko TA, Bhagwat JG, Zou KH, Ohno-Machado L (2005) The use of receiver operating characteristic curves in biomedical informatics. J Biomed Inform, 38(5):404-415.
39. Lee F, Teich JM, Spurr CD, Bates DW (1996) Implementation of physician order entry: User satisfaction and self-reported usage patterns. J Am Med Inform Assoc, 3(1):42-55.
40. Lehmann TM, Guld MO, Thies C, Fischer B, Spitzer K, Keysers D, Ney H, Kohnen M, Schubert H, Wein BB (2004) Content-based image retrieval in medical applications. Methods Inf Med, 43(4):354-361.
41. Lewis JR (1995) IBM computer usability satisfaction questionnaires: Psychometric evaluation and instructions for use. Intl J Human-computer Interaction, 7(1):57-78.
42. Limbourg Q, Vanderdonckt J (2003) Comparing task models for user interface design. In: Diaper D, Stanton N (eds) The Handbook of Task Analysis for Human-Computer Interaction, pp 135-154.
43. Lindgaard G, Chattratichart J (2007) Usability testing: What have we overlooked? Proc SIGCHI Conf Human Factors in Computing Systems pp 1415-1424.
44. Loh WY, Shih YS (1997) Split selection methods for classification trees. Statistica Sinica, 7:815-840.
45. Long LR, Antani S, Deserno T, Thoma GR (2009) Content-based image retrieval in medicine: Retrospective assessment, state of the art, and future directions. Intl J Healthcare Information Systems and Informatics, 4(1):1-17.
46. Maclure M (1991) The case-crossover design: A method for studying transient effects on the risk of acute events. Am J Epidemiol, 133(2):144-153.

47. Mayhew DJ (1999) The Usability Engineering Lifecycle: A Practitioner's Handbook for User Interface Design. Morgan Kaufmann Publishers, San Francisco, Calif.
48. Metz CE (2006) Receiver operating characteristic analysis: A tool for the quantitative evaluation of observer performance and imaging systems. J Am Coll Radiol, 3(6):413-422.
49. Militello LG, Hutton RJB (1998) Applied cognitive task analysis (ACTA): A practitioner's toolkit for understanding cognitive task demands. Ergonomics, 41(11):1618-1641.
50. Morton SC, Adams JL, Suttorp MK, Shanman R, Valentine D, Rhodes S, Shekelle PG (2004) Meta-regression approaches: What, why, when, and how? (Technical Review 04-0033). Agency for Healthcare Research and Quality, Rockville, MD.
51. Müller H, Clough P, Hersh B, Geissbühler A (2007) Variation of relevance assessments for medical image retrieval. In: Marchand-Maillet S, Bruno E, Nurnberger A, Detyniecki M (eds) Adaptive Multimedia Retrieval: User, Context, and Feedback (LNCS). Springer, pp 232-246.
52. Müller H, Deselaers T, Deserno T, Kalpathy-Cramer J, Kim E, Hersh W (2007) Overview of the ImageCLEF 2007 medical retrieval and annotation tasks. Advances in Multilingual and Multimodal Information Retrieval: Proc 8th Workshop Cross-Language Evaluation Forum (CLEF), Budapest, Hungary, pp 472-491.
53. Müller H, Michoux N, Bandon D, Geissbuhler A (2004) A review of content-based image retrieval systems in medical applications-clinical benefits and future directions. Int J Med Inform, 73(1):1-23.
54. Müller H, Rosset A, Vallée J, Terrier F, Geissbuhler A (2004) A reference data set for the evaluation of medical image retrieval systems. Comp Med Imaging and Graphics, 28(6):295-305.
55. Murff HJ, Kannry J (2001) Physician satisfaction with two order entry systems. J Am Med Inform Assoc, 8(5):499-509.
56. Nielsen J (1993) Usability Engineering. Academic Press, Boston.
57. Nielsen J (1994) Heuristic evaluation. In: Nielsen J, Mack RL (eds) Usability Inspection Methods. Wiley, New York.
58. Obuchowski NA (2003) Receiver operating characteristic curves and their use in radiology. Radiology, 229(1):3-8.
59. Obuchowski NA (2005) ROC analysis. Am. J. Roentgenol., 184(2):364-372.
60. Pampel FC (2000) Logistic Regression: A Primer Sage Publications, Thousand Oaks, CA.
61. Quinlan JR (1986) Induction of decision trees. Machine Learning, 1(1):81-106.
62. Quinlan JR (1996) Improved use of continuous attributes in C4.5. J Artificial Intelligence, 4:77-90.
63. Rose AF, Schnipper JL, Park ER, Poon EG, Li Q, Middleton B (2005) Using qualitative studies to improve the usability of an EMR. J Biomedical Informatics, 38(1):51-60.
64. Rosenberger WF, Lachin JM (2002) Randomization in Clinical Trials: Theory and practice. Wiley, New York, NY.
65. Salton G, Lesk M (1965) The SMART automatic document retrieval systems - An illustration. Communications of the ACM, 8(6):391-398.
66. Salton G, Wong A, C.S. Y (1975) A vector space model for automatic indexing. Communications of the ACM, 18(11):613-620.
67. Schamber L, Eisenberg M, Nilan M (1990) A re-examination of relevance: Toward a dynamic, situational definition. Information Processing and Management, 26(6):755-776.
68. Shneiderman B, Plaisant C (2004) Designing the User Interface: Strategies for Effective Human-Computer Interaction. 4th edition. Pearson/Addison Wesley, Boston.
69. Shyu CR, Brodley C, Kak A, Kosaka A, Aisen A, Broderick L (1999) ASSERT: A physician-in-the-loop content-based retrieval system for HRCT image databases. Computer Vision and Image Understanding, 75(1-2):111-132.

70. Sittig DF, Kuperman GJ, Fiskio J (1999) Evaluating physician satisfaction regarding user interactions with an electronic medical record system. Proc AMIA Symp:400-404.
71. Snyder C (2006) Bias in usability testing. http://www2.stc.org/edu/54thConf/data-Show.asp?ID=65. Accessed February 19, 2009.
72. Stein C (1945) A two-sample test for a linear hypothesis whose power is independent of the variance. Ann Math Stat, 16:243-258.
73. Stoicu-Tivadar L, Stoicu-Tivadar V (2006) Human-computer interaction reflected in the design of user interfaces for general practitioners. Int J Med Inform, 75(3-4):335-342.
74. Tagare H, Jaffe C, Duncan J (1997) Medical image databases: A content-based retrieval approach. J Am Med Inform Assoc, 4:184-198.
75. Talmon J, Enning J, Castaneda G, Eurlings F, Hoyer D, Nykanen P, Sanz F, Thayer C, Vissers M (1999) The VATAM guidelines. Int J Med Inform, 56(1-3):107-115.
76. Tang Z, Johnson TR, Tindall RD, Zhang J (2006) Applying heuristic evaluation to improve the usability of a telemedicine system. Telemed J E Health, 12(1):24-34.
77. Taylor RS (1962) The process of asking questions. American Documentation, 13(4):391-396.
78. Tractinsky N, Katz AS, Ikar D (2000) What is beautiful is usable. Interact Comp, 13(2):127-145.
79. Vicente KJ (1999) Cognitive Work Analysis: Toward Safe, Productive, and Healthy Computer-based Work. Lawrence Erlbaum Associates, Mahwah, NJ.
80. Virzi RA (1992) Refining the test phase of usability evaluation: How many subjects is enough? Human Factors, 34(4):457-468.
81. Wittes J, Brittain E (1990) The role of internal pilot studies in increasing the efficiency of clinical trials. Stat Med, 9:65-72.

Index

D

E

F

G

M

N

P

Q

R

S

T

U

V

W

X

CPSIA information can be obtained at www.ICGtesting.com
Printed in the USA
LVOW03*1440051213

364039LV00002B/5/P

9 781441 903846